Managed Aquifer Recharge for Water Resilience

Managed Aquifer Recharge for Water Resilience

Editors

Peter Dillon
Enrique Fernández Escalante
Sharon B. Megdal
Gudrun Massmann

MDPI • Basel • Beijing • Wuhan • Barcelona • Belgrade • Manchester • Tokyo • Cluj • Tianjin

Editors
Peter Dillon
CSIRO Hon Fellow
Australia

Enrique Fernández Escalante
Specialist R&D Tragsa Group
Spain

Sharon B. Megdal
The University of Arizona
USA

Gudrun Massmann
Carl von Ossietzky Universität Oldenburg
Germany

Editorial Office
MDPI
St. Alban-Anlage 66
4052 Basel, Switzerland

This is a reprint of articles from the Special Issue published online in the open access journal *Water* (ISSN 2073-4441) (available at: https://www.mdpi.com/journal/water/special_issues/ISMAR10_2019).

For citation purposes, cite each article independently as indicated on the article page online and as indicated below:

LastName, A.A.; LastName, B.B.; LastName, C.C. Article Title. *Journal Name* **Year**, *Volume Number*, Page Range.

ISBN 978-3-03943-042-0 (Hbk)
ISBN 978-3-03943-043-7 (PDF)

Cover image courtesy of Enrique Fernández Escalante.
Water is piped 21km from Cega river to eleven infiltration basins at El Carracillo, Los Arenales, Spain. The cover image shows water discharging through an outlet structure into one of these, Chatún basin. MAR increases resilience of groundwater supplies for irrigation and cattle feeding in this area including on farms seen behind the basin.

© 2021 by the authors. Articles in this book are Open Access and distributed under the Creative Commons Attribution (CC BY) license, which allows users to download, copy and build upon published articles, as long as the author and publisher are properly credited, which ensures maximum dissemination and a wider impact of our publications.

The book as a whole is distributed by MDPI under the terms and conditions of the Creative Commons license CC BY-NC-ND.

Contents

About the Editors ix

Preface to "Managed Aquifer Recharge for Water Resilience" xi

Peter Dillon, Enrique Fernández Escalante, Sharon B. Megdal and Gudrun Massmann
Managed Aquifer Recharge for Water Resilience
Reprinted from: *Water* 2020, 12, 1846, doi:10.3390/w12071846 1

Enrique Fernández Escalante, Jon San Sebastián Sauto and Rodrigo Calero Gil
Sites and Indicators of MAR as a Successful Tool to Mitigate Climate Change Effects in Spain
Reprinted from: *Water* 2019, 11, 1943, doi:10.3390/w11091943 13

Mohammad Faiz Alam, Paul Pavelic, Navneet Sharma and Alok Sikka
Managed Aquifer Recharge of Monsoon Runoff Using Village Ponds: Performance Assessment
of a Pilot Trial in the Ramganga Basin, India
Reprinted from: *Water* 2020, 12, 1028, doi:10.3390/w12041028 31

Prahlad Soni, Yogita Dashora, Basant Maheshwari, Peter Dillon, Pradeep Singh and Anupama Kumar
Managed Aquifer Recharge at a Farm Level: Evaluating the Performance of Direct Well
Recharge Structures
Reprinted from: *Water* 2020, 12, 1069, doi:10.3390/w12041069 51

Roksana Kruć, Krzysztof Dragon and Józef Górski
Migration of Pharmaceuticals from the Warta River to the Aquifer at a Riverbank Filtration Site
in Krajkowo (Poland)
Reprinted from: *Water* 2019, 11, 2238, doi:10.3390/w11112238 71

Janie Masse-Dufresne, Paul Baudron, Florent Barbecot, Marc Patenaude, Coralie Pontoreau, Francis Proteau-Bédard, Matthieu Menou, Philippe Pasquier, Sabine Veuille and Benoit Barbeau
Anthropic and Meteorological Controls on the Origin and Quality of Water at a Bank Filtration
Site in Canada
Reprinted from: *Water* 2019, 11, 2510, doi:10.3390/w11122510 83

Marc Patenaude, Paul Baudron, Laurence Labelle and Janie Masse-Dufresne
Evaluating Bank-Filtration Occurrence in the Province of Quebec (Canada) with a
GIS Approach
Reprinted from: *Water* 2020, 12, 662, doi:10.3390/w12030662 105

Cristina Valhondo, Jesús Carrera, Lurdes Martínez-Landa, Jingjing Wang, Stefano Amalfitano, Caterina Levantesi and M. Silvia Diaz-Cruz
Reactive Barriers for Renaturalization of Reclaimed Water during Soil Aquifer Treatment
Reprinted from: *Water* 2020, 12, 1012, doi:10.3390/w12041012 123

Robert W. Van Kirk, Bryce A. Contor, Christina N. Morrisett, Sarah E. Null and Ashly S. Loibman
Potential for Managed Aquifer Recharge to Enhance Fish Habitat in a Regulated River
Reprinted from: *Water* 2020, 12, 673, doi:10.3390/w12030673 141

Jana Sallwey, Robert Schlick, José Pablo Bonilla Valverde, Ralf Junghanns,
Felipe Vásquez López and Catalin Stefan
Suitability Mapping for Managed Aquifer Recharge: Development of Web-Tools
Reprinted from: *Water* **2019**, *11*, 2254, doi:10.3390/w11112254 163

Peter Dahlqvist, Karin Sjöstrand, Andreas Lindhe, Lars Rosén, Jakob Nisell, Eva Hellstrand
and Björn Holgersson
Potential Benefits of Managed Aquifer Recharge MAR on the Island of Gotland, Sweden
Reprinted from: *Water* **2019**, *11*, 2164, doi:10.3390/w11102164 175

Anthony Knapton, Declan Page, Joanne Vanderzalm, Dennis Gonzalez, Karen Barry,
Andrew Taylor, Nerida Horner, Chris Chilcott and Cuan Petherem
Managed Aquifer Recharge as a Strategic Storage and Urban Water Management Tool in
Darwin, Northern Territory, Australia
Reprinted from: *Water* **2019**, *11*, 1869, doi:10.3390/w11091869 187

Jean-Christophe Maréchal, Madjid Bouzit, Jean-Daniel Rinaudo, Fanny Moiroux,
Jean-François Desprats and Yvan Caballero
Mapping Economic Feasibility of Managed Aquifer Recharge
Reprinted from: *Water* **2020**, *12*, 680, doi:10.3390/w12030680 197

Andreas Lindhe, Lars Rosén, Per-Olof Johansson and Tommy Norberg
Dynamic Water Balance Modelling for Risk Assessment and Decision Support on MAR
Potential in Botswana
Reprinted from: *Water* **2020**, *12*, 721, doi:10.3390/w12030721 211

Girma Y Ebrahim, Jonathan F. Lautze and Karen G. Villholth
Managed Aquifer Recharge in Africa: Taking Stock and Looking Forward
Reprinted from: *Water* **2020**, *12*, 1844, doi:10.3390/w12071844 225

Tatsuo Shubo, Lucila Fernandes and Suzana Gico Montenegro
An Overview of Managed Aquifer Recharge in Brazil
Reprinted from: *Water* **2020**, *12*, 1072, doi:10.3390/w12041072 247

Mary Belle Cruz-Ayala and Sharon B. Megdal
An Overview of Managed Aquifer Recharge in Mexico and Its Legal Framework
Reprinted from: *Water* **2020**, *12*, 474, doi:10.3390/w12020474 269

Peter Dillon, Declan Page, Joanne Vanderzalm, Simon Toze, Craig Simmons, Grant Hose,
Russell Martin, Karen Johnston, Simon Higginson and Ryan Morris
Lessons from 10 Years of Experience with Australia's Risk-Based Guidelines for Managed
Aquifer Recharge
Reprinted from: *Water* **2020**, *12*, 537, doi:10.3390/w12020537 293

Ido Negev, Tamir Shechter, Lilach Shtrasler, Hadar Rozenbach and Avri Livne
The Effect of Soil Tillage Equipment on the Recharge Capacity of Infiltration Ponds
Reprinted from: *Water* **2020**, *12*, 541, doi:10.3390/w12020541 313

Pieter J. Stuyfzand and Javier Osma
Clogging Issues with Aquifer Storage and Recovery of Reclaimed Water in the Brackish
Werribee Aquifer, Melbourne, Australia
Reprinted from: *Water* **2019**, *11*, 1807, doi:10.3390/w11091807 325

Shuai Liu, Weiping Wang, Shisong Qu, Yan Zheng and Wenliang Li
Specific Types and Adaptability Evaluation of Managed Aquifer Recharge for Irrigation in the North China Plain
Reprinted from: *Water* **2020**, *12*, 562, doi:10.3390/w12020562 . . . . **349**

Nasanbayar Narantsogt and Ulf Mohrlok
Evaluation of MAR Methods for Semi-Arid, Cold Regions
Reprinted from: *Water* **2019**, *11*, 2548, doi:10.3390/w11122548 . . . . **365**

Yasmin Adomat, Gerit-Hartmut Orzechowski, Marc Pelger, Robert Haas, Rico Bartak, Zsuzsanna Ágnes Nagy-Kovács, Joep Appels and Thomas Grischek
New Methods for Microbiological Monitoring at Riverbank Filtration Sites
Reprinted from: *Water* **2020**, *12*, 584, doi:10.3390/w12020584 . . . . **379**

About the Editors

Peter Dillon is a research engineer who led teams on groundwater quality protection and water recycling in CSIRO Land and Water from 1985 to 2014. He is the author or co-author of 150 journal papers and 200 reports, and has edited nine books or journal Special Issues largely on MAR. He was a founding co-chair of the IAH Commission on Managing Aquifer Recharge, from 2001 to 2019, and co-founded the Australian Water Association's Water Recycling Special Interest Group, and is a Fellow of Engineers Australia. He has supervised 24 postgraduates, run 25 MAR training courses in 14 countries and also consults. His research interests include: hydrogeology; water quality protection; stormwater management; water recycling; and water policy.

Enrique Fernández Escalante has a major interest in hydrogeological interventions, especially for managed aquifer recharge (MAR) and has implemented projects in 16 countries during his 30 years of professional experience. While his activity is more directed to the practical deployment of MAR systems than in academia, he does have research interests including related aspects of environmental hydrogeology, unsaturated zone studies and integrated water resources management (IWRM). Dr. Escalante has been a co-chair of the IAH MAR commission since 2014. He is the author, co-author or editor of 27 books, most of which are MAR-related, and more than 60 peer-reviewed articles. Dr. Escalante also teaches a Masters course at the Technical University of Madrid. Over many years he has been or currently serves as a member of 18 competitive R&D projects on IWRM and MAR, including as a coordinator for four of these.

Sharon B. Megdal holds a Ph.D. in Economics from Princeton University and is the Director of the Water Resources Research Center at the University of Arizona, USA. She also serves as a Professor, at the Department of Environmental Science, a C.W. & Modene Neely Endowed Professor, and a Distinguished Outreach Professor. Dr. Megdal works at geographic scales, ranging from local to international, and her research efforts include the comparative evaluation of water management, policy, and governance in water-scarce regions, aquifer recharge, and transboundary aquifer assessment. As the author of many articles, she is also the editor of Shared Borders, shared Waters: Israeli–Palestinian and Colorado River Basin Water Challenges and multiple special journal issues. Dr. Megdal teaches the multi-disciplinary graduate course "Water Policy in Arizona and Semi-Arid Regions".

Gudrun Massmann is a professor at the Carl von Ossietzky University, Oldenburg, where she has led the Hydrogeology & Landscape Hydrology group since 2010. The research interests of the group include the fate of organic trace pollutants in groundwater, managed aquifer recharge, surface water–groundwater interaction, coastal hydrogeology and ecohydrology. She has co-authored 79 peer-reviewed journal papers that that are well cited (¿2000 citations with a corresponding h-index of 26 based on theWeb of Science) and is presently involved in projects dealing with water recycling, groundwater salinisation following sea-level rise, the fate of freshwater lenses in view of climate change and the biogeochemistry of subterranean estuaries.

Preface to "Managed Aquifer Recharge for Water Resilience"

Managed aquifer recharge (MAR) is part of the palette of solutions to water shortage, water security, water quality decline, falling water tables, and endangered groundwater-dependent ecosystems. It can be the most economic, most benign, most resilient, and most socially acceptable solution, but it has frequently not been implemented due to a lack of awareness, the inadequate knowledge of aquifers, the immature perception of risk, and incomplete policies for integrated water management, including linking MAR with demand management. MAR can achieve much towards solving the myriad local water problems that have collectively been termed "the global water crisis". This Special Issue strives to elucidate the effectiveness, benefits, constraints, limitations, and applicability of MAR, together with its scientific advances, to a wide variety of situations that have global relevance. This Special Issue was initiated by the International Association of Hydrogeologists Commission on Managing Aquifer Recharge to capture and extend from selected papers at the 10th International Symposium on Managed Aquifer Recharge (ISMAR10) held in Madrid, Spain, 20–24 May 2019.

Peter Dillon, Enrique Fernández Escalante, Sharon B. Megdal, Gudrun Massmann
Editors

Editorial

Managed Aquifer Recharge for Water Resilience

Peter Dillon [1,2,*], Enrique Fernández Escalante [3], Sharon B. Megdal [4] and Gudrun Massmann [5]

1. CSIRO Land and Water, Waite Laboratories, Waite Rd, Urrbrae, SA 5064, Australia
2. National Centre for Groundwater Research and Training (NCGRT) & College of Science and Engineering, Flinders University, Adelaide, SA 5001, Australia
3. Grupo Tragsa, Subdirección de Innovación, 28006 Madrid, Spain; efernan6@tragsa.es
4. Water Resources Research Center, University of Arizona, Tucson, AZ 85721, USA; smegdal@arizona.edu
5. Department of Biology and Environmental Sciences, Carl von Ossietzky University of Oldenburg, D-26111 Oldenburg, Germany; gudrun.massmann@uni-oldenburg.de
* Correspondence: pdillon500@gmail.com

Received: 4 June 2020; Accepted: 8 June 2020; Published: 28 June 2020

Abstract: Managed aquifer recharge (MAR) is part of the palette of solutions to water shortage, water security, water quality decline, falling water tables, and endangered groundwater-dependent ecosystems. It can be the most economic, most benign, most resilient, and most socially acceptable solution, but frequently has not been implemented due to lack of awareness, inadequate knowledge of aquifers, immature perception of risk, and incomplete policies for integrated water management, including linking MAR with demand management. MAR can achieve much towards solving the myriad local water problems that have collectively been termed "the global water crisis". This special issue strives to elucidate the effectiveness, benefits, constraints, limitations, and applicability of MAR, together with its scientific advances, to a wide variety of situations that have global relevance. This special issue was initiated by the International Association of Hydrogeologists Commission on Managing Aquifer Recharge to capture and extend from selected papers at the 10th International Symposium on Managed Aquifer Recharge (ISMAR10) held in Madrid, Spain, 20–24 May 2019.

Keywords: groundwater recharge; water quality; water banking; managed aquifer recharge; water crisis

---

1. Introduction

The papers in this special issue explain how managed aquifer recharge (MAR) addresses water resilience challenges across the globe. A key water management objective is increasing the security of water supplies in droughts and emergencies. Another is improving water quality so that sources of water are able to supply drinking water or buffer against water quality decline due to ingress of saline or polluted waters. MAR is also used for ecological restoration of wetlands and stream habitats that have been impacted by surface water and groundwater extraction. Well-conceived and executed MAR projects therefore offer water managers the opportunity to realize water resilience benefits.

This collection of papers goes beyond enumerating these benefits in various climatic, geological and social settings. It also addresses the supportive measures to enhance the ability of MAR to proceed sustainably and effectively to achieve these benefits. Identifying suitable sites for MAR is one fundamental prerequisite. In recent years, a systematic way of doing this has been by overlaying layers of relevant variables within a geographic information system and taking combinations of these with predetermined weights and criteria for likelihood of success (multi-criteria decision analysis). Examples and a synthesis of this approach are presented in this special issue. In addition to aquifer suitability mapping, there is also a need to know where sources of water are available for recharge and where there are existing or projected demands for recovered water. The composite is known as opportunity assessment and examples are given. Time series modelling of water availability is also used in one paper to determine when recharge is possible and when recovery is needed to help with integrating MAR into a national water supply system.

Creating awareness of MAR, especially where it is an underutilised tool in water management, is an important step to increase its effective deployment and impacts. Hence, overviews of MAR practices at the national and continental scales help develop understanding of the relevant conditions where MAR has proven effective. Awareness of the policies and guidelines relating to MAR at the national and state scales, at which water is commonly managed, also helps water regulators determine the regulations warranted for effective implementation of MAR. Examples are presented where policies have had positive and unintended negative impacts on the usefulness of MAR.

Concerns by operators over chronic operational issues, such as clogging, must be addressed to avoid MAR projects becoming unsustainable and therefore not producing the water resilience intended over time. The largest cause of failure of MAR systems is that methods to manage clogging have been insufficient at some sites. Two papers focus on clogging—one in infiltration basins and one in injection wells. They show how well-constructed investigations and research can provide necessary information for the long-term successful operation of projects where recycled water is recharged.

Finally, the future of MAR is enhanced through innovation in MAR methods and monitoring. Several papers reveal highly innovative MAR methods. One paper describes a variety of ways to harness surface water irrigation canals to recharge aquifers where irrigation can draw from canals and aquifers. Another paper initiates an exploration of a method to simplify monitoring of microbiota in aquifers used for bank filtration, which has implications for pathogen removal.

Table 1 maps each paper to water resilience themes and the discussion of this introductory paper. The thematic categories include water security improvement, water quality improvement and environmental protection and restoration. Following these are some cross-lapping supportive themes referenced above: mapping of suitable MAR sites and identifying opportunities; continental-scale and national overviews of MAR practices and policies; operational issues including management of clogging; and innovation in MAR methods and monitoring. Table 1 shows the papers in order of mention. It highlights the section of this introductory paper where each paper is featured and also includes information on the type of source of water used; type of target aquifer involved; type of recharge method; end use of recovered water, and represented geographic area.

Table 1. Directory to the matters addressed and the characteristics of managed aquifer recharge (MAR) sites for each paper; Highlighting shows the introductory paper section assignment.

Reference Number	Authors of Paper	Improve Water Security	Improve Water Quality	Improve Environment	Mapping/ Opportunity Assessment	National Summary/ Legislation/Policy	Clogging/ Operational Issues	Innovative MAR Methods	Source Water *	Aquifer Type	Recharge Method #	End Use	Geographic Area
[1]	Fernández et al. (2019)	Y	y	y				y	all	all	all	irrigation	Spain
[2]	Alam et al. (2020)	Y					y	y	N	alluvial	hybrid-basin & wells	irrigation	India
[3]	Soni et al. (2020)	Y	y						N	hardrock	dug wells	irrigation	India
[4]	Kruć et al. (2019)		Y						N	alluvial	river bank filtration	potable	Poland
[5]	Masse-Dufresne et al. (2019)		Y						N	alluvial	bank filtration	potable	Canada
[6]	Patenaude et al. (2020)		Y		y				N	all	bank filtration	potable	Canada
[7]	Valhondo et al. (2020)		Y					y	R	alluvial	SAT	potable	Spain
[8]	Van Kirk et al. (2020)	y	y	Y		y		y	N	alluvial	infiltration	fishery, agric	USA
[9]	Sallwey et al. (2019)				Y				all	all	all	all	universal
[10]	Dahlqvist et al. (2019)	y			Y				N	limestone	infiltration	potable	Sweden
[11]	Knapton et al. (2019)	y			Y				N	laterite	all	all	Australia
[12]	Maréchal et al. (2020)				Y				N	alluvial	infiltration	non-potable	France
[13]	Lindhe et al. (2020)				Y				N	alluvial	all	potable	Botswana
[14]	Ebrahim et al. (2020)	y	y		y	Y		y	all	all	all	all	Africa
[15]	Shubo et al. (2020)	y	y			Y		y	N	all	all, incl. novel	any	Brazil
[16]	Cruz-Ayala and Megdal (2020)	y	y			Y			all	all	all	all	Mexico
[17]	Dillon et al. (2020)		y	y		Y	Y	y	all	all	all	all	Australia
[18]	Negev et al. (2020)	y					Y	y	R	sandstone	SAT	non-potable	Israel
[19]	Stuyfzand and Osma (2019)		y					y	R	siliclastic	injection wells	non-potable	Australia
[20]	Liu et al. (2020)	y			y			Y	N	alluvial	all, incl. novel	irrigation	China
[21]	Narantsogt and Mohrlok (2019)	y			y			Y	N	alluvial	infiltration	potable	Mongolia
[22]	Adomat et al. (2020)		y					Y	N	alluvial	river bank filtration	potable	Hungary

Notes: * N = natural water; R = recycled water; # SAT = soil aquifer treatment; Y = primary contribution of paper; y = additional contribution of paper.

2. Synopsis of Contents of This Special Issue

2.1. Water Security Improvement

Most papers reported on water supply security improvements, with three of the papers providing an assessment of benefits. The broadest range of benefits is reported for a diversity of MAR projects in Spain. Fernández et al. [1] explains how additional storage enables adaptation to climate change by buffering water availability during reduced rainfall and extended droughts. For these cases, the additional storage has been quantified. In Los Arenales aquifer, Santiuste Basin, this is sufficient to supply farmers for three years with no rainfall. Another benefit is the quantified reduced energy demand for the pumping of groundwater, which itself is a step to reduce carbon emissions and mitigate climate change. Furthermore, the aquifer acts as a reticulation system to deliver water without pumping to farmers wells. The integration of treated wastewater in several projects enhanced groundwater recharge and its reliability and further increased storage.

In monsoonal North India, imbalance between supply and demand is an annual and interannual problem. MAR has been proposed by Alam et al. [2] as a possible solution to both. They conducted the first systematic, multi-year assessment of the performance of pilot-scale MAR designed to harness village ponds to replenish alluvial aquifers in an intensively groundwater-irrigated, flood-prone area of the Indo-Gangetic Plain. In Ramganga Basin, adjacent to an irrigation canal, an unused village pond in clay soil was equipped with 10 recharge wells, and volumes and levels were measured over each wet season for three years. Recharge averaged 44,000 m^3 $year^{-1}$ at a rate of 580 m^3 day^{-1} (221 mm day^{-1}) during up to 3 months each year, enough to irrigate 8–18 ha dry season crop. This was up to 9 times the recharge without wells. Significant reductions in recharge rates occurred during each wet season due to clogging of the annular sand filters surrounding recharge wells and due to hydraulic connection with the aquifer. Authors conclude that the pilot has a beneficial impact on water security for village supplies but would need widespread replication to have an observable impact on flooding.

Another multi-year pilot-scale trial, also in India but using gravel filters to filter field runoff before recharging farmers open dug wells in hard-rock terrain in Rajasthan, was undertaken by Soni et al. [3]. A total of 11 wells were recharged between 1 and 3 years, and depth to water level was monitored weekly for 5 years for all recharge wells and for two control wells near each. In this case, volumes of water recharged were too small to produce sufficient additional crop to justify the cost of recharge infrastructure. This is unlike check dams on streams in the same catchment that have a benefit to cost ratio greater than 4. Water sampling suggested lowered salinity and fluoride in recharged wells but increased turbidity and *Escherichia coli*. An unexpected finding of this study was that no sampled open dug well met drinking water standards. Hence, wellhead water quality protection measures, including parapet walls and covers and prevention of direct recharge, were recommended for wells used for drinking water supplies. Testing of larger-scale field infiltration pits is now planned.

2.2. Water Quality Improvement

Improving the quality of drinking water supplies through bank filtration was the focus of three papers. Kruć et al. [4] studied the fate of 25 pharmaceuticals in the Warta River at a bank filtration site in Poland. Thirteen compounds were detected in bank filtrate and removal increased with distance from the stream. Some chemicals were completely removed at distances less than 38 m, while a few known persistent chemicals were still present but at greatly reduced concentrations for wells up to 250 m from the river. At the most distant well, only carbamazepine and sulfamethoxazole were detected. Average removal of most parameters was 70–80% even at less than 100 m distance from the river, demonstrating the additional value of bank filtration in the drinking water treatment train.

Masse-Dufresne et al. [5] studied the quality of water at a bank filtration site near Montreal, Canada, where two lakes contributed to the supply, and the mixing ratios were dynamic depending on relative lake levels and the pumping regime for wells. Salinity contrasts between lakes and seasonal differences in iron and manganese concentrations allowed an understanding of how to modify pumping to improve the quality of water pumped.

In the same area of south east Canada that contains many streams and lakes and a huge number of municipal water supply wells, Patenaude et al. [6] posed the question "which of these are in fact induced river bank filtration wells that may require greater protection from potential surface water pollution?" They used a GIS with multi-criteria decision analysis (MCDA) to categorise the likelihood of wells inducing infiltration from surface water. Minimum distance of wells from lakes or streams and type of aquifer were the variables selected for categorising wells. It was found that almost one million people are supplied from wells within 500 m of either streams or lakes. The method is seen by authors as a starting point for a risk-based analysis that takes account of water quality, environmental tracers and contaminants in source waters.

Water quality improvement is also an objective of soil aquifer treatment systems that intermittently infiltrate recycled water. Valhondo et al. [7] tested the use of several types of organic-rich reactive layers placed at the bottom of infiltration basins to enhance water quality improvement during soil passage. Field tests were performed at two sites in Spain. Results showed that the reactive layers in most cases enhanced the removal of the selected organic chemicals analysed (pharmaceuticals and personal care products). Candidate mechanisms for removal were proposed but not evaluated, so further research is needed to discuss persistence and resilience. The reactive layer did not increase the removal of *E.coli* (a bacterial pathogen indicator) beyond the 2–4 $\log_{10}$ removals observed in controls.

An aquifer affected by seawater intrusion in Barcelona (Spain) has been preserved by a hydraulic barrier created by MAR, in a study by Fernández et al. [1], which demonstrated improved water quality by mitigating and preventing further water quality deterioration.

2.3. Environmental Protection and Restoration

In a novel case study in the Snake River catchment of Idaho, USA, Van Kirk et al. [8] used a groundwater model and stream and aquifer water temperature data to assess potential benefits of MAR to protect a trout fishery. Winter and spring MAR operations 8 km from the river supplement recharge incidental to irrigation and were calculated to increase streamflow in 2019 by 4–7% during the driest and warmest time of year by increasing cool groundwater discharge, rather than by reducing stream losses. This lowered the stream temperature from approximately 19 °C, where trout are under heat stress, to give cool refuges adjacent to springs at 14 °C, which is optimal for trout. This habitat improvement is an additional benefit of MAR that also supports agricultural irrigation. Well-developed water rights and water transaction systems in Idaho and other western states enable MAR. However, the authors note that there remain legal and administrative hurdles to using MAR for cold-water fisheries conservation in Idaho, where conservation groups so far are unable to engage directly in water transactions.

In Spain, wetland restoration has also been achieved through MAR in Castilla y León to restore water levels and maintain a geochemical equilibrium vital for bacteria, vegetation and refuge for aquatic birds (Fernández et al. [1]). Since 1995, a deep recharge well in a karstic aquifer capable of accepting 1000 L/s has been used in Lliria (Valencia) for flood mitigation while also enhancing irrigation water security [1]. In Neila, Burgos, Spain, 15–40% of flow in streams is directed via constructed channels into contour bunds in forested areas to enhance diffuse source recharge while also increasing forest production [1].

2.4. Mapping of Suitable MAR Sites and Identifying Opportunities

A number of papers made use of geographic information systems (GIS) with multi-criteria decision analysis (MCDA) to identify suitable locations for MAR operations. Sallwey et al. [9] undertook a review of such studies and out of this developed two open-source web-based tools, a query tool and a tool to help standardise weight assignment and criteria. These will help users to make mapping of MAR site suitability more structured and assist in collaboration among multiple partners. Site suitability focuses on the presence of an aquifer capable of storage and recovery of water, as well as information on the unsaturated zone characteristics to indicate viability of infiltration type methods. Data availability and quality are important in the mapping process and the tools still depend on the assessor's expertise in choosing relevant datasets for each specific study.

Although not discussed in any of the GIS-MCDA papers, modern remote sensing methods, particularly those that are satellite-based provide a dense raster of data relevant to site selection. Spatial correlation ranges can be determined using geostatistics to suggest more robust predictors than possible from sparse point-scale measurements, such as aquifer parameters from pumping tests, although these are valuable to help ground-truth predicted aquifer suitability. It is hoped that in future, greater effort will be put into parameter selection for parsimonious and robust mapping of MAR suitability, and into validation of predictions.

MAR site suitability mapping is a foundational layer in assessing MAR opportunity, where the proximity of such aquifers to sources of water such as streams, dams and water recycling plants is also considered. One example is the Island of Gottland, Sweden, where Dahlqvist et al. [10] determined the role for MAR to contribute to future water supplies. They found that 7.5% of the area of Gotland was suitable for MAR compared with 3.3% suitable for surface water supplies through new dams. Although lacking detailed site-specific studies, which they recommend, they claim MAR to be a viable option. They estimated that the unit cost of MAR was four times that of expansion of conventional groundwater supplies where this was possible. However, MAR was comparable in unit cost and yield of expanded surface water supplies and approximately one-quarter of the unit cost of seawater desalination.

Knapton et al. [11] studied MAR options using a partially calibrated groundwater model for the Darwin rural area of northern Australia. The unconfined aquifer is characterised as a lateritic aquifer that refills each wet season and was previously presumed unsuitable for MAR. However, in specific areas, some wet season storage capacity remains, with potential for up to 1.2 Mm^3/year recharge. A confined part of this aquifer was identified to have up to 5 Mm^3 storage opportunity for water banking for Darwin's water security if a 20 m head increase is acceptable in the aquifer.

Maréchal et al. [12] aim to advance GIS-MCDA mapping approaches by adding an economic evaluation for siting a MAR facility anywhere on an aquifer. They assess the levelised unit cost of recharge from an infiltration basin, including capital and operating costs, implementing a GIS-tool in order to build maps of levelised costs at the aquifer scale. The method was tested in simplified form, with assumptions declared and dependent sensitivity analysis, for an alluvial aquifer in Southern France. Authors propose that this approach be integrated into a broader analysis of soil and aquifer parameters that would influence costs and refine the consequent maps.

GIS-MCDA was also used to map zones suitable for different types of innovative recharge operations on the North China Plain (as mentioned later by Liu et al. [20]).

A different type of opportunity assessment is not based on mapping, but instead uses time series analysis of water supply and demand to determine the need for MAR and the extent to which it can contribute to security of national water supplies. Such an analysis is performed by Lindhe et al. [13] for the north–south water carrier in Botswana. This combines large shallow dams that only irregularly fill, well fields that have small and reliable supplies but only low rates of natural replenishment, and possible future MAR systems of different capabilities. The water supply security model uses monthly time steps over 23 years to relate supply with demand and simulate the magnitude and probability of water supply shortages. Implementing large-scale MAR can be shown to improve the supply reliability from 88% to 95%. The model reveals system properties that constrain the effectiveness of MAR and suggest how to further improve its benefits for an integrated system.

2.5. Continental-Scale and National Overviews of MAR Practices and Policies

Awareness of existing, relevant MAR practices alerts water managers to the possibilities and is reassuring to those contemplating undertaking a MAR project. This special issue contains a summary of MAR practice in the African continent and at the national level in Brazil and Mexico for both practice and policies. These cover a wealth of experience that is, to date, underreported in international literature. A decade of experience in Australia with MAR guidelines for health and environment protection is also reported. These accounts each have unique and highly advanced elements that will be of interest not only to these geographic areas but also globally.

Ebrahim et al. [14] review and synthesize MAR experience in Africa from 52 reported cases in 9 countries, dating back to the 1960s and covering all main types of MAR. Cases were classified under 13 characteristics including objective of the MAR, hydrogeology and climate. It was found that MAR occurred most commonly in areas of high interannual variability in water availability. The most common objective for projects is to secure and augment water supply and balance variability in supply and demand, in both urban and rural areas. Results revealed a wide diversity of applications including reservoir releases (Morocco), surface spreading/infiltration (Algeria, Tunisia, Egypt, South Africa and Nigeria), riverbank filtration (Egypt), in-channel modifications (Kenya, Tunisia and Ethiopia) and recharge wells (South Africa). Africa also contains several of the world's most sophisticated MAR projects, including aquifer injection of highly treated recycled water into crystalline rock to secure city drinking water supplies (Windhoek in Namibia) and recycling of stormwater and treated sewage via infiltration basins for town water supplies (Atlantis in South Africa). In total, the estimated annual recharge volume is 158 Mm^3/year or 0.4% of the continent's annual groundwater extraction. Advancing MAR in Africa requires fostering awareness of existing MAR projects, mapping suitability of aquifers for MAR (as performed in South Africa) and informing account of MAR in water allocation and water quality protection policies.

A study of national advance in the practice and governance of MAR in Brazil is reported by Shubo et al. [15]. Community level and government-level programs have been implemented at many sites to address dry season and drought supplies. The Barraginhas Project alone has seen construction of more than 500,000 infiltration ponds in north east Brazil up to 2013. Another Brazilian MAR design, Caixa Seca (or 'dry box') is widely used to recharge road runoff and would also have international application. More than 90 in-channel modifications for MAR have been recorded. Urban drainage public policies have stimulated urban aquifer recharge initiatives mostly aimed to reduce runoff peak flows. Concerning MAR policies, Brazil has been progressive at the federal level since 2001, when the Water Resources National Council Resolution n° 15 encouraged municipalities to adopt MAR. By 2008, its Resolution n° 92 made prior authorization and mandatory monitoring a condition of aquifer recharge. At the subnational level, regulations in all states mention MAR ('artificial recharge') and two, Pernambuco and Ceará, give incentives and prescriptions for community- and company-established MAR projects. The authors also note where improvements could be made in the reporting, monitoring, and systematic appraisal of opportunities and water quality risk management aspects.

Cruz-Ayala and Megdal [16] reviewed the occurrence and legal framework for MAR in Mexico. They found seven documented operational projects, five pilot projects and five research activities since the 1950s involving natural waters, recycled water and stormwater. Their combined recharge restores depleted aquifers, reduces land subsidence, increases water availability and mitigates floods. There are also very significant opportunities to expand MAR. Regulations are discussed that involve at least three levels of governance from national to basin and user level. There are also Mexican National Standards (NOMs) that create a specific regulatory framework for water allocation and water quality standards that MAR projects must fulfill. These specify the information needed to obtain a permit. Some gaps in regulations are identified, such as on entitlements to recover recharged water, that, if addressed, would help to motivate new MAR projects to address critical needs.

Dillon et al. [17] reviewed the consequences of the Australian MAR guidelines for health and environment protection after 10 years of implementation. They found that, in those states where MAR is progressing, the guidelines are welcomed as giving certainty and objectivity to approvals and fitting broader risk management approaches to water quality. In the other states, there has been no progress, although the need for MAR is just as great. Only minor adjustments are suggested to the guidelines, such as taking specific account of temperature change as a hazard in geothermal settings, referencing advances in environmental genomics, and accounting more explicitly for cumulative impacts of multiple MAR projects. In the entry level section of the guidelines, more explicit water entitlement arrangements for sourcing water, recharging aquifers, recovering from aquifers and end uses are suggested for basins where groundwater management policies need to be strengthened to be effective in securing MAR entitlements. Its relevance for application in other countries depends on capabilities to monitor, sample and analyse water quality. If such capabilities are scant, other forms of guideline are more appropriate, and India's is given as an example.

2.6. Operational Issues Including Management of Clogging

Two papers focus on operational procedures to manage clogging in MAR projects utilising recycled water. The first of these describes changes to tilling operations in intermittent infiltration basins (soil aquifer treatment) at the Dan Region Reclamation Project (Shafdan) near Tel Aviv, Israel (Negev et al. [18]). After 20 years of stable operation, infiltration rates declined over a 3 year period due to changing water quality, reduction in drying periods, and seasonal effects. Tillage changes introduced in a replicated full-scale trial increased recharge capacity up to 95% for deep ploughing and 15% for chisel knife cultivator treatments, both with improved tractor power and depth control systems. Measurements included infiltration rates and soil compaction depths. Minimising compaction by allowing complete drainage before tillage is important for sustaining higher infiltration rates.

Stuyfzand and Osma [19] evaluated clogging at a pilot-scale recycled water aquifer storage and recovery (ASR) well in a confined siliclastic aquifer near Melbourne Australia. They recorded head build up during an injection and recovery trial, analysed water quality and purged solids, and developed some novel tests to predict clogging by suspended particles and biofouling. These revealed that additional water treatment would be needed and reduced rates of injection, requiring more wells to achieve the injection volume target.

2.7. Innovation in MAR Methods and Monitoring

Two novel recharge systems called a "well–canal combination mode" and an "open channel–underground perforated pipe–shaft–water saving irrigation system" as practiced in the Yellow River irrigation district on the north China Plan are described by Liu et al. [20]. These are among the described numerous types of agricultural MAR practiced since the 1970s in the North China Plain. Adaptive measures to compensate for diversion of more Yellow River water to cities, to sustain conjunctive use during irrigation efficiency improvement, and to prevent clogging by fine sediment are described. Further, a GIS system using a multi-criteria decision analysis (MCDA) was developed to identify zones for sustainable development of MAR projects in the Yellow River Irrigation District of Shandong Province. Mapping revealed highest opportunities for MAR systems in the western part of Liaocheng City irrigation area, where deeper water table and greater sand thickness gave more storage potential.

Non-conventional methods for recharge enhancement are proposed by Narantsogt and Mohrlok [21] to secure the depleting water supply for Ulaanbaatar, the capital city of Mongolia. They modelled several configurations for ice storage and melting in the dry season, when groundwater levels are low and there is no river flow in this cold semi-arid area. Combining these with recharge releases from a dam is predicted to meet the ongoing water needs for the city.

Advances in monitoring methods are important to the efficient operation of MAR and protection of public health and the environment. Microbially-mediated processes in porous media are important for the removal of viruses, bacteria, and protozoa that are pathogenic to humans. Measuring bacterial biomass concentration and enzymatic activity is an innovative way of improving understanding of these processes. Adomat et al. [22] did this using flow cytometry and two precise enzymatic detection methods to monitor dynamic fluctuations in bacterial biomass at three riverbank filtration sites. They also performed online flow cytometry in an ultrafiltration pilot plant. The method showed promise as a rapid, easy and sensitive future alternative to traditional labour-intensive methods for assessing the microbial quality of RBF water, but this is still an early stage of development. Findings of bacterial regrowth on membranes reinforce the value of river bank filtration as a pretreatment for ultrafiltration for drinking water supplies.

3. Summary

The water security improvement is, perhaps, the biggest target and advantage of MAR, understood as an integrated water resources management (IWRM) technique. Examples from more than ten countries demonstrate MAR as a climate change adaptation mechanism (no regret technique) towards securing long-term groundwater availability. It also facilitates the transmission of water throughout an aquifer to users' wells; a reduction in pumping energy costs due to higher groundwater levels; and, in karst areas, can assist flood mitigation by infiltrating water to the aquifer, thereby increasing the concentration time of the storm and reducing erosion.

Another major motivation to conduct MAR is the fact that the water quality of the source water generally improves during subsurface passage. Bank filtration for drinking water production is discussed in several papers in this special issue. These cover aspects from studies of passive removal of organic chemicals, modifying mixing of waters by adjusting pumping rates, and efforts to differentiate unintentional bank filtration from other groundwater supplies for managing water quality risk. Several studies mentioned MAR in the form of hydraulic barriers to prevent saltwater intrusion, and one used reactive organic-rich layers placed at the bottom of intermittent infiltration basins to accelerate degradation of constituents of recycled water.

To date, less in the spotlight are the benefits of MAR for environmental protection and restoration. Examples herein show that MAR may increase streamflow by increased groundwater discharge to adjacent rivers, hence decreasing water temperatures in summer to benefit a trout fishery. Moreover, wetland restoration can be achieved by increasing water levels following MAR, as shown for cases in Spain.

Mapping of aquifer suitability for MAR has used geographic information systems (GIS) with multi-criteria decision analysis (MCDA) in several areas and web-based tools are emerging aiming to simplify the process. Overlaying aquifer suitability maps with maps of water demand and water sources has allowed MAR opportunity maps to be produced. A pioneering attempt to extend these to levelised cost mapping for MAR has been included. Modelling of the historical operation of a national water grid with and without MAR has helped to define how to integrate MAR with the existing infrastructure for maximum benefit.

The enormous diversity of experience in MAR globally means that many different countries have undertaken initiatives to advance aspects of MAR. Two examples are Brazil's 500,000 Barraginhas (infiltration ponds) and the injection of highly treated recycled water into crystalline rock in Namibia for supplying drinking water. Countries with emerging and growing economies see real opportunities for making precious water supplies resilient and are moving ahead to do so at a faster pace than some countries with more established economies. A number of countries, including Mexico, Brazil and Australia, have for some time recognised the need for MAR regulations to protect aquifers. In Australia, regulations have assured sustainable operations and assisted uptake of MAR. There remain some challenges. An example is in Mexico, where existing water allocation policies give no incentive for investing in MAR to increase water supply.

Clogging, the most frequently encountered threat to successful MAR, has been addressed by a range of methods to maintain the infiltration rates and to improve water quality. In this special issue, field studies of infiltration basins and injection wells have been combined with innovative measurement methods and solutions to trial or recommend strategies. These start with problem avoidance, by improving the quality of source water, monitoring to detect and reject water with key parameters exceeded, removing residual drilling fluids and preventing air entrainment. Then small cycle improvements become the focus, such as improved practices with valves, metering, drying times and flow rates. Finally, longer-term maintenance and remedial measures such as changing the method of purging of injection wells or tillage in infiltration basins are tested and adopted.

Innovations in MAR methods and monitoring are continuously developed and implemented, and a number of very diverse approaches are presented herein. These include non-conventional methods for recharge enhancement, such as melting of ice storage and underground ice dams in semi-arid cold Mongolia, innovative MAR types for agricultural irrigation in the "open channel–underground perforated pipe–shaft–water saving irrigation system" developed for the North China plain, and new methods to monitor dynamic fluctuations in bacterial biomass at riverbank filtration sites applied in Europe.

The analyses in this special issue strongly indicate that MAR innovations will continue, as will the sharing of results, as exemplified by the following 22 papers. With increased project diversity, and changing climate leading to tighter constraints on availability of surface water and increased needs for recovery of stored water, MAR operations will themselves need to become increasingly resilient and efficient. Good monitoring of demonstration projects to improve process understanding, and good site selection, as illustrated in this special issue, will be foundational for achieving reliable MAR systems that produce resilient water supplies.

Author Contributions: Each guest editor contributed as an author to the formulation of the concept and structure of this paper, to writing various sections and to revising the text and table. All authors have read and agreed to the published version of the manuscript.

Funding: This research received no external funding.

Acknowledgments: This special issue was initiated jointly by the Commission on Managing Aquifer Recharge of the International Association of Hydrogeologists, the organising committee of the 10th International Symposium on Managed Aquifer Recharge (ISMAR10) held in Madrid, Spain, in the period 20–24 May 2019, and MDPI Journal Water. This is the third special issue in *Water* following those for ISMAR8: Policy and Economics of MAR and Water Banking, and ISMAR9: Water Quality Considerations for Managed Aquifer Recharge Systems. The editors thank all those authors who contributed to this special issue and to the many reviewers whose comments have helped to improve the papers it contains. Finally, thank you to Rachel Lu and the staff of MDPI, who expertly and efficiently managed the publication of this special issue.

Conflicts of Interest: The authors declare no conflict of interest.

References

1. Fernández Escalante, E.; San Sebastián Sauto, J.; Calero Gil, R. Sites and indicators of MAR as a successful tool to mitigate climate change effects in Spain. *Water* **2019**, *11*, 1943. [CrossRef]
2. Alam, M.F.; Pavelic, P.; Sharma, N.; Sikka, A. Managed aquifer recharge of monsoon runoff using village ponds: Performance assessment of a pilot trial in the Ramganga Basin, India. *Water* **2020**, *12*, 1028. [CrossRef]
3. Soni, P.; Dashora, Y.; Maheshwari, B.; Dillon, P.; Singh, P.K. Managed aquifer recharge at a farm level: Evaluating the performance of direct well recharge structures. *Water* **2020**, *12*, 1069. [CrossRef]
4. Kruć, R.; Dragon, K.; Górski, J. Migration of pharmaceuticals from the Warta River to the aquifer at a riverbank filtration site in Krajkowo (Poland). *Water* **2019**, *11*, 2238. [CrossRef]
5. Masse-Dufresne, J.; Baudron, P.; Barbecot, F.; Patenaude, M.; Pontoreau, C.; Proteau-Bédard, F.; Menou, M.; Pasquier, P.; Veuille, S.; Barbeau, B. Anthropic and meteorological controls on the origin and quality of water at a bank filtration site in Canada. *Water* **2019**, *11*, 2510. [CrossRef]
6. Patenaude, M.; Baudron, P.; Labelle, L.; Masse-Dufresne, J. Evaluating bank-filtration occurrence in the Province of Quebec (Canada) with a GIS approach. *Water* **2020**, *12*, 662. [CrossRef]
7. Valhondo, C.; Carrera, J.; Martínez-Landa, L.; Wang, J.; Amalfitano, S.; Levantesi, C.; Diaz-Cruz, M.S. Reactive barriers for renaturalization of reclaimed water during soil aquifer treatment. *Water* **2020**, *12*, 1012. [CrossRef]
8. Van Kirk, R.W.; Contor, B.A.; Morrisett, C.N.; Null, S.E.; Loibman, A.S. Potential for managed aquifer recharge to enhance fish habitat in a regulated river. *Water* **2020**, *12*, 673. [CrossRef]
9. Sallwey, J.; Schlick, R.; Bonilla Valverde, J.P.; Junghanns, R.; Vásquez, F.L.; Stefan, C. Suitability mapping for managed aquifer recharge: Development of web-tools. *Water* **2019**, *11*, 2254. [CrossRef]
10. Dahlqvist, P.; Sjöstrand, K.; Lindhe, A.; Rosén, L.; Nisell, J.; Hellstrand, E.; Holgersson, B. Potential benefits of managed aquifer recharge on the island of Gotland, Sweden. *Water* **2019**, *11*, 2164. [CrossRef]

11. Knapton, A.; Page, D.; Vanderzalm, J.; Gonzalez, D.; Barry, K.; Taylor, A.; Horner, N.; Chilcott, C.; Petheram, C. Managed aquifer recharge as a strategic storage and urban water management tool in Darwin, Northern Territory, Australia. *Water* **2019**, *11*, 1869. [CrossRef]
12. Maréchal, J.-C.; Bouzit, M.; Rinaudo, J.-D.; Moiroux, F.; Desprats, J.-F.; Caballero, Y. Mapping economic feasibility of managed aquifer recharge. *Water* **2020**, *12*, 680. [CrossRef]
13. Lindhe, A.; Rosén, L.; Johansson, P.-O.; Norberg, T. Dynamic water balance modelling for risk assessment and decision support on MAR potential in Botswana. *Water* **2020**, *12*, 721. [CrossRef]
14. Ebrahim, G.; Lautze, J.; Villholth, K. Managed aquifer recharge in Africa: Taking stock and looking forward. *Water* **2020**, *12*, 1844. [CrossRef]
15. Shubo, T.; Fernandes, L.; Montenegro, S.G. An overview of managed aquifer recharge in Brazil. *Water* **2020**, *12*, 1072. [CrossRef]
16. Cruz-Ayala, M.B.; Megdal, S.B. An overview of managed aquifer recharge in Mexico and its legal framework. *Water* **2020**, *12*, 474. [CrossRef]
17. Dillon, P.; Page, D.; Vanderzalm, J.; Toze, S.; Simmons, C.; Hose, G.; Martin, R.; Johnston, K.; Higginson, S.; Morris, R. Lessons from 10 years of experience with Australia's risk-based guidelines for managed aquifer recharge. *Water* **2020**, *12*, 537. [CrossRef]
18. Negev, I.; Shechter, T.; Shtrasler, L.; Rozenbach, H.; Livne, A. The effect of soil tillage equipment on the recharge capacity of infiltration ponds. *Water* **2020**, *12*, 541. [CrossRef]
19. Stuyfzand, P.J.; Osma, J. Clogging issues with aquifer storage and recovery of reclaimed water in the brackish Werribee aquifer, Melbourne, Australia. *Water* **2019**, *11*, 1807. [CrossRef]
20. Liu, S.; Wang, W.; Qu, S.; Zheng, Y.; Li, W. Specific types and adaptability evaluation of managed aquifer recharge for irrigation in the North China Plain. *Water* **2020**, *12*, 562. [CrossRef]
21. Narantsogt, N.; Mohrlok, U. Evaluation of MAR methods for semi-arid, cold regions. *Water* **2019**, *11*, 2548. [CrossRef]
22. Adomat, Y.; Orzechowski, G.-H.; Pelger, M.; Haas, R.; Bartak, R.; Nagy-Kovács, Z.Á.; Appels, J.; Grischek, T. New methods for microbiological monitoring at riverbank filtration sites. *Water* **2020**, *12*, 584. [CrossRef]

© 2020 by the authors. Licensee MDPI, Basel, Switzerland. This article is an open access article distributed under the terms and conditions of the Creative Commons Attribution (CC BY) license (http://creativecommons.org/licenses/by/4.0/).

Article

Sites and Indicators of MAR as a Successful Tool to Mitigate Climate Change Effects in Spain

Enrique Fernández Escalante [1,*], Jon San Sebastián Sauto [2] and Rodrigo Calero Gil [1]

[1] Tragsa, Innovation Subdirectorate, Maldonado 58, 28006 Madrid, Spain; rcalero@tragsa.es
[2] Tragsatec, Architecture and Enginering Area, Julián Camarillo 6b, 28037 Madrid, Spain; jsss@tragsa.es
* Correspondence: efernan6@tragsa.es; Tel.: +34-91-3226106

Received: 6 August 2019; Accepted: 15 September 2019; Published: 18 September 2019

Abstract: In this article, the authors will support Managed Aquifer Recharge (MAR) as a tool to combat Climate Change (CC) adverse impacts on the basis of real sites, indicators, and specific cases located Spain. MAR has been used in Spain in combination with other measures of Integrated Water Resources Management (IWRM) to mitigate and adapt to Climate Change (CC) challenges. The main effects of CC are that the rising of the average atmospheric temperature together with the decreasing average annual precipitation rate cause extreme weather and induce sea level rise. These pattern results in a series of negative impacts reflected in an increase of certain events or parameters, such as evaporation, evapotranspiration, water demand, fire risk, run-off, floods, droughts, and saltwater intrusion; and a decrease of others such as availability of water resources, the wetland area, and the hydro-electrical power production. Solutions include underground storage, lowering the temperature, increasing soil humidity, reclaimed water infiltration, punctual and directed infiltration, self-purification and naturalization, off-river storage, wetland restoration and/or establishment, flow water distribution by gravity, power saving, eventual recharge of extreme flows, multi-annual management and positive barrier wells against saline water intrusion. The main advantages and disadvantages for each MAR solution have been addressed. As success must be measured, some indicators have been designed or adopted and calculated to quantify the actual effect of these solutions and their evolution. They have been expressed in the form of volumes, lengths, areas, percentages, grades, euros, CO_2 emissions, and years. Therefore, MAR in Spain demonstrably supports its usefulness in battling CC adverse impacts in a broad variety of environments and circumstances. This situation is comparable to other countries where MAR improvements have also been assessed.

Keywords: Managed Aquifer Recharge; MAR; climate change; water management; IWRM; adaptation measures; indicators; Spain

1. Introduction

In a world of arising concern for the effects of Climate Change (CC), the search for practical solutions to mitigate undesirable consequences implies a global change of mentality in the management of water resources. Beyond overexploitation of water bodies, it is mandatory to build models that take into account the current effects of CC, especially in those countries with arid or a semiarid climate, such as the Mediterranean area, where the annual rain scarcity overlaps with punctual extreme precipitations. These accepted phenomena are indeed heightened according to the prevailing CC models.

The main manifestations of CC shown in this paper on which the Managed Aquifer Recharge (MAR) techniques can incise are an increase in the average temperature, a decrease in the annual precipitation, recurrent extreme weather and a rise in the sea level [1].

The key problems and impacts of CC whose figures are globally rising are the evaporation rate, water demand, fire risk, and run-off. On the other hand, decreasing figures are found, at least, in the water supply, wetland surface and hydro-electric energy production [2].

Managed Aquifer Recharge (MAR) can provide with a large array of technical solutions to mitigate those adverse CC effects by not only managing groundwater, but also showing an integrated vision of water resources and their associated wetlands, following the EU Water Framework Directive approach (2000/60/CE) [3]. This concept has been put into practice all over the world, facing different CC challenges, from building extreme run-off reservoir systems to fighting sea level intrusion. The monitoring of these devices shows a bunch of indicators of successful recharge and simultaneous local CC mitigation effects.

Some out-of-Spain models to assess the potential impact of future climate conditions on groundwater quantity and quality have been performed, e.g., in the Central Huai Luang Basin of Thailand. There, four different cases were developed to study the spreading saline groundwater and saline soils in this basin as a consequence of MAR activities, concluding that for all future climate conditions, the depths of the groundwater water table not only will increase, but also the salinity distribution areas will follow this trend by about 8.08% and 56.92% in the deep and shallow groundwater systems, respectively [4].

On the Spanish Mediterranean coast—an area with severe salinization over the last 40 years due to intensive exploitation of groundwater—piezometric levels and chloride concentrations have been monitored. Dry periods and their associated increases in pumping caused the advance of seawater intrusion. The sharp reduction in groundwater withdrawals over the last decade has pushed the saline wedge backwards, although the ongoing extraction and the climate conditions mean that this retreat is quite slow, and could be enhanced by means of MAR applications [5].

On the Southern Italian Mediterraneas coast, over four models were tested to foresee saline water intrusion threatening fresh groundwater resources. Among all the processes taken in consideration for the simulations, authors remarked the importance of a detailed statigraphic reconstruction and geomorphological settings. The results of the validated model indicated the strong link between surface water bodies (specially affected by CC impacts) and the coastal aquifer, with a slight salinization increase for the horizon 2050 [6].

In a recent study performed by the International Association of Hydrogeologists (IAH) considering inputs from a vast variety of countries with special focus on Brazil and South Africa, authors claimed the excellent groundwater drought resilience and how it provided a 'natural solution' for the deployment in CC adaptation, by means of 'strategic rethinking', conjunctive use, and quality protection. These actions should be applied on storage availability, supply productivity, natural quality and pollution vulnerability. They advised that though uncertainty remains over the long-term effects of CC on groundwater recharge, a higher impact on shallow aquifers is still expected. Nevertheless, they also remarked the necessity for more studies to be undertaken due to the current lack of definitive data [7].

2. Materials and Methods

The main objective of this paper was to collect figures and examples that could illustrate that MAR, as well as in combination with other techniques, could successfully contest the effects of CC by adaptation and even mitigation strategies.

The methodological approach used below pairs the problems and solutions together. Main CC impacts and risks were going to be matched with the available MAR techniques that could be used to mitigate them. Effects, impacts and corresponding MAR solutions were organized in 3 different columns in Figure 1.

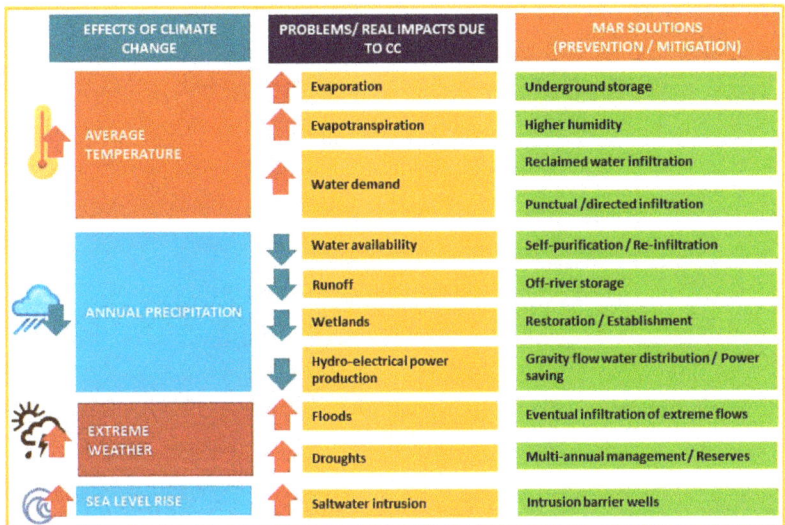

Figure 1. Relationships between the main manifestations of climate change (CC) and their main problems and impacts, and the technological solutions that can be implemented as adaptation measures.

3. Results

Results will be explained in the following pages, where some indicators will be proposed to measure the quantitative impact on CC mitigation as related to other usual techniques.

Examples of initiatives to combat climate change have been organised into four groups, as shown in Figure 1:

1. Examples of technological solutions to palliate rising temperatures (Section 3.1).
2. Examples of technological solutions to palliate decreasing annual precipitation rates (Section 3.2).
3. Examples of technological solutions to manage extreme phenomena (Section 3.3).
4. Examples of technological solutions to reduce the rising of the sea level and saline water intrusion (Section 3.4).

Spanish examples are located for every group (Figure 2).

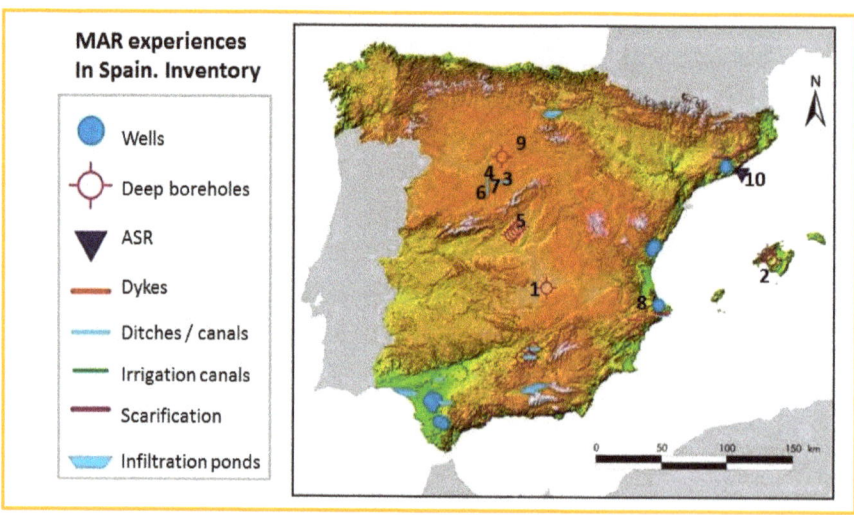

Figure 2. Map of Spain containing the Managed Aquifer Recharge (MAR) sites to fight CC adverse impacts studied and exposed. Numbers in the map follow in brackets after the header of the corresponding example in the next paragraphs.

3.1. Examples of Technological Solutions to Palliate Rising Temperatures

The International Panel for Climate Change (IPCC) in the 5th Assessment Report (2014) declared that the global temperature will rise more than 1.5 °C during the 21st century in all the possible scenarios and probably 2 °C in two of the highest emission sceneries [8]. Evaporation, evapo-transpiration and water demand are expected to follow this trend, but MAR has its own means to counteract those effects.

The indicators for each of the following examples have been gathered in Table 1.

Table 1. Selected Spanish MAR sites and indicators to track their relationship with CC adverse impacts.

MAR SITE (*)	CC IMPACT	INDICATOR/S
Guadiana Canal, Castilla-La Mancha (1)	PALLIATE RISING TEMPERATURE	Capability to recharge peak flows: Intermittent underground water storage. Total recharge in 48 supplementary hm^3/year
Parc Bit Majorca Island (2)	Palliate rising temperature	Lower surface temperature according to thermographic photographs
Gomezserracín, Castilla y León (3)	Palliate rising temperature	Increase in soil humidity during MAR cycles
Alcazarén, Castilla y León (4)	Strategic water storage/Palliate rising temperature	Increases of 0.4 hm^3 per year recharging reclaimed water
CYII Madrid (5)	Strategic water storage/Palliate rising temperature	Capability to recharge peak flows: Increases up to 5 hm^3 per year by recharging potable water excess
Santiuste, Castilla y León (6)	PALLIATE DECREASING PRECIPITATION RATES	Strategic reserve for drought periods +/−12–53% in water physical and chemical parameters

Table 1. *Cont.*

MAR SITE (*)	CC IMPACT	INDICATOR/S
Santiuste, Castilla y León (6)	Wetlands Restoration	5% recharge volume dedicated to alkaline lake restoration
El Carracillo, Castilla y León (7)	Gravity flow water distribution	Transport length without pumpage: 40.7 km of pipes and channels by gravity. Supplied irrigated area: 3500 ha
El Carracillo, Castilla y León (7)	Energy efficiency through Managed Aquifer Recharge	Saving in terms of kW-h is between 12 and 36% thanks to water level rise
Arnachos, Valencia (8)	MANAGE EXTREME CC PHENOMENA	Reduce precipitation peak thank to a high recharge capacity borehole (up to 1000 L/s)
Neila, Castilla y León (9)	Forested watersheds	Forest is capable of retaining and channelling 15%–40% of the volume of surface runoff
Santiuste, Castilla y León (5)	Multiannual management by means of Off-river storage	2.62 hm^3/year stored out of Voltoya River would allow groundwater extractions for irrigation during 3 years with no rain
El Prat de Llobregat, Cataluña (10)	REDUCE SEA LEVEL RISE AND SALINE WATER INTRUSION	Evolution of seawater intrusion by iso-chlorides lines evolution

(*) The positions of these MAR sites have been exposed in the Figure 2.

3.1.1. Underground Water Storage. Canal del Guadiana, Castilla-La Mancha (1 in Figure 2)

The high temperatures over 40 °C in August, that can be found in the historic record of the Castilla-La Mancha Region [9], and the shallow streams multiply the evaporation rate in summer. A net of wells close to the canal of Guadiana was built by the Guadiana River Water Authority (CHG) in Castilla-La Mancha (Figure 3), for rural development and mitigation of the overexploitation of the groundwater body (known as aquifer #23). This MAR system can increase the total storage volume by means of intentional recharge in about 48 supplementary hm^3 per year [10].

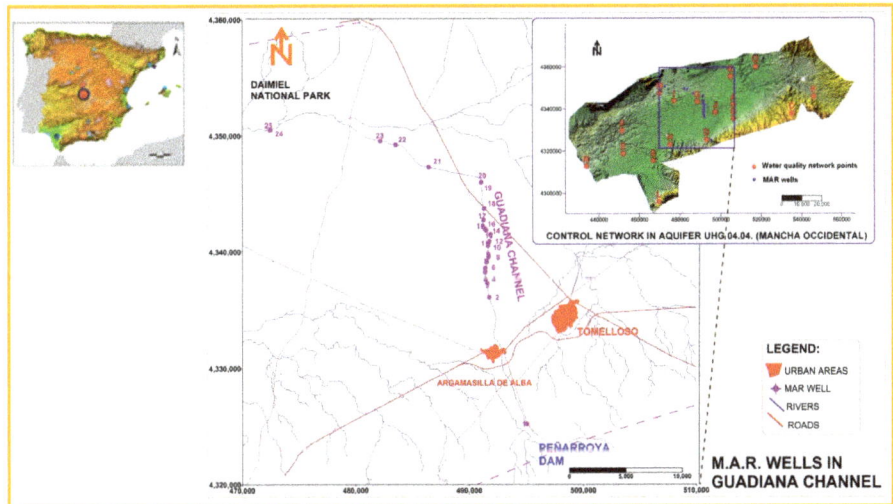

Figure 3. General sketch of the MAR system of wells near Canal del Guadiana for irrigation and environmental purposes.

3.1.2. Temperature Reduction. Parc Bit, Palma de Mallorca, I. Balears (2)

The Sustainable Drainage Urban Systems (SDUS) consisted in a group of building items that was integrated into the urban architecture [11], with the goal to increase the city water permeability by means of rising run-off infiltration into the aquifers under the town surface. At the same time, they could also combat Urban Heat Island (UHI) through the development of water stores and green areas within the city landscape.

A good example of this practice can be found in Parc Bit (Palma de Mallorca, Figure 4), where the vegetated roofs, fed by rain collection, were able to reduce the air temperature in the range of 1.5 to 6 °C. Thermographs were able to establish a clear and quick difference displayed in the pattern of colours when areas with or without a canopy were compared [12]. A square meter of green cover could evaporate more than half a litre of water per day.

Figure 4. Sustainable drainage urban systems (SUDS) to reduce the urban heat island (UHI). Model and development of green roof on the Parc Bit building, Palma de Mallorca, Spain. Example of thermography to track the UHI evolution.

3.1.3. Increase in Soil Humidity. Gomezserracín, Castilla y León (3)

Los Arenales aquifer in Gomezserracín provided an example of increased soil moisture and a rise in the phreatic surface brought about by underground storage through a system of canals and streams (Figure 5). Artificial recharge operations, initiated in 2003, resulted in an average rise in the phreatic surface of more than 2 m, even though it was a passive system since it did not require any electrical power to work. This additional storage in the unsaturated zone increased soil moisture by 15%–20% according to datasets obtained from the MARSOL ZNS-3 station [9–17], equipped with a set of sensors which captured measures in both, the saturated and the unsaturated zones. Humidity evolution has been the main assessed indicator after taking into account the natural precipitation. The costs, appart from the initial investment, were due to cleaning and maintenance, with an average of about 30,000 €/year, contributed by the irrigators´ association.

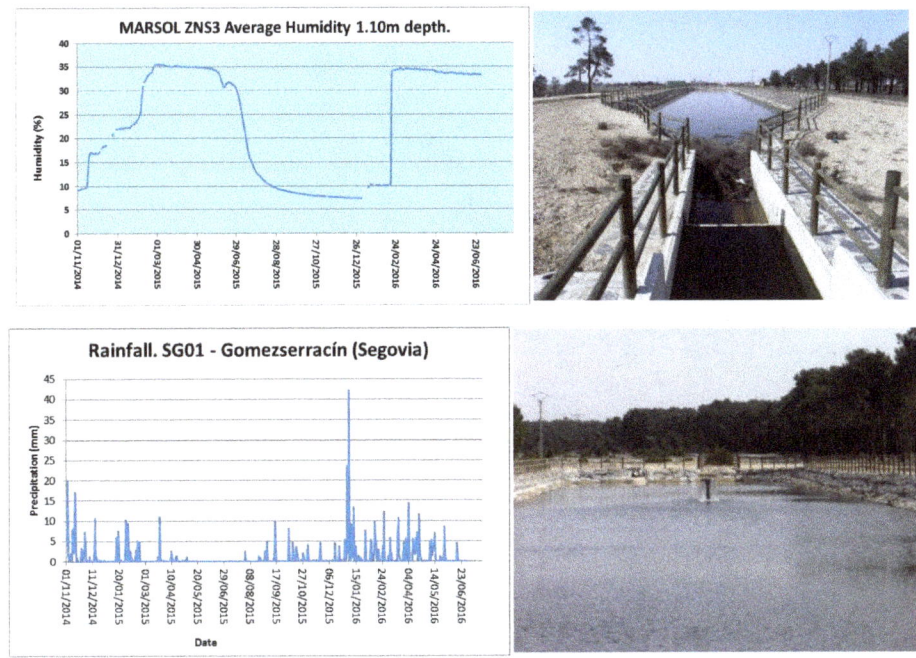

Figure 5. Infiltration ponds in the Los Arenales aquifer: Gomezserracín, soil humidity evolution from the so called MARSOL ZNS-3 station datasets (02/11/2014–30/06/2016) and natural precipitation evolution [9].

3.1.4. Reclaimed Water Infiltration. Alcazarén, Castilla y León (4)

The recharge system (Figure 6) began operating in 2012, with an estimated annual recharge of 0.6 hm^3 for the whole working period, with scarce variations, thus the main indicator remained constant [14–23]. In the case of Alcazarén, the recharge water came from an advanced secondary treatment at Pedrajas de San Esteban Waste Water Treatment Plant (WWTP). It was convenient to perform post-treatment actions on the treated water (filter beds, geofabrics, reactive filters, and tests with disinfectants or Disinfection By Products (DBP), thus that its quality was more appropriate to make MAR without causing damage to either the environment or the consumers' health. These waters were subsequently used for irrigation and agro-industry supply.

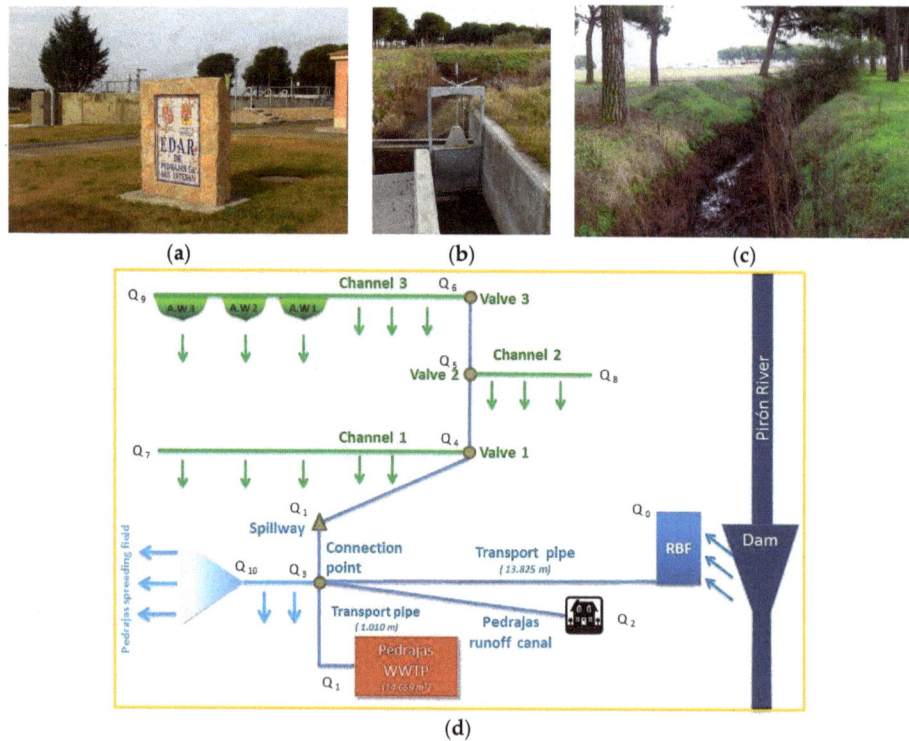

Figure 6. Alcazarén Area and its MAR components, Valladolid (Spain). Photos of some key points (**a**) Pedrajas waste water treatment plant (WWTP); (**b**). Connecting junction of water from WWTP, Pirón River abstraction and urban run-off channel; (**c**). Run-off canal from Pedrajas Village to the connection point; (**d**) General scheme.

3.1.5. Punctual Infiltration. Canal de Isabel II, Madrid (5)

Canal de Isabel II or CYII is the public enterprise in charge of water purification, supply and wastewater management in Madrid. This company has built a system of deep injection in a semi-confined aquifer in the aquifer under the city. Punctual recharge takes advantage of low surface need and high capability to recharge peak flows.

This MAR device was mainly used during drought alerts for potable water supply, increasing resources in the city of Madrid with up to 5 hm^3 per year [11]. Thus, the indicator remained about this figure along that time period.

3.2. Examples of Technological Solutions to Palliate Decreasing Annual Precipitation Rates

The impact of CC over the last decades has been connected to changes on a large scale in the hydrological cycle. Changes in the precipitation pattern were subjected to a significant variability in space and time. During the 20th century, precipitation had risen in inland areas and northern latitudes, while it had fallen between 10° S and 30° N from the 70s [16].

3.2.1. Self-Purification by Natural Biofilters and Nature Based Solutions. Santiuste, Castilla y León (6)

The Wastewater Treatment Plant (WWTP) of Santiuste pours the treated water into four lagooning purifications ponds, and then in the East MAR canal, with two different stretches: The first section works as a natural filter and as a MAR canal, and occupies more than one kilometre in length; while the second has a scarce filtering section, and extends for 1.5 km up to the mouth of the Sanchón artificial wetlands complex. Natural vegetation is respected in this stretch, as it acts as a biofilter, until it reaches the Sanchón spillway or is sent to the wetlands for post-processing actions. After the third (2b) artificial wetlands (AW), water returns to the East MAR canal with improved quality. Sunlight and plant growth play a crucial role in the purifying processes of the resulting water, combining one part from the Voltoya River and another from the treatment plant. Indicators assess the evolution of the main parameters, e.g., nitrate concentration was reduced by almost 30%, turbidity by 34%, and copper ions by more than 60% [9].

3.2.2. Wetlands Restoration. Santiuste, Castilla y León (6)

La Iglesia Lagoon is an alkaline wetland (salt-lake with basic salts of very high pH), which was rehabilitated by means of a solution specifically designed to take advantage of MAR facilities in the area. The recovery of the mineralization fundamental to maintain the characteristics of this type of water bodies, which was thus unique, was achieved through the interaction between the recharge water interacting with the biological and saline sediments deposited in the beach of the lagoon. This allowed the maintenance of a colony of endemic bacteria and the protection of vegetation of high ecological value. It was also an important refuge for aquatic birds. Finally MAR contributed in the preservation of minerals and biominerals considered "rare", thanks to about 5% of the total MAR volume being diverted to La Iglesia Wetland from the Santiuste West MAR Canal [17–22] by gravity (passive system). The amount of water used for environmental purposes has been adopted as the indicator for wetlands restoration.

3.2.3. Gravity Flow Water Distribution. El Carracillo, Castilla y León (7)

A "passive" MAR system is one that does not require electrical energy to operate. They generally function by gravity. Once the behaviour of the aquifer is known, it is possible to infiltrate the recharge water concentrated in a given area, relying on water resources being reused simply by gravity and the quality improvement by naturalization thanks to the aquifer. This technique makes it possible to reduce pipe layouts, with its consequent environmental benefits and cost savings. An example is the MAR artificial recharge at the head of the Carracillo, with distribution of the recharge waters from the storage area throughout the irrigable area, where most of the wells in the region are scattered. The volume of recharge water in the headwaters (East) is naturally directed through the aquifer to the discharge into the Pirón River and its tributary Malucas (West), and can be intercepted throughout the circuit by the irrigation and agro-industry wells, thus avoiding the laying of pipes.

The gravity distribution system (Figure 7) covers up to 40.7 km in length between canals and pipes from the dam to the final discharge area, serving an area of 3500 ha irrigated within 7586 ha of agricultural area [14]. Consequently, the adopted indicator is the length of the network divided by the number of irrigated hectares.

The system represents an important energy saving, which can be added to the savings involved in pumping water from shallower groundwater levels.

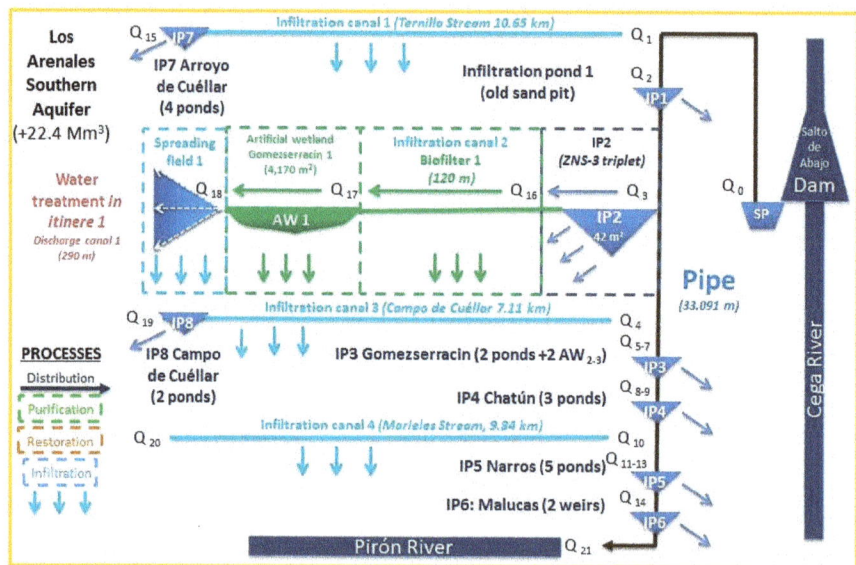

Figure 7. El Carracillo Area and its MAR sites. Devices, junctions, and functions, Segovia (Spain).

3.2.4. Energy Efficiency Increase through Managed Aquifer Recharge (El Carracillo, Castilla y León)

The recharge in El Carracillo contributes to the rise of the water table. Monitoring of the pumping costs of 314 extraction wells, with an average pumping of 9957 m^3 per well per annum, and the rise in the average phreatic level from 6.30 m to 4.00 m since 2003 to 2015 represented a rise of +2.30 m (Figure 8).

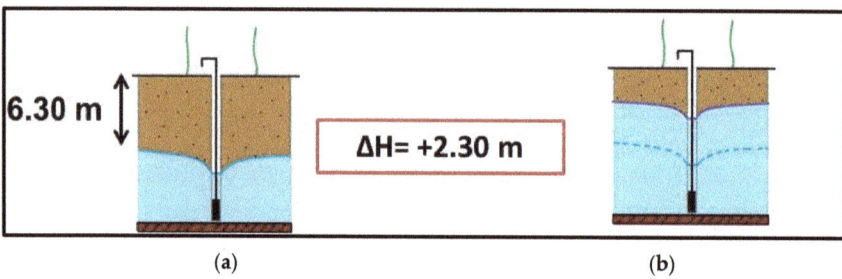

Figure 8. Mock-up with the rise in water level resulting from artificial recharge operations in El Carracillo aquifer. Groundwater levels before (**a**) and after MAR operations (**b**).

The next task was to calculate how much a rise in water level of 2.30 m represented in energy terms. The saving for the irrigators' community, over a calculation of about 0.16 kW·h/m^3 as an average for water extractions, was between 12% and 36% depending on the area, the equivalent of 3000 €/annum as a maximum. This situation is very beneficial for irrigation and for shallow water ecosystems [9].

The volume of CO_2 emitted annually in the El Carracillo irrigation community had fallen by 10,780 kg, which was proportional to the rise in the phreatic surface, without taking into account upgrading, energy efficiency initiatives, etc.

The indicator adopted was either the cost savings or the reduction of emissions thanks to groundwater level rise caused by MAR actions.

3.3. Examples of Technological Solutions to Manage Extreme Phenomena

Extreme situations characterised by an abundance of water, such as floods, "cold drop" events, etc., GIAE [12] can be used, to a certain extent, for MAR. For this purpose, it was necessary to create a system to detain the fast-flowing water and channel it towards recharge devices.

3.3.1. Infiltration of Extreme Flows. Lliria, Valencia (8)

Since 1995, the Basic Civil Protection Guidelines for flood risk included safety procedures for preventing and limiting potential damage arising from this risk. An outstanding example was "Arnachos", a 300 m deep borehole drilled in Losa del Obispo (Valencia) with an extremely high recharge capacity. This was located just a few metres from the irrigation pond of the Tarragó Irrigation Community (Figure 9). It enabled the extraction of a signification fraction of clean water from the irrigation pond in times of heavy rain. Therefore, this recharge system acted as a safety system, reducing the water excess during floods with zero electricity consumption.

Figure 9. Deep borehole "Arnachos" at Balsa del Campo, Valencia (UTM 685,744/4,391,256) located in the margin of an irrigation pond and used as both, a safety and recharge element. Photos courtesy of J.M. Montes and FEGA.

In 2014, it was used twice to reduce the peak-flow in a flood and to recharge the karstified aquifer with an infiltration rate of almost 1000 L/s for a period of 14 h (0.0504 hm^3), a significant amount of water that otherwise would have worsened the devastation caused by the flooding.

3.3.2. Forested watersheds. Neila, Castilla y León (9)

Many examples can be given of mechanical soil preparation for the purpose of increasing the infiltration rate: Channelling of river water to forests conditioned to store the water for a period and facilitate infiltration, as well as forests "organised" to receive "ordered" runoff and facilitate infiltration (Figure 10), etc. According to the DINAMAR project, the main share of artificial recharge in Spain comes from these kinds of devices and is estimated to be 200 hm^3 per year.

Figure 10. Mechanical initiatives to minimise runoff, facilitate recharge, and subsequent plantation (**a**) and infrastructure to channel and level excess runoff water (**b**) Neila (Castilla y León).

An example of this sort was found in Neila, Burgos, where a canal had been constructed to channel water from a road towards a forest adequately prepared for this purpose. This forest was capable of retaining and channelling 15%–40% of the volume of surface runoff [10–18], therefore the indicator adopted is: Percentage of trapped water out of the total runoff.

3.3.3. Multi-Annual Management. Santiuste Basin (CyL) (6)

On certain occasions, conditioned by the potential storage volume of the receiving medium, multi-annual management actions may be performed on the recharge waters. This situation is possible either in areas of high volume available and any demand, or in areas of low potential storage volume and low demand.

In previous sections, situations of inter-annual water management have been described, including nodes of return to aquifers in topological schemes and strategic storage as a preventive measure of adaptation to hypothetical future adverse situations. In this same context, it is worth mentioning the multi-year management of reserves. This is a basic water management technique that considers water as a mining resource, renewable in years of favourable weather conditions, for use in years of prolonged drought.

For example, in the Los Arenales aquifer, Santiuste Basin (Figure 11), the storage of water during several winter periods, in addition to that previously existing when the aquifer was provisionally declared overexploited, could cushion short-term drought situations with almost no repercussions for farmers, as the system was passive too. According to data from the DINAMAR R&D project, the economic activity of the region could be maintained for a period of three years with zero rainfall during all this time, thanks to the reserves stored in the different underground basins that the aquifer presents [10–17].

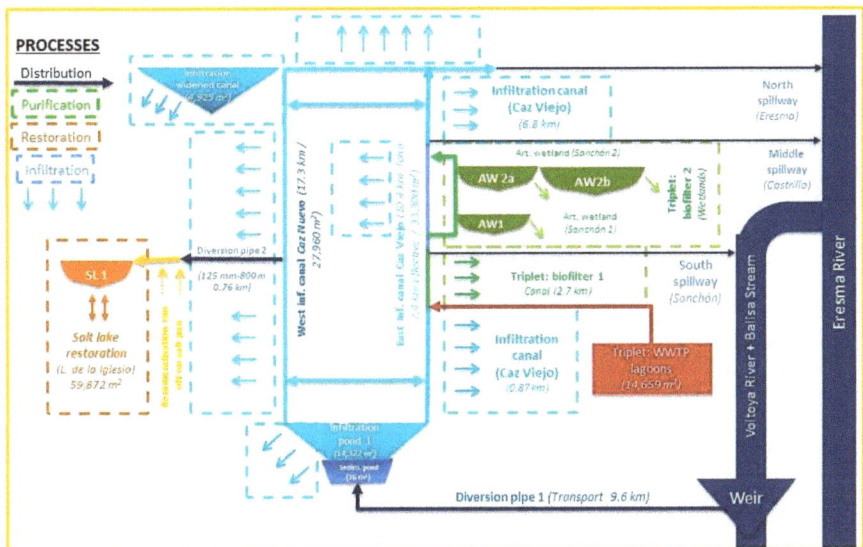

Figure 11. Managed underground water storage for use at annual intervals. Santiuste Basin. MAR devices and functions [15].

3.4. Examples of Technological Solutions to Reduce Sea Level Rise

Positive Hydraulic Barrier. El Prat de Llobregat, Barcelona (10)

One of the most emblematic examples of a water barrier against sea water intrusion is located in the surroundings of Barcelona city´s airport. It is a system of recharging wells injecting water from El Prat WWTP, a positive hydraulic barrier (Figure 12). According to the mathematical models the recovery of the preoperational state previous to the sea intrusion should take around 30 years [20]. The main disadvantage was the huge electricity consumption, thus the activity was eventually stopped during the global economic crisis affecting Spain from 2008.

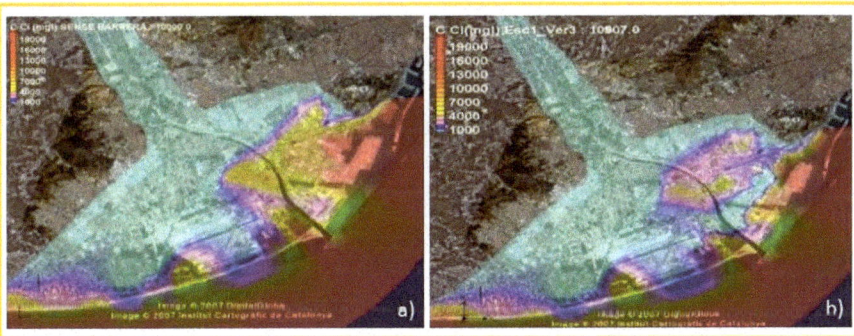

Figure 12. Intrusion barrier in the Llobregat River delta. Hydraulic barrier at Llobregat River delta. Iso-chlorides evolution graphic model for 2035 horizon: Evolution of seawater intrusion without (**a**) and with (**b**) operative recharges.

Specific Spanish MAR sites and proposal/examples of indicators to monitor and track their relationship with CC adverse impacts are exposed in Table 1.

4. Discussion

Analysing one by one some of the MAR solutions with a direct connection to CC impacts, some outcomes are obtained, according to the different groups established for disaggregated studies. These groups are underground water storage, temperature reduction, soil humidity increase, reclaimed water Infiltration, punctual infiltration, self-purification, off-river storage, restoration of key elements, ground-water distribution by gravity, savings/Lower emissions, infiltration of a part of extreme flows, forested watersheds techniques, multiannual management and intrusion barrier wells.

Table 2 summarizes the main pros and cons for MAR solutions regarding CC adverse impacts.

Table 2. Advantages and disadvantages of MAR technical solutions as mitigation measures of CC negative effects.

MAR SOLUTIONS	ADVANTAGES	DISADVANTAGES
Underground water storage	Water recharge can help to restore wetlands associated with overexploited aquifers, especially when winter extraordinary flows are used as a recharging source.	Run-off abstraction can change recharge into negative impact, considering downstream ecosystems
Temperature reduction	Broad array of possibilities in SDUS, from parking lots to roofs, from rain storage to high evaporation systems	Risk of accidental pollution through run off on contaminated areas
Soil humidity increase	Maintenance of micro-flora and fauna in the soil, increase in fertility, low infiltration with small investment and good purification	High soil humidity can facilitate flooding by water table rising or freezing in cold climates. Balance between unsaturated and saturated areas should be searched
Reclaimed water infiltration	Decreasing offer of primary sources (rain and run-off) and increasing offer of secondary ones (WWTP, desalination, storm reservoirs). Chance to change a split into a resource	Reclaimed water involves unbalance between recharging water quality and receptor aquifer quality, clogging during infiltration and legal limits to recharge (EIA) or to use (authorization)
Punctual infiltration	High potential to manage peak flows in constrained areas with filtering systems and possibility of deep recharge as a safety measure in open aquifers	Decantation processes can get clogged. Forced refill can reduce the availability of extreme flows from unexpected storms
Self-purification	Possibility of design according to characteristics of the spillage parameters, combining depth or development of vegetation that allow the development of physical, chemical and biological phenomena depending on draft, type of background, speed of flow, entry of light. Manageable characteristics also to accommodate different types of habitats	The mixture with poor quality waters can affect the infiltrating capacity of the aquifer by either clogging the unsaturated zone or compromising the possibilities and authorizations to use the final mixture. The development of certain vegetation can favour a greater infiltration through the roots or, on the contrary, encourage surface clogging by the formation of bacterial biofilms
Restoration	Slow infiltration into areas where sufficient surface is available for infiltration ponds allows temporary wetlands permanence requiring only a fraction of the total rechargeable volume and, at the same time, fulfils relevant ecosystem functions as a refuge for wild fauna and flora	The establishment of free water sheets may limit the use of reclaimed water due to possible health risks
Gravity flow water distribution	The greater the knowledge of the aquifer is, the greater is the established systems that take advantage of the hydraulic characteristics of the terrain	Detailed hydrology and geotechnics knowledge play a fundamental role in order to take advantage of the potential distribution of water along the aquifer by simple gravity. Precise studies are essential
Savings/Lower emissions	In this context, new lines of action are being considered to improve energy efficiency, such as the replacement of diesel engines by electric motors, the use of alternative energies to reduce pumping costs, such as solar panels, wind energy, and greater use of biomass	The improvement of the economic conditions allowing energy consumption can become a dangerous stimulus for the excessive increase in agricultural demand, thus, it is necessary to establish regulations for general resources management

Table 2. Cont.

MAR SOLUTIONS	ADVANTAGES	DISADVANTAGES
Infiltration of a part of extreme flows	High capacity to manage overfloods and peak flows in reduced spaces with the application of measures that decrease solid load. Ability to redirect flows to deep aquifers to avoid flooding in certain sectors of unconfined aquifers	Overfloods must be previously laminated to be partially infiltrated. The enhanced infiltration in the aquifer might reduce the soil capacity to absorb extreme precipitation by infiltration
Forested watersheds	Watersheds erosion control and promotion of forest hydrological restoration thanks to detention/retention devices to form soil and reduce slope. Development of deep soil botanical species with greater terrain stability (the retention of solids allows to increase the useful life of dams)	Water retention in the heading of the basin reduces downstream runoff, enhances soil formation and has a direct effect on associated wetlands
Multiannual management	Underground reserves do not require certain precautionary measures such as winter water releases, but might need to divert certain volumes to deep aquifers for exploitation in emergencies	Multi-year management implies a very good organization of uses with a great cohesive spirit among stakeholders. Despite their advantages over dams, they also require precautions against water table excessive rises
Intrusion barrier wells	Acceptable use of low-quality sources (high NaCl or NO_3 concentration) carefully combined for MAR	Collateral effects of pollutants in the recharged volumes on the aquifer's potential storage

Advantages and disadvantages should be considered before selecting the best fitted or a combination of techniques.

Regarding results, Figure 13 summarizes the relationship between CC current impacts and MAR solutions with their specific site mention and the assessed indicators of positive achievements against CC. Trends and evolutions of the different indicators are explained in each one of the references, but the figures exposed represent an accurate approach for the present MAR-CC binomial.

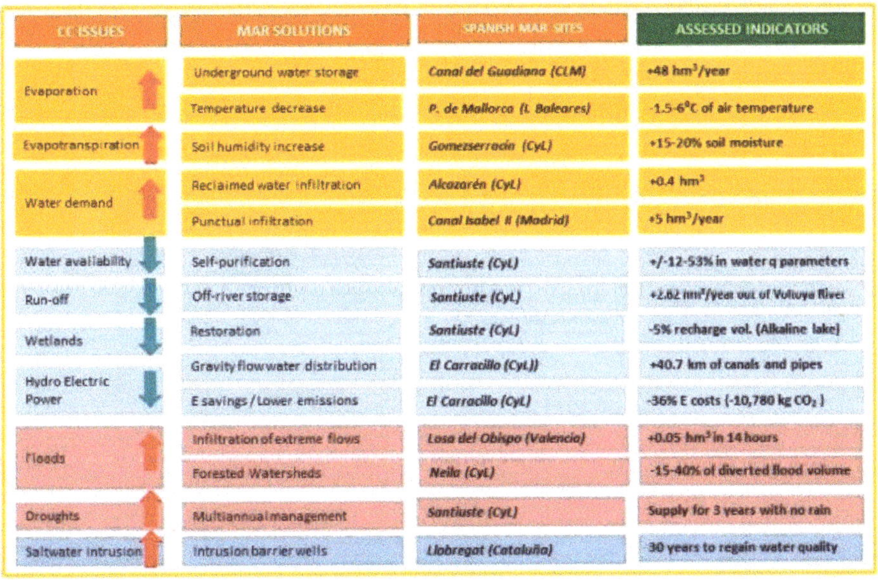

CC ISSUES	MAR SOLUTIONS	SPANISH MAR SITES	ASSESSED INDICATORS
Evaporation	Underground water storage	Canal del Guadiana (CLM)	+48 hm^3/year
	Temperature decrease	P. de Mallorca (I. Baleares)	-1.5-6°C of air temperature
Evapotranspiration	Soil humidity increase	Gomezserracin (CyL)	+15-20% soil moisture
Water demand	Reclaimed water infiltration	Alcazarén (CyL)	+0.4 hm^3
	Punctual infiltration	Canal Isabel II (Madrid)	+5 hm^3/year
Water availability	Self-purification	Santiuste (CyL)	+/-12-53% in water q parameters
Run-off	Off-river storage	Santiuste (CyL)	+2.82 hm^3/year out of Voltoya River
Wetlands	Restoration	Santiuste (CyL)	-5% recharge vol. (Alkaline lake)
Hydro Electric Power	Gravity flow water distribution	El Carracillo (CyL)	+40.7 km of canals and pipes
	E savings / Lower emissions	El Carracillo (CyL)	-36% E costs (-10,780 kg CO_2)
Floods	Infiltration of extreme flows	Losa del Obispo (Valencia)	+0.05 hm^3 in 14 hours
	Forested Watersheds	Neila (CyL)	-15-40% of diverted flood volume
Droughts	Multiannual management	Santiuste (CyL)	Supply for 3 years with no rain
Saltwater intrusion	Intrusion barrier wells	Llobregat (Cataluña)	30 years to regain water quality

Figure 13. Relations between CC impacts and MAR solutions with their site locations and indicators of positive achievements.

These results have been compared with the referenced international parallel cases. All the studied demo-sites reflect a homogeneous evolution:

Most of the published articles pay special attention to modelling and saline water intrusion and groundwater salinity evolution, though they have scarce data and results about indicators to monitor most of the identified impacts.

Those sites affected by saline water intrusion and salinity increase do not show the groundwater resilience that the other MAR pilots expose, where indicators are showing a better reaction regarding water storage, soils humidity and extreme water related events response.

Monitoring water quality is being a pendant issue, as there are more indicators facing quantity constraints than quality issues.

Reclaimed water infiltration is a first-row topic under permanent revision. In the future and for the studied sites, this kind of MAR will not be an option but a priority.

Dykes play a key role regarding runoff capture and floods, extending the concentration time and enlarging the volume peak, therefore reducing the flood´s devastation capacity.

Wetlands are under permanent support. Most of the studied cases invest about 5% of available water for environmental purposes. A general regeneration is achieved to a certain extent in both, water availability and biodiversity.

5. Conclusions

Climate change effects and their associated impacts have been related to 10 successful MAR sites in Spain through a series of indicators (Figure 13), that let us assess the efficacy and efficiency of the MAR technique as a multifunctional technique that can simultaneously achieve several purposes.

The list of climate change effects in Spain has been accompanied by several fruitful cases of MAR. This success is economically sustainable as most of them are passive systems (do not require electricity to work). The data associated with these monitored cases have enabled the establishment of status indicators, whilst demonstrating the proficiency of MAR to face frontally CC adverse impacts, not only within the context of the case-studies in Mediterranean areas (Figure 13), but also in parallel circumstances all around the world.

The exposed examples affirm that management schemes featuring intentional aquifer recharge constitute an important set of climate change adaptation measures, while providing guarantees with respect to future water supply. These examples are aligned with other international cases consulted in the references, where despite isolated actions, the response to CC appears to be collegiated [24]. Some of the exposed technical solutions also serve to palliate the adverse effects of CC as mitigation measures. According to indicators, some progress is achieved in replenished aquifers where pumping costs save electricity due to a higher water level with an attached CO_2 emission reduction. The attention paid on water and anergy efficiency is also a general asset found in the whole MAR cases under study.

The exposed examples and their comparable potential may have a high practical value for MAR constructions, specially adapted to combat CC in Mediterranean countries [25] where droughts have dramatic effects.

Author Contributions: Investigation, E.F.E.; methodology, J.S.S.S.; project administration, R.C.G.

Funding: This research was funded by H2020 Marie Skłodowska-Curie Actions, CA 814,066.

Acknowledgments: This paper has been edited by members of the TRAGSA Group in the MARSolut European Project (Managed Aquifer Recharge Solutions Training Network) with the support of the European Commission, H2020 MSCA). The information and views set out in this article are those of the authors and do not necessarily reflect the official opinion of the European Union. Authors wish to express their gratitude for the collaboration and support of TRAGSA Group technical staff, IAH-MAR Commission as a source of knowledge and inspiration, and two anonymous reviewers who kindly helped us to improve the quality of this article. The authors wish also to thank Miren San Sebastián and James Haworth who assisted in the proof-reading of the manuscript.

Conflicts of Interest: The authors declare no conflict of interest. The founding sponsors had no role in the design of the study; in the collection, analyses, or interpretation of data; in the writing of the manuscript and in the decision to publish the results.

Abbreviations

The following abbreviations are used in this manuscript:

AW	Artificial Wetlands
MAR	Managed Aquifer Recharge
CC	Climate Change
NBS	Nature Based Solutions
SUDS	Sustainable Drainage Urban Systems
UHI	Urban Heat Island
WWTP	Waste Water Treatment Plant
MSCA	Marie Sklodowska Curie Action

References

1. Taylor, R.G.; Scanlon, B.; Döll, P.; Rodell, M.; Van Beek, R.; Wada, Y.; Longuevergne, L.; Leblanc, M.; Famiglietti, J.S.; Edmunds, M.; et al. Ground water and climate change. *Nat. Clim. Chang.* **2013**, *3*, 322–329. [CrossRef]
2. Mimikou, M.A.; Baltas, E.; Varanou, E.; Pantazis, K. Regional impacts of climate change on water resources quantity and quality indicators. *J. Hydrol.* **2000**, *234*, 95–109. [CrossRef]
3. The EU Water Framework Directive—Integrated River Basin Management for Europe. Available online: http://ec.europa.eu/environment/water/water-framework/index_en.html (accessed on 9 March 2019).
4. Pholkern, K.; Saraphirom, P.; Srisuk, K. Potential impact of climate change on groundwater resources in the Central Huai Luang Basin, Northeast Thailand. *Sci. Total Environ.* **2018**, *633*, 1518–1535. [CrossRef] [PubMed]
5. García-Menéndez, O.; Morell, I.; Ballesteros, B.J.; Renau-Prunonosa, A.; Renau-Llorens, A.; Esteller, M.V. Spatial characterization of the seawater upconing process in a coastal Mediterranean aquifer (Plana de Castellón, Spain): Evolution and controls. *Environ. Earth Sci.* **2016**, *75*, 728. [CrossRef]
6. Mastrocicco, M.; Busico, G.; Colombani, N.; Vigliotti, M.; Ruberti, D. Modelling Actual and Future Seawater Intrusion in the Variconi Coastal Wetland (Italy) Due to Climate and Landscape Changes. *Water* **2019**, *11*, 1502. [CrossRef]
7. Foster, S.; Tyson, G. Climate-Change Adaptation & Groundwater, IAH Strategic Overview Series. 2019. Available online: https://iah.org/wp-content/uploads/2019/07/IAH_Climate-ChangeAdaptationGdwtr.pdf (accessed on 1 September 2019).
8. IPCC. Resumen para responsables de políticas. In *Cambio Climático 2013: Bases Físicas. Contribución del Grupo de trabajo I al Quinto Informe de Evaluación del Grupo Intergubernamental de Expertos sobre el Cambio Climático*; Stocker, T.F., Qin, D., Plattner, G.-K., Tignor, M., Allen, S.K., Boschung, J., Nauels, A., Xia, Y., Bex y, V., Midgley, P.M., Eds.; Cambridge University Press: Cambridge, UK; New York, NY, USA, 2013.
9. Valores Climatológicos Extremos Medidos en Siguientes Periodos. Temperatura: 955 a 2010. Temperatura Máxima Absoluta (°C). Available online: http://www.castillalamancha.es/sites/default/files/clima_valores_extremos.pdf (accessed on 19 March 2019).
10. DINAMAR. La Gestión de la Recarga Artificial de Acuíferos en el Marco del Desarrollo Sostenible. Desarrollo Tecnológico. Coord. Enrique Fernández-Escalante. Serie Hidrogeología Hoy, No 6. Método Gráfico. 2010. Available online: http://goo.gl/6lp4m (accessed on 2 June 2019).
11. Zhou, Q. A Review of Sustainable Urban Drainage Systems Considering the Climate Change and Urbanization Impacts. *Water* **2014**, *6*, 976–992. [CrossRef]
12. GIAE. *The Comprehensive Management of Rainwater in Built-up Areas*; Legal Deposit M-9957-2015; Tragsa Group: Madrid, Spain, 2015.
13. MARSOL; Fernández Escalante, E.; Calero Gil, R.; Villanueva Lago, M.; San Sebastián Sauto, J.; Martínez Tejero, O.; Valiente Blázquez, J.A. *Appropriate MAR Methodology and Tested Know-How for the General Rural Development*; MARSOL deliverable 5-3; MARSOL: Darmstadt, Germany, 2016; Available online: http://marsol.eu/35-0-Results.html (accessed on 15 May 2019).

14. IMTA. *Manejo de la Recarga de Acuíferos: Un enfoque hacia Latinoamérica*; IMTA: Progreso, México, 2017; p. 977. Available online: https://www.imta.gob.mx/biblioteca/libros_html/manejo-recarga-acuiferos-ehl.pdf (accessed on 15 May 2019).
15. López-Camacho, B.; Iglesias, J.A.; Muñoz, A.; Sánchez, E.; Cabrera, E. Gestión sostenible de los recursos hídricos en el sistema de abastecimiento de la Comunidad de Madrid. *Equip. Serv. Munic.* **2007**, *133*, 78–88.
16. Bates, B.C.; Kundzewicz, Z.W.; Wu, S.; Palutikof, J.P. (Eds.) El Cambio Climático y el Agua. In *Documento Técnico del Grupo Intergubernamental de Expertos Sobre el Cambio Climático*; Secretaría del IPCC: Ginebra, Switzerland, 2008.
17. Fernández Escalante, E. 2002–2012, una década de recarga gestionada. In *Acuífero de la Cubeta de Santiuste (Castilla y León)*; Tragsa, Ed.; Tragsa Group: Madrid, Spain, 2014; ISBN 84-616-8910-0.
18. Fernández Escalante, E. Mecanismos de "detención-infiltración" para la recarga intencionada de los acuíferos como estrategia de adaptación al cambio climático. In *Revista IDiAgua, Número 1, Junio de 2018 Fenómenos Extremos y Cambio Climático*; Plataforma Tecnológica Española del Agua: Madrid, Spain, 2018.
19. San Sebastián Sauto, J.; Fernández Escalante, E.; González Herrarte, F.D.B. La Recarga Gestionada en Santiuste: 13 Años de Usos y Servicios Múltiples para la Comunidad Rural. *Rev. Tierras* **2015**, *234*, 78–85.
20. CETaqua, Centro Tecnológico del Agua. *Enhancement of Soil Aquifer Treatment to Improve the Quality of Recharge Water in the Llobregat River Delta Aquifer Life+ ENSAT project 2010–2012 Layman's Report*; CETaqua, Centro Tecnológico del Agua: Barcelona, Spain, 2013.
21. Del Barrio, V. The activity seen from the Duero river basin and the RBMP. In Proceedings of the MAR4FARM Training Workshop Presentation, Gomezserracín, Segovia, Spain, 29–30 October 2014; Available online: http://www.dina-mar.es/post/2014/11/17/mar4farm-presentaciones-presentations-available-freely-on-the-internet.aspx (accessed on 15 August 2019).
22. MARSOL; Fernández Escalante, E.; Calero Gil, R.; González Herrarte, B.; San Sebastián Sauto, J.; Del Pozo Campos, E. *Los Arenales Demonstration Site Characterization. Report on the Los Arenales Pilot Site Improvements*; MARSOL deliverable 5-1; MARSOL: Darmstadt, Germany, 2015; Available online: http://marsol.eu/35-0-Results.html (accessed on 10 April 2019).
23. IWMI en GRIPP; Escalante, E.F.; San Sebastián Sauto, J. The Alcazarén-Pedrajas MAR scheme in Central Spain. 2018. Available online: https://bit.ly/2LnLXzT (accessed on 3 June 2019).
24. Ebi, K.; Lim, B.; Aguilar, I. *Scoping and Designing an Adaptation Project in Adaptation Policy Frameworks for Climate Change: Developing Strategies, Policies and Measures*; Lim, B., Spanger-Siegfried, E., Eds.; United Nations Development Programme; Cambridge University Press: Cambridge, UK, 2004; pp. 33–47. ISBN 0 521 61760 X.
25. Alvarez, J.; Estrela, T. Regionalisation and identification of droughts in Mediterranean countries of Europe. In *Tools for Drought Mitigation in Mediterranean Regions, Water Science and Technology Library Volume 44*; Rossi, G., Cancelliere, A., Pereira, L.S., Oweis, T., Shatanawi, M., Zairi, A., Eds.; Kluwer Academic Publishers: Dordrecht, The Netherlands, 2003; pp. 123–146. ISBN 1-4020-1140-7.

© 2019 by the authors. Licensee MDPI, Basel, Switzerland. This article is an open access article distributed under the terms and conditions of the Creative Commons Attribution (CC BY) license (http://creativecommons.org/licenses/by/4.0/).

Article

Managed Aquifer Recharge of Monsoon Runoff Using Village Ponds: Performance Assessment of a Pilot Trial in the Ramganga Basin, India

Mohammad Faiz Alam [1,*], Paul Pavelic [2], Navneet Sharma [1] and Alok Sikka [1]

1. International Water Management Institute, New Delhi 110012, India; navneetsharma.mit@gmail.com (N.S.); a.sikka@cgiar.org (A.S.)
2. International Water Management Institute, Vientiane 4199, Laos; p.pavelic@cgiar.org
* Correspondence: m.alam@cgiar.org; Tel.: +91-7042-035-620

Received: 25 November 2019; Accepted: 31 March 2020; Published: 4 April 2020

Abstract: The managed aquifer recharge (MAR) of excess monsoonal runoff to mitigate downstream flooding and enhance groundwater storage has received limited attention across the Indo-Gangetic Plain of the Indian subcontinent. Here, we assess the performance of a pilot MAR trial carried out in the Ramganga basin in India. The pilot consisted of a battery of 10 recharge wells, each 24 to 30 m deep, installed in a formerly unused village pond situated adjacent to an irrigation canal that provided river water during the monsoon season. Over three years of pilot testing, volumes ranging from 26,000 to 62,000 m^3 were recharged each year over durations ranging from 62 to 85 days. These volumes are equivalent to 1.3–3.6% of the total recharge in the village, and would be sufficient to irrigate 8 to 18 hectares of rabi season crop. High inter-year variation in performance was observed, with yearly average recharge rates ranging from 430 to 775 m^3 day^{-1} (164–295 mm day^{-1}) and overall average recharge rates of 580 m^3 day^{-1} (221 mm day^{-1}). High intra-year variation was also observed, with recharge rates at the end of recharge period reducing by 72%, 88% and 96% in 2016, 2017 and 2018 respectively, relative to the initial recharge rates. The observed inter- and intra-year variability is due to the groundwater levels that strongly influence gravity recharge heads and lateral groundwater flows, as well as the source water quality, which leads to clogging. The increase in groundwater levels in response to MAR was found to be limited due to the high specific yield and transmissivity of the alluvial aquifer, and, in all but one year, was difficult to distinguish from the overall groundwater level rise due to a range of confounding factors. The results from this study provide the first systematic, multi-year assessment of the performance of pilot-scale MAR harnessing village ponds in the intensively groundwater irrigated, flood prone, alluvial aquifers of the Indo-Gangetic Plain.

Keywords: managed aquifer recharge; Underground Transfer of Floods for Irrigation; droughts; floods; groundwater depletion

1. Introduction

The inter- and intra-annual variability of water availability, manifesting in extreme flood and drought events, presents a considerable challenge to ensuring water security globally [1,2]. This variation in water availability, separated by time and space, co-exists in most river basins globally [2]. The impact of water variability is magnified in the agriculture sector due to its strong dependence on climate. This is exemplified by the fact that of the total loss of USD 80 billion in crop and livestock production in 67 countries between 2003 and 2013, due to 140 medium-to-large-scale disasters (including non-water related events), 83% was caused by flood or drought [3]. With climate change increasing rainfall variability and inducing more and severe extreme weather events, the predictability of water availability will further reduce in coming years [4,5], prompting the need for urgent attention to adaptation.

Groundwater, with its high buffering capacity due to relatively large storage [6], presents a potential opportunity to resolve the temporal and spatial imbalances in water supply and availability. The extensive use of groundwater, in many places leading to overexploitation [7], also creates additional depleted storage. This additional storage capacity could be used, similarly to dams, to capture excess monsoonal runoff in the wet season, making it available during dry periods, and thus mitigating both flood and drought hazards [8].

One novel way of operationalizing this concept is "Underground Transfer of Floods for Irrigation (UTFI)" [8,9]. UTFI is a form of managed aquifer recharge (MAR) that involves the targeted recharge of excess monsoonal runoff that potentially poses a flood risk downstream, in aquifers at the basin scale through the strategic establishment of groundwater recharge infrastructure to mitigate flooding and enhance groundwater storage [8,10]. Enhanced groundwater storage increases water availability so that the water can be used during the dry season for domestic, livestock or irrigation use, or, if retained, can support dry season inflows, enhancing ecosystem services [11].

Efforts to test UTFI started in the Ganges river basin, with its high population density, cropping intensity, recurring floods and droughts due to the concentrated monsoon season [11,12] along with extensive aquifer systems (underlain by highly productive alluvial aquifers of the Indo-Gangetic Plain) used intensively for irrigation [13]. These characteristics present both the favorable conditions and challenges UTFI aims to solve. A GIS-based multi-criteria analysis revealed high suitability across the Ganges basin [9].

However, to successfully implement UTFI at the basin scale requires thoughtful planning and staged testing and development to minimize the potential environmental, social and financial risks [8]. Though there are some MAR pilot studies in porous alluvial aquifers of the Indo-Gangetic Plain [14–16], they lack the long term comprehensive and systematic approaches required to assess how UTFI would perform if upscaled. This is unlike the case in hard rock settings in India where experience is much more extensive [17–19]. Therefore, the piloting and testing of UTFI was carried out to generate the body of knowledge necessary to establish the scope for wider implementation across similar settings. This paper presents the learnings gained from piloting in hydraulic- and hydrological-related aspects. Detailed information on site selection and setting up the pilot is covered in [8], and these are briefly covered here. A broader perspective on the findings from the piloting can be gained from related studies on water quality, and environmental and socio-institutional aspects [20,21].

2. Study Area

Pilot testing was carried out in Jiwai Jadid village, located in Rampur district, Uttar Pradesh, India (Figure 1). The climate of the area is sub-humid and characterized by hot, humid summers and cold winters. The average annual rainfall of Rampur district is 933 mm, and about 85% of the rainfall is received during the south-western monsoon between June and September. Agriculture is the primary means of livelihood in the district, with about 60% of the working population reliant on agriculture. This is reflected in the land use of the district, where 81% of the 2357 km^2 area is under cropping [22,23]. The major cropping pattern of the district is paddy and wheat, grown in two major seasons known as the kharif (coinciding with the monsoon season: June to November) and rabi (November to March), respectively.

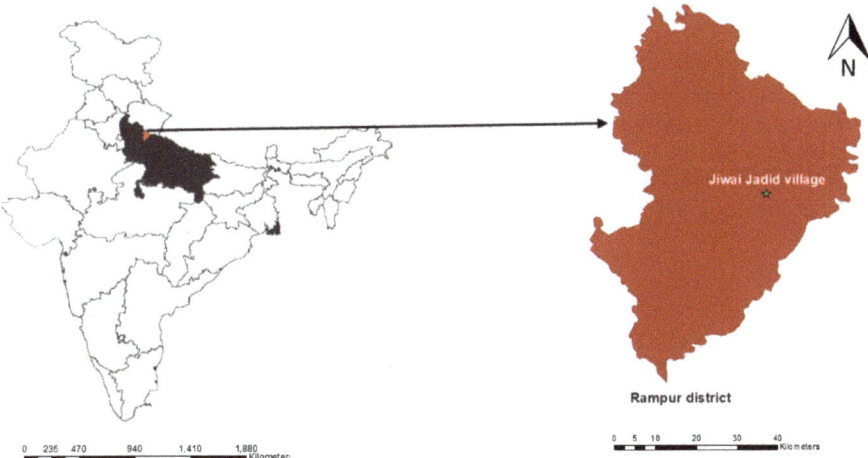

Figure 1. The location of the Underground Transfer of Floods for Irrigation (UTFI) pilot study area in Jiwai Jadid village, Rampur District, Uttar Pradesh, India.

Concentrated rainfall during the monsoon months leads to excess water/floods, followed by limited water availability during the non-monsoon season. High flows in the Ramganga river and its tributaries (e.g., the Kosi and Pilakhar rivers) during the monsoon season makes flooding a recurrent and major problem in Rampur and the surrounding districts. Major flooding experienced in Rampur in 1973 affected almost 238,165 people across 561 villages and impacted a total crop area of 144,836 ha. In recent years, floods have been reported in 2010, 2013 and 2015, affecting 15, 18 and 207 villages, respectively [24–26].

Rampur district is intensively irrigated (~99.8% of the cropped area), with groundwater being the main source of irrigation (~97%) [27]. Around 80,000 shallow tube wells equipped with diesel-powered centrifugal pumps account for most of the groundwater pumped. Intensive groundwater irrigation in the district is made possible by the highly productive Indo-Gangetic Plain alluvial aquifers that underlie much of northern India, as well as several neighboring countries [13]. There are four aquifer groups present in the area down to 440 metres below ground level (mbgl). The first aquifer group is unconfined and extends down to depth of 60 to 90 mbgl [28], and is utilized mostly for irrigation.

Average groundwater table depths in Rampur district range from 4–8 m during the monsoon and 8–12 m in the non-monsoon months [28]. However, as a result of the intensive demand for groundwater year-round, groundwater overexploitation is a risk, with groundwater tables falling across the district. The number of 'dark' administrative blocks, a Government of India term representing a high level of overexploitation of groundwater resources, has increased from only one block in 2003 to four blocks in 2013 (out of a total of six blocks in the district) [29].

Intra seasonal water availability, variability, recurrent floods and high irrigation demand with intensive groundwater irrigation, leading to groundwater overexploitation in the region, provides the challenging conditions well-suited to piloting UTFI.

3. Pilot Trial Design

The UTFI pilot features 10 gravity-fed recharge wells of 150 mm diameter (PVC pipe) installed in the village pond (Figure 2). Recharge wells were drilled to depths of 24–30 m with gravel packing, and the lowest 18 m of pipe was screened. The village pond was used to install UTFI infrastructure, as land availability is a serious constraint owing to high population density and intensive year-round cultivation in the region. The pond (75 × 35 m) was dewatered and excavated to a depth of 2.5 m, creating a maximum storage capacity of 5250 m^3. To ensure a common reference point for the measurement of the water table in the piezometers and the pond, the top of the recharge well RW1 (Figure 2), which was 2.1 m above the base of the pond, was selected. With the bottom of the pond at 2.1 mbgl and the lowest pond level that allowed water to enter recharge wells at 1.1 mbgl, the height of dead storage in the pond was 1.0 m. Water, after the pond water level was above dead storage, entered the wells only through a pea-gravel-filled chamber, to filter out suspended silts. Nine piezometers were installed with P1, P2, P3, P6 and P7 positioned along a transect in the direction of groundwater flow from north to south, whilst P4 and P5 (deep (D) and shallow (S) pairs for each) were positional along a shorter north-south transect.

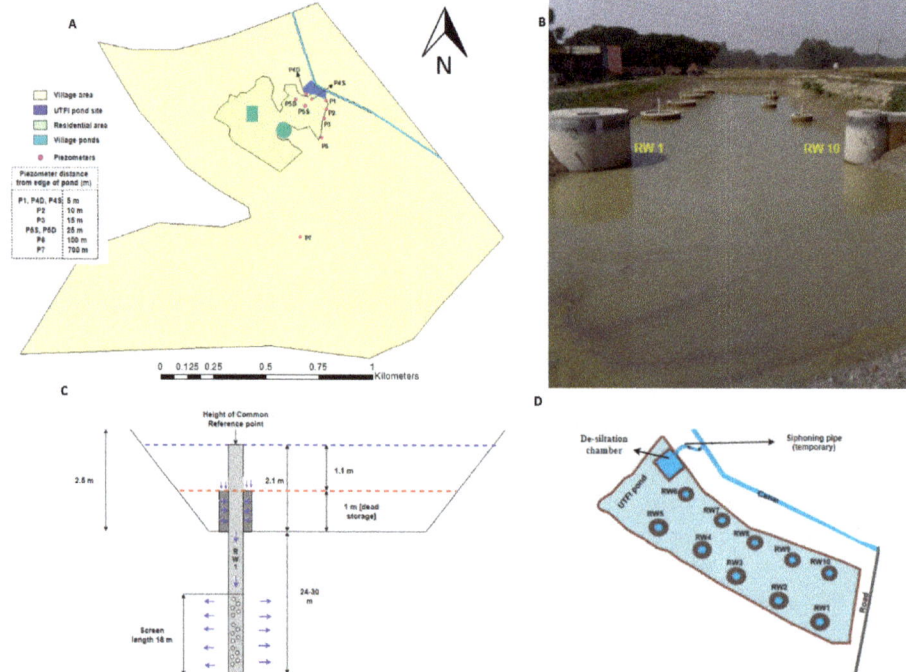

Figure 2. (**A**): The location of the UTFI pond and the installed piezometers at the pilot site (P1 to P7 = piezometers); (**B**): The completed UTFI pilot site; (**C**): A simplified vertical cross section of the UTFI pond; (**D**): A schematic of the pond showing the locations of recharge wells within the pond (RW1 to RW10 = recharge wells).

The source water (i.e., excess rain water/flood water) is brought to the site through a canal (Left Pilakhar canal) that is situated next to the pond. This canal carries water from the Pilakhar river and provides water for irrigation during the summer and winter seasons, though the crops in the pilot village were not irrigated by canal water. However, during the monsoon season, the canal flow was more than the irrigation demand, and excess water flowed downstream to the river/canal. A de-siltation chamber (2 × 2 × 1.5 m) was also built at the intake of the pond (Figure 2).

Operation and Maintenance

Pilot operations during the trial period were handled with the support of an appointed village representative, who was provided training on the protocols. Recharge was done only during the monsoon season when the water level in the canal was above a pre-defined level of 0.6 m. The water from the canal was initially pumped into the pond over an embankment using a diesel-operated pump for 15 min, and thereafter, it flowed naturally under gravity (siphon flow) due to the water level difference.

Maintenance of the UTFI system to enhance the recharge rates included three main types of activities:

a. Recharge well cleaning: recharge wells were cleaned using a compressor to remove silt deposited inside the recharge wells and fine particles that had blocked the slots of the recharge well.
b. Recharge well filter cleaning: pea-gravels in the recharge filters packed in brick structures were cleaned by removing and washing by hand.
c. Desilting pond bottom: silt deposited at the pond bottom was scraped off, and the embankments and side slopes were restabilized.

4. Methods

4.1. Monitoring and Data Collection

Table 1 gives an overview of the different parameters measured, methodologies applied and frequencies of observations in accordance with a monitoring protocol prepared to guide the monitoring of the physical dimensions of the pilot system.

Table 1. The monitored parameters, methods of measurement and frequencies.

Parameter	Measurement Method	Frequency
Groundwater levels	Nine piezometers were installed (Figure 2). All groundwater levels of piezometers are given relative to the reference point, RW1 (in mbgl).	Measured weekly using a portable water level meter during recharge operations, and every 2 weeks during non-recharge periods
UTFI pond infiltration rate	(A) Single ring infiltration test up to 8 h was conducted at the bottom of pond at four locations after pond cleaning/development (B) Days taken for pond dead storage to dry out after stopping recharge operations	(A) 6 h test at 4 locations at 45 cm depth on the 11 and 18 September, 2015 (B) At the end of the recharge operations each year
Rainfall	Rain gauge at Krishi Vigyan Kendra (KVK), Rampur city, situated approximately 20 km from the pilot site	Daily from 24 June 2016
Canal water levels	Measuring scale was marked in canal wall near road bridge	Daily basis during the recharge operation and at 15 day intervals during non-recharge periods
UTFI pond storage volume	Relationship between depth of water in pond and volume of water in pond was developed	After pond development
UTFI Pond water level measurement	Measuring scale was marked at recharge well (RW1 in Figure 2) to record pond water level. All depths are relative to the RW1 reference point.	Pond water level was recorded on a daily basis during the recharge operations
Source water silt content	Water samples analyzed for total solid solids (TSS); Mass of silt accumulated at pond bottom after recharge seasons (tonnes)	Monthly; After recharge season of 2016 and 2017
Socio-economic survey	Socio-economic survey of 120 farmers within a 1 km radius of the UTFI site	At the start of the pilot trial in 2016

4.2. Groundwater Recharge from the UTFI Pond

Groundwater recharge from the UTFI pond (V_{UTFI}), consisting of recharge from 10 gravity recharge wells and infiltration from the pond bottom, was estimated from the observed decrease in volume of water stored in the pond over a period of time (Equation (1)). To estimate the change in volume, a specific depth-storage relationship representing the volume of water relative to the corresponding water depth in the pond was developed based on the geometry of the structure. Groundwater recharge tests were conducted over durations of about 12 h at approximately 10 day intervals during the recharge periods.

$$V_{UTFI(d)} = \frac{\Delta V_{vp}}{t_i - t_{i+1}} \quad (1)$$

where $V_{UTFI(d)}$ = the recharge rate of UTFI pond (m^3 h^{-1}) at day d from the start of the recharge period; ΔV_{vp} is the change in the volume of water in the pond (m^3) between the start (t_i) and end (t_{i+1}) of the monitoring period based on the depth-storage function, and $t_i - t_{i+1}$ = the duration of the test period (hours). The evaporation rate (6–9 mm day^{-1} during the recharge period, or 3–4.5 mm day^{-1} during the recharge test) was assumed to be negligible in the calculation.

The total recharge for the whole season was then calculated by summing up the recharges between the two tests at days d and d + i (Equation (2)). The recharge rates at the start and end of the trial were taken to be equal to the first and last measured recharge rates, respectively.

$$V_{UTFI} = \sum_{i=1}^{n} \left(\frac{(R_{UTFI(d)} + R_{UTFI(d+i)}) \times 24}{2} \right) \times ((d+i) - (d)) \quad (2)$$

where V_{UTFI} is the volume of water recharge (m^3), n is the total number of recharge tests conducted, and $R_{UTFI(d)}$ and $R_{UTFI(d+i)}$ are the recharge rates on days d and $d + i$ respectively.

The average recharge rate (R_{UTFI}) for the UTFI pond for the whole recharge season in m^3 day^{-1} is calculated by dividing the total recharge by the number of recharge days.

4.3. Groundwater Level Response to UTFI Recharge

Groundwater level changes during the recharge periods in response to UTFI recharge was analyzed from monitored groundwater levels in the piezometers and relative mound formation, which is the difference in groundwater levels at distance d from a reference piezometer caused by UTFI recharge. Relative mound height provides a key measure of the hydraulic impact of UTFI recharge, on the assumption that other factors impacting groundwater levels will influence all of the piezometers uniformly. Groundwater level comparisons before and after UTFI were not possible as this required historical data in the village that were lacking, as monitoring only started with the commencement of the pilot.

Concentrated recharge over the small pilot area would lead to the formation of a relative mound, and, as time progressed and groundwater spreaded horizontally, the relative mound would flatten and eventually dissipate after recharge operations ceased [30,31]. The relative mound (Equation (3)) at distance r from the UTFI pond and days d after recharge started ($H_{mound(r,d)}$) was estimated against the groundwater level of the referenced piezometer ($GWL_{(ref,d)}$). The selection of the specific reference piezometer, which is expected to be unaffected (or the least affected) by recharge operations, was based on the groundwater level observations covered in the results section.

$$H_{mound(r,d)} = GWL_{(r,d)} - GWL_{(ref,d)} \quad (3)$$

where $H_{mound(r,d)}$ is the mound height at distance r from the pond at day d from the start of recharge period, and $GWL_{(r,d)}$ and $GWL_{(ref,d)}$ are the groundwater levels at distance r and at the reference piezometer at day d after the start of the recharge period.

The Hantush Analytical Solution

The Hantush analytical solution [31] was used to model mound formation in order to study the impact of UTFI recharge alone on relative mound formation, thus distinguishing it from other potentially confounding factors (such as rainfall, groundwater pumping, recharge from canal and river and regional groundwater flow) and to substantiate the field observations from Equation (3). The Hantush solution gives the growth and decay of groundwater mounds beneath rectangular or circular infiltration basins [31] and has been used in numerous studies to provide insights into mounding behavior [32,33]. In this analysis, the ten recharge wells are approximated as a uniform recharge source over the rectangular area of the pond.

To apply the Hantush solution, a spreadsheet solution of the equation was used [34]. To simulate mounding with decreasing recharge rates, mound height was estimated at 5 day intervals (up to 80 days) from the start of recharge operations, and the corresponding 5 day average recharge rates were calculated (based on Equation (2)) and given as input. Aquifer characteristics (a hydraulic conductivity of 20 m day^{-1} and a specific yield of 0.1) were assumed to be uniform over the study area [35].

4.4. Groundwater Bbalance at the Village Scale

The magnitude of UTFI recharge (V_{UTFI}) relative to the total groundwater recharge over the village area (Figure 2) is estimated to draw inferences on how additional recharge from UTFI pond contributes to overall groundwater recharge. The selection of the village boundary as the scale for analysis is largely for demonstrative purposes. It was deemed an appropriate scale as UTFI is designed as a village-level intervention, and its zone of influence is expected to be largely within village boundaries.

Village total recharge (m^3) ($V_{village}$) is calculated using the water table fluctuation method [36] (Equation (4)). Post–monsoon recharge is considered negligible as more than 90% of the rainfall in all 3 years took place during the monsoon season [35]. In Equation (4), $V_{village}$ is made up of recharge from the UTFI pond (V_{UTFI}), as well as recharge from canal seepage and other village ponds. Also, storage change due to any net inflow across village boundaries is accounted for.

$$V_{village} = \Delta GW_S + VA_{monsoon} = \Delta H \times S_y \times A_{village} + VA_{monsoon} \quad (4)$$

where ΔGW_S (m^3) = the change in village groundwater storage, ΔH(m) = the rise in water level in the monsoon season (estimated in any year from the average rise in all piezometers from the start of rainfall period to the end of September), $A_{village}$ (m^2) = the area for the computation of recharge (village area), S_y = specific yield, and $VA_{monsoon}$ (m^3) = net groundwater abstraction in the monsoon season (taken to be equal to net abstraction for paddy in the kharif season).

Groundwater abstraction for irrigation was estimated based on the proportion of area irrigated with groundwater (based on information derived from the socio-economic survey), combined with the irrigation requirements of paddy as taken from [37]. Net abstraction was estimated from gross groundwater irrigation abstraction following groundwater resource estimation guidelines [35], which recommend that for groundwater depths of less than 10 metres, a return flow of 45% should be taken for paddy.

5. Results

5.1. Groundwater Recharge from UTFI

Table 2 summarizes UTFI recharge, average recharge rates and recharge duration from pilot testing for each year (2016–2018). On average, the recharge period (i.e., the number of days when the UTFI system was operated for recharge) was 75 days, recharging 44,415 m^3 of water at an average recharge rate of 580 m^3 day^{-1} (or 221 mm day^{-1} over the ponded area). The total volume recharged was 5.0–11.8 fold higher than the total storage capacity of the pond. The quantities of water recharged in 2016, 2017 and 2018 would have been sufficient to irrigate 12.8 ha, 17.7 ha and 7.5 ha of crop land (rabi wheat with an irrigation requirement of ~350 mm), respectively.

Table 2. UTFI recharge volumes (V_{UTFI}) and recharge rates (R_{UTFI}) for 2016, 2017 and 2018.

Year	Start Date–End Date	Number of Recharge Days [a]	V_{UTFI} (m^3)	R_{UTFI} (m^3 day^{-1}) (mm day^{-1}) [b]
2016	15 July–7 October	85	45,070	537 (204)
2017	17 July–5 October	78	61,969	775 (295)
2018	6 August–6 October	62	26,207	430 (164)
Average	-	75	44,415	580 (221)

[a] During monsoon season after the recharge period started, despite dry periods, the pond had water above dead storage height, so recharge continued. [b] Recharge rate in mm day^{-1} = ((Recharge rate m^3 day^{-1})/(pond area m^2)) × 1000.

There is a high inter- and intra-year variation in the UTFI recharge and recharge rates, with the highest recharge (average) observed in 2017 and the lowest in 2018. The lowest recharge in 2018 was due to both the low recharge rates observed and the relatively shorter recharge period, with recharge lasting for only 62 days, in comparison to 85 and 78 days in 2016 and 2017, respectively. The short recharge season in 2018 was due to the delayed onset of the monsoon, which shortened the duration of recharge operations.

In all years, recharge rates started high and gradually declined during the recharging period (Figure 5). Starting recharge rates in 2016, 2017 and 2018 were 996, 2499, 1978 m^3 day^{-1} respectively, which by the end of recharge period (when water storage in the pond was effectively dead storage) decreased to 274, 289 and 85 m^3 day^{-1}, i.e., there were reductions of 72.4%, 88.4% and 95.7%, respectively, from the starting conditions.

5.2. Groundwater Levels and the Response to UTFI Recharge

Figure 3 shows the monitored groundwater levels (relative to the reference point) for the piezometers and in recharge wells (RW1 and RW10), together with the pond water levels and daily rainfall over three years. The shallowest depth is observed in recharge wells with the highest depth in P6 and P7, which is as per the groundwater flow direction in the village. However, as all recharge wells (including RW1 and RW10) were used for recharge during recharge periods, they quickly filled and mainly indicated water levels in the pond (Figure 3). Thus, groundwater levels directly beneath the pond (beyond the recharge wells) could not be ascertained during recharge periods, and thus RW1 and RW10 were not used further in the analysis.

The pond water level during the entire recharge season remained above the minimum threshold level for recharge (dashed line in Figure 5), but shows some variation, with water highest during mid-season, which was also reflected in recharge well readings. As pond water levels are influenced by rainfall, the rate of siphoning from canal and also recharge rates, the precise reason for the variation remains unclear. The last readings of the pond level after each recharge period in Figure 5 show the times when the pond dried out. The pond water level readings indicate that for two discrete events, in September 2016 and August 2018, the water level rose above the recharge well heights, and thus potentially allowed unfiltered pond water to bypass the pea gravel filter and enter at the top of recharge well casings through small openings that serve to purge entrapped air. Though this happened for short time periods during high intensity rainfall events, it warrants building some margin of safety into recharge well heights in the future.

Despite extensive groundwater abstraction for the supplemental irrigation of paddy that takes place in the village, groundwater levels show a consistent rise in all piezometers during the monsoon season in all three years, coinciding with and then falling gradually over the non-monsoon season. Average groundwater level buildup in piezometers (ΔH) during the monsoon seasons in 2016, 2017 and 2018 was 2.74 m, 2.75 m and 3.95 m, respectively indicating high total recharge.

For the years 2016 and 2017, the average groundwater levels were lowest just before the start of the monsoon on 15 June and 24 June, at 6.45 mbgl and 6.51 mbgl, respectively. Meanwhile, in 2018, groundwater levels kept receding until 7 July, reaching 6.87 m due to the delayed onset of the monsoon rains (rains starting in mid-July, versus the last week of June when monsoon is expected).

However, despite this delay, 2018 was characterized by both high rainfall and high intensity rainfall events compared to 2016 and 2017 (Table 3), leading to substantially higher groundwater level buildup, with observed average groundwater levels reaching up to 2.22 mbgl. On the other hand, the highest groundwater levels in 2016 and 2017 were 3.25 and 3.67 mbgl, respectively.

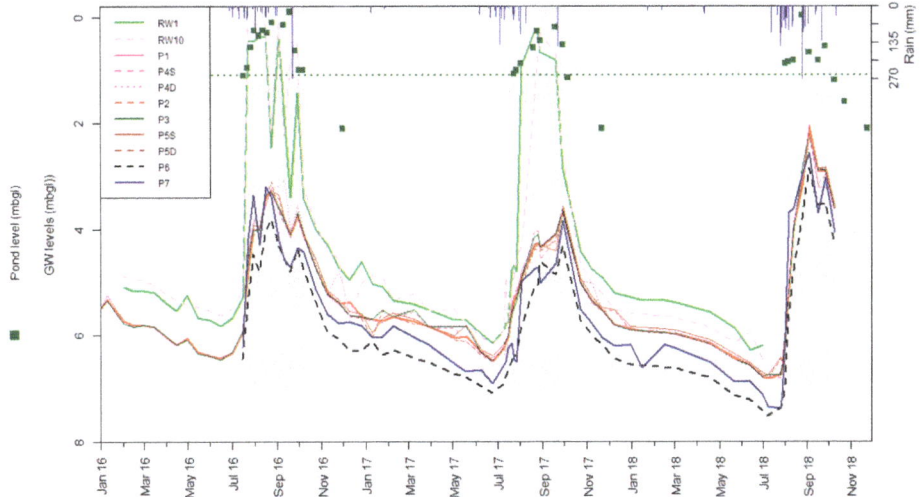

Figure 3. Observed groundwater levels (depth below ground level in mbgl) for piezometers and recharge wells (RW1 and RW10) with pond water levels over the period from January 2016 to November 2018. The grey shaded areas represent the periods of UTFI recharge operations. The dashed line shows the minimum pond level for which well recharge takes place.

Table 3. Rainfall and rainy days for 2016, 2017 and 2018.

Parameter	Time Period	2016	2017	2018
Rainfall (mm)	Annual	857	905	1812
	Monsoon (28 June–22 September)	857	874	1708
	Recharge period	737	472	992
Rainy days	Annual	23	22	27
	Monsoon (28 June–22 September)	23	20	22
	Recharge period	14	8	12

The Hantush Analytical Solution

The relative mound height was calculated (Figure 3) in order to distinguish the groundwater level response due to UTFI recharge from that due to other factors (e.g., other recharge mechanisms and pumping). By taking the mounding relative to a reference well, and not the absolute mounding, (Equation (3)), background differences in the groundwater levels of the piezometers were accounted for. Also, as the groundwater levels monitored in recharge wells were not representative of the groundwater conditions beneath the pond during the recharge season (as discussed above), the closest piezometers P1 and P5 were used instead for the analysis. Piezometer P6 was chosen as a reference well for this purpose. P7, the piezometer furthest from the pond, was not used for reference purposes, as closer inspection of water levels showed that it is more sensitive to rainfall events and subject to abrupt changes, likely due to water ponding in the local area.

Analysis showed that the observed relative mound height was subject to noise (high fluctuations) in the years 2016 and 2018 due to a range of confounding factors (e.g., rainfall recharge, pumping and canal recharge). Only in the year 2017 was a distinct signature resembling the expected theoretical relative mound, as modelled by the Hantush model, clearly visible (Figure 4). A more distinct—though still small and with noise—signature in 2017 with respect to the signatures in 2016 and 2018 could be attributed to low rainfall during the UTFI recharge operations (Table 3) when groundwater levels naturally increased (Figure 3), limiting rainfall recharge, and the higher recharge rates in 2017 relative to those in 2016 and 2018 (Table 2). The observed relative mound in 2017 shows the expected dynamics of mound formation, with relative mound decreasing as the distance from UTFI pond increases (i.e., mound P1 > P5), and flattening and becoming insignificant as time progresses, with groundwater spreading horizontally by the end of the season. High noise in the years 2016 and 2018, and overall low relative observed and modelled mound values in all years illustrate that the impact of UTFI recharge alone on groundwater levels was small—due to the high specific yield and transmissivity of the aquifer in the pilot area—which is difficult to distinguish from fluctuations in groundwater levels due to other confounding factors.

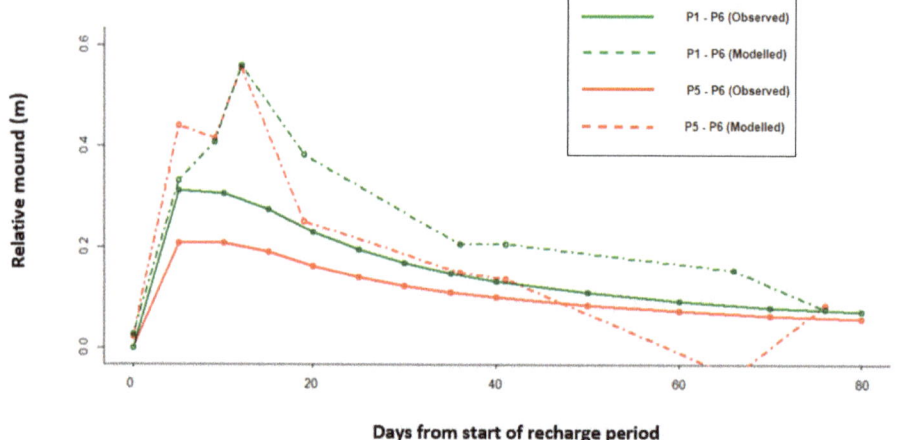

Figure 4. Observed and modelled relative mound heights, relative to P6 that was used as the reference piezometer, for piezometers at P1 (i.e., P1–P6) and P5 (i.e., P5–P6) for the year 2017.

5.3. UTFI Contribution to Recharge

Table 4 gives the estimated groundwater recharge at the village scale and the contribution of UTFI to overall recharge. Recharge in 2018 is 14% more than in both 2016 and 2017 due to higher groundwater level buildup. For years 2016, 2017 and 2018, the UTFI contribution to total recharge is 2.6%, 3.6% and 1.3% of the total recharge, respectively. Low values relative to both overall recharge is a reflection of the limited scale of the pilot relative to the village, the high recharge and the storage capacity of the groundwater systems in the region. However, the UTFI pond, with only 0.12% of the village area (indicative of the limited scale of pilot), contributed, on average, to about 2.5% of the village recharge. This shows that recharge per unit area within the pond is ~21 times higher than that occurring in other parts of the village.

Table 4. Annual village and UTFI recharge with recharge components (Equation (4)) for 2016–2018.

Year	ΔGW_S [a]	$VA_{monsoon}$ [b]	$V_{Village}$	V_{UTFI}	UTFI (% of Recharge)
		($\times 10^3$ m^3)			
2016	581	1158	1739	45.07	2.6
2017	583	1158	1741	61.97	3.6
2018	835	1158	1993	26.21	1.3

[a] ΔGW_S (m^3) = $\Delta H \times S_y \times A_{village}$ with ΔH for 2016, 2017 and 2018 is 2.74 m, 2.75 and 3.95 m, respectively; S_y = 0.1; $A_{village}$ = 212 hectares. [b] Gross irrigation applied for paddy in the kharif season in western Uttar Pradesh is 1100 mm [37]. Considering 45% return flow, net irrigation is 605 mm. Considering a crop area of village of 191.2 ha and 100% irrigation by GW, $VA_{monsoon}$ (m^3) = (605/1000) × (191.5 × 10000). Irrigation applied is not scaled to rainfall, as irrigation is influenced by both the distribution of rainfall and farmers' decisions on the sowing date. For this reason, only the average irrigation applied is used.

5.4. Clogging Effects on Recharge

The clogging of recharge wells due to the presence of silt and clay particles in the infiltration water has been identified as one of the main challenges for the sustainable operation of managed aquifer recharge (MAR) schemes [38,39]. High silt content in the source water entering the UTFI pond reduces recharge rates by clogging recharge wells, reducing the filtering capacity of filters and silting the bottom (reducing infiltration) as the recharge season progresses.

Analysis and measurement of the particulate matter deposited in recharge wells and filters during maintenance operations showed the appearance of only physical clogging [40]. Recharge water quality, measured as total suspended solids (mg/L), in source water showed that the average total suspended solids (TSS) of source water was in the range of 260–340 mg/L over the three monsoon seasons. The range is well above the limit of 10 mg/l for which clogging has been found to be moderate-to-severe in sand and gravel aquifer types [41], and above admissible guideline values for both direct injection and indirect infiltration recharge (20–60 mg/L in EU countries [42]. However, the relationship between TSS and recharge performance is site-specific [43] and is not considered in detail here.

In addition to the measurement of the TSS of the source water, the total silt entering and, in turn, accumulated at the bottom of pond, was estimated using gridded sampling of silt depth on the pond surface at the end of the 2016 and 2017 recharge seasons [40]. For 2016 and 2017, the total accumulated silt (Table 5) was estimated to be equivalent to 3.4 mm and 5.8 mm depth of silt at the pond surface, respectively. The development of a 3–6 mm clogging layer during recharge seasons increases hydraulic resistance. This results in the much lower infiltration rate observed at the end of recharge of 14.4–17.0 mm day^{-1} (Table 5) in comparison to the infiltration rate that was observed at the pond bottom in 2015 after cleaning and deepening (indicator of recharge from the infiltration pond without clogging), of 106 mm day^{-1}.

The high silt accumulation was expected as the canal also carries runoff generated by high monsoon rainfall, which consists of high particulate load [44]. For this reason, for the piloting, it was decided to start the recharge operation after the first few rainfall events, which were expected to carry the maximum sediment loads, to ensure these loads were not recharged.

In addition, regular yearly maintenance operations were carried out. In between recharge seasons, maintenance involving the cleaning of the recharge well and filters, as well as the desilting of the pond bottom, were carried out to enhance the recharge rates (Table 5).

The starting recharge rate in 2016 of 996 m^3 day^{-1} was much lower than the initial recharge rate tested in 2015 of 3150 m^3 day^{-1}. As a result, the cleaning of recharge wells and the filter box, and the de-silting of the pond bottom took place before the 2017 recharge season. In addition to well cleaning, which removes any clogging, the process also develops a well cavity that also leads to higher recharge rates. The effect of this was clearly visible with much higher (by more than a factor of two) starting recharge rates in 2017 relative to in 2016. Prior to the 2018 recharge season, recharge wells were not cleaned, and only the pond was desilted and the filters were cleaned. Despite overall low average recharge rates, the starting recharge rate of 1978 m^3 day^{-1} in 2018 suggests that the effect of recharge well cleaning persisted and that the recharge wells hadn't clogged. Overall, average recharge rates post maintenance operations in 2017 increased by 44.4%, whereas in 2018, average recharge rates—despite filter cleaning and pond desiltation—were reduced by 44.5% relative to 2017, and were lower by 19.9% relative to 2016. The lower recharge rates in 2018, despite starting with high recharge rates similar to 2017, suggest the influence of factors other than clogging alone on recharge rates. Relative recharge from recharge wells and by infiltration through the basin alone is discussed in Section 6.2.1.

Table 5. A summary of groundwater levels, recharge rates (R_{UTFI}) and maintenance operations over the recharge season from 2016–2018.

Year	R_{UTFI} (m^3 day^{-1})			Recharge Well Cleaning	Filter Cleaning	Desilting Pond Bottom	Gravity Head for Recharge (m) [a]	Avg. water Quality Entering Pond (TSS: mg/L) [b]	Total Silt Accumulated [c] (Tonnes)	Days Taken for Dead Storage to Dry [d]	End Season Infiltration Rate (mm day^{-1}) [e]
	Start	End	Avg.								
2016	996	274	537	No	No	No	4.0	340	12.2	55	14.4
2017	2499	289	775	yes	yes	Yes	4.7	260	21.2	48	17.0
2018	1978	85	430	No	Yes	Yes	3.5	282	-	50	16.2

[a] Taken as elevation difference between the surface water level in the pond and the elevation of the water table in the nearest piezometer (P1). [b] Number of readings were limited: monthly in 2016 and 2018; much more in 2017. [c] Estimated based on silt load accumulated at the pond bottom, determined based on gridded sampling. [d] Total number of days taken for the pond to dry out from 1 m of dead storage (no recharge of recharge wells. Indicated by the last pond water level reading in Figure 5) by infiltration + evaporation. [e] Estimated from the time taken for dead storage to dry (dead storage (volume)/days) and subtracting the evaporation rate calculated using $ET_c = K_c \times ET_O$ and taking $K_c = 1$ for open water [45]. Reference evapotranspiration (ET_O) is taken as the average of the months of October and November) and is 3.82 mm day^{-1} [46].

6. Discussion

6.1. Factors Influencing Recharge Rates

For all the years, recharge rates start at their highest values and decrease as the season progresses (Figure 5). Similar hydrologic trends have been observed at MAR sites with surface runoff containing high levels of particulate matter [38,47]. These studies tend to suggest that the reduction in recharge rates over time is a function of two major processes: (i) groundwater levels linked to rainfall and (ii) clogging linked to recharge water quality.

The increase in recharge rates post maintenance cleaning operations in 2017 shows the impact of physical clogging on recharge rates (Table 5). A high starting recharge rate in 2018, similar to that in 2017, suggests the cleaning operations post the 2017 recharge season (excluding recharge well cleaning) maintained similar clogging as in 2017. However, recharge rates in 2018 dropped steeply in line with the observed steep rise in groundwater levels, whereas in 2016 and 2017, the decline in recharge rates was gradual, in line with the rising trend in groundwater levels (Figure 5). This points to the influence of groundwater levels on recharge rates as there was no apparent difference in average source water TSS (Table 5, Figure 5). However, the limited readings of TSS add some uncertainty to the analysis, as TSS in source water show high variability associated with rainfall events.

Groundwater levels influence the recharge rates as they change the gravity head (i.e., the elevation difference between the surface water level in the pond and the groundwater level), on which gravity recharge depends [38,48]. As the groundwater levels rise during the monsoon, the gravity head decreases, resulting in declining recharge rates over the recharge season (Figure 5). The highest and lowest recharge rates in the years 2017 and 2018, respectively, are associated with the highest and lowest average available gravity heads (Table 5). Steep rises in groundwater levels towards the middle of the 2018 season brought groundwater levels up to ~2 mbgl in the nearest piezometer(P1). This shows the potential hydraulic connection taking place in between groundwater and the pond base (2.1 mbgl), which would have reduced the recharge rates [48]. This is reflected in a steep decline in recharge rates, and recharge rate values reduced to <100 m^3 day^{-1} by the end of recharge season in 2018 in comparison to end of season recharge rates of ~300 m³ day⁻¹ in 2017 and 2016, where groundwater levels remained deeper than 3 m. The pond water level remained above the minimum threshold level for recharge (Figure 5). No direct correlation between recharge rates and pond levels could be made out, which shows that the pond water level, which has a source from the canal, is not the limiting factor in recharge rates, and that clogging and the groundwater level beneath the pond when this becomes hydraulically connected to the pond remain the leading factors influencing recharge rates.

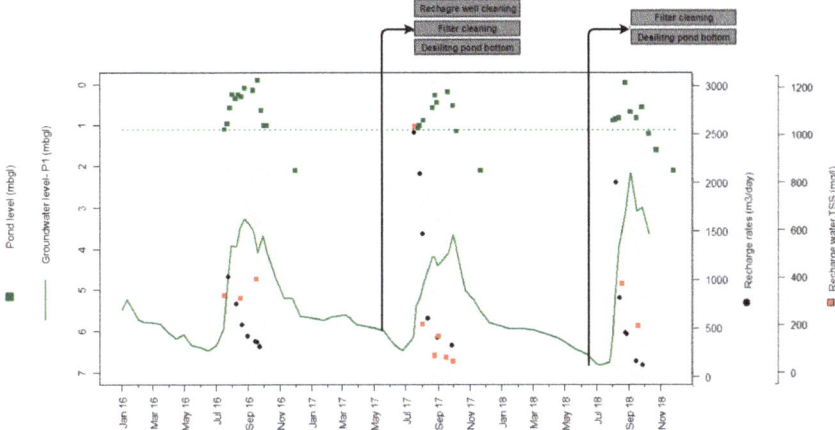

Figure 5. UTFI recharge rates, the groundwater level of P1 (meters below ground level (mbgl)), the pond water level (mblg) and the TSS of source water UTFI from 2016 to 2018. The dashed line shows the minimum pond level for which well recharge takes place. Text boxes show the desilting activities. The grey shaded areas represent the periods of recharge.

UTFI Recharge Variation Implications for Flood Mitigation

The high inter-year variation of recharge rates carries high significance for the dual aims of UTFI in the region: flood mitigation and enhancing groundwater storage. UTFI recharge is highest in low rainfall years (61,969 m³ in 2017 with rainfall of 905 mm) and lowest in high rainfall years (26,207 m³ in 2018 with rainfall of 1811 mm). This is because high rainfall years lead to high groundwater levels, thereby decreasing the available gravity head, which reduces UTFI recharge rates. At the same time, high rainfall intensity years, such as 2018, may also reduce the number of recharge days, which are already constrained by the monsoon season.

Thus, on one hand, higher recharge in low rainfall years shows that UTFI could play an important positive role in addressing groundwater storage depletion in dry years when recharge is limited. However, lower recharge in wet years, which are also the years when more intense flooding is expected to take place, shows the diminishing returns of UTFI if upscaled specifically for flood mitigation. This influence of rainfall and groundwater levels is critical and needs to be taken into consideration for planning purposes when flood mitigation is a central objective.

Several multi-year studies analyzing MAR recharge performance in other parts of India have observed recharge behaviors quite different to those in this study. Those studies indicate high recharge in high rainfall years and vice versa [19,49]. Recharge-rainfall relationships in those cases would appear to be characterized by a different set of limiting factors for recharge. For example, in the case of check dam recharge reported by [19,49], all of the monitored check dams showed higher recharge in high rainfall years, as low rainfall years result in less recharge due to low inflow into the dams. However, study [19] was later extended to include a wetter year [50], where recharge was observed to be less than for the average year, indicating recharge limited by aquifer capacity. This is similar to the UTFI pilot case where source water was not a limitation (supplied by a canal at flow rates far higher than what could be recharged). Instead, recharge is limited by the available gravity head and the infrastructure performance.

6.2. Performance of the UTFI Pilot

6.2.1. Comparing Recharge Wells to the Infiltration Pond

The pilot chose to employ sub-surface recharge methods using recharge wells to maximize recharge rates to overcome land availability constraints, as the region is one of the most densely populated places in the world [51]. To assess the performance, UTFI system recharge rates (R_{UTFI}) were compared against infiltration rates observed from the pond bottom alone. Recharge rates from the base of the pond were estimated in two ways at two different times: first, during 6 h ring infiltration tests during preliminary recharge testing in 2015, and later on, during recharge operations, by observing the time taken for the dead storage to empty. Assuming the same decreasing temporal trends in infiltration from the pond bottom as overall recharge rates, rates from the infiltration ring are comparable with recharge rates at the start of season. The decline in dead storage (corrected for evaporative losses), can be compared with the end of season recharge rates.

The average infiltration rate from the pond bottom in 2015 after cleaning and deepening the pond (an indicator of recharge from the infiltration pond only) was 106 mm day^{-1}. In comparison, the UTFI system recharge rate at the start of season was 379 mm day^{-1} in 2016, 933 mm day^{-1} in 2017 and 752 mm day^{-1} in 2018; these rates were, on average, higher by factors of 3.6, 8.8 and 7.1, respectively. Similarly, infiltration from the pond bottom at the end of the pilot was 14.4, 17.0 and 16.2 mm day^{-1} in 2016, 2017 and 2018, respectively (Table 5). In comparison, end-of-season UTFI recharge rates for the corresponding years were 104.4, 110.1 and 32.4 mm day^{-1}, which were, on average, higher by factors of 7.2, 6.5 and 2.0 respectively. These high recharge rates justify the use of sub-surface methods, without which far lower volumes would have been recharged.

6.2.2. Comparing UTFI to Comparable Studies in the Region

The limited existing studies of MAR using sub-surface methods in the Indo-Gangetic Plains having similar aquifer characteristics as the study area were reviewed to compare UTFI performance. In one study in the neighboring state of Haryana, recharge wells were used to recharge canal water during the rainy season (July–October), with reported average recharge rates from a single well of 794–989 m^3 day^{-1} in the first year with no typical trend, whereas in the second year, recharge rates reduced from 1088 m^3 day^{-1} at the start to 798 m^3 day^{-1} by the end of the season [14]. In other studies from an alluvial area in the state of Punjab, all of which used canal water as the source, reported recharge rates ranged from 302 to 3784 m^3 day^{-1} [15], and from 588 to 1766 m^3 day^{-1} well^{-1} in three case studies reported by [16]. The Master Plan for MAR in India gave an expected average design recharge rate from a recharge shaft in alluvial areas of Uttar Pradesh of 1000 m^3 day^{-1}, running on about 60 operational days during monsoon [52]; however, no supporting data are provided on how these estimates were derived.

Large differences in recharge rates among the studies presented above point to the number of factors influencing recharge rates such as the aquifer characteristics, groundwater levels, source water, design and methods used to estimate recharge. Contrasts in system design ranged from pressure injection in Ghaggar, Punjab [15] to a battery of 20 recharge wells installed in 20 trenches in Patiala [16]. The depth of recharge wells ranged from 26 to 81 m, and groundwater levels were >10–15 m, in comparison to the shallow groundwater levels in the UTFI pilot case (2–7 m). Given the large number of differences and the general lack of studies in the region that investigated recharge performance comprehensively over multi-year time periods, along with the limited details on monitoring provided, any systematic comparisons are difficult.

6.3. Dependence of Aquifer Type on Groundwater Mounding

Most MAR case studies in India report a visible increase in groundwater levels or longer availability of water in wells [17,18]. For example, [18] reviewed six case studies, all in hard rock areas, which reported increases in groundwater levels of up to 4–7 m due to MAR interventions. In contrast, for the UTFI pilot, modelled and observed mounds are small (~0.4–0.8 m). In addition, the expected mound is difficult to distinguish from background water level variations.

The low mound formation observed during the pilot relative to high values reported in other MAR studies in India, most of which come from hard rock areas, reflects the contrast in hydrogeological characteristics such as porosity, transmissivity and lithology, and the thickness of the aquifer. To analyze the expected difference due to hydrogeology, the Hantush solution was run with typical hard rock aquifer (basaltic) characteristics (a hydraulic conductivity (k) of 5 m day^{-1} and a specific yield (Sy) of 0.02) [35] and compared with the UTFI pilot site (Figure 6). The difference in mound height is apparent, with hard rock aquifers showing a mound height much higher than the pilot (an average of 2.8 m and 2.1 m in hard rock vs. 0.7 m and 0.5 m in the pilot case) at distances of 0 m (i.e., the center of the pond) and at 5 m (i.e., P1), respectively for the same recharge volume. In the Hantush model, mound height was calculated from the center of the rectangular basin in the x-direction with the basin edge at a distance of 17.5 m (i.e., (35 m)/2). Therefore, a piezometer at a distance of 5 m from the edge of pond is at a distance of 5 + 17.5 = 22.5 from the center of the rectangular basin. The contrast with the hard rock aquifers is a reflection of high aquifer transmissivity and specific yield of the alluvial aquifer in the village. Though this is as expected, it underlies that the same recharge performance would have had a very different impact on groundwater levels, depending on the hydrogeological conditions. In alluvial aquifers, as is the case in the UTFI pilot, the storage changes are more subdued and would require detailed monitoring to discern the change in groundwater levels in response to recharge.

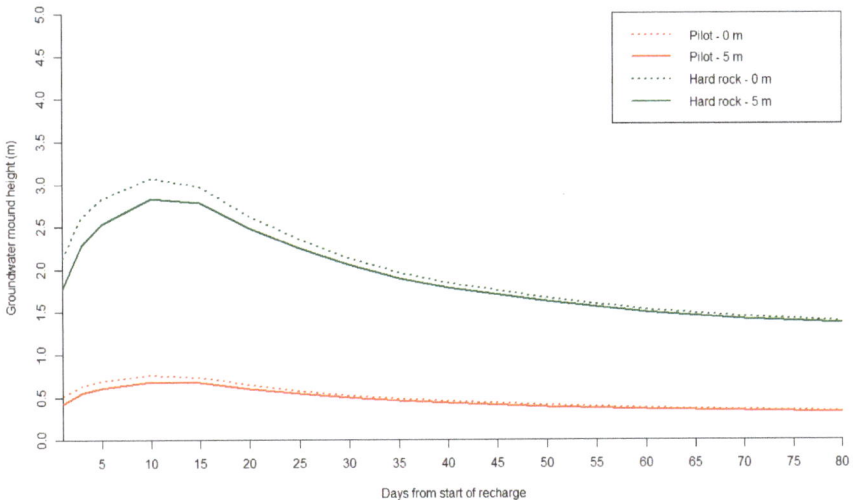

Figure 6. Modelled mound heights at the center of pond (0 m) and P1 (5 m) for the pilot (S_y = 0.1 and k = 20 m day^{-1}) and general hard rock aquifers (S_y = 0.02 and k = 5 m day^{-1}) under the observed UTFI recharge rates of the year 2017.

6.4. Scaling up of UTFI

The scaling up of UTFI to the basin scale requires a sound assessment of the availability of flows for recharge, the capacity of available groundwater storage to store runoff and the demand for recharge water. In addition, studies on the costs and benefits are required to ascertain the economic rationale for upscaling. Previous research by [11] has shown that the average monsoon runoff in the Ramganga basin during the monsoon months is 5782 million m^3. Recharging 10–50% of these runoff volumes can reduce flooding (with a 5 year return period) by 5–27%, increase groundwater recharge by 11–56%, and increase groundwater levels, on average, by 1.2–4.6 m with respect to a no-UTFI scenario. Assessment of the economic benefits (based on increasing crop intensity by using recharge water) by [53] shows that upscaling would require huge capital investment in infrastructure, but on average have a benefit to cost ratio of >1.

However, the above study is based on major assumptions [53], nor does it consider the actual performance of the UTFI system that this pilot study explored. However, one UTFI pilot could not be considered representative of the entire basin, but can help to identify lessons that are critical when developing any upscaling strategy. For example, the insights on diminishing UTFI recharge rates during high rainfall years, due to high recharge leading to hydraulic connection, warrant the need to look more critically at the available groundwater storage and the variability in recharge rates. The need for routine operation and maintenance must be clearly accounted for while doing any economic analysis. In addition, any scaling up would require the consideration of socioeconomic, institutional and related issues, which are not analyzed in this study, but are touched upon in [20,21].

7. Conclusions

The first UTFI pilot trial in the Indo-Gangetic Plain in India was capable of recharging an average annual volume of 44,000 m^3 with the recharge volume over three years varying from 26,000 to 62,000 m^3 (i.e., 430 to 775 m^3 day^{-1} during the recharge periods). These volumes are 5.0–11.8 times the total storage capacity of the pond and would be sufficient to irrigate 8–18 hectares of rabi crop. High intra-year variation, reflected in recharge rate reductions ranging from 72.4% to 95.7% relative to the starting conditions, was observed. This is linked to the roles of: (a) source water quality (TSS of 260–340 mg/L) leading to clogging, and (b) groundwater levels influencing gravity heads as the recharge season progresses, and hydraulic connection further reducing recharge in the wetter year (2018) when recharge from other sources is high. Annual maintenance activities involving desilting basins, and cleaning filters and recharge wells appear to be effective in controlling clogging and restoring recharge rates. Overall, the UTFI design achieved much higher recharge rates (2–9 times) than what would have been achieved from village ponds alone through infiltration. The results show that the relative mounding in nearby wells w.r.t to the reference well at 100 m distance due to UTFI is limited, due to the high specific yield and transmissivity of the alluvial aquifer, and is influenced by a range of confounding factors that make the delineation of small mound heights difficult. The results provide the first systematic, multi-year assessment of the performance of UTFI systems at the individual pond scale in the flood prone, intensively irrigated, alluvial aquifer regions of the Indo-Gangetic Plain.

Author Contributions: Conceptualization, M.F.A. and P.P.; Data curation, M.F.A. and N.S.; Formal analysis, M.F.A.; Funding acquisition, P.P.; Investigation, M.F.A., P.P. and N.S.; Methodology, M.F.A. and P.P.; Project administration, A.S. and P.P.; Supervision, P.P. and A.S.; Visualization, M.F.A.; Writing—original draft, M.F.A.; Writing—review & editing, P.P. All authors have read and agree to the published version of the manuscript.

Funding: This research received no external funding.

Acknowledgments: It was undertaken as part of the CGIAR Research Program on Climate Change, Agriculture and Food Security (CCAFS). It also contributes to the CGIAR Research Program on Water, Land and Ecosystems (WLE). The contributions and support provided by IWMI colleagues including Prasun Gangopadhyay (IWMI), Krishi Vigyan Kendra (KVK), Rampur and the Central Soil Salinity Research Institute (CSSRI), Lucknow for carrying out UTFI pilot are gratefully acknowledged. Critical reviews by two anonymous reviewers and the editor of this Special Issues helped to improve the quality of this paper.

Conflicts of Interest: The authors declare no conflict of interest.

References

1. Hall, J.; Grey, D.; Garrick, D.; Fung, F.; Brown, C.; Dadson, S.; Sadoff, C.W. Water Security. Coping with the curse of freshwater variability. *Science* **2014**, *346*, 429–430. [CrossRef]
2. Hoekstra, A.; Mekonnen, M.; Chapagain, A.; Mathews, R.; Richter, B. Global Monthly Water Scarcity: Blue Water Footprints versus Blue Water. *PLoS ONE* **2012**, *7*, e32688. [CrossRef] [PubMed]
3. FAO. *The Impact of Disasters on Agriculture and Food Security*; Food and Agriculture Organization of the United Nations (FAO): Rome, Italy, 2015.
4. United Nations. *Climate Change and Water: UN-Water Policy Brief*; United Nations: Geneva, Switzerland, 2019.
5. IPCC. Summary for Policymakers. In *Managing the Risks of Extreme Events and Disasters to Advance Climate Change Adaptation, A Special Report of Working Groups I and II of the Intergovernmental Panel on Climate Change*; Field, C.B., Barros, V., Stocker, T.F., Qin, D., Dokken, D.J., Ebi, K.L., Mastrandrea, M.D., Mach, K.J., Plattner, G.-K., Allen, S.K., et al., Eds.; Cambridge University Press: Cambridge, UK; New York, NY, USA, 2012; pp. 1–19.
6. Van der Gun, J. *Groundwater and Global Change: Trends, Opportunities and Challenges*; UNESCO: Paris, France, 2012.
7. Döll, P.; Schmied, H.M.; Schuh, C.; Portmann, F.T.; Eicker, A. Global-scale assessment of groundwater depletion and related groundwater abstractions: Combining hydrological modeling with information from well observations and GRACE satellites. *Water Resour. Res.* **2014**, *50*, 5698–5720. [CrossRef]
8. Pavelic, P.; Brindha, K.; Amarnath, G.; Eriyagama, N.; Muthuwatta, L.; Smakhtin, V.; Gangopadhyay, P.K.; Malik, R.P.S.; Mishra, A.; Sharma, B.R.; et al. *Controlling Floods and Droughts through Underground Storage: From Concept to Pilot Implementation in the Ganges River Basin*; IWMI Research Report; International Water Management Institute: Colombo, Sri Lanka, 2015; Volume 165, 33p. [CrossRef]
9. Brindha, K.; Pavelic, P. Identifying priority watersheds to mitigate flood and drought impacts by novel conjunctive water use management. *Environ. Earth Sci.* **2016**, *40*, 407–431. [CrossRef]
10. Pavelic, P.; Srisuk, K.; Sarap, P.; Nadee, S.; Pholkern, K.; Chusanathas, S.; Sitisak, M.; Theerasak, T.; Teerawash, I.; Vladimir, S. Balancing-out floods and droughts: Opportunities to utilize floodwater harvesting and groundwater storage for agricultural development in Thailand. *J. Hydrol.* **2012**, *470–471*, 55–64. [CrossRef]
11. Chinnasamy, P.; Muthuwatta, L.; Eriyagama, N.; Pavelic, P.; Lagudu, S. Modeling the potential for floodwater recharge to offset groundwater depletion: A case study from the Ramganga basin, India. *Sustain. Water Resour. Manag.* **2018**. [CrossRef]
12. Khan, R.M.; Voss, C.I.; Yu, W.; Michael, H.A. Water resources management in the Ganges Basin: A comparison of three strategies for conjunctive use of groundwater and surface water. *Water Resour. Manag.* **2014**, *28*, 1235–1250. [CrossRef]
13. MacDonald, A.M.; Bonsor, H.C.; Taylor, R.; Shamsudduha, M.; Burgess, W.G.; Ahmed, K.M.; Mukherjee, A.; Zahid, A.; Lapworth, D.; Gopal, K.; et al. *Groundwater Resources in the Indo-Gangetic Basin: Resilience to Climate Change and Abstraction*; British Geological Survey Open Report OR/15/047; British Geological Survey: Nottingham, UK, 2015.
14. Kaledhonkar, M.J.; Singh, O.P.; Ambast, S.K.; Tyagi, N.K.; Tyagi, K.C. Artificial groundwater recharge through recharge tube wells: A case study. *J. Inst. Eng. (India)* **2003**, *84*, 5.
15. Sakthivadivel, R. The groundwater recharge movement in India. In *The Agricultural Groundwater Revolution: Opportunities and Threats to Development*; Giordano, M., Villholth, K., Eds.; CAB International Publishing: Colombo, Sri Lanka, 2007.
16. Gupta, S. Groundwater Management in Alluvial Areas, Incidental Paper. CGWB, Ministry of Water Resources, Government of India; 2011. Available online: http://cgwb.gov.in/documents/papers/incidpapers/Paper%2011-%20sushil%20gupta.pdf (accessed on 12 November 2019).
17. Glendenning, C.J.; van Ogtrop, F.F.; Mishra, A.K.; Vervoort, R.W. Balancing watershed and local scale impacts of rain water harvesting in India: A review. *Agric. Water Manag.* **2012**, *107*, 1–13. [CrossRef]
18. Prathapar, S.; Dhar, S.; Rao, G.T.; Maheshwari, B. Performance and impacts of managed aquifer recharge interventions for agricultural water security: A framework for evaluation. *Agric. Water Manag.* **2015**, *159*, 165–175. [CrossRef]

19. Dashora, Y.; Dillon, P.; Maheshwari, B.; Soni, P.; Dashora, R.; Davande, S.; Purohit, R.C.; Mittal, H.K. A simple method using farmers' measurements applied to estimate check dam recharge in Rajasthan, India. *Sustain. Water Resour. Manag.* **2018**, *4*, 301–316. [CrossRef]
20. Reddy, V.R.; Pavelic, P.; Hanjra, M.A. Underground Taming of Floods for Irrigation (UTFI) in the river basins of South Asia: Institutionalizing approaches and policies for sustainable water management and livelihood enhancement. *Water Policy* **2017**, *20*, wp2017150. [CrossRef]
21. Reddy, V.R.; Rout, S.K.; Shalsi, S.; Pavelic, P.; Ross, A. Managing underground transfer of floods for irrigation: A case study from the Ramganga Sub-Basin (Ganges Basin), India. *J. Hydrol.* **2020**. [CrossRef]
22. Census of India. *District Census Handbook, Rampur*; Village and town wise primary census abstract; Directorate of Census Operations: Uttar Pradesh, India, 2011.
23. DoES. *Total Area and Classification of Area in Each District of Uttar Pradesh State for the Year Ending 2014-15 (Hectare)*; Special Data Dissemination Standard Division, Directorate of Economics & Statistics (DoES), Ministry of Agriculture and Farmers Welfare; Govt. of India: New Delhi, India, 2015. Available online: https://aps.dac.gov.in/LUS/Index.htm (accessed on 22 October 2019).
24. Government of Uttar Pradesh. *District Disaster Management Plan*; DDMA: Rampur, India, 2010.
25. Government of Uttar Pradesh. *District Disaster Management Plan*; DDMA: Rampur, India, 2013.
26. Government of Uttar Pradesh. *Report of Officer-in-Charge, District Flood Protection and Relief*; Government of Uttar Pradesh: Rampur, India, 2015.
27. Government of Uttar Pradesh. *District Statistical Handbook*; Government of Uttar Pradesh: Rampur, India, 2014.
28. Tripathi, A.K. *District Ground Water Brochure of Rampur District*; U.P. Central Ground Water Board (CGWB): Rampur, India, 2009.
29. CGWB. *Dynamic Groundwater Resources of India (As on 31st March, 2013)*; Central Ground Water Board, Ministry of Water Resources, River Development & Ganga Rejuvenation Government of India: Faridabad, India, 2017.
30. Bouwer, H. Artificial recharge of groundwater: Hydrogeology and engineering. *Hydrogeol. J.* **2002**, *10*, 121–142. [CrossRef]
31. Hantush, M.S. Growth and decay of groundwater-mounds in response to uniform percolation. *Water Resour. Res.* **1967**, *3*, 227–234. [CrossRef]
32. Yihdego, Y. Simulation of Groundwater Mounding Due to Irrigation Practice: Case of Wastewater Reuse Engineering Design. *Hydrology* **2017**, *4*, 19. [CrossRef]
33. Poeter, E.P.; McCray, J.E. Modeling water table mounding to design cluster and high-density wastewater soil absorption systems. *J. Hydrol. Eng.* **2008**, *13*, 702–709. [CrossRef]
34. Carleton, G.B. *Simulation of Groundwater Mounding Beneath Hypothetical Stormwater Infiltration Basins*; U.S. Geological Survey: Reston, VA, USA, 2010.
35. Sharma, B.; Gulati, A.; Mohan, G.; Manchanda, S.; Ray, I.; Amarasinghe, U. *Water Productivity Mapping of Major Indian Crops*; NABARD and ICRIER: New Delhi, India, 2018.
36. MoWR, RD & GR, GoI. Ministry of Water Resources, River Development & Ganga Rejuvenation Government of India. In *Report of the Ground Water Resource Estimation Committee (GEC-2015)*; MoWR, RD & GR, GoI: New Delhi, India, 2017.
37. Healy, R.W.; Cook, P.G. Using groundwater levels to estimate recharge. *Hydrogeol. J.* **2002**, *10*, 91–109. [CrossRef]
38. Holländer, H.M.; Mull, R.; Panda, S.N. A concept for managed aquifer recharge using ASR-wells for sustainable use of groundwater resources in an alluvial coastal aquifer in Eastern India. *Phys. Chem. Earth Parts A/B/C* **2009**, *34*, 270–278. [CrossRef]
39. Martin, R. (Ed.) *Clogging Issues Associated with Managed Aquifer Recharge Methods*; IAH Commission on Managed Aquifer Recharge: Adelaide, Australia, 2013.
40. CSRRI. *Technological Assessment of Underground Taming of Floods for Irrigation*; Progress report prepared for the International Water Management Institute for project: Piloting and Up-scaling an Innovative Approach for Mitigating Urban Floods and Improving Rural Water Security in South Asia; ICAR-Central Soil Salinity Research Institute, Regional Research Station: Lucknow, India, 2018.

41. NRMMC, ERHC and NHMRC. *Australian Guidelines for Water Recycling: Managed Aquifer Recharge (National Water Quality Management Strategy No. 24)*; Natural Resource Management Ministerial Council (NRMMC), Environment Protection and Heritage Council (ERHC) and National Health and Medical Research Council (NHMRC): Canberra, Australia, 2009.
42. Miret, M.; Vilanova, E.; Molinero, J.; Sprenger, C. The Management of Aquifer Recharge in the European Legal Framework (DEMEAU Project, European Commission). Available online: https://demeau-fp7.eu/sites/files/D121%20legal%20framework%20and%20MAR%20DEMEAU%20project_1.pdf (accessed on 12 October 2019).
43. Hutchinson, A.; Milczarek, M.; Banerjee, M. Clogging phenomena related to surface water recharge facilities. In *Clogging Issues Associated with Managed Aquifer Recharge Methods*; Martin, R., Ed.; IAH Commission on Managed Aquifer Recharge: Adelaide, Australia, 2013; pp. 95–106.
44. Kumar, A.; Bahadur, Y. Water quality of River Kosi and Rajera system at Rampur (India): Impact Assessment. *J. Chem.* **2012**, *2013*, 618612. [CrossRef]
45. Kohli, A.; Frenken, K. *Evaporation from Artificial Lakes and Reservoirs (FAO AQUSTAT Reports)*; FAO: Rome, Italy, 2015.
46. IWP. India Water Portal Met Data [WWW Document]. Available online: https://www.indiawaterportal.org/met_data/ (accessed on 22 November 2019).
47. Dillon, P.J.; Vanderzalm, J.; Page, D.; Barry, K.; Gonzalez, D.; Muthukaruppan, M.; Hudson, M. Analysis of ASR clogging investigations at three Australian ASR sites in a Bayesian context. *Water* **2016**, *8*, 442. [CrossRef]
48. Dillon, P.J.; Liggett, J.A. An ephemeral stream aquifer interaction model. *Water Resour. Res.* **1983**, *19*, 621–626. [CrossRef]
49. Glendenning, C.J.; Vervoort, R.W. Hydrological impacts of rainwater harvesting (RWH) in a case study catchment: The Arvari River, Rajasthan, India. Part 1: Field-scale impacts. *Agric. Water Manag.* **2010**, *98*, 331–342. [CrossRef]
50. Dashora, Y.; Dillon, P.; Maheshwari, B.; Soni, P.; Mittal, H.; Dashora, R.; Singh, P.K.; Purohit, R.C.; Katara, P. Hydrologic and cost benefit analysis at local scale of streambed recharge structures in Rajasthan (India) and their value for securing irrigation water supplies. *Hydrogeol. J.* **2019**, *27*, 1889–1909. [CrossRef]
51. FAO. Irrigation in Southern and Eastern Asia in Figures—AQUASTAT Survey—2011. 2011. Available online: http://www.fao.org/nr/water/aquastat/basins/gbm/gbm-CP_eng.pdf (accessed on 12 November 2019).
52. GoI. *Master Plan. for Artificial Recharge to Ground Water in India*; Ministry of Water Resources and Central Groundwater Board, Government of India: New Delhi, India, 2013.
53. Alam, M.F.; Pavelic, P. *Underground Transfer of Floods for Irrigation (UTFI): Exploring the Potential at the Global Scale*; IWMI Research Report; International Water Management Institute: Colombo, Sri Lanka, in press.

 © 2020 by the authors. Licensee MDPI, Basel, Switzerland. This article is an open access article distributed under the terms and conditions of the Creative Commons Attribution (CC BY) license (http://creativecommons.org/licenses/by/4.0/).

Article

Managed Aquifer Recharge at a Farm Level: Evaluating the Performance of Direct Well Recharge Structures

Prahlad Soni [1], Yogita Dashora [1], Basant Maheshwari [2,*], Peter Dillon [3,4], Pradeep Singh [5] and Anupama Kumar [3]

1. Vidya Bhawan Krishi Vigyan Kendra, Udaipur 313001, India; prahladsoni.baif@gmail.com (P.S.); dashora.yogita@gmail.com (Y.D.)
2. Western Sydney University, Penrith, Locked Bag 1797, Penrith, NSW 2767, Australia
3. CSIRO Land and Water, Waite Laboratories, Waite Rd, Urrbrae, SA 5064, Australia; pdillon500@gmail.com (P.D.); Anupama.Kumar@csiro.au (A.K.)
4. National Centre for Groundwater Research and Training (NCGRT) & College of Science and Engineering, Flinders University, Adelaide, SA 5001, Australia
5. Maharana Pratap University of Agriculture and Technology, Udaipur 313001, India; pksingh35@yahoo.com
* Correspondence: b.maheshwari@westernsydney.edu.au

Received: 10 October 2019; Accepted: 7 April 2020; Published: 9 April 2020

Abstract: A field study evaluated the performance of direct well recharge structures (DWRS) in order to harvest and filter farm runoff and its discharge into open dug wells to augment groundwater recharge. This was undertaken between 2016 and 2018 using a total of 11 wells in the Dharta watershed, situated in a semi-arid hardrock region of Udaipur district, Rajasthan, India. The depth to water level in each DWRS well was monitored weekly for 1 to 3 years before and after the DWRS was established, and water samples were taken for water quality analysis (pH, electrical conductivity (EC), total dissolved solids (TDS), turbidity, fluoride, and *Escherichia coli*) before and during the monsoon period. For each DWRS well, two control wells in close proximity were also monitored and sampled. Five of the DWRS established in 2018 also had flow meters installed in order to measure discharge from the filter to the well. The volume of water recharged through DWRS into individual wells during the 2018 monsoon ranged from 2 to 176 m^3 per well. Although the mean rise in water levels over the monsoon was higher in DWRS wells than in nearby control wells, the difference was not significant. Values of pH, EC, TDS, and F decreased in DWRS and control wells as each monsoon progressed, whereas the turbidity of wells with DWRS increased slightly. There was no significant difference between DWRS and control wells for pH, EC/TDS, turbidity, or fluoride. The presence of *E. coli* in DWRS wells was higher than in control wells, however, *E. coli* exceeded drinking water guidelines in all sampled wells. On the basis of this study, it is recommended that rural runoff should not be admitted to wells that are used for, or close to, wells used for drinking water supplies, even though salinity and fluoride concentrations may be reduced. For this study, none of the 11 DWRS wells produced sufficient additional recharge to potentially increase dry season irrigation supplies to justify expenditure on DWRS. This even applies to the DWRS well adjacent to a small ephemeral stream that had a significantly larger catchment area than those drawing on farmers' fields alone. An important and unexpected finding of this study was that no sampled open dug well met drinking water standards. This has led to a shift in local priorities to implement well-head water quality protection measures for wells used for drinking water supplies. It is recommended that parapet walls be built around the perimeter of such dug wells, as well as having covers be installed.

Keywords: groundwater recharge; water quality; water level monitoring; recharge performance; rainwater harvesting; India

1. Introduction

Water scarcity has become a major problem, especially in most of the arid regions of the world. It ultimately affects food security, natural ecosystems, and plant and human health (Seckler et al., 1999) [1]. Water scarcity arises due to the various anthropogenic factors and one of them is the depletion of groundwater resource. Farmers in semi-arid parts of India use groundwater to save rainfed crops from failure and to increase yields. As it is a relatively cheap and easily accessible water resource for individual farmers, irrespective of their farm size, annual groundwater use often reaches or exceeds the average annual natural recharge. Depth to watertable in hard rock terrain fluctuates considerably during the year, and shallow aquifers become depleted where the use of groundwater has increased; thus, tubewells are drilled to allow pumping from deeper down (in the same or different aquifers), in some areas rendering marginal quality water (Shah, 2009) [2]. The extensive use of groundwater resources by farmers all over the country pumping out water in an unregulated manner creates its own sets of complex management and sustainability issues.

According to a report of CGWB (2017) [3], almost the whole of India shows declining groundwater levels, with the largest declines observed in parts of Rajasthan, Haryana, Punjab, Gujarat, Telangana, and Maharashtra. Water harvesting and recharge enhancement at micro-watershed level have been identified as means to benefit farmers at the village level to address water scarcity (Cavelaars et al., 1994) [4]. However, groundwater levels are declining despite water harvesting measures to conserve water and enhance aquifer recharge, supported on a large scale by watershed development programmes. It is therefore crucial to increase our understanding of the capability and constraints of managed aquifer recharge (MAR) to overcome the threat of groundwater scarcity in the future (Massuel et al., 2014) [5]. Equally important is the understanding of the potential for managing or influencing the new patterns of use (Burke and Moench, 2000) [6], patterns that are often highly dispersed and individualized. To cope with lowering groundwater level, MAR has become an important complementary measure along with demand management to cope with groundwater scarcity (Dillon et al. (2012) [7].

The MARVI project, Managing Aquifer Recharge and Sustaining Groundwater Use through Village-level Intervention (www.marvi.org.in), has demonstrated that it is important to monitor and manage groundwater at the village level, particularly in hard rock areas of India (Maheshwari et al., 2014 [8]; Jadeja et al., 2018 [9]). This approach involves the training of village volunteers and developing a participatory process to assist cooperative management of groundwater. The methods include groundwater data collection at the village level; a methodology to estimate groundwater recharge from simple measurements on check dams (Dashora et al., 2018) [10]); and a smart phone app (MyWell) for collecting and visualising groundwater, rainfall, and check dam data. This approach supports village level decision-making for groundwater use and management. This has been field tested and is considered ready for extended out-scaling across India.

In this study area, village groundwater cooperatives are being formed to help achieve sustainable groundwater supplies. These have informed rabi (winter) crop decision making based on measured groundwater levels. They can also support maintenance of watershed measures for soil and water conservation, including maintenance of streambed recharge structures, as well as encouraging uptake of other options when proven. There is a watershed development program at the state level to increase groundwater recharge through the construction of check dams.

2. Why This Study?

Roof-top rainwater harvesting to recharge dug wells has been widely practiced in India with a varying degree of success (CGWB (2007) [11]; Rainwater Harvesting Association (2020) [12]). However, the use of harvested runoff from farmers' fields to recharge dug wells has been practiced mostly on a trial and error basis (e.g., examples reported in Bali Water Protection Program (2020) [13]), but with relatively rare monitoring. One exception is the work of Pendke et al. (2017) [14], in a study in Maharashtra from 2011 to 2015, who reported 64% removal efficiency of silt in the entry pit containing a preliminary filter rising to 93% removal at the end of the main filter before water is discharged to an open well. This was at a research site with a catchment area of 1.8 ha where runoff was estimated using an uncalibrated model. In 2015, the study was expanded to involve 10 recharge wells and two wells as controls. The size of the catchment areas for these was not reported. In 2015, water table rise was reported to be significantly larger in recharged wells than control wells. Aside from measurement of suspended silt at the pilot site, there was no evaluation of water quality that might impact on the safe use of well water.

The overall aim of this study was to understand the effectiveness of direct well recharge structures (DWRS) to improve groundwater supplies and quality at the local level. The activity reported in this paper arose because some farmers, who were at a considerable distance from streams, perceived that check dams in their catchment were not directly benefiting them as much as farmers whose wells were closer to those check dams. They sought an alternative way of increasing recharge at their wells. They were intending to harvest runoff from fields close to their wells, and divert this into their wells. Researchers from the MARVI team became involved due to well-founded concerns over potential for groundwater contamination. They evaluated wells proposed for direct recharge by farmers to avoid wells used for drinking water supplies, insisted on a filtration step and on monitoring the impacts on levels and quality, and developed a water quality laboratory in the village to enable analyses to be performed. The results of this investigation were to be reported back to farmers before considering any possible ongoing operation. Without these precautionary interventions, this approach could not be considered MAR.

3. Study Area

The study was carried out in the Dharta watershed, which is situated in Bhindar block of Udaipur district of southern Rajasthan, India. This area lies between 24°30′ and 24°37′ N latitude and 73°05′ to 73°15′ E longitude. Four adjoining villages were selected within a radius of 4 km, these being Badgaon, Dharta, Hinta, and Varni, for evaluating the performance of direct well recharge structures (Figure 1). Topography is often undulating with slope up to 2.7%. The ground elevation of the area is 470 m above the mean sea level. The average annual rainfall of the area is about 665 mm (Dashora et al. 2018) [10] and the temperature ranges from 19 to 48 °C in summer and 3 to 29 °C during winter.

The occurrence of groundwater in the watershed is mainly controlled by the topographic and structural features present in the Proterozoic gneisses and schists underlying the area. Groundwater in these rocks occurs in the zone of weathering and in fractures, joints, and foliation plains. When schists are inter-mixed with gneisses, they form a better aquifer (CGWB, 2013) [15]. The depth of dug wells ranges from 14 to 38 m. The major crops grown in the area are maize, wheat, mustard, cluster bean (guar), chickpea, and barley. About 25% of the total land area in the watershed is irrigated by dug wells and tube wells.

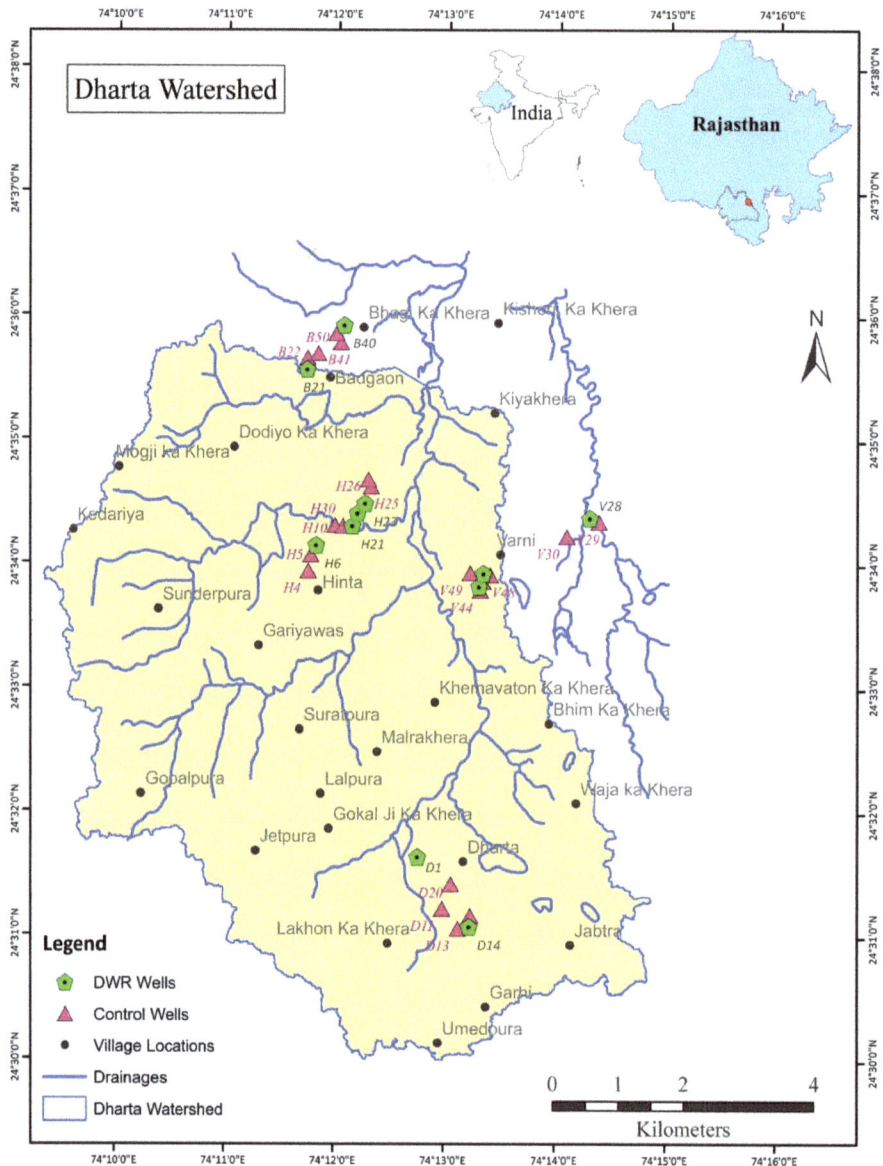

Figure 1. Location map of direct well recharge structures sites in the Dharta watershed and adjacent control wells.

4. Methodology

The study was carried out during 2016–2018. The steps followed in this study were (i) selecting the dug wells for implementing DWRS and nearby control wells, (ii) identifying suitable locations for pits, (iii) building pits and filters to reduce sediment discharge into wells, (iv) installing flow meters, (v) calculating the cost of construction, (vi) monitoring rainfall, (vii) monitoring groundwater levels, and (viii) water quality sampling and analysis.

4.1. Selection of the Dug Wells

With a view to evaluating the performance of direct well recharge at a farm level, a number of dug wells were selected and marked with the code numbers for identification. In the year of 2016, a total of 18 wells were selected, out of which 6 wells were selected for direct well recharge and 12 control wells were selected, with 2 separate wells in close proximity to each DWRS well. In 2018, an additional 15 wells were selected, and out of these, 5 wells were used as DWRS wells and 10 as control wells, again with 2 controls close to each DWRS well. Only the DWRS wells constructed in 2018 were fitted with flow meters to estimate the annual recharge volume. Hence, in 2018, there were a total of 11 DWRS wells and 22 wells as controls (Table 1). All the control wells were in close proximity to their recharge wells. Further, the wells in Table 1 are identified by whether they have parapet walls, overhanging trees and rotten plant debris, or whether they are fitted with flow meters for measuring runoff discharge into the wells. All wells are used for irrigation supplies.

Table 1. Total well depths of direct well recharge structures (DWRS) and control wells.

	DWRS Well	Total Well Depth, m	Control Well (1)	Total Well Depth, m	Control Well (2)	Total Well Depth, m
2016	H6 [ab]	19.60	H4 [a]	24.50	H5	17.65
	H21	28.90	H30 [b]	29.20	H10 [a]	25.40
	B21	20.50	B22 [a]	23.20	B44	20.60
	B40	18.45	B41	23.20	B50	27.90
	V43	30.45	V44 [a]	35.80	V45 [a]	33.10
	V47 [ab]	27.10	V48 [a]	28.45	V49 [ab]	30.10
2018	H22 [a] *	21.20	H30	29.20	H10 [a]	25.40
	H23 *	18.30	H25	24.30	H26	21.80
	D1 *	32.10	D11 [a]	18.95	D20	19.60
	D14 *	31.20	D13 [a]	22.80	D15	31.00
	V28 [ab] *	19.20	V29 [a]	19.10	V30 [ab]	22.70
	Average depth (m)	24.3	-	25.3	-	25.0

* = DWRS wells established in 2018 were fitted with flow meters. [a] = well with parapet wall; [b] = wells without overhanging trees and rotten plant debris. All wells were infested with birds.

4.2. Identification of Suitable Locations for Pits

It was considered important that the recharge pit (details described below) was located close to the recharge well to reduce the cost, and it was also located such that the runoff could easily flow towards the pit. For this, the important consideration was the general slope of the runoff contributing area. An earthen channel was constructed to guide runoff towards the pit. The catchment area was a secondary consideration, and subsequently this was identified as constraining the measured benefits. If the pit filled during a rainfall event, excess flow diverted along natural drainage lines and did not enter the well.

4.3. Pit and Filter Constructions and Pipe Installations

The pits were dug near the recharge wells with the help of earth moving machinery. The size of the pits varied slightly due to construction method. The median length, width, and depth of pits were 1.40, 1.55, and 1.15 m, respectively (Table 2). Once the pit was dug to the required dimensions, the masonry work was done on the four sides of the pit walls to maintain the stability of the pits. The bottoms of the pits were cemented, incorporating stones from a local quarry. The pit was divided into two sections by a brick wall constructed in the middle with a height of about two-thirds of the pit depth. This division was done to allow extra deposition time of sediments in the pit, as reported useful by Pendke et al. (2017) [14]. Runoff from pits was discharged from the pit into the recharge well through one or more 50 mm diameter high-density polyethylene (HDPE) pipes, which were laid in a trench to allow gravitational flow and perforated the well perimeter through an aperture just large enough to contain the pipe(s). The pipe inlets were installed about 0.2–0.3 m above the bottom of the pit to minimize clogging of the inlets (Figure 2). In some wells, two or even three pipes of 50 mm diameter were used in order to increase the proportion of runoff that entered the pit and well. After pipe installation, the trench was backfilled and compacted.

Table 2. Design details of DWRS pits.

DWRS Code	Length, m	Width, m	Depth, m	Volume (m³)	Catchment Area (m²)	Vol as mm over Catchment Area
H6	3.35	1.90	1.00	6.40	1131	5.6
H21	0.90	1.35	1.15	1.40	585	2.4
B21	1.15	1.20	0.70	1.00	2343	0.4
B40	2.30	2.30	1.10	5.80	1155	5.0
V43	1.90	1.90	1.20	4.30	304	14.3
V47	1.90	1.80	0.85	2.90	263	11.1
H22 *	1.20	1.40	1.10	1.80	3200	0.6
H23 *	1.40	1.55	1.20	2.60	662	3.9
D1 *	1.00	1.37	1.22	1.70	2860	0.6
D14 *	1.34	1.13	1.22	1.80	11,954	0.2
V28 *	1.40	2.40	1.30	4.40	2,902,300	0.0
Median	1.40	1.55	1.15	2.60	1155	2.4

* DWRS well established in 2018.

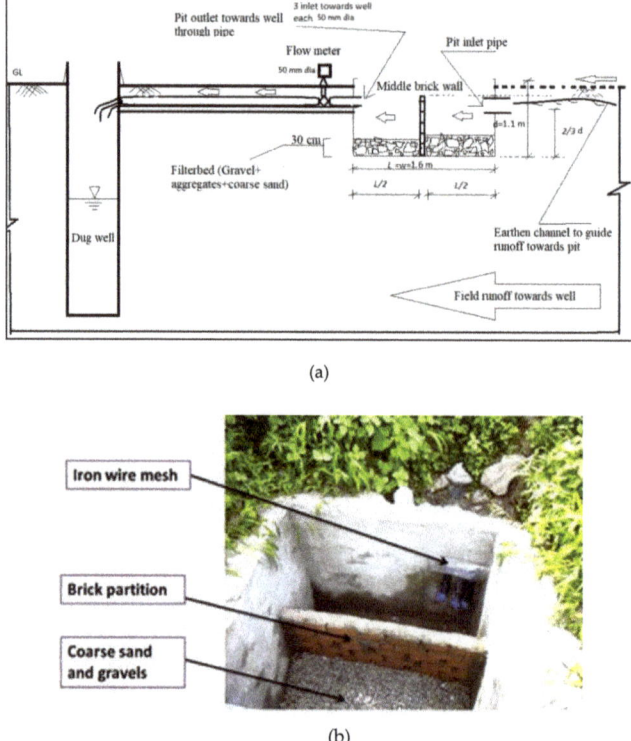

(a)

(b)

Figure 2. View of DWRS installed in the study: (**a**) a cross-sectional view of DWRS (not to scale); and (**b**) photograph of a sample structure constructed in the study area.

4.4. Reducing Sediment Discharge into Wells

The runoff carries suspended sediment particles throughout the rainy season, although the concentration was expected to be highest at the beginning of the monsoon season, when the ground was parched and there was no vegetation cover. It was considered important to prevent the discharge of sediments into the DWRS well in order to reduce the likelihood of turbid water clogging the fractures that allowed natural ingress of groundwater. A simple and cheap roughing filter was devised in which coarse sand and stone aggregates were placed in the pit on both sides of the dividing wall and covered with net cloth to help make suspended sediments settle in the pit and allow easy removal of detritus. Table 2 reports gross volume of pits, not accounting for filter material; hence, the holding capacity for water was quite small (<6.4 m^3) in relation to typical monsoon rainfall events, which could exceed 60 mm in a day.

4.5. Installation of Flow Meters

In the five DWRSs constructed in 2018, a flow meter was installed between the pit and recharge well to monitor the cumulative volume of water discharged into those wells. Flow meters with 50 mm diameter were used to measure the total volume of the runoff water discharged in a single pipe. If there were more than one pipe, it was assumed that other pipes discharged the same volume as the metered pipe. For additional protection of water meters from clogging due to plant debris in runoff water, iron wire meshes were placed at the inlet of pipes. A schematic diagram of field settings of components of the recharge structure are shown in Figure 2a, and a photo of a typical structure (one of 11) is shown in Figure 2b, whereas Figure 3 shows the discharge of runoff into a well after it has passed through the filter. The dial pad reading of the flow meter was recorded photographically at the time of installation, and subsequently after every runoff event.

Figure 3. Runoff discharge into well after it was collected in the pit and had passed through the filter.

4.6. Managed Aquifer Recharge Operations

For the DWRSs constructed in 2016, managed aquifer recharge (MAR) commenced in July 2016 and continued through the monsoons of 2017 and 2018, generally over the months of July to October. For DWRSs constructed in 2018, MAR commenced in July 2018. The systems were shut down at the end of the 2018 monsoon. DWRS and control well water levels were measured weekly from January

2013 to December 2018 for the wells of Hinta, Dharta, and Badgaon village, whereas for the Varni village, monitoring was done from December 2013 to December 2018.

4.7. Calculating the Cost of Construction

The cost of construction of the recharge pits varied on the basis of the location and material used. Locally available construction material was used, and well owners were engaged throughout the construction process. All the cost components starting from digging the pit to installing water-meter and outlet pipes were recorded. The cost of construction and installation depended on access to the site, distance between pit and recharge well, and construction of runoff collection field channel (wherever necessary). Only existing wells were used, and thus these are regarded as a sunk cost. The site specific average estimate of cost for installing a DWRS is given in Table 3, in Indian rupees at 2018 costs.

Table 3. Installation cost of a DWRS structure at field site (for conversion USD 1 = INR 70 in 2018).

Items	Quantity	Cost, INR	Cost, USD
Hiring cost for earth moving equipment	1 h	800	11
Stones	1 trolley load	1300	19
Coarse sand	$\frac{1}{4}$ trolley load	600	9
Cement bag	2	600	9
Bricks for partition	50	250	4
Stone aggregates	$\frac{1}{4}$ trolley load	300	4
Pipes (m)	3	600	9
Builder and labour	1 + 1	1600	23
Flow meter *	1	4500	64
Total cost without flow meter		6050	86
Total cost with flow meter		10,550	151

* Installed for flow measurement.

4.8. Rainfall Monitoring

Rainfall monitoring was done on a daily basis by farmer volunteers, known as BJs (Bhujul Jaankaars or "groundwater informed"). Rain gauges were installed in all four villages, and annual rainfalls were recorded (Figure 4) by BJs. To evaluate the effect of the runoff on the water level fluctuation of the wells, the rainfall data obtained were used to correlate with the water table level and the influence of the recharge pit for specific rainfall events.

4.9. Groundwater Level Monitoring

Groundwater level monitoring was done at a weekly interval and commenced a few weeks before the monsoon, continuing until after the end of the monsoon season when levels had peaked and were in decline. An ordinary measuring tape with a float at its end was used for monitoring the depth to water level in each DWRS and control well, below a datum that was marked on the well head with the well identification number. Readings were taken by the farmer BJs who had been trained to undertake such measurements and had considerable experience. The water level data obtained during weekly monitoring were used to plot well hydrographs.

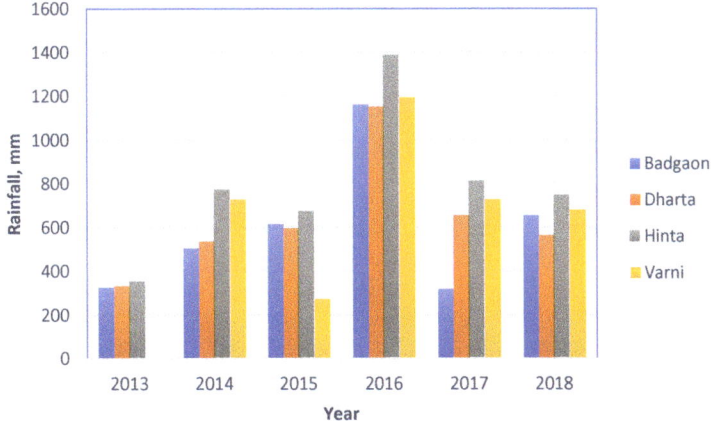

Figure 4. Annual rainfall in study villages during the study period, 2013–2018.

4.10. Water Quality Monitoring

4.10.1. Sampling

Water samples were taken on five occasions for analysis of pH, EC/TDS, and turbidity—July 2015, July 2017, June 2018, August 2018, and October 2018. Samples were analysed for fluoride on three of these occasions—July 2017, June 2018, and August 2018. *Escherichia coli* analysis was conducted on the water samples collected in August, September, and October 2018.

4.10.2. Physical and Chemical Analyses

The water samples were collected in order to analyse pH, EC, TDS, turbidity, and fluoride. They were analysed in the field for these physico-chemical parameters using an Aquaread instrument (https://www.aquaread.com/portofolio/ap-5000/) to test pH, EC, TDS, and turbidity. A HACH DR/890 portable colorimeter (https://www.hach.com/dr-890-portable-colorimeter/product?id=7640439041) was used to measure fluoride (F) concentration. *E. coli* samples were collected and taken to a laboratory in the Hinta village for analysis within 8–24 h, and samples were stored in a refrigerator for the time period between sampling and laboratory analysis. On each day of sampling before testing of water samples, the instruments were calibrated using distilled water and stock solutions. On one occasion, a split set of 10 samples was provided to an independent university laboratory for analysis of TDS (by EC) and fluoride. The coefficient of determination (R^2) for TDS was 0.82, and in terms of fluoride, R^2 was 0.98 for samples within the prescribed range of <2–2.5 mg/L for the colorimeter. To establish the reliability of the measurements, the testing of duplicate water samples was carried out. The results indicated average differences for 10 samples for pH, EC, TDS, and F and for 9 samples for turbidity of between 2.5% and 5% of the range in observed values. Hence, these field data are considered reliable for the purposes of the investigation.

4.10.3. Bacteriological Analysis

The MacConkey Agar (MAC) method was used to grow *Escherichia coli* bacteria. For the bacteriological analysis, standard lab procedure was used—the MAC flasks, spreader, and Petri dishes were sterilized in an autoclave at 120 °C at 15 psi for 15 min, after which spreading of field samples was done under laminar flow conditions. The MAC was poured into sterilized Petri dishes on which *E. coli* was cultured. This agar provides a solid medium on which selected bacteria are able to decompose agar. MAC is a selective and differential medium designed to selectively isolate Gram-negative bacteria such as *E. coli* and enteric bacilli on their ability to ferment lactose. Groundwater samples of DWRS wells and control wells were tested for microorganisms that would ferment lactose to produce end products that react with the pH indicator neutral red and would produce a pink colour colony. Results were reported as *E. coli* log colony-forming units (CFU)/mL.

5. Results and Discussion

The results of the evaluation of DWRS at a farm level are presented and discussed below.

5.1. Recharge in DWRS Wells

The metered volume of water recharging wells could only be determined at three DWRS wells in 2018 due to meter failures at two sites. Failures were thought to be caused by detritus clogging the impellors on mechanical flow meters in spite of the precautions taken. For the two sites representative of the catchment areas for 10 of the 11 DWRS wells, the average recharged proportion of monsoon rainfall on the catchment areas was 1.17%. This is considerably lower than the estimated 17% runoff generated from rainfall in 10 Maharashtra DWRS catchments (Pendke et al. 2017) [14]. It was observed that pits filled in heavy storms and subsequent runoff bypassed DWRS. Applying the average proportion of catchment rainfall recharged from these two wells to all other DWRS wells in all years since they were established gives the volumetric recharge estimates shown in Table 4.

Table 4. Observed and estimated recharge through DWRS pits.

Well ID	Year of Pit Establishment	Recharge Volume Metered, m^3	Recharge as mm over Catchment	As % of Rainfall on Catchment	Estimated Recharge, m^3		
					2016	2017	2018
H6	2016			*	13	7	10
H21	2016			*	7	4	5
B21	2016			*	22	6	18
B40	2016			*	11	3	9
V43	2016			*	3	2	2
V47	2016			*	3	2	2
H22	2018	27	8.44	1.13%			27
H23	2018			*			4
D1	2018			*			13
D14	2018	81	6.78	1.20%			81
V28	2018	176	0.06	0.01%			176
Mean				1.17%*	14	6	32
		Total (pits established in 2016)			59	24	46
		Total (pits established in 2018)			0	0	309
		Total			59	24	355

* The mean value for H22 and D14 was applied to all unmetered sites and sites where meters failed to register. V28 represents a DWRS well besides a stream with a catchment area three orders of magnitude larger than the median of the DWRS sites, and hence was excluded from estimation of recharge at other wells.

The volumes of recharge are very low, in part due to the small catchment area of farm fields, in part by the low proportion of runoff diverted into wells due to the very small volumes of pits (Table 2) with respect to typical monsoon rainfall events, and possibly in part due to under-estimation of recharge by under-performing flow meters.

5.2. Head Rise Comparison between DWRS Wells and Control Wells

Six DWRS wells were constructed in 2016 and another five in the year 2018 (Table 1), and in this catchment that had been intensively monitored since 2013 in the MARVI project (Maheshwari et al. 2014) [8], we calculated the head rise in each well by subtracting the depth to water level at the end of the monsoon from that at the beginning of the monsoon. The ratio of head rise of each DWRS well to the mean of its adjacent control wells was calculated for each year (2013 to 2018). Subsequently, the change in these ratios was analysed to compare head rises before and after construction of DWRS for both construction years (2016 and 2018). Table 5 shows the mean and standard deviation of the head rise ratios.

Table 5. Statistical analysis of ratio of mean head rise of each DWRS and nearby control wells.

DWR	2012	2013	2014	2015	2016	2017	2018	Mean before Construction	Mean after Construction	Mean (after Minus before)
H6 ##	0.72	0.54	0.12	0.88	0.85	1.63	0.72	0.56	1.07	0.51
H21 ##	1.35	0.71	0.70	2.06	0.86	0.86	0.89	1.20	0.87	−0.33
B21 ##	0.88	0.55	0.66	1.72	0.91	1.02	1.24	0.95	1.06	0.11
B40 ##	0.52	0.73	0.63	1.30	0.75	0.78	0.64	0.80	0.72	−0.07
V43 ##			0.79	1.20	0.76	2.39	1.17	1.00	1.44	0.44
V47 ##			0.25	0.24	0.82	1.06	1.20	0.25	1.02	0.78
H22 *	0.65	0.17	0.31	0.36	0.49	0.77	0.91	0.37	0.91	0.54
H23 *	0.37	0.08	0.02	0.48	0.60	0.81	0.76	0.23	0.76	0.53
D1 *	1.81	2.21	4.39	2.95	1.78	1.31	2.41	2.41	3.49	1.08
D14 *		1.33	1.36	1.24	1.17	1.14	2.15	1.31	2.15	0.84
V28 *			0.26	0.51	1.26	0.91	2.12	0.38	2.12	1.74
Summary statistics of head rise ratio by year										
Mean	**0.90**	**0.79**	**0.86**	**1.18**	**0.93**	**1.15**	**1.39**	**0.86**	**1.42**	**0.56**
SD	**0.51**	**0.69**	**1.23**	**0.82**	**0.36**	**0.49**	**0.87**	**0.64**	**0.85**	**0.57**
CoV	**0.57**	**0.87**	**1.42**	**0.70**	**0.38**	**0.42**	**0.62**	**0.74**	**0.60**	**1.01**
Values below are for DWR wells commencing in 2016 only ##										
Mean	0.87	0.63	0.53	1.23	0.83	1.29	0.98	0.79	1.03	0.24
SD	0.35	0.10	0.27	0.64	0.06	0.62	0.26	0.34	0.24	0.41
CoV	0.41	0.16	0.52	0.52	0.08	0.48	0.27	0.43	0.23	1.73
Values below are for DWR wells commencing in 2018 only *										
Mean	0.94	0.95	1.27	1.11	1.06	0.99	1.89	0.94	1.89	0.94
SD	0.77	1.02	1.82	1.09	0.53	0.23	1.11	0.92	1.11	0.50
CoV	0.81	1.07	1.44	0.98	0.50	0.23	0.59	0.98	0.59	0.53

DWRS constructed in 2016; * DWRS constructed in 2018; Bold is summary for all DWRS wells.

The statistical analysis of ratio of mean head rise of DWRS and control wells indicated that the effect of DWRS to raise water level in DWRS was not statistically significant at $p < 0.05$. This is not surprising due to the fact that the natural recharge in the area is considerably larger than the generally small additional volumes of water recharged through DWRS. This, combined with the local factors such as geology, topography, and rainfall intensity variations, can mask the DWRS contribution to the aquifer. The maximum increase in head rise ratio was observed at DWRS V28 (which had the highest recharge volume, more than three times the next highest measured or estimated value (at D14)) (Table 4).

Pendke et al. (2017) [14] studied direct well recharge at 10 sites in the Maharashtra state of India and observed that the difference between the post-monsoon (September) and pre-monsoon (June) water level depths was greater when compared with those of two controls. However, the catchment areas were more than 10 times the median in the Dharta case study, but inflow volumes were not recorded. It is expected that head rise in individual wells is unlikely to be an effective diagnostic of DWRS recharge effectiveness. Variations in transmissivity and specific yield in the aquifer could even suggest the reverse is true where for the same recharge volume the groundwater mound would be higher for aquifers with low transmissivity and low specific yield. Reliable measurements of recharge are the most decisive information on which to assess recharge effectiveness, as found for check dams in the same catchment by Dashora et al. (2018, 2019) [10,16].

5.2.1. Water Quality

The water quality information for the various wells in four villages is summarised in Table 6.

Table 6. Water quality parameter values DWRS and control wells for different villages during the study period.

Parameter	Badgaon		Dharta		Hinta		Varni		All Villages		All Villages
	DWRS Wells	Control Wells	DWRS Wells	Control Wells	DWRS Wells	Control Wells	DWRS Wells	Control Wells	DWRS Wells	Control Wells	DWRS Wells − Control Wells
No. of wells	2	4	2	4	4	6	3	6	11	20	
No. of samples	6	12	6	12	12	18	9	18	33	60	
pH (mean)	7.92	7.85	7.76	8.04	7.8	7.87	8.06	8.07	7.89	7.96	−0.07
pH (standard deviation)	0.32	0.24	0.43	0.18	0.21	0.29	0.42	0.37	0.33	0.28	
TDS (mean), mg/L	1772	2676	1649	2031	3081	2937	2571	2041	2444	2435	9
TDS (standard deviation), mg/L	690	1705	319	415	1604	1504	1591	970	1201	1166	
Turbidity (mean), NTU	36.02	61.71	57.72	82.7	80.99	76.16	37.13	30.98	57	61	−4
Turbidity (standard deviation), NTU	20.15	48.67	51.34	113.42	63.81	57.58	28.83	29.55	44	59	
No. of samples fluoride	4	8	4	8	8	12	6	12	22	40	
Fluoride (mean), mg/L	1.12	0.77	0.74	1.06	0.98	0.94	0.95	1.03	0.95	0.96	0.00
Fluoride (std. deviation), mg/L	0.41	0.38	0.37	0.36	0.49	0.37	1.03	0.21	0.60	0.32	
No. of samples *Escherichia coli*	8	10	7	11	16	20	12	12	43	53	
E. coli (mean), log number CFU/mL	3.26	2.83	3.04	2.55	3.06	3.06	2.81	2.48	3.02	2.78	0.24
E. coli (standard deviation), log number CFU/mL	0.46	0.47	0.69	0.65	0.45	0.38	0.32	0.31	0.45	0.44	

5.2.2. pH

Water samples were collected and tested for pre-monsoon (Jun 2018), during monsoon (July 2015, July 2017, and August 2018), and post-monsoon (October 2018) periods. The mean pH values of most of the DWRS and less of their control wells were found to be between the permissible limits (6.5–8.5) of the Bureau of Indian Standards (BIS; 2004) [17]. Figure 5 shows the percentage of samples that met the (BIS) criteria. Both in July 2015 (before any DWRS recharge) and October 2018 (post-monsoon), all the DWRS wells met the criteria, whereas half of the control wells had a pH greater than 8.5. In 2017, only about 26% of samples of both DWRS and control wells met the criteria due to elevated pH. That is, the introduction of DWRS made little difference to the acceptability of the pH of the water for drinking.

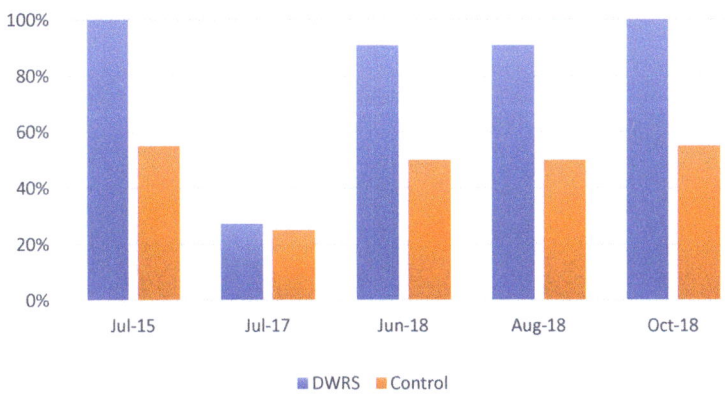

Figure 5. Percentage of samples meeting Bureau of Indian Standards (BIS) guidelines for pH in drinking water with or without an alternative supply (BIS acceptance range pH: 6.5–8.5).

5.2.3. TDS

In July 2015, about 82% samples of DWRS met BIS criteria (TDS (2000 mg/L), compared with 55% for the control wells (Figure 6). Although a higher proportion of DWRS wells than control wells had TDS less than 2000 mg/L, before and during occurrence of DWRS recharge, it is evident that these proportions can increase during the monsoon for both DWRS and control wells due to dilution with fresh natural recharge. However, the volume of DWRS recharge in the DWRS wells is so small that it does not make a marked benefit if wells were to be used for drinking, and it will be seen that other parameters relevant for drinking are adversely impacted by DWRS.

5.2.4. Fluoride

The average values of fluoride of DWRS and control wells ranged from 0.75 to 1.13 mg/L and 0.83 to 0.94 mg/L, respectively. The proportion meeting the BIS criteria (<1.5 mg/L in the absence of an alternative supply) of DWRS was 73% in July 2017, compared with 55% for control wells (Figure 7). Between June 2018 (before monsoon) and August 2018 (mid monsoon), the proportion of DWRS wells with $F < 1.5$ mg/L increased with respect to control wells. This is not surprising because of the generally lower ambient TDS and F of DWRS wells than in control wells, and thus rainfall recharge is expected to have a greater diluting influence in DWRS wells.

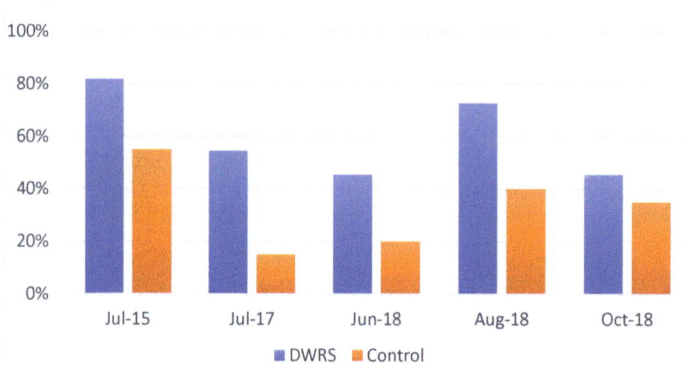

Figure 6. Percentage of samples meeting BIS guideline for TDS in drinking water in the absence of an alternative supply (BIS threshold < 2000 mg/L).

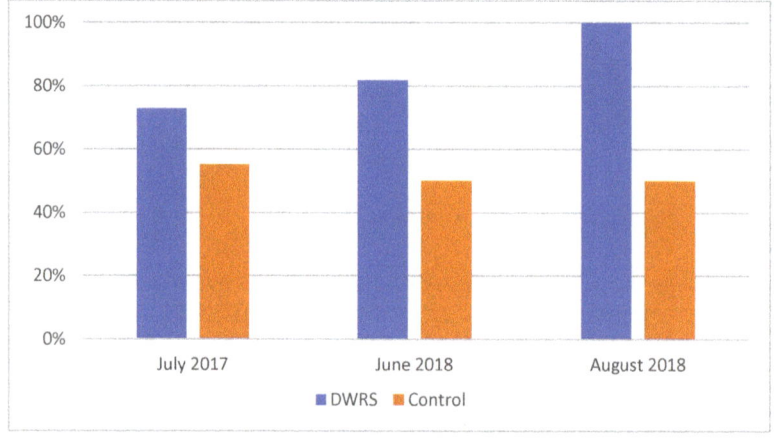

Figure 7. Percentage of groundwater samples that meet BIS guidelines for fluoride in drinking water in the absence of an alternative supply (BIS threshold < 1.5 mg/L).

5.2.5. Turbidity

As indicated in Table 6, the mean values of the turbidity of DWRS and control wells ranged from 30 to 65 and 29 to 66 NTU (Nephelometric Turbidity Units), respectively. As illustrated in Figure 8, from the years 2015 to 2018, none of the samples met the BIS criteria (10 NTU in the absence of an alternative supply) except in June 2018 (DWRS 27% and control 20%) before the monsoon broke, as well as in October 2018 (only control 10%). It was found that wells with parapet wall (45 NTU) had less turbidity when compared to wells without parapet wall (54 NTU). This suggests that a parapet wall alone may be insufficient in providing adequate protection for drinking water wells in this area.

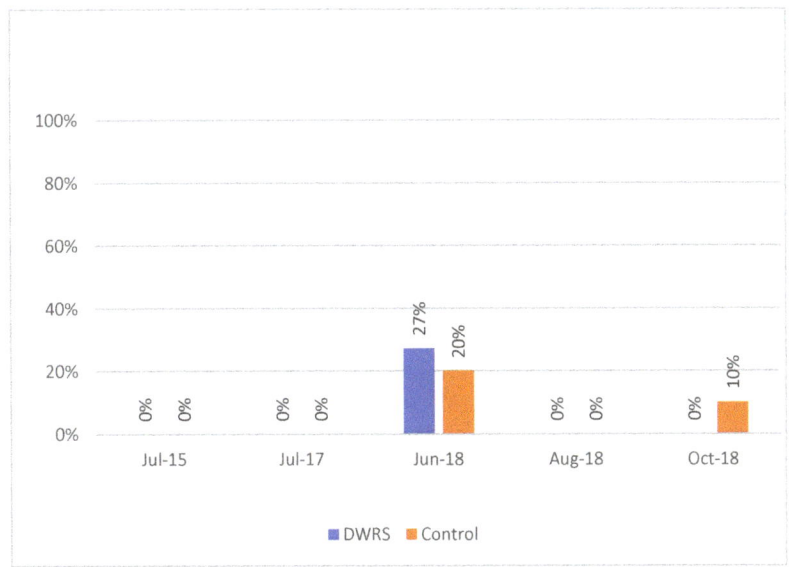

Figure 8. Percentage of samples of samples meeting BIS criteria for turbidity in drinking water in the absence of an alternative supply (<10 NTU).

5.3. E. coli

The presence of *E. coli* bacteria in any 100 mL sample of water indicates that the water is contaminated and unfit for drinking (BIS standards). The water samples for both DWRS and control wells were tested and found that not only the wells that were recharged but also control wells showed the presence of *E. coli*. Table 7 shows the mean of DWRS wells was between 0.12 and 0.68 log CFU/mL higher than the mean of control wells; however, in relation to standard deviations, this departure was not significantly different.

Table 7. *E. coli* log number colony-forming units (CFU)/mL of DWRS and control wells.

Date	Number of Wells		Mean Value of *E. coli*, log CFU/mL			Standard Deviation, log CFU/mL	
	DWRS	Control	DWRS	Control	DWRS - Control	DWRS	Control
16-08-2018	10	0	2.77	-		0.39	-
25-08-2018	10	12	3.03	2.35	0.68	0.54	0.70
25-09-2018	11	19	3.15	2.70	0.45	0.57	0.65
06-10-2018	9	16	3.22	3.10	0.12	0.46	0.48
All samples	40	47	3.04	2.75	0.29	0.49	0.61

The data revealed that both DWRS and control wells were found to be infected by *E. coli*. It was also noticed that the control wells that did not have a well-constructed parapet were affected by the bird droppings and rotten plant debris in creating the possibility of the *E. coli*. No wells had covers, and only 15 wells out of a total of 31 wells monitored had a parapet wall. It was found that wells with parapet walls had a lower average number of *E. coli* (2.47 log CFU/mL) than wells without parapet walls (2.85 log CFU/mL) (Table 6). The wells with over hanging trees and bird activities inside wells had *E. coli* 2.92 log CFU/mL, whereas without hanging trees showed *E. coli* 2.09 log CFU/mL. There were no wells with covers to keep out birds and bats from well heads, and thus it was possible these were the source of *E. coli* found in all wells.

The Water Quality Guide for Managed Aquifer Recharge in India (Dillon et al. 2014) [18] allows for a very simple approach to accepting natural water to recharge an aquifer if the recharge mechanism does not bypass the unsaturated zone. If the unsaturated zone is bypassed, as is the case in DWRS, the guide then refers proponents to the Australian Guidelines for MAR (NRMMC, EPHC, NHMRC (2009) [19]. These require a monitoring regime to ensure that the aquifer is not polluted, which could have an adverse impact on human health or the environment. Although the monitoring effort undertaken in this study did not cover all potentially present contaminants, such as agricultural organic chemicals, nutrients, and other types of microorganisms such as viruses and protozoa, the selection of parameters is sufficiently convincing in order to demonstrate the fact that improved treatment is required if any well influenced by the water introduced via DWRS is used for drinking water supplies.

5.3.1. Performance of Filters and Potential for Fracture Clogging

The runoff water was filtered before redirecting it into the recharge well to retain suspended sediments and thereby reduce the blockages of fractures (see Figure 9) and improve the groundwater quality. During the first two to three rainfall events in the study, we observed that the surface water carried with it considerable amounts of suspended fine silt particles and organic plant materials, including rotten leaves and plant debris. The filter bed made up of coarse sand and gravels retained much of the suspended silt. It was also observed that timely manual cleaning of the pit, namely, the removal of the silt and plant debris, was an important activity to reduce any blockage of the discharge pipe inlets. During the monitoring, on some occasions, the water meters were observed as being clogged by plant debris, and thus to overcome this problem, we installed a wire mesh at each flow meter inlet.

Figure 9. Recharge pit with filters. and clogging of the flow meter inlet.

For the long-term success of DWRS structures, removal of any suspended material through filtering is important before runoff water is discharged into wells to avoid potential clogging of aquifer fractures. Clogging has been observed a significant issue in Australia when stormwater runoff and treated municipal waste-water effluent are injected into aquifers to produce water for irrigation (NRMMC, EPHC, NHMRC 2009) [19]. Baveye et al. (1998) [20] reported that the main problem in infiltration systems for enhancing recharge of groundwater is clogging of the infiltrating surface (basin bottoms, walls of trenches and vadose-zone wells, and well-aquifer interface in recharge wells), resulting in reduced infiltration rates. Silt removal is done mechanically with scrapers, front-end loaders, and graders, or manually with shovels and rakes.

5.3.2. Costs and Benefits of DWRS

The costs of establishing a DWRS without and with a flow meter were shown in Table 3 to be INR 6050 and 10,550 (USD 86 and 151), respectively. Benefits of additional water were determined to be 2.36 INR/m^3 (0.034 USD/m^3) (Dashora et al. 2019) [16] in this same catchment using the net value of increased production per cubic metre of additional water available from check dam recharge. Assuming the life of the DWRS infrastructure was either 10 or 30 years, and following the procedure laid out by Dashora et al. (2019) [16] using the same discount rate of 8%, we found that an annual volume of 382 or 250 m^3, respectively, would need to be recharged and used productively for agricultural irrigation in order to warrant the capital expense (and including flow meter (666 or 416 m^3)). These calculated economic recharge volumes are under-estimates because they neglect annual maintenance costs, such as scraping out the pit. The lowest of these numbers exceeds the maximum annual recharge recorded in 2018 and suggests that none of the DWRSs evaluated would be economically feasible (i.e., present value of benefits exceed the present value of costs). The mean recharge in 2018 was 32 m^3, suggesting that, if this was representative of mean annual recharge, the B/C ratio would be between 0.05 and 0.13 depending on the assumed life of the infrastructure and absence or presence of meters. Even the DWRS harvesting from a large catchment (V28) failed to reach this feasibility criterion. This was quite a different result than found for check dams that had a benefit/cost ratio of 4.1 [16], and therefore remain a preferred approach to recharge enhancement in this area.

6. Concluding Remarks

In this study, we evaluated the effect of direct well recharge structures (DWRS) on the groundwater level rise over the monsoon season and the quality of water in recharged wells as compared to nearby control wells. This was the first micro scale (farm level) evaluation in a semi-arid region of Rajasthan state, which is facing the problem of groundwater over-exploitation. Water quality observations were made to determine whether groundwater quality was protected.

The volume of water recharged through DWRS into individual wells during the monsoon season varied with catchment area, rainfall amount, and intensity, and in 2018, in three wells where water flow meters did not clog, these were 27, 81, and 176 m^3 per well. Using the same ratio of recharge to rainfall over the catchment area, in the same year, the other eight wells were estimated to recharge between 2 and 19 m^3. The value of average recharge for all the wells monitored in 2018 was 32 m^3. The mean rise in well water levels over the monsoon season was higher in wells with DWRS than in nearby control wells, but not significantly different. The study revealed that some wells with DWRS have shown a larger increase in water level than in control wells, and this was particularly true for one well (V28) that accounted for 50% of the total recharge to 11 wells in 2018.

Similarly, monitoring of water quality revealed no significant difference between DWRS and control wells for pH, EC/TDS, turbidity, or fluoride. The presence of E. coli in DWRS wells was higher than in control wells, however, E. coli exceeded drinking water guidelines in all sampled wells. Values of pH, EC/TDS, and F decreased in DWRS and control wells as each monsoon progressed, whereas the turbidity of wells with DWRS increased slightly. The turbidity and E. coli values suggest that DWRS should not be attempted in or near wells that could be used for drinking water supplies.

The high proportion of both DWRS and control wells that failed to meet BIS criteria for drinking water suggests that well-head protection measures are needed, such as parapet walls and covers, in order to reduce these contaminant loads for wells that are used as a source of drinking water. As a result of this study, trials are commencing to monitor the changes in water quality due to well-head protection measures in the treated wells and control wells, in order to provide the evidence base necessary to inform appropriate actions by the village communities.

The volume of water recharged by DWRS was too small to warrant the expenditure on DWRS, even for the system with a very large catchment, on the basis of a present value analysis and assuming the asset life of the DWRS system is between 10 and 30 years and neglecting maintenance costs.

It is anticipated that pit filters would need to be removed, cleaned, and replaced periodically to enable DWRS to remain operational. Diverting the first flush runoff in a monsoon before water enters the filter pit, until after vegetation cover is established and turbidity reduces, would be expected to reduce maintenance needs at the cost of a reduced harvest. It is also expected that improved watershed management such as contour banking will improve quality of runoff and reduce the needed frequency of desilting of filters. It may also be a more effective form of increasing recharge than DWRS, but it would be difficult to measure recharge increase as a result of such dispersed recharge methods.

Author Contributions: Conceptualization, B.M., P.D. and P.K.S.; methodology, B.M., P.D. and P.K.S., P.S. and Y.D.; formal analysis, P.S., Y.D., P.D. and B.M.; resources, B.M.; data curation, P.S., Y.D., P.D., B.M., P.K.S. and A.K.; writing— P.S., Y.D., B.M., and P.D.; writing—review and editing, P.D., B.M., P.S., Y.D. and A.K.; supervision, B.M. and P.K.S.; project administration, B.M.; funding acquisition, B.M. All authors have read and agreed to the published version of the manuscript.

Funding: This research was funded by Australian Centre for International Agricultural Research, Canberra (Project LWR2012/15) and Australian Water Partnership (Project no. 660147.93).

Acknowledgments: This research was conducted under the Managing Aquifer Recharge and Sustaining Groundwater Use through Village-level Intervention (MARVI) project, which was funded by the Australian Centre for International Agricultural Research (ACIAR) and the Australian Water Partnership (AWP). The support of International Centre of Excellence in Water Resources Management (ICE WaRM), Adelaide, and Darryl Day is acknowledged for the training provided to the first two authors of this article. Authors also thank Rameshwar Lal Soni, Hari Ram Gadri, and other Bhujal Jaankaar for their valuable support in data collection in the study villages.

Conflicts of Interest: The authors declare no conflict of interest.

References

1. Seckler, D.; David, M.; Randolph, B. *Water Scarcity in the 21st Century, Research Report*; International Water Management Institute: Colombo, Sri Lanka, 1999.
2. Shah, T. *Taming the Anarchy: Groundwater Governance in South Asia*; Routledge: New Delhi, India, 2009.
3. CGWB. *Report on Ground Water Scenario in India (Premonsoon 2017)*; Central Ground Water Board: New Delhi, India, 2017; pp. 1–142.
4. Cavelaars, J.C.; Vlotman, W.F.; Spoor, G. Subsurface Drainage System. In *ILRI Publication 16*; Ritzema, H.P., Ed.; International Institute for Land Reclamation and Improvement: Wageningen, The Netherlands, 1994; p. 827.
5. Massuel, S.; Perrin, J.; Mascre, C.; Mohamed, W.; Boisson, A.; Ahmed, S. Managed aquifer recharge in South India: What to expect from small percolation tanks in hard rock? *J. Hydrol.* **2014**, *512*, 157–167. [CrossRef]
6. Burke, J.J.; Moench, M.H. *Groundwater and Society: Resources, Tensions and Opportunities. Themes in Groundwater Management for the Twenty-First Century*; Department of International Economic and Social Affairs, Statistical Office, United Nations: New York, NY, USA, 2000.
7. Dillon, P.; Fernandez, E.E.; Tuinhof, A. Management of Aquifer Recharge and Discharge Processes and Aquifer Storage Equilibrium. IAH Contribution to GEF-FAO Groundwater Governance Thematic Paper 4. 2012. Available online: www.groundwatergovernance.org/resources/thematic-papers/en/ (accessed on 21 February 2020).
8. Maheshwari, B.; Varua, M.; Ward, J.; Packham, R.; Chinnasamy, P.; Dashora, Y.; Dave, S.; Soni, P.; Dillon, P.; Purohit, R.; et al. The role of transdisciplinary approach and community participation in village scale groundwater management: Insights from Gujarat and Rajasthan, India. *Int. Open Access J. Water* **2014**, *6*, 3386–3408. Available online: http://www.mdpi.com/journal/water/special_issues/MAR (accessed on 31 March 2020). [CrossRef]
9. Jadeja, Y.; Maheshwari, B.; Packham, R.; Hakimuddin, B.; Purohit, R.; Thaker, B.; Dillon, P.; Oza, S.; Dave, S.; Soni, P.; et al. Managing aquifer recharge and sustaining groundwater use: Developing a capacity building program for creating local groundwater champions. *Sustain. Water Resour. Manag.* **2018**, *4*, 317–329. Available online: https://recharge.iah.org/thematic-issues-journals (accessed on 31 March 2020). [CrossRef]

10. Dashora, Y.; Dillon, P.; Maheshwari, B.; Soni, P.; Dashora, R.; Davande, S.; Purohit, R.C.; Mittal, H.K. (2018). A simple method using farmers' measurements applied to estimate check dam recharge in Rajasthan, India. *Sustain. Water Resour. Manag.* **2018**, *4*, 301–316. Available online: https://recharge.iah.org/thematic-issues-journals (accessed on 31 March 2020). [CrossRef]
11. CGWB. *Manual on Artificial Recharge of Ground Water*; Government of India, Ministry of Water Resources, Central Ground Water Board: Shillong, India, 2007.
12. Rainwater Harvesting Association. Available online: www.rainwaterharvesting.org (accessed on 23 February 2020).
13. Bali Water Protection Program-Penyelamatan Air Tanah Bali–Annex 2-Aquifer Recharge Technical Overview. (no date). Available online: https://www.idepfoundation.org/images/idep/how-you-can-help/support-a-project/bali-water-protection-program/bwp_annex2_technical_brief_overview.pdf (accessed on 23 February 2020).
14. Pendke, M.S.; Asewar, B.V.; Waskar, D.P.; Gore, A.K.; Samindre, M.S. Openwell and Borewell Recharge Technology. In *All-India Co-ordinated Research Project for Dryland Agriculture*; Vasantrao Naik Marathwada Krishi Vidyapeeth Parbhani: Maharashtra, India, 2017; p. 20. ISBN 978-93-85456-19-0. Available online: https://www.researchgate.net/publication/319553655_Openwell_and_Borewell_Recharge_Technology (accessed on 23 February 2020).
15. CGWB. Report on Groundwater Scenario, Udaipur District, Rajasthan, Ministry of Water Resources, Central Ground Water Board, Western Region Jaipur. 2013. Available online: http://cgwb.gov.in/District_Profile/Rajasthan/Udaipur.pdf (accessed on 5 November 2019).
16. Dashora, Y.; Dillon, P.; Maheshwari, B.; Soni, P.; Mittal, H.K.; Dashora, R.; Singh, P.K.; Purohit, R.C.; Katara, P. Hydrologic and cost benefit analysis at local scale of streambed recharge structures in Rajasthan (India) and their value for securing irrigation water supplies. *Hydrogeol. J.* **2019**, *27*, 1889–1909. [CrossRef]
17. BIS (Bureau of Indian Standards). *Specification for Drinking Water IS 10500: 2004*; BIS: New Delhi, India, 2004. Available online: https://www.indiawaterportal.org/sites/indiawaterportal.org/files/drinking_water_standards_bis_10500_2004_by_bis.pdf (accessed on 31 March 2020).
18. Dillon, P.; Vanderzalm, J.; Sidhu, J.; Page, D.; Chadha, D. (2014). A Water Quality Guide to Managed Aquifer Recharge in India. CSIRO Land and Water and UNESCO Report of AusAID PSLP Project ROU 14476. Available online: https://publications.csiro.au/rpr/pub?pid=csiro:EP149116 (accessed on 23 February 2020).
19. NRMMC, EPHC, NHMRC. *Australian Guidelines for Water Recycling, Managing Health and Environmental Risks, Volume 2C–Managed Aquifer Recharge*; Natural Resource Management Ministerial Council, Environment Protection and Heritage Council National Health and Medical Research Council, NWQMS: Canberra, Australia, 2009; p. 237. Available online: http://www.waterquality.gov.au/guidelines/recycled-water (accessed on 31 March 2020).
20. Baveye, P.; Vandevivere, P.; Hoyle, B.L.; De Leo, P.C.; de Lozada, D.S. Environmental impact and mechanisms of the biological clogging of saturated soils and aquifer materials. *Crit. Rev. Environ. Sci. Technol.* **1998**, *28*, 123–191. [CrossRef]

© 2020 by the authors. Licensee MDPI, Basel, Switzerland. This article is an open access article distributed under the terms and conditions of the Creative Commons Attribution (CC BY) license (http://creativecommons.org/licenses/by/4.0/).

Article

Migration of Pharmaceuticals from the Warta River to the Aquifer at a Riverbank Filtration Site in Krajkowo (Poland)

Roksana Kruć *, Krzysztof Dragon and Józef Górski

Institute of Geology, Department of Hydrogeology and Water Protection, Adam Mickiewicz University, ul. Bogumiła Krygowskiego 12, 61-680 Poznań; Poland; smok@amu.edu.pl (K.D.); gorski@amu.edu.pl (J.G.)
* Correspondence: roksana.kruc@amu.edu.pl

Received: 7 June 2019; Accepted: 24 October 2019; Published: 26 October 2019

Abstract: Studies on the presence of pharmaceuticals in water were carried out on the riverbank filtration site, Krajkowo–Poznań (Poland). A preliminary investigation conducted in 3 sampling points showed the presence of pharmaceuticals in both surface water and bank filtrate. Based on the above, an extended analysis was made in July, August and October 2018 and included surface water and wells located at a different distance (5–250 m) and travel time (1–150 days) from source water (Warta River). Firstly, 75 compounds (antibiotics, anti-inflammatory and analgesic drugs, psychotropic drugs, x-ray agents and β-blockers) were tested and 25 of them were detected in the river or bank filtrate. The highest concentrations were observed in source water and then were reduced along the flow path. The sampling points located close to the river (<38 m) are characterized by low removal. Higher removal is visible in wells located 64–82 m away from the river, while 250 m from the river most compounds are completely attenuated. Carbamazepine, gabapentin, tramadol, oxypurinol, fluconazole, and lamotrigine are the most common compounds. Some of the tested parameters occur only in the river water, e.g., iopromide, diclofenac, iohexol, clindamycin, fexofenadine and valsartan. The research shows that at the site, a significant attenuation of pharmaceuticals can be achieved at travel times of 40–50 days and distances of 60–80 m, although higher values are ensured when the well is located more than 250 m away.

Keywords: riverbank filtration; pharmaceuticals in groundwater; removal of pharmaceuticals

1. Introduction

Riverbank filtration (RBF) systems are widely used for drinking water supplies. RBF, by forcing the infiltration of surface water into the groundwater systems, allows relatively large amounts of water to be obtained, especially in the alluvial aquifers located in the European lowland areas in river valleys and ice-marginal valleys [1,2]. The infiltration of surface water to groundwater systems and water passage through the aquifer media causes improvements in water quality by a set of processes including: sorption, redox processes and biodegradation [3,4]. The mixing of bank filtrates with ambient, usually unpolluted groundwater, also takes place [5,6]. Nevertheless, the quality of bank filtrate is strongly dependent on surface water quality. Currently, this dependency is extremely important due to the detection of contaminants (e.g., pharmaceuticals) in the river (source) water. The occurrence of pharmaceuticals (such as antibiotics, analgesics, blood lipid regulators, contrast agents) has been studied all over the world in surface and also in groundwater [7–9]. The occurrence of micropollutants was documented in Chinese rivers [10,11], Japanese rivers [11], Korean rivers [11], Kenyan rivers [12], USA rivers [13,14] and also European rivers [1,15,16] and has also been previously documented in the Warta River [17]. In cases of heavily polluted surface water or temporary occurrences of peak constituent concentrations in rivers (e.g., during extreme weather conditions [18]), the contaminants can migrate to production wells in reduced concentration [4,19]. These remaining residues necessitate removal by the use of engineering techniques in treatment plants. However, a properly constructed RBF system can also be used as a natural water treatment method [16]. This can be achieved if the travel time (i.e., time of water passage from surface water to wells) is long enough to remove or considerably reduce the contaminants from the bank filtrate [1,4,16].

The goals of the research presented here are (i) to report the occurrence of a large number of pharmaceuticals in both river and bank filtrate and (ii) the investigation of their attenuation during bank filtrations. The data was analysed at points at different distances (and likewise travel times) from the river, as well as in various types of wells (vertical and horizontal), as according to the literature [4,7] the removal of pharmaceuticals increases with increasing distance (as well as travel time) from the source water.

2. Materials and Methods

2.1. Site Description

For the investigation of pharmaceuticals in river and bank filtrate water, the Krajkowo well field was selected. This well field is located 30 km from Poznań City. The well field is composed of the following (Figure 1): (1) a well group located on the floodplain along the Warta River (RBF-c) at a distance of 60–80 m from the riverbank; (2) a group of 56 wells situated on a higher terrace located 400–1000 m from the river (RBF-f); (3) one horizontal well (HW) with 8 radial drains situated 5 m below the river bottom. In the Krajkowo well field, one additional well group is recharged from artificial ponds. This part of the well field was not considered in this study. A detailed description of the well fields is presented in previous work [20].

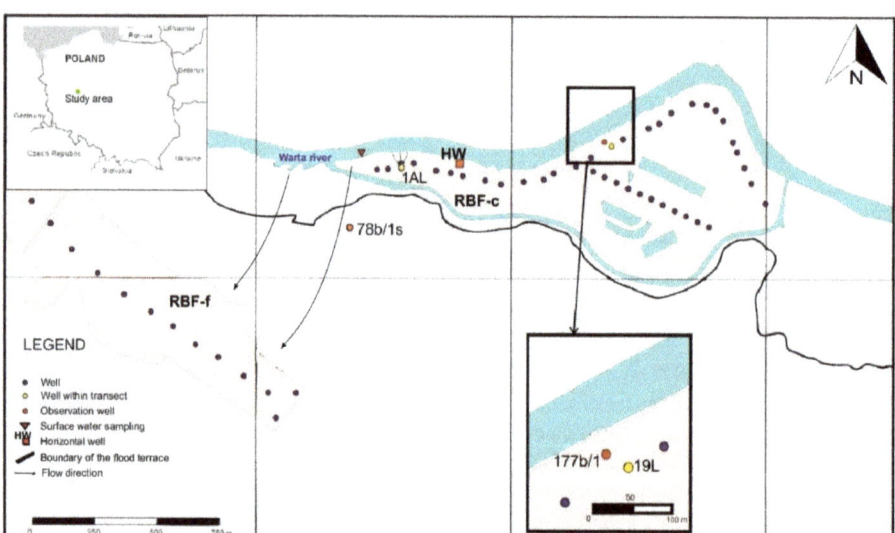

Figure 1. Situation map of the study area. RBF: riverbank filtration; RBF-c: wells on the flood terrace; RBF-f: wells on the higher terrace; and HW: horizontal well. [2] modified.

The Krajkowo well field is located in a region of favourable hydrogeological conditions. The total thickness of the aquifer is up to 40 m. In the upper part of the aquifer, there are sediments of the Warsaw-Berlin ice-marginal valley. Deeper sediments of the Wielkopolska Buried Valley are present. In the profile of aquifer sediments, there are fluvial fine and medium-grained sands and fluvioglacial coarse-grained sands with gravels. The total well field production is approximately 70,000–120,000 m^3/day.

2.2. Methods

For the investigation of pharmaceutical behaviour along flow paths from the river to the wells, 6 sampling points were selected, source water (the Warta River) and the wells located at different distances from the river (Table 1). Three production wells were selected for the research: HW, 19L, and 1AL. The closest sampling point is HW. Observation well 177b/1 is located between the river and well 19L. Observation well 78b/1s is the furthest away sampling point. The RBF-f wells shown in Figure 1 were in continuous operation during 2 years, including the period of our investigations. This situation enabled the observation of bank filtrate in well 78b/1s (Figure 1). The water balance and residence time were estimated based on the analyses of the hydrochemical data and the results of the mathematical modelling of groundwater flow. The well field monitoring data performed by the water company were also used for the interpretation.

Table 1. Characterization of sampling points.

Sampling Points	Location	Distance from the River Bank (m)	Depth of the Well Screen (m)	Contribution of River Water to Total Water Balance in Well (%)	Residence Time (days)
Warta River	-	-	-	-	-
Horizontal well-HW	Drains under river bottom	-	5 m below river bottom	100	1
Observation well 177b/1	Floodplain	38	12.5–14.5	100	24
Vertical well 19L	Floodplain	64	24.0–32.0	65–85	40
Vertical well 1AL	Floodplain	82	16.5–32.5	65–85	50
Observation well 78b/1s	Higher terrace	250	18.0–28.0	60	150

For the preliminary investigation of pharmaceuticals, 3 sampling points were selected (surface water, 1AL, and 78b/1s). Three sampling sessions were performed on September 2017, May 2018 and June 2018. The laboratory measurements addressing 13 constituents were performed in the ALS Laboratory in Prague. Based on this investigation, consecutive sampling campaigns were planned. The next investigations were performed in July, August and October 2018 and included six sampling points (surface water, HW, 177b/1, 1AL, 19L, and 78b/1s). The measurements of 75 constituents were performed in the Laboratory of Povodí Vltavy VHL Plzeň. (Table 2).

Table 2. List of substances tested in extended investigation (July, August, October 2018).

Parameters	LOQ	Parameters	LOQ	Parameters	LOQ
Carbamazepine	<10	Saccharin	<50	Alfuzosin	<10
Erythromycin	<10	Gabapentin	<10	Bisoprolol	<10
Sulfamethoxazol	<10	Tramadol	<10	Celiprolol	<10
Iopromide	<50	Clarithromycin	<10	Citalopram	<20
Ibuprofen	<20	Roxithromycin	<10	Clindamycin	<10
Diclofenac	<20	Azithromycin	<10	Cyclophosphamide	<10
Iopamidol	<50	Carbamazepine-DH	<10	Diltiazem	<10
Atenolol	<10	Oxcarbazepine	<10	Fexofenadine	<10
Caffein	<100	Ibuprofen-2-hydroxy	<30	Fluconazole	<10
Ketoprofen	<10	Ibuprofen-carboxy	<20	Fluoxentine	<10
Metoprolol	<10	Diclofenac-4-hydroxy	<20	Iomeprol	<50
Peniciline G	<10	Naproxene-O-desmeth	<20	Irbesartan	<10
Sulfamerazine	<10	Venlafaxine	<10	Ivermectin	<10
Sulfamethazin	<10	Sertraline	<10	Lamotrigine	<10
Sulfapyridin	<10	Ranitidine	<10	Lovastatin	<10
Trimetoprim	<10	Iohexol	<50	Memantine	<20
Furosemide	<10	Carbamazepine-2-hydr	<10	Mirtazapine	<10
Gemfibrozil	<50	Clofibric acid	<10	Phenazone	<10
Hydrochlorothiazide	<10	Cotinine	<20	Primidone	<10

Table 2. Cont.

Parameters	LOQ	Parameters	LOQ	Parameters	LOQ
Naproxene	<50	Paraxanthine	<100	Propranolol	<10
Triclocarban	<10	Bisfenol B	<50	Propyphenazone	<10
Triclosan	<20	Bisfenol S	<50	Simvastatin	<10
Chloramphenicol	<20	Oxypurinol	<50	Sotalol	<10
Bezafibrate	<10	Tiamulin	<10	Telmisartan	<20
Warfarin	<10	Acebutolol	<10	Valsartan	<10

The sampling collection took one day. The samples were taken from surface water, observation and production wells. The observation wells were pumped using a portable pump (MP-1, Grundfos, Bjerringbro, Denmark). The production wells were pumped continuously before and during the sampling periods. The water samples were stored in glass bottles and transported in a refrigerated container and frozen. After 5 days of storage at −18 °C temperature, the samples were delivered to the laboratory. The investigation of pharmaceuticals in the ALS Laboratory in Prague was performed using liquid chromatography (LC-MS/MS). The extended investigations in the laboratory of Povodí Vltavy VHL Plzeň were carried out using liquid chromatography (LC-MS/MS) and ultra-high-performance liquid chromatography (UHPLC MS/MS). A 1200 Ultra-High-Performance Liquid Chromatograph (UHPLC) tandem with 6495 Triple Quad Mass Spectrophotometer (MS/MS) of Agilent Technologies was used in ESI mode. The separation was carried out on an X-bridge C18 analytical column (100 × 4.6 mm, 3.5 µm particle size). The mobile phase consisted of (A) methanol and (B) water with 0.02% acetic acid and 5 mM ammonium fluoride as mobile phase additives. The flow rate was 0.5 mL min^{-1}. The injection volume was 0.050 mL.

3. Results

Preliminary investigations performed in September 2017 and, May and June 2018 at three sampling points allowed the determination of occurrences of pharmaceuticals in the surface and bank filtrate water (Table 3). Among the 13 measured parameters, antibiotics, anti-inflammatory and analgesic drugs, psychotropic drugs, X-ray agents and β-blockers were detected. The highest pharmaceutical concentrations and the largest variety of substances were detected in the Warta River (max. 485 ng/L). The investigation showed that the concentrations in bank filtration wells were considerably lower (max. 184 ng/L). Some of the pharmaceuticals were detected only in the river water (iomeprol (max. 156 ng/L), iopromide (max. 413 ng/L), metoprolol (max. 26 ng/L), metformin (max. 88 ng/L) and 1H-Benzotriazole (140 ng/L)). In well 1AL, located 82 m away from the river, 5 substances were detected (carbamazepine (max. 145 ng/L), sulfamethoxazole (max. 20 ng/L), diclofenac (max. 99 ng/L), naproxen (max. 21 ng/L) and iohexol (max. 146 ng/L)). In observation well 78b/1s that is located 250 m from the river, only 2 constituents were detected (carbamazepine (max. 81 ng/L), iohexol (max. 184 ng/L)). The results documented the occurrence of pharmaceuticals in both surface water and bank filtrates.

In July, August and October 2018, the analyses involving 75 different compounds at 6 sampling points were conducted. The analyses included antibiotics, anti-inflammatory and analgesic drugs, psychotropic drugs, X-ray agents, β-blockers, sweeteners and drugs, such as caffeine. A total of 25 of the 75 tested pharmaceuticals were detected (Table 4).

Table 3. Concentrations of pharmaceuticals in ng/L: The preliminary investigation. <LOQ - below limit of quantification. (Measurements performed in ALS Laboratory in Prague).

		LOQ	September 2017			May 2018			June 2018		
			Warta	1AL	78b/1s	Warta	1AL	78b/1s	Warta	1AL	78b/1s
Antibiotics	Sulfamethoxazole	<10	43	15	<LOQ	306	20	<LOQ	24	16	<LOQ
	Iopromide	<30	<LOQ	<LOQ	<LOQ	413	<LOQ	<LOQ	79	<LOQ	<LOQ
X-ray agents	Iohexol	<10	120	<LOQ	<LOQ	217	<LOQ	<LOQ	485	146	184
	Iomeprol	<39	<LOQ	<LOQ	<LOQ	156	<LOQ	<LOQ	<LOQ	<LOQ	<LOQ
Psychotropic	Carbamazepine	<10	110	145	81	208	73	9	91	77	75
Beta-blockers	Metoprolol	<100	<LOQ	<LOQ	<LOQ	26	<LOQ	<LOQ	<LOQ	<LOQ	<LOQ
Anti-inflammatory	Diclofenac	<10	43	99	<LOQ	<LOQ	<LOQ	<LOQ	<LOQ	<LOQ	<LOQ
	Naproxene	<10	39	<LOQ	<LOQ	31	21	<LOQ	<LOQ	<LOQ	<LOQ
Antidiabetic	Metmorfina	<50	88	<LOQ	<LOQ	79	<LOQ	<LOQ	55	<LOQ	<LOQ
Benzotriazole	1H-Benzotriazole	<80	140	<LOQ	<LOQ	<LOQ	<LOQ	<LOQ	<LOQ	<LOQ	<LOQ

Ketoprofen, iopamidol, and ibuprofen were never detected.

In general, the highest concentration of pharmaceuticals was detected in the river water (Table 4). However, the concentrations decrease along the flow path from the river to the wells (Figure 2). The distance and travel time have an impact on the decrease in concentrations. Some of the substances occurred only in the river water (iopromide (max. 149 ng/L), diclofenac (max. 37.4 ng/L), metoprolol (max. 19.6 ng/L), penicillin G (max. 17.1 ng/L), saccharine (max. 360 ng/L), iohexol (max. 120 ng/L), cotinine (max. 50.8 ng/L), clindamycin (max. 12.7 ng/L), fexofenadine (max. 40.7 ng/L), valsartan) others also in the closest wells, HW and 177b/1 (caffeine, paraxanthine, sulfapyridine, sotalol, telmisartan) or just there (primidone). Carbamazepine, sulfamethoxazole, gabapentin, tramadol, oxypurinol, fluconazole and lamotrigine, are the most common compounds from all sampling sessions and sampling points, being episodically detected also in the farthest production wells: 19L and 1AL.

The concentration of some pharmaceuticals in the Warta River and the nearest well, HW, are similar (e.g., carbamazepine, sulfamethoxazole, tramadol, fluconazole, lamotrigine (Table 4)). This result is due to the short distance (5 m) and short travel time (1 day) between the river and this well. Most of the substances found in the HW well were also observed in well 177b/1, but at lower concentrations. The significant decreases in concentrations occurred in production wells 19L and 1AL, where most of the parameters were below LOQ. This finding is due to the longer distances (64–82 m) and travel times (40–50 days) for these wells. In well 78b/1s, which is located 250 m away from the Warta River with a travel time of 150 days, only two parameters, carbamazepine and gabapentin, were detected and were at relatively low concentrations. This is the result of water mixing (Figure 2 and Table 4).

The detected parameter concentrations in the river water range from 10.8 ng/L (sulfapyridine) to 1470 ng/L (paraxanthine). The highest concentrations in river water occurred in the August 2018 sampling session. Oxypurinol presented high concentrations in river water that persisted (even at higher values) in nearby wells (HW) and also in more distant ones (1AL). Carbamazepine also persisted at high concentrations (135 ng/L in river water and 179 ng/L in HW).

Figure 3 shows the concentration of individuals groups of parameters. The groups were established based on the use of the substances. Nine groups were separated: antibiotics; X-ray agents; psychotropics, anticonvulsants, and antiepileptics; beta-blockers and cardiac drugs; drugs like caffeine; analgesics and anti-inflammatories; antifungals and antibacterials; antihistamines; and xanthine oxidase inhibitors. The highest concentrations show xanthine oxidase inhibitors, although there is only one substance in this group (oxypurinol). Psychotropics, anticonvulsant and antiepileptic drugs and drugs like caffeine also reach high concentrations. On the lower level antibiotics were detected: X-ray agents; beta-blockers and cardiac drugs; analgesic and anti-inflammatory; as well as antifungal and antibacterial.

Table 4. Concentrations of pharmaceuticals in ng/L: Extended investigations. HW - Horizontal well, <LOQ - below limit of quantification. Measurements performed in VHL Plzeň.

		LOQ	July 2018						August 2018						October 2018				
			Warta	HW	177b/1	19L	1AL	78b/1s	Warta	HW	177b/1	19L	1AL	78b/1s	Warta	HW	177b/1	19L	1AL	78b/1s
Antibiotics	Clindamycin	<10	12.7	<LOQ	<LOQ	<LOQ	<LOQ	<LOQ	<LOQ	<LOQ	<LOQ	<LOQ	<LOQ	<LOQ	12.2	<LOQ	<LOQ	<LOQ	<LOQ	<LOQ
	Penicillin G	<10	<LOQ	<LOQ	<LOQ	<LOQ	<LOQ	<LOQ	13	<LOQ	<LOQ	<LOQ	<LOQ	<LOQ	17.1	<LOQ	<LOQ	<LOQ	<LOQ	<LOQ
	Sulfamethoxazole	<10	29.3	27.1	15.9	<LOQ	15	<LOQ	18.8	17.1	12.4	<LOQ	<LOQ	<LOQ	37.7	21.8	10.1	<LOQ	10.5	<LOQ
X-ray Agents	Iohexol	<50	120	<LOQ	<LOQ	<LOQ	<LOQ	<LOQ	<LOQ	<LOQ	<LOQ	<LOQ	<LOQ	<LOQ	90	<LOQ	<LOQ	<LOQ	<LOQ	<LOQ
	Iopromide	<50	149	<LOQ	<LOQ	<LOQ	<LOQ	<LOQ	59.8	<LOQ	<LOQ	<LOQ	<LOQ	<LOQ	105	<LOQ	<LOQ	<LOQ	<LOQ	<LOQ
Psychotropic, Anticonvulsant, Antiepileptic	Carbamazepine	<10	130	179	161	112	110	83.1	132	131	131	88.6	99	63.6	135	134	148	135	123	80.3
	Gabapentin	<10	97	53.3	18.7	13	14	21.3	55.6	27	25.2	12.6	13.8	15.6	81.5	61.7	13.5	<LOQ	10.2	24
	Lamotrigine	<10	35.8	54	29.1	15	21	<LOQ	36.1	44.9	26.7	15.6	16.6	<LOQ	45.1	42.6	38.3	24.6	25.2	<LOQ
	Primidone	<10	<LOQ	12.4	<LOQ	<LOQ	<LOQ	<LOQ	<LOQ	10.4	<LOQ	<LOQ	<LOQ	<LOQ	<LOQ	10.2	11.6	<LOQ	<LOQ	<LOQ
Beta-blockers, Cardiac Drugs	Metoprolol	<10	11.9	<LOQ	<LOQ	<LOQ	<LOQ	<LOQ	<LOQ	<LOQ	<LOQ	<LOQ	<LOQ	<LOQ	19.6	<LOQ	<LOQ	<LOQ	<LOQ	<LOQ
	Sotalol	<10	23.3	<LOQ	<LOQ	<LOQ	<LOQ	<LOQ	14.3	<LOQ	<LOQ	<LOQ	<LOQ	<LOQ	50.3	14.1	<LOQ	<LOQ	<LOQ	<LOQ
	Telmisartan	<20	140	62.5	<LOQ	<LOQ	<LOQ	<LOQ	132	52	<LOQ	<LOQ	<LOQ	<LOQ	136	60.5	<LOQ	<LOQ	<LOQ	<LOQ
	Valsartan	<10	61.1	<LOQ	<LOQ	<LOQ	<LOQ	<LOQ	23	<LOQ	<LOQ	<LOQ	<LOQ	<LOQ	28.9	10.2	11.6	<LOQ	<LOQ	<LOQ
Drugs, e.g., Caffeine	Caffeine	<100	154	<LOQ	<LOQ	<LOQ	<LOQ	<LOQ	1350	<LOQ	140	<LOQ	<LOQ	<LOQ	<LOQ	<LOQ	<LOQ	<LOQ	<LOQ	<LOQ
	Continine	<20	30.9	<LOQ	<LOQ	<LOQ	<LOQ	<LOQ	50.8	<LOQ	<LOQ	<LOQ	<LOQ	<LOQ	<LOQ	<LOQ	<LOQ	<LOQ	<LOQ	<LOQ
	Saccharin	<50	111	<LOQ	<LOQ	<LOQ	<LOQ	<LOQ	360	<LOQ	<LOQ	<LOQ	<LOQ	<LOQ	<LOQ	<LOQ	<LOQ	<LOQ	<LOQ	<LOQ
	Paraxanthine	<100	163	<LOQ	<LOQ	<LOQ	<LOQ	<LOQ	1470	<LOQ	104	<LOQ	<LOQ	<LOQ	<LOQ	<LOQ	<LOQ	<LOQ	<LOQ	<LOQ
Analgesics, anti-Inflammatory	Diclofenac	<20	24.5	<LOQ	<LOQ	<LOQ	<LOQ	<LOQ	<LOQ	<LOQ	<LOQ	<LOQ	<LOQ	<LOQ	37.4	<LOQ	<LOQ	<LOQ	<LOQ	<LOQ
	Tramadol	<10	76.1	73.7	35.9	19	22	<LOQ	52	38.1	27.4	17	20.5	<LOQ	83.8	64.4	35.3	24.4	27.9	<LOQ
Antifungal and Antibacterial	Fluconazole	<10	35.6	48.4	21.5	12	21	<LOQ	32.5	42.1	20.2	10.4	19.6	<LOQ	51.7	51.6	29.2	15	21.5	<LOQ
	Sulfapyridine	<10	<LOQ	10.7	14.2	<LOQ	<LOQ	<LOQ	<LOQ	<LOQ	11.7	<LOQ	<LOQ	<LOQ	10.8	11.2	13	<LOQ	<LOQ	<LOQ
Antihistamine	Fexofenadine	<10	40.7	<LOQ	<LOQ	<LOQ	<LOQ	<LOQ	28.9	<LOQ	<LOQ	<LOQ	<LOQ	<LOQ	33.2	<LOQ	<LOQ	<LOQ	<LOQ	<LOQ
Xanthine Oxidase Inhibit	Oxypurinol	<50	388	1350	503	237	345	<LOQ	610	1100	486	130	228	<LOQ	1050	1010	652	260	317	<LOQ

Iomeprol (94.7 ng/L) and venlafaxine (12.1 ng/L), were detected once and only in the Warta River.

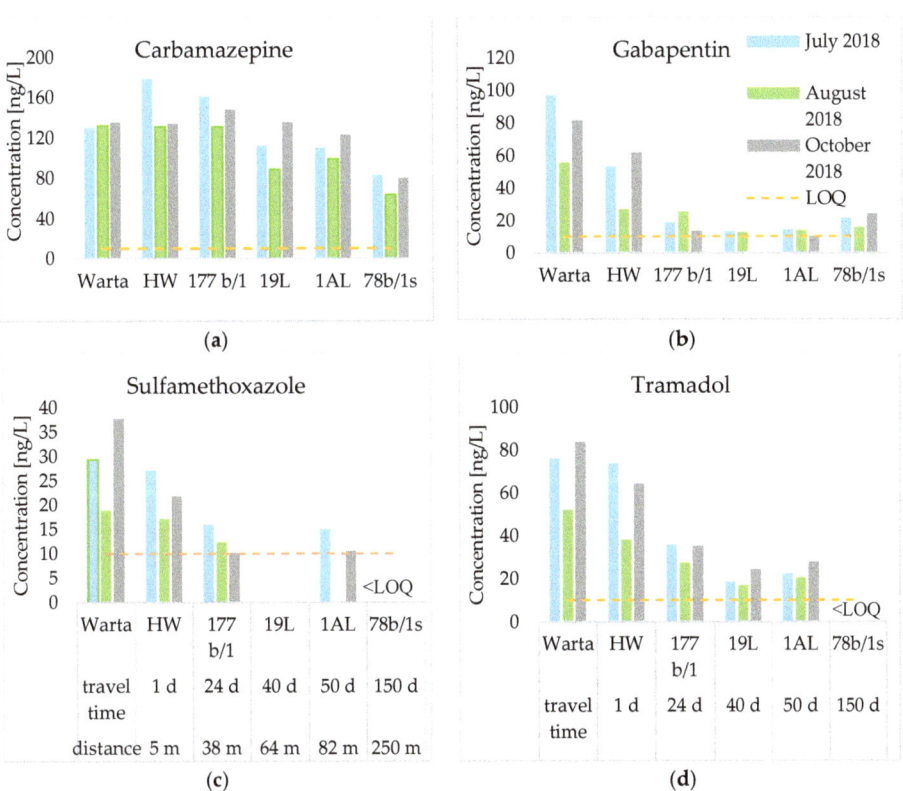

Figure 2. Concentrations of carbamazepine (**a**), gabapentin (**b**), sulfamethoxazole (**c**), and tramadol (**d**) for 3 sampling sessions.

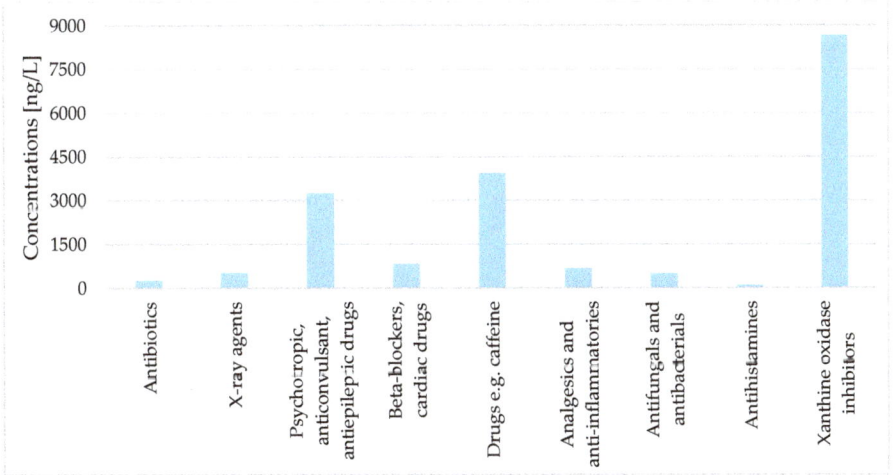

Figure 3. Sum of concentrations of all pharmaceuticals in each series in categories of pharmaceutical according to the application.

Table 5 shows the percentage of removal for pharmaceuticals at sampling points located at different distances from the river. The removal was calculated using the formula:

$$Removal\ (\%) = \frac{\text{concentration in river} - \text{concentration in well}}{\text{concentration in river}} \times 100\% \qquad (1)$$

Table 5. Removal of pharmaceuticals in %. HW—Horizontal well.

	HW			177b/1			19L			1AL			78b/1s		
	VII	VIII	X	VII	VIII	X	VII	VIII	X	VII	VIII	X	VII	VIII	X
	2018			2018			2018			2018			2018		
Carbamazepine	−37.7	0.8	0.7	−23.8	0.8	−9.6	13.8	32.9	0	15.4	25.0	26.7	36.1	51.8	40.5
Sulfamethoxazole	7.5	9.0	42.2	45.7	34.0	73.2	100	100	100	48.8	100	100	100	100	100
Gabapentin	45.1	51.4	24.3	80.7	54.7	83.4	86.5	77.3	100	85.2	75.2	83.1	78.0	71.9	70.6
Tramadol	3.2	26.7	23.2	52.8	47.3	57.9	75.6	67.3	70.9	70.6	60.6	75.5	100	100	100
Oxypurinol	−247.9	−80.3	3.8	−29.6	20.3	37.9	38.9	78.7	75.2	11.1	62.6	78.3	100	100	100
Fluconazole	−36.0	−29.5	0.2	39.6	37.8	43.5	65.7	68.0	71.0	40.7	39.7	62.1	100	100	100
Lamotrigine	−50.8	−24.4	5.5	18.7	26.0	15.1	58.1	56.8	45.5	40.5	54.0	63.2	100	100	100

The removal is calculated on detected values only and mixing was not accounted for.

The lowest removal was observed in the HW. In the HW, some of the parameters increase, which probably occurs because there were higher concentrations in the Warta River before the sampling periods. In observation well 177b/1, removal varies over a range of −29.6–100% depending of the compound. The removal in two production wells, 19L and 1AL, show similar values. At the furthest sampling point, 78b/1s, most parameters reduced by 100%. The removal probably depends on the location of the sampling point (distance and travel time from the river) but is also different for specific compounds. The evaluation of the lowest removal shows that carbamazepine (a psychotropic drug) is found at the farthest points (78b/1s – 250 m from the river) and decreases by 36.1–51.8%, whereas sulfamethoxazole (an antibiotic), gabapentin (an anti-epileptic drug) and tramadol (an analgesic drug) reach similar values at a distance of 38 m (177b/1s). Carbamazepine is a difficult compound to remove in spite of long distances and travel times. Gabapentin attains the highest removal but is not completely removed, even at the farthest point.

The total reductions of some (Table 5) pharmaceuticals (sulfamethoxazole, tramadol, oxypurinol, fluconazole, lamotrigine) are achieved in wells 19L, 1AL and an observation well 78b/1s, while this did not occur in HW and 177b/1. The results indicate that at the given conditions, significant reductions in pharmaceutical concentrations can be achieved at travel times of 40–50 days and distances of 60–80 m, although higher values of the reduction can be achieved when the well is located more than 250 m away.

The degree of removal of pharmaceuticals at sampling points depends not only on the travel time in the subsurface, but also on the diverse impact of sorption and biodegradation, and the influence of temperature and redox conditions on those processes [21]. The assessment of the impact of these factors was not analyzed in detail in this study. However, based on well field monitoring data, it can be assumed that in wells located close to the river (HW, 177b/1, 19L, 1AL), the biodegradation and oxidation occur because of oxic conditions. The following data confirmed this: oxygen 1–6.2 mg/L, nitrate 0.5–18 mg/L and a lack of hydrogen sulfide. In the well located further away from the river (78b/s), there are trace concentrations of nitrates (0.08–0.26 mg/L) and a lack of oxygen, however, the presence of hydrogen sulfide (0.024–0.066 mg/L) is noted. It can also be added that the redox processes and biodegradation in wells located close to the river are also favored by higher temperatures in summer (15–17 °C). Whereas, further away from the river (78b/s well), the temperatures are leveled in the range of (8–12 °C), similar to ambient groundwater.

4. Discussion

The concentrations of pharmaceuticals in the Warta River were found at levels previously documented in European rivers and lakes [1,7,22]. Carbamazepine concentrations in the Warta River

(130–135 ng/L) are at a similar level as in the Nairobi River (Kenya) [23] 100 ng/L and in the Leine River (Germany) 144 ng/L [24]. However, carbamazepine concentrations in the Warta River are much lower than in Lake Tegel (510 ng/L) and Lake Wennsee (310 ng/L) [19]. Similar concentrations also show Sulfamethoxazole in the Warta River is 18.8–37.7 ng/L and in the Lake Maggiore (Italy) 10ng/L [25], in the Douro River (Portugal) 53.3 ng/L [26]. Among 75 substances, 25 were detected in the river. Nonsteroidal anti-inflammatory drugs (diclofenac) previously measured in the Warta River were documented at lower concentrations in the current research than in 2007 [15], while ibuprofen and benzafibrate documented earlier were not detected in the current research [2,15].

The research presented confirms high percentages of removal for organic micropollutants at the RBF sites [2,7,8,19,22,27–30]. Among 25 substances measured in the Warta River, 12 were not detected in the RBF site in Krajkowo (valsartan, fexofenadine, clindamycin, saccharin, iopromide, diclofenac, cotinine, iohexol, metoprolol, penicillin G, iomeprol and venlafaxine). In the case of the organic micropollutants research at two sites in Budapest, out of the 36 analyzed micropollutants, 12 were present in almost all the samples [22]. It is documented in the literature [3,4,27] that the transport and removal of organic micropollutants during subsurface movement from rivers to wells depends highly on the prevailing hydrochemical conditions along the flow path. As a result, different degradation behaviour can be seen for individual sites. The percentage of removal of carbamazepine varied between 37.7 and 51.8%, which was relatively persistent during subsurface flow as was observed previously at other sites [4,22,27,28]. Carbamazepine was also detected in well 78b/1s, where the travel time is 5 months. The result is comparable to findings from Berlin, where carbamazepine occurs in the well where the travel time is 2.8–4.3 months [19]. In the 78b/1s well, Gabapentin was also detected but was characterized by a relatively high percentage of removal (>70%). Oxypurinol was not removed along short distances (relatively high concentrations were seen in HW and 177b/1), but in production wells (distance 64–82 m), the percentage of the removal increased to a range of 11–78% and at distances of 250 m (78b/1s), and the complete removal was achieved. These analyses confirm earlier findings, documenting carbamazepine as a persistent constituent, while gabapentin and oxypurinol are subjects to primary degradation during filtration [27].

The high percentages of removal are achieved for the remaining substances that occur in bank filtrates (Figure 2, Table 5). The remaining substances detected in bank filtrates show a relatively high percentage of removal (typically more than 70%) in production wells located 64–82 m from the river. A similar reduction was observed in the Rhine River in wells located at 70 m, where the removal was >51% [8] and Lake Tegel in Berlin where the wells, located at 90 m distance from a lake, were removed >51% (Table 5) [29]. A total of 12 substances were detected in the Warta River that did not occur in bank filtrates, showing the complete removal even at short distances.

The negative removal observed in the case of HW and 177b/1 (the sampling points located at the nearest distance to the river) inaccurately suggest an increase in concentrations during subsurface flow and is probably due to unrecognized fluctuations in concentrations in the source water before sampling (carbamazepine, oxypurinol, lamotrigine, fluconazole). A similar situation was encountered at the RBF site in Austria. The higher concentrations of some substances appear in the wells at higher distances [24]. The same effect is responsible for fluctuations in the removal during the investigation periods (e.g., 11.1–78.3% for the case of oxypurinol in well 1AL). It is also possible due to the transformation from other compounds.

5. Conclusions

The research carried out on the Krajkowo riverbank filtration site (Poland) contained 75 different compounds, including antibiotics, anti-inflammatory and analgesic drugs, psychotropic drugs, X-ray agents, β-blockers and sweeteners. A total of 25 of these have been detected. The highest concentrations were found in the Warta River.

In the bank filtrates, 13 compounds were detected. Their concentrations declined along the flow path. The number of detected pharmaceuticals at each sampling point decreased with increasing distances. The lowest removal was noticed in the horizontal well. In wells 1AL and 19L (distances from the river of 64 to 82 m, respectively), the removal of most parameters was approximately 70–80%. For the observation well 78b/1s (at a distance of 250 m from the river), only 2 compounds were detected.

This research shows the significant role of bank filtration in the removal of pharmaceuticals. Under similar hydrogeological conditions, wells should be located at least 60 m from the river. Higher removal can be achieved at distances of 250 m from the source water. However, the results obtained emphasize the need for further monitoring studies to recognize the factors that determine the variability of micropollutants in the river, as well as in the production wells (hydrological conditions and seasons of the year). It is also necessary to identify processes that condition the migration and removal of micropollutants. Future research should focus on fewer compounds and their metabolites.

Author Contributions: R.K. prepared the manuscript and all authors read and approved the manuscript, R.K. took part in fieldwork and performed graphical and statistical interpretations; K.D., J.G. and R.K. interpreted the data and were involved in discussing the study; J.G., K.D. and R.K., were responsible for the overall coordination of the research team.

Funding: This research was completed with support from the AquaNES project. This project has received funding from the European Union's Horizon 2020 Research and Innovation Program under grant agreement no. 689450.

Acknowledgments: The authors would like to thank Aquanet SA (Poznań Waterworks operator) for their contribution.

Conflicts of Interest: The authors declare no conflicts of interest.

Abbreviations

The following abbreviations are used in this manuscript:

BF bank filtrate
HW horizontal well

References

1. Kovačević, S.; Radišić, M.; Laušević, M.; Dimkić, M. Occurrence and behavior of selected pharmaceuticals during riverbank filtration in The Republic of Serbia. *Environ. Sci. Pollut. Res.* **2017**, *24*, 2075–2088. [CrossRef] [PubMed]
2. Dragon, K.; Górski, J.; Kruć, R.; Drożdżyński, D.; Grischek, T. Removal of Natural Organic Matter and Organic Micropollutants during Riverbank Filtration in Krajkowo, Poland. *Water* **2018**, *10*, 1457. [CrossRef]
3. Hiscock, K.M.; Grischek, T. Attenuation of groundwater pollution by bank filtration. *J. Hydrol.* **2002**, *266*, 139–144. [CrossRef]
4. Maeng, S.K.; Ameda, E.; Sharma, S.K.; Grutzmacher, G.; Amy, G.L. Organic micropollutant removal from wastewater effluent-impacted drinking water sources during bank filtration and artificial recharge. *Water Res.* **2010**, *44*, 4003–4014. [CrossRef] [PubMed]
5. Forizs, T.; Berecz, Z.; Molnar, Z.; Suveges, M. Origin of shallow groundwater of Csepel Island (south of Budapest. Hungary. River Danube): Isotopic and chemical approach. *Hydrol. Process.* **2005**, *19*, 3299–3312. [CrossRef]
6. Lasagna, M.; De Luca, D.A.; Franchino, E. Nitrates contamination of groundwater in the western Po Plain (Italy): The effects of groundwater and surface water interactions. *Environ. Earth Sci.* **2016**, *75*, 240. [CrossRef]
7. Heberer, T.; Mechlinski, A.; Fanck, B.; Knappe, A.; Massmann, G.; Pekdeger, A.; Fritz, B. Field Studies on the Fate and Transport of Pharmaceutical Residues in Bank Filtration. *Groundw. Monit. Remediat.* **2004**, *24*, 70–77. [CrossRef]
8. Schmidt, C.K.; Lange, F.T.; Brauch, H.J. Characteristics and evaluation of natural attenuation processes for organic micropollutant removal during riverbank filtration. *Water Supply* **2007**, *7*, 1–7. [CrossRef]
9. Maeng, S.K.; Salinas Rodriguez, C.N.A.; Sharma, S.K. Removal of Pharmaceuticals by Bank Filtration and Artifical Recharge and Recovery. *Compr. Anal. Chem.* **2013**, *62*, 435–451.
10. Wang, L.; Ying, G.G.; Zhao, J.L.; Yang, X.B.; Chen, F. Occurrence and risk assessment of acidic pharmaceuticals in the Yellow River, Hai River and Liao River of north China. *Sci. Total Environ.* **2010**, *408*, 3139–3147. [CrossRef]

11. Zheng, Q.; Zhang, R.; Wang, Y.; Pan, X. Occurrence and distribution of antibiotics in the Beibu Gulf, China: Impacts of river discharge and aquaculture activities. *Mar. Environ. Res.* **2012**, *78*, 26–33. [CrossRef] [PubMed]
12. Zhou, X.F.; Dai, M.C.; Zhang, Y.L.; Surampalli, R.Y. A preliminary study on the occurrence and behavior of carbamazepine (CBZ) in aquatic environment of Yangtze River Delta, China. *Environ. Monit. Assess.* **2011**, *173*, 45–53. [CrossRef] [PubMed]
13. Tamtam, F.; Mercier, F.; Le Bot, B.; Eurin, J.; Dinh, Q.T. Occurrence and fate of antibiotics in the Seine River in various hydrological conditions. *Sci. Total Environ.* **2008**, *393*, 84–95. [CrossRef] [PubMed]
14. Shala, L.; Foster, G.D. Surface water concentrations and loading budgets of pharmaceuticals and other domestic-use chemicals in an urban watershed (Washington, DC, USA). *Arch. Environ. Contam. Toxicol.* **2010**, *58*, 551–561. [CrossRef]
15. Szymonik, A.; Lach, J.; Malińska, K. Fate and removal of pharmaceuticals and illegal drugs present in drinking water and wastewater. *Ecol. Chem. Eng. S* **2017**, *24*, 65–85. [CrossRef]
16. Hamann, E.; Stuyfzand, P.J.; Greskowiak, J.; Timmer, H.; Massmann, G. The fate of organicmicropollutants during long-term/long-distance river bank filtration. *Sci. Total Environ.* **2016**, *545–546*, 629–640. [CrossRef]
17. Kasprzyk-Hordern, B.; Dąbrowska, A.; Vieno, N.; Kronberg, L.; Nawrocki, J. Occurrence of Acidic Pharmaceuticals in the Warta River in Poland. *Chem. Anal.* **2007**, *52*, 289–303.
18. Górski, J.; Dragon, K.; Kaczmarek, P. Nitrate pollution in the Warta River (Poland) between 1958 and 2016: Trend and causes. *Environ. Sci. Pollut. Res.* **2019**, *26*, 2038–2046. [CrossRef]
19. Massmann, G.; Sultenfuß, J.; Dunnbier, U.; Knappe, A.; Taute, T.; Pekdeger, A. Investigation of groundwater residence times during bank filtration in Berlin: A multi-tracer approach. *Hydrol. Process.* **2008**, *22*, 788–801. [CrossRef]
20. Górski, J.; Dragon, K.; Kruć, R. A comparison of river water treatment efficiency in different types of wells. *Geologos* **2018**, *24*, 245–251. [CrossRef]
21. Massmann, G.; Greskowiak, J.; Dunnbier, U.; Zuehlke, S.; Knappe, A.; Pekdeger, A. The impact of variable temperatures on the redox conditions and the behaviour of pharmaceutical residues during artificial recharge. *J. Hydrol.* **2006**, *328*, 141–156. [CrossRef]
22. Nagy-Kovács, Z.; László, B.; Fleit, E.; Czihat-Mártonné, K.; Till, G.; Börnick, H.; Adomat, Y.; Grischek, T. Behavior of Organic Micropollutants during River Bank Filtration in Budapest, Hungary. *Water* **2018**, *10*, 1861. [CrossRef]
23. Koreje, K.O.; Demeestere, K.; De Wispelaere, P.; Vergeynst, L. From multi-residue screening to target analysis of pharmaceuticals in water: Development of a new approach based on magnetic sector mass spectrometry and application in the Nairobi River basin, Kenya. *Sci. Total Environ.* **2012**, *437*, 153–164. [CrossRef] [PubMed]
24. Nödler, K.; Licha, T.; Fischer, S.; Wagner, B. A case study on the correlation of micro-contaminants and potassium in the Leine River (Germany). *Appl. Geochem.* **2011**, *26*, 2172–2180. [CrossRef]
25. Loos, R.; Wollgast, J.; Huber, T.; Hanke, G. Polar herbicides, pharmaceutical products, perfluorooctanesulfonate (PFOS), perfluorooctanoate (PFOA), and nonylphenol and its carboxylates and ethoxylates in surface and tap waters around Lake Maggiore in Northern Italy. *Anal. Bioanal. Chem.* **2007**, *387*, 1469–1478. [CrossRef] [PubMed]
26. Madureira, V.T.; Barreiro, J.C.; Rocha, M.J.; Rocha, E. Spatiotemporal distribution of pharmaceuticals in the Douro River estuary (Portugal). *Sci. Total Environ.* **2010**, *408*, 5513–5520. [CrossRef]
27. Burke, V.; Schneider, L.; Greskowiak, J.; Zerball-van Baar, P.; Sperlich, A.; Dünnbier, U.; Massmann, G. Trace Organic Removal during River Bank Filtration for Two Types of Sediment. *Water* **2018**, *10*, 1736. [CrossRef]
28. Driezum, I.H.; Derx, J.; Oudega, T.J.; Zessner, M.; Naus, F.L.; Saracevic, E.; Kirschner, A.K.T.; Sommer, R.; Farnleitner, A.H.; Blaschke, A.P. Spatiotemporal resolved sampling for the interpretation of micropollutant removal during riverbank filtration. *Sci. Total Environ.* **2019**, *649*, 212–223. [CrossRef]
29. Grünheid, S.; Amy, G.; Jekel, M. Removal of bulk dissolved organic carbon (DOC) and trace organic compounds by bank filtration and artificial recharge. *Water Res.* **2005**, *39*, 3219–3228. [CrossRef]
30. Pedeger, A.; Massmann, G.; Ohm, B.; Pühringer, S.; Richter, D.; Engemann, N.; Gruß, S. *Hydrogeological-Hydrogeochemical Processes during Bank Filtration and Groundwater Recharge Using a Multi-Tracer Approach*; NASRI: Berlin, Germany, 2006.

© 2019 by the authors. Licensee MDPI, Basel, Switzerland. This article is an open access article distributed under the terms and conditions of the Creative Commons Attribution (CC BY) license (http://creativecommons.org/licenses/by/4.0/).

Article

Anthropic and Meteorological Controls on the Origin and Quality of Water at a Bank Filtration Site in Canada

Janie Masse-Dufresne [1],*, Paul Baudron [1,2], Florent Barbecot [3], Marc Patenaude [1], Coralie Pontoreau [1], Francis Proteau-Bédard [1], Matthieu Menou [4], Philippe Pasquier [1], Sabine Veuille [1] and Benoit Barbeau [1]

1. Polytechnique Montréal, Department of Civil, Geological and Mining Engineering, Montreal, QC H3T 1J4, Canada
2. Institut de Recherche pour le Développement, UMR G-EAU, 34090 Montpellier, France
3. Geotop-UQAM, Department of Earth and Atmospheric Sciences, Montreal, QC H2X 3Y7, Canada
4. Loire sécurité risques, Direction Départementale des Territoires de la Nièvre (Ministère de la Transition Écologique et solidaire), 58020 Nevers CEDEX, France
* Correspondence: janie.masse-dufresne@polymtl.ca

Received: 2 October 2019; Accepted: 20 November 2019; Published: 28 November 2019

Abstract: At many bank filtration (BF) sites, mixing ratios between the contributing sources of water are typically regarded as values with no temporal variation, even though hydraulic conditions and pumping regimes can be transient. This study illustrates how anthropic and meteorological forcings influence the origin of the water of a BF system that interacts with two lakes (named A and B). The development of a time-varying binary mixing model based on electrical conductivity (EC) allowed the estimation of mixing ratios over a year. A sensitivity analysis quantified the importance of considering the temporal variability of the end-members for reliable results. The model revealed that the contribution from Lake A may vary from 0% to 100%. At the wells that were operated continuously at >1000 m^3/day, the contribution from Lake A stabilized between 54% and 78%. On the other hand, intermittent and occasional pumping regimes caused the mixing ratios to be controlled by indirect anthropic and/or meteorological forcing. The flow conditions have implications for the quality of the bank filtrate, as highlighted via the spatiotemporal variability of total Fe and Mn concentrations. We therefore propose guidelines for rapid decision-making regarding the origin and quality of the pumped drinking water.

Keywords: anthropic forcing; meteorological forcing; lake bank filtration; mixing ratios; environmental tracer; time-varying mixing model; sensitivity analysis

1. Introduction

Bank filtration (BF) is known as a cost-effective treatment step to produce drinking water [1,2]. This natural or artificially induced process occurs as surface water infiltrates into the aquifer from the banks and/or bed of a lake or a river and is subsequently intercepted by a pumping well [3]. During subsurface passage, water is exposed to physical, chemical, and biological processes, which may attenuate contaminants initially present in the surface water but also release unwanted minerals [4,5]. BF systems have proven to be efficient for the removal of turbidity [6–8], pathogens [9–11], and organic compounds [12–15]. The efficiency of BF systems to attenuate contaminants is strongly controlled by travel times [13,16] and redox conditions [17–19], which in turn depend on numerous site-specific natural and engineered parameters. Natural parameters include the hydrological and hydrogeological conditions, surface and groundwater quality, and prevailing physico–chemical conditions [20]. Engineered parameters refer to the number of wells, the distance between wells and surface water, well spacing, the well type, depth, radius, location, and screen length [21,22].

Most BF systems are in the vicinity of rivers, where the bank filtrate is a mixture of surface water and ambient groundwater [4,23]. Numerous studies have shown that the dilution of contaminants by high-quality groundwater can also help to attenuate contaminants, enhancing the efficiency of a BF system. For instance, Kvitsand et al. [24] reported that dilution with ambient groundwater was significant enough to lower concentrations of natural organic matter. Derx et al. [25] numerically studied the effects of flooding on virus removal by bank filtration. They reported that a rapid decrease in river water level can lead to a hydraulic gradient towards the river and a dilution of virus concentrations by regional groundwater. In addition, some BF systems are placed in hydrogeological contexts with low-quality groundwater but can still achieve high-quality raw water with adequate regulation of mixing ratios [26–29]. Hence, when assessing the performance of a BF system, estimating mixing ratios is crucial to: (1) correctly differentiate between dilution and removal mechanisms and (2) control the occurrence of groundwater-borne contaminants [30]. BF systems typically show spatial variability of mixing ratios at the pumping wells, since they are affected by the distance to the surface water body [22]. Another factor governing the mixing ratios is the drawdown at the pumping wells [27]. The latter is subject to spatial and temporal variations, since BF systems are rarely operated under steady-state hydraulic conditions (e.g., river stage) and/or pumping regimes. However, when calculating mixing ratios, authors rarely discuss the temporal variations and the factors controlling this variability, even though erroneous estimation of the mixing ratios can lead to misinterpretation of the performance of the BF system.

This study aims to provide a better understanding of the relationship between anthropic (i.e., pumping regimes) and meteorological (i.e., hydraulic gradients) effects on the origin of bank filtrate. To this end, we investigated the spatiotemporal variability of flow patterns and mixing ratios at a two-lake BF site, where two surface water types (Lake A and Lake B) contribute to seven pumping wells. A time-varying mixing model based on electrical conductivity (EC) was developed in order to quantify the contributions of Lake A and Lake B (i.e., two water sources and further referred to as end-members) over a one-year period. A sensitivity analysis was conducted in order to test the assumptions concerning the definition of the end-members.

2. Site Description

2.1. Hydrogeological Context

2.1.1. Description of the Bank Filtration and Aquifer System

The studied BF system supplies drinking water to more than 18,000 people in a town near Montreal, Canada (Figure 1a). A total of eight pumping wells are located between two artificial lakes (Figure 1a,b), which were created by sand dredging activities. The exploitation stopped a few decades ago at Lake B, while Lake A is still in operation. As described by Ageos [31], the aquifer is a buried valley embedded in the Champlain Sea clays (Figure 1b,d). The aquifer is mainly composed of alluvial fine to medium sands. A small lens (≤3.45 m thick) of alluvial gravel (with a sandy matrix in places) lies between the Champlain Sea clays and the alluvial sands near pumping wells P4 and P5 (see Figure 1d). The aquifer is fully unconfined. Hydraulic conductivity was estimated as 2.7×10^{-3} m/s [31].

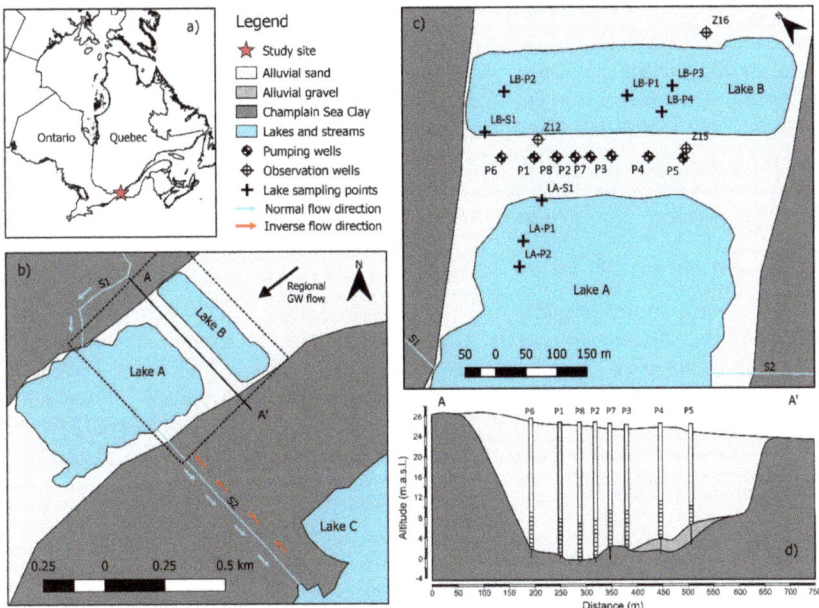

Figure 1. Study site location maps (**a**–**c**) and schematic lithological cut along pumping wells (**d**) (adapted with permission from Ageos, 2010 [31]).

The maximum thickness of the aquifer is 26 m and the static water level is about 4 m below the ground surface. The sandy bank is 100 m to 120 m wide and approximately 500 m long. All the pumping wells are screened at the base of the aquifer over an 8 m long section, except for pumping well P5, which only has a 4 m long screen due the shallower depth of the aquifer at this location. The distance between Lake A and the well cluster is 70 m to 80 m, whereas a distance of 30–35 m separates the wells from Lake B. Finally, the wells are spaced 30–60 m from one another.

2.1.2. Lake A and Lake B

Lake A (2.8×10^5 m^2) is fed by a stream named S1, which discharges from the North with a mean annual rate of 0.32 m^3/s [31]. It drains a small watershed (14.4 km^2), where land use is mostly industrial and agricultural. A 1 km long channeled stream named S2, located at the southeastern bank, allows water to exit Lake A and flow towards Lake C. The flow direction between Lake A and Lake C can be temporally reversed (Figure 1b) when the surface water level of Lake C exceeds both the elevation of Lake A and a topographic threshold at 22.12 m.a.s.l. [31]. Under these hydraulic conditions, Lake A receives surface water inputs from Lake C. This process typically occurs during spring (from April to May) and more occasionally during autumn (from October to December) due to snowpack melting and/or abundant precipitations. Ageos [31] reported that surface water input into Lake A seems to control its geochemistry, as it features a Ca-HCO$_3$ water type.

Lake B (7.6×10^4 m^2) is a groundwater-fed lake without any inlet stream. An artificial outlet channel can drain Lake B water towards the town's stormwater collection system (when Lake B elevation is above approximately 21.8 m.a.s.l.). A NaCl water type is found in Lake B [31]. Pazouki et al. [32] stated that the salinity of Lake B originates from de-icing road salts that are applied during wintertime. This is supported by the fact that a regional and widely used road is located less than 100 m from the study site. Precipitations are approximately 1000 mm/year and contribute to the water mass balance of both lakes. Runoff is likely a negligible contribution to the water mass balances of the lakes, considering the nearly flat topography. The maximum observed depths at Lake A and Lake B are 20 m and 19 m, respectively (at LA-P2 and LB-P2). Based on the lithological cross sections at the pumping wells and observation wells [31], it is believed that lake bottoms roughly correspond to the elevation of the marine clay sediments. In this geological context, no or only minor groundwater flow could occur beneath the lake bottom. The sediments at the bottom of the lakes were not sampled and no quantitative information concerning clogging is available. However, while sampling for surface water, relatively high turbidity (denoted by the color and the milky appearance of water) was observed at Lake A, which indicates that the sediments at the lake–aquifer interface are susceptible to clogging [33,34]. Sampling of the sediments would be needed to confirm this hypothesis.

2.2. Hydraulics of the Two-Lake BF System

A water table monitoring program was performed by Ageos [35] from 2012 to 2015. This study reported that, prior to the activation of the BF system in October 2012, surface water levels of Lake B were higher than in Lake A. Such conditions forced surface water to infiltrate and flow naturally through the sandy bank from Lake B to Lake A (Figure 2a). For instance, during summer 2012, the water level difference was about 0.1 m, which created a natural hydraulic gradient of approximately 0.001 between the lakes. Based on Darcy's law, the mean residence time of the water in the bank was approximately one month.

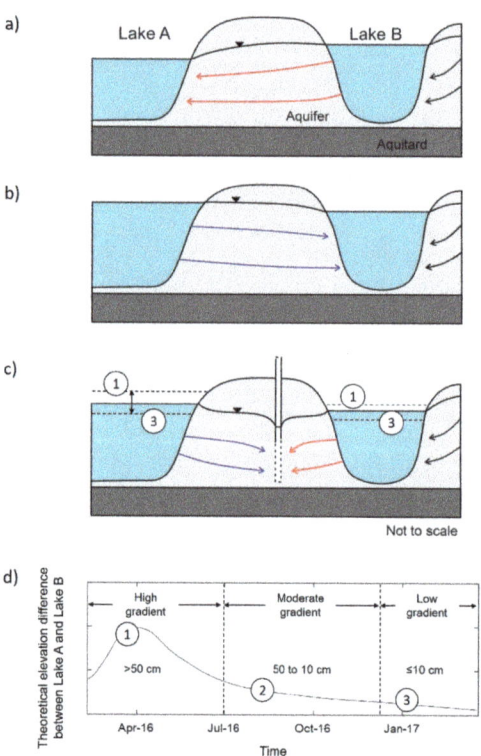

Figure 2. Schematic representation of the flow patterns and directions at the study site when (**a**) elevation of Lake B > elevation of Lake A, (**b**) elevation of Lake A > elevation of Lake B, and (**c**) the pumping wells are in operation. Black, blue, and red arrows refer to regional groundwater and water originating from Lake A and Lake B, respectively. Theoretical elevation difference between Lake A and Lake B in (**d**). Numbers 1 to 3 correspond to different hydraulic conditions, namely high, moderate, and low hydraulic gradients between Lake A and Lake B.

The above-mentioned water table monitoring program also demonstrated that the water level in Lake A was significantly higher than in Lake B during springtime from 2012 to 2016 (i.e., up to 1 m water level difference). This is due to the intermittent hydraulic connection between Lake C and Lake A and supporting surface water inputs into Lake A (see Section 2.1). Under such hydraulic conditions, the direction of groundwater flow into the bank is reversed, i.e., from Lake A to Lake B (Figure 2b).

Since the implementation of the BF system (on 3 October 2012), the relative surface water elevations of Lake A and Lake B have not been the only controlling factors on the direction and intensity of groundwater flow through the sandy bank. Drawdown of the water table in the vicinity of the active pumping wells induces an artificial hydraulic gradient, forcing surface water from both lakes (A and B) to infiltrate into the sandy bank and travel toward the pumping wells (Figure 2c). A schematic representation of the theoretical elevation difference between Lake A and Lake B is shown in Figure 2d. When analyzing the data from the monitoring program conducted by Ageos [35], we depicted three typical hydraulic conditions recurring each year. First, a high hydraulic gradient between Lake A and Lake B develops in response to the hydraulic connection between Lake A and Lake C (as explained above). Second, in summertime, the hydraulic connection between Lake A and Lake C stops and water demand increases. This leads to a moderate hydraulic gradient between the lakes. Finally, in wintertime, a low hydraulic gradient is expected, as surface water inputs into Lake A are very limited and municipal water demands are reduced. In sum, the lake dynamics and the pumping regimes both influence the relative surface water elevations of Lake A and Lake B and allow for a gradual transition from high (during springtime) to low (during wintertime) hydraulic gradient between the lakes.

3. Materials and Methods

3.1. Surface and Groundwater Sampling

Monitoring of surface water and groundwater was conducted on a monthly basis and included measurements of physico–chemical parameters and water sampling for geochemical analyses. Surface water sampling was performed near the shore (see location of lake sampling points LA-S and LB-S in Figure 1c). Additional sampling campaigns were conducted at Lake A (on 15 February 2017 at LA-P1 and LA-P2) and Lake B (on 9 September 2016 at LB-P1 and LB-P2 and on 3 March 2017 at LB-P3 and LB-P4) to assess for vertical heterogeneity. Physico–chemical parameters were measured along vertical profiles at 1 to 2 m intervals and water was sampled at multiple depths (e.g., 3 m, 7 m, and 12 m) with a submersible pump. Groundwater sampling was conducted at the pumping wells via a bypass faucet, as submersible pumps permanently regulate flow rate at each well. Water sampling was conducted at least 30 min after pumping started, allowing the stagnant water to be purged. In the case of observation wells, a submersible pump (WSP-12V-5 Tornado, Proactive Environmental Products, Bradenton, FL, USA) with a 30 m long polyvinyl chloride (PVC) tube was used and sampling was conducted after purging at least three well volumes and stabilizing the physico–chemical parameters.

Measurements of temperature, pH, electrical conductivity (EC), and redox potential (Eh) were performed with a multiparameter probe (YSI Pro Plus 6051030 and Pro Series pH/ORP/ISE and Conductivity Field Cable 6051030-1, YSI Incorporated, Yellow Springs, OH, USA) installed in an airtight cell connected to the pump. Samples for major ions and alkalinity were collected in 50 mL low-density polyethylene (LDPE) containers and were filtered through a 0.45 µm hydrophilic polyvinylidene fluoride (PVDF) membrane (Millex-HV, Millipore, Burlington, MA, USA) prior to analysis. Water samples were transported and stored at 4 °C. The same sampling and transport procedures were applied for total and dissolved metals analysis (Fe and Mn). Following on-site filtration, acidification with HNO_3 (in order to lower pH < 2) was performed in the laboratory within a 24 h delay.

3.2. Analytical Techniques

Major ion quantification was performed via either atomic absorption (Aanalyst 200 Atomic Absorption Spectrometer, Perkin Elmer, Waltham, MA, USA) or ion chromatography (ICS 5000 AS-DP

DIONEX Thermo Fisher Scientific, Saint-Laurent, QC, Canada) for all surface water samples and groundwater samples collected at observation wells, depending on the availability of the equipment. Total Fe and Mn concentrations were measured via atomic absorption for all surface water and observation wells samples. Inductively coupled plasma mass spectrometry was used for the quantification of major ions and total and dissolved Fe and Mn concentrations for the water samples collected at the pumping wells. The limit of detection (LOD) was 0.2 mg/L for all major ions and 0.01 mg/L or 0.05 mg/L for total and dissolved Fe and Mn, depending on the quantification method. For subsequent calculations and interpretations, all results ≤ LOD will be considered equal to LOD/2. Duplicates were analyzed to confirm the repeatability of the quantification methods. Bicarbonate concentrations were derived from alkalinity, which was measured manually in the laboratory according to the Gran method [36]. On samples with measured alkalinity (n = 98), the ionic balance errors were all below 10%. The mean and median ionic balance errors were 1% and the standard deviation was 3%.

3.3. Estimating Mixing Ratios

The mixing between two end-members can be quantified via a binary mixing model which can be described by the following equations:

$$f_A + f_B = 1, \qquad (1)$$

$$X_A f_A + X_B f_B = X_W, \qquad (2)$$

where f represents the fraction of the different sources and X the concentration (or value) of the tracer. A and B correspond to the two water sources, whereas W represents the water sampled at the well.

Tracer-based approaches can be used to estimate mixing ratios and travel times, as long as the tracer presents conservative or predictable behavior [37,38]. Various natural tracers, such as chloride (Cl$^-$), electrical conductivity (EC), and stable isotopes of waters (δ^{18}O-δ^2H), have been applied in numerous BF or alluvial aquifer contexts [39–44]. In this paper, we used EC values as a quantitative mass balance tracer for the application of the mixing model, with the assumption that it behaves conservatively. Violation of this assumption was unlikely at the study site, considering that the aquifer matrix is alluvial sands (mainly siliceous with no calcite). Good correlation (R^2 = 0.95) between EC values and Cl$^-$ (a conservative tracer) was also observed. The advantages of using EC instead of Cl$^-$ are that measurements can be done at a low-cost, as well as remotely and continuously.

4. Results and Discussion

4.1. Highly Transient Pumping Schemes

In this section, we (1) identify typical pumping schemes and (2) depict the seasonal variability of the total pumped volume.

Figure 3a–d shows the pumping rates for P1, P3, P5, and P6 over a typical one-week period (from 16 January 2017 to 23 January 2017). P1 was mainly active during daytime for 1–12 h (Figure 3a). A similar pumping scheme was applied to P2, P7, and P8 during summertime (data not shown). P3 and P6 were operated at rates ranging from 1000 m^3/day to 3000 m^3/day. Both were typically active on a daily basis, although P3 was turned off during night time (for less than 6 h) as water demand diminished (Figure 3b,d). P5 and P4 were typically activated on a monthly basis for monitoring and sampling procedures (Figure 3c). Three general pumping schemes emerged from this analysis of pumping rates and made it possible to distinguish three groups: (1) wells operated at nearly continuous rates (P3 and P6); (2) wells operated intermittently (P1, P2, P7 and P8); and (3) wells operated occasionally (P4 and P5).

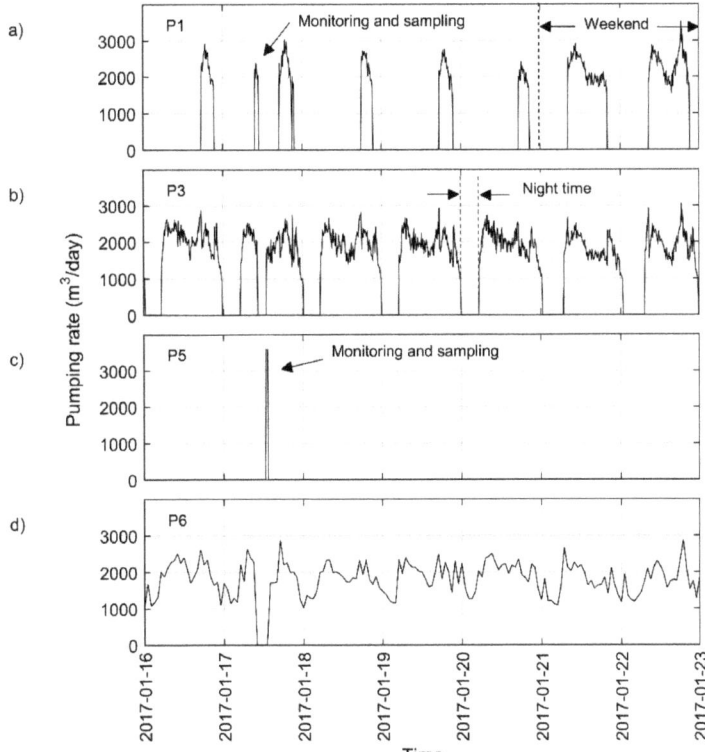

Figure 3. Pumping rates for wells (**a**) P1, (**b**) P3, (**c**) P5, and (**d**) P6 during a typical one-week period (from 16 January 2017 to 23 January 2016). Monitoring and water sampling were conducted on 17 January 2017 at all the pumping wells.

Figure 4 illustrates the monthly mean total pumping rate for all wells from March 2016 to March 2017. The mean pumping rate was about 4400 m^3/day, excluding summertime (May 2016 to September 2016), during which it was approximately 7000 m^3/day. Throughout most of the year, with the exception of summer months, 71% to 83% of the total daily pumped volume was provided by the continuously pumping wells. The intermittently pumping wells provided 16% to 29% of the pumped volume. The remaining volume (<1%) was supplied by the occasionally pumping wells. In summertime, pumping rates increased at all wells, except for P5. Continuously pumping wells were operated at mean rates of approximately 2000 m^3/day, representing from 52% to 63% of the total pumping rate. The intermittently pumping wells together supported 36% to 46% of the total pumped rate and the occasionally pumping wells supplied together the remaining 3%.

Over the study period, the total pumped volume fluctuated daily and seasonally to accommodate the municipal water demand. Indeed, higher pumping rates prevailed during (1) mornings and evenings, (2) weekends, and (3) summertime. This well field is typically operated with a hierarchical system, giving priority to the continuously pumping wells. If the water demand increases, intermittently pumping wells are subsequently activated. Lastly, the occasionally pumping wells can be solicited. This implies that anywhere from one to eight pumping wells were solicited to fulfill the water demand and accommodate for the daily and seasonal water demand fluctuations.

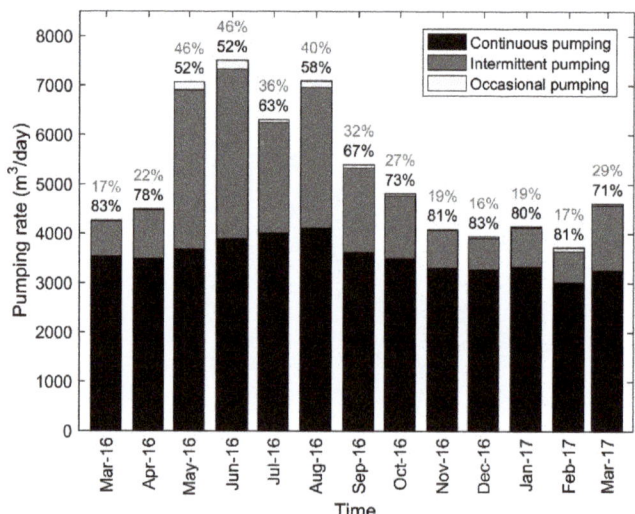

Figure 4. Monthly mean total pumping rate from March 2016 to March 2017. Above each bar are the proportions of the total pumped volume supplied by the continuously (in black) and intermittently (in grey) pumping wells. The occasionally pumping wells supply only <1–2% of the total pumped volume.

4.2. Geochemistry as a Proxy of the Hydrosystem Dynamics

The objective of this section was to examine the geochemistry of Lake A, Lake B, regional groundwater, and the bank filtrate in order to identify the contributing water sources to the pumping wells.

Box plots of the temperature, EC, pH, and Eh at Lake A (<1 m depth), Lake B (<1 m depth), the pumping wells, and the observation wells Z12, Z15, and Z16 are illustrated in Figure 5. Concerning Lake A and Lake B, note that the presented data correspond to measurements at the surface of the lakes (i.e., <1 m depth). Hence, the medians and the 25th and 75th quartiles values may not be representative of the entire water column. Observed temperatures at Lake A and Lake B ranged from 1.3 °C to 27.5 °C and from 3.9 °C to 27.5 °C, respectively. For the pumping wells, box plots are spatially sorted (P6; P1; P8; P2; P7; P3; P4; P5) from the northwest to the southeast ends of the well field (see location of the pumping wells in Figure 1c). Temperatures ranged from 3.4 °C to 16.2 °C, with minimum and maximum values being observed in occasionally and continuously pumping wells, respectively. EC values at the pumping wells ranged from 491 µS/cm (at P5) to 895 µS/cm (at P8), which is in between observed EC values in Lake A and Lake B. Note that the EC values for Lake A and Lake B in Figure 5 are associated with water sampled at <1 m depth. Higher EC values were measured in situ in Lake B at >12 m deep (further details in Section 4.3). Observed EC values at Z12 were similar to those in Lake B, whereas Z15 showed lower values, similar to Lake A. The highest EC values were observed at Z16, which is representative of regional groundwater. Measured pH values at the pumping wells tended to increase spatially from NW to SE (P6 to P5). Redox conditions also varied spatially and decreased from NW to SE. A H_2S odor was noticed when sampling at P4, P5, Z15, and Z16, which is consistent with Eh measurements that indicate more reduced conditions.

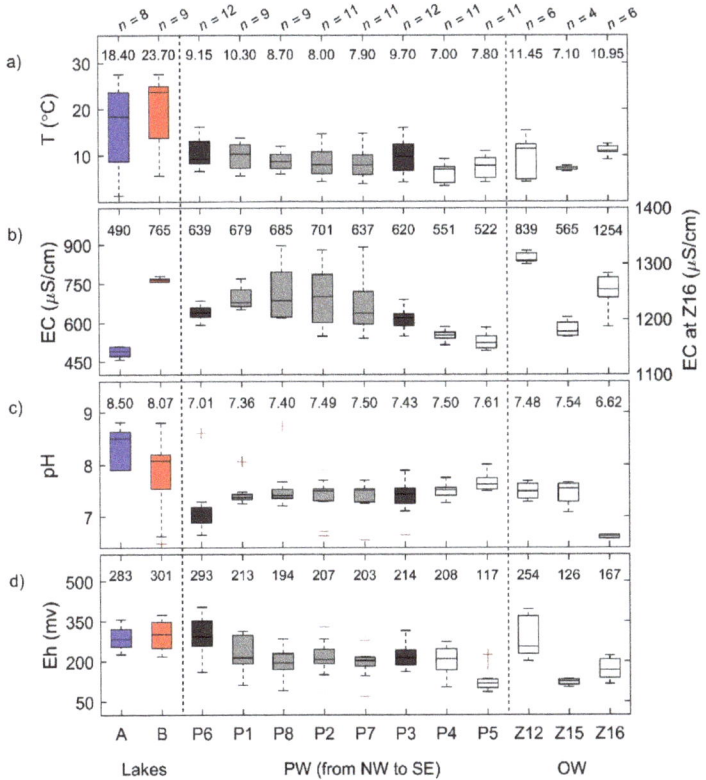

Figure 5. Boxplots of (**a**) temperature, (**b**) electrical conductivity (EC), (**c**) pH, and (**d**) redox potential (Eh) at the lakes, pumping wells (PW), and observation wells (OW). Blue and red boxes are associated with Lake A (<1 m depth) and Lake B (<1 m depth), whereas dark, medium, and light grey boxes correspond to continuously, intermittently, and occasionally pumping wells, respectively. Numeric values above each box correspond to the median.

Figure 6 shows the spatial variability of total Fe and Mn concentrations at Lake A (<1 m depth), Lake B (<1 m depth), the pumping wells, and the observation wells Z12, Z15, and Z16. Concentrations in total Fe ranged from <0.01 mg/L to 1.28 mg/L at the pumping wells, with median concentrations increasing from NW to SE. Median total Fe concentrations at P4 and P5 were high relative to Canada's aesthetic objective for total Fe in drinking water (i.e., 0.3 mg/L) [45]. Analyses also reported high concentrations (from 0.05 mg/L to 2.12 mg/L) at Z15 (near P5). The highest total Fe concentrations were observed at Z16. Total Mn concentrations ranged from 0.1 mg/L to 1.3 mg/L at the pumping wells, which exceeded the aesthetic objective for total Mn in drinking water in Canada (i.e., 0.02 mg/L) [46]. The highest concentrations were measured at the intermittently pumping wells. Total Mn concentrations at the surface of Lake A and Lake B were relatively low (i.e., typically ≤0.03 mg/L). However, it is important to note that 1.06 mg/L was observed at 6 m depth in Lake B (see red circle in Figure 6b). This result highlights that total Mn concentrations may be more important at greater depths in Lake B. Release of Fe and Mn to the water column in Lake A may potentially occur from sand dredging activities, as this lake is still actively mined for sand. However, no data are available to discuss the evolution of Fe and Mn concentrations in relation to these anthropic activities. Dissolved Fe concentrations were all <0.05 mg/L, while dissolved Mn concentrations were similar to the total ones.

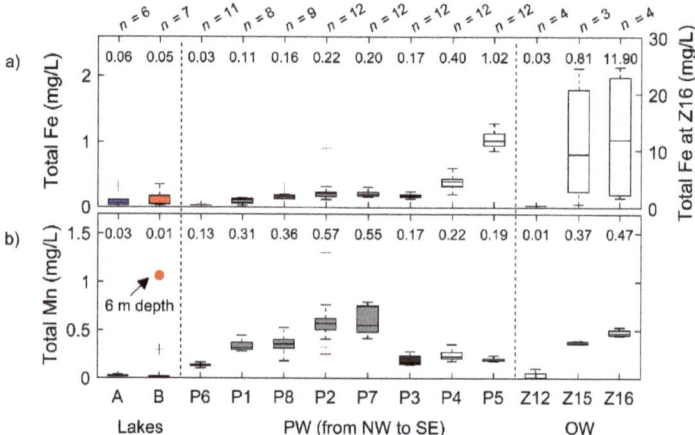

Figure 6. Boxplots of (**a**) total Fe, and (**b**) total Mn concentrations at the lakes, pumping wells, and piezometers. Blue and red boxes are associated with Lake A (<1 m depth) and Lake B (<1 m depth), whereas dark, medium, and light grey boxes correspond to continuously, intermittently, and occasionally pumping wells, respectively. Numeric values above each box correspond to the median. Red circle represents the maximal observed total Mn in Lake B (at 6 m depth).

Figure 7 shows the relationship between $(Ca^{2+} + Mg^{2+})/Na^+$ and cationic content (i.e., sum of major cations) for Lake A, Lake B, pumping wells, and regional groundwater. Potassium (K^+) was excluded from these calculations since only a few samples were analysed for K^+ and concentrations in K^+ only represent a small fraction of the total cation content (i.e., approximately 1.5%). Lake A and Lake B samples are plotted in opposing regions of the graph, Lake A having a high $(Ca^{2+} + Mg^{2+})/Na^+$ ratio and low cationic content and Lake B having a low $(Ca^{2+} + Mg^{2+})/Na^+$ ratio and high cationic content. Concerning the samples from the pumping wells, they are mostly plotted in the area extending from the Lake A to Lake B regions. Occasionally pumping wells had a geochemical signature similar to Lake A, whereas continuously and intermittently pumping wells spread between both lake signatures. Regional groundwater samples were sampled from one observation well, namely Z16, located on the NE side of Lake B (see Figure 1c). These samples were characterized by the lowest $(Ca^{2+} + Mg^{2+})/Na^+$ ratios and highest cationic content. It is believed that direct contribution to the pumping wells from regional groundwater is not likely at this site, due to the hydrogeological context (see Section 2.1). Hence, we hypothesize that the spreading of pumping well samples relative to the Lake A–Lake B mixing line is potentially due to an indirect contribution from regional groundwater, which discharged into Lake B. Only three wells (i.e., P2, P7, and P8) were affected from November 2016 to February 2017. During this period, the three wells together supplied <10% of the total pumped volume. Based on these observations, we propose that the mixing between Lake A and Lake B is the dominant process governing the geochemical facies of the pumping wells.

4.3. EC Time-Varying Mixing Model

It was discussed in Section 4.2 that the geochemical facies at the pumping wells are controlled by mixing between Lake A and Lake B. Hence, we used the binary mixing model of Equations (1) and (2) to estimate the relative contributions of each lake to the pumping wells. In this section, we first present the temporal and vertical variability of EC at Lake A and Lake B in order to define the end-member values. Then, estimations of the mixing ratios are evaluated with respect to a reference scenario, and spatiotemporal evolution is discussed. We also provide a sensitivity analysis, which helps strengthen the conclusions of the model.

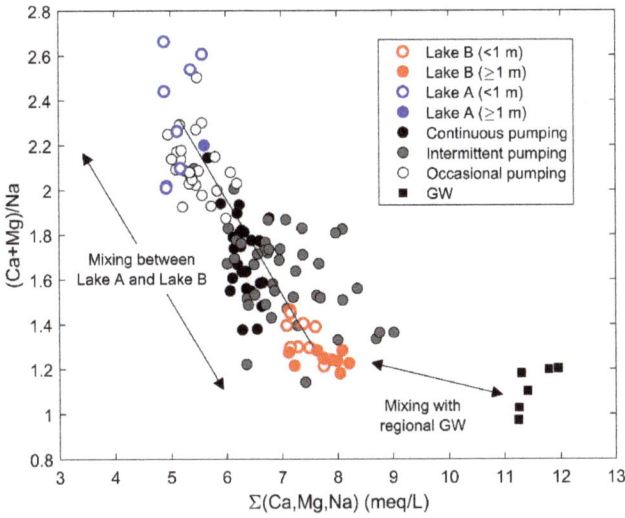

Figure 7. Comparison of the geochemical facies of Lake A, Lake B, pumping wells, and the regional groundwater (GW). The solid black line represents the mixing line between Lake A and Lake B mean values.

4.3.1. Temporal and Vertical EC Variability at Lake A and Lake B

Temporal variability in EC values at the surface (<1 m depth) of Lake A and Lake B are illustrated in Figure 8a,b. At the surface of Lake A, minimal and maximal EC values were observed in springtime and wintertime, respectively. Low EC values were expected for springtime, since it corresponds to the period of hydraulic connection between Lake A and Lake C. During this period, surface water with low EC is discharged into Lake A from streams S1 and S2 with inverted flow direction (see Figure 1b). In Lake B, EC values at the surface (<1 m depth) were also found to be variable over time. During springtime and summertime, EC values were relatively constant. A significant increase in EC values was observed in autumn–winter. Figure 8b also depicts the EC time series at an observation well (namely Z12) which was located between pumping well P1 and Lake B. It is screened at the bottom of the aquifer over a 9.14 m long section. EC measurements at Z12 are thus representative of the mixing between multiple flow lines originating from various depths in Lake B. The mean EC value at Z12 was 848 µS/cm and values were typically higher than at the surface of Lake B. These results reveal that (1) the EC measurements at the surface of Lake B are not representative of the infiltrating water and (2) considering the vertical variability in EC in Lake B is important.

Figure 9 shows vertical EC profiles for Lake A and Lake B. EC was measured at depth in winter (on 15 February 2017) at Lake A and in summer and winter (on 9 September 2016 and 2 March 2017) at Lake B. For each campaign, at least two vertical profiles were conducted in order to assess the horizontal variability (see Figure 1c for location of the vertical profiles). Maximal EC differences (at the same depth) were 6 µS/cm and 25 µS/cm for Lake A and Lake B, respectively, suggesting no significant horizontal variability at both lakes.

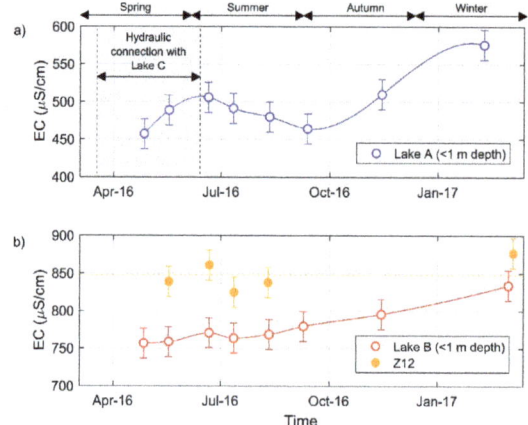

Figure 8. Time series of electrical conductivity (EC) at (**a**) Lake A (<1 m depth), and (**b**) Lake B (<1 m depth) and Z12. The grey shaded area represents the timing for hydraulic connection between Lake A and Lake C. Error bars (±20 µS/cm) represent the maximal expected measurement error on EC. The yellow dashed line corresponds to the mean EC value at Z12.

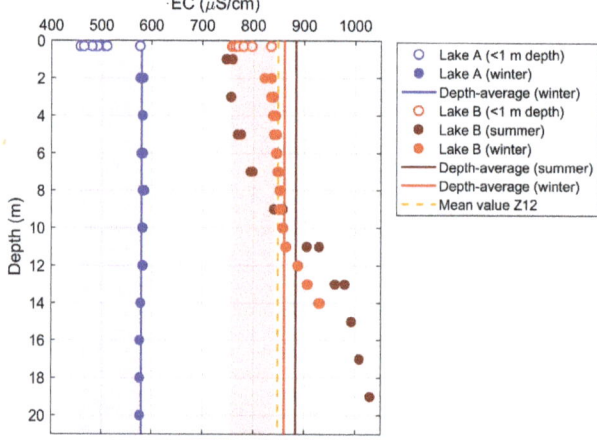

Figure 9. Electrical conductivity (EC) measurements against depth at Lake A and Lake B in winter and summer. Solid lines represent depth-average value at the lakes, while the yellow dashed line is associated with mean EC values at observation well Z12. Blue and red shaded areas illustrate the variability in EC at the surface (<1 m depth) of Lake A and Lake B, respectively.

Concerning Lake A, no significant vertical variability in EC values was observed. This suggests that Lake A was vertically well mixed in wintertime. Given that Lake A receives surface water from a stream and that some industrial activity (i.e., sand dredging) takes place in the lake during the ice-free period (typically form early May to late October), it is likely that some currents in Lake A are stimulating mixing of the water column. Hence, we assumed that Lake A is fully mixed and does not develop any significant vertical EC stratification over a hydrological year. Temporal variability in EC at the surface of Lake A (at <1 m depth) is presumably representative of the evolution of the entire water body.

At Lake B, observed EC values increased with depth for all vertical profiles. Higher EC at greater depth could be induced by regional groundwater inputs into Lake B. In Canada, groundwater inputs are typically found at the bottom of lakes, due to thermal (and density) contrast [47]. Smaller vertical variability was observed in wintertime (in comparison to summertime). However, both summertime and wintertime depth-averaged values were similar (861 µS/cm and 884 µS/cm; a difference of 23 µS/cm being barely significant). It is important to note that the wintertime vertical profiles were conducted in a shallower zone of the lake and could explain the discrepancy between the depth-averaged values. Additionally, it is interesting to note that the depth-averaged EC values were similar to the Z12 mean EC value. This suggests that the depth-averaged EC value was adequate to depict the EC signal originating from Lake B.

4.3.2. Reference Scenario

Based on the assumption that Lake A is well mixed, we considered that EC measurements at <1 m depth were representative of the Lake A end-member. Hence, the Lake A end-member is a time-varying EC signal. Temporal interpolation between discrete measurements was done with the cubic spline method (using the *spline* function in MATLAB). The result of this calculation is represented in Figure 8a by the solid blue line and gives the best available estimate of the Lake A end-member from 27 April 2016 to 9 February 2017. No temporal shifting was considered, since travel times were expected to be much smaller than the observed changes in EC in Lake A. Concerning Lake B, a constant value of 873 µS/cm (i.e., mean of the wintertime and summertime depth-average values) was considered to correctly represent this second end-member. There was no need to consider temporal variation for Lake B end-member as both depth-averaged values were found to be similar.

The results of the mixing model are shown in Figure 10. By considering the relative pumping rates and the estimated contributions from Lake A at each well, we calculated that 62% of the annual pumped volume originated from Lake A. The continuously pumping wells are characterized by 54% to 78% of water originating from Lake A, with the highest contributions from Lake A occurring from April to July, i.e., during the highest hydraulic gradient period. The lowest contributions from Lake A were observed from July to September, i.e., during the moderate hydraulic gradient period. This is likely related to an increase in the total pumped volume during summertime (see Section 4.1).

The intermittently pumping wells showed the widest distribution in mixing ratios, with contributions from Lake A ranging from 0% to 87%. Similar to the continuously pumping wells, the highest contributions from Lake A were observed during the high hydraulic gradient period, while its contribution decreased as the hydraulic forcing became less important. It was estimated that during the low hydraulic gradient period, the fraction of Lake B can reach up to 100% for the intermittently pumping wells. In fact, the mixing model yields such mixing ratios when the measured EC at the pumping wells is greater or equal to the Lake B end-member (i.e., 873 µS/cm). This condition was observed four times and EC measurements at the concerned pumping wells ranged from 879 µS/cm to 895 µS/cm. However, expectations were that Lake A and Lake B would always contribute to the pumping wells, since a radial depression cone normally develops in the vicinity of an active pumping well, forcing water to infiltrate from both sides of the sandy bank. Hence, we considered that a calculated contribution of 100% of Lake B depicts a limit of the developed mixing model. In reality, this result could be an indication that, during winter, water preferentially infiltrates from the bottommost zone (≥12 m) of Lake B (where EC >873 µS/cm), leading to higher EC values at the intermittently pumping wells. Controls on the development of such preferential flow paths are not within the scope of this paper and, thus, will not be further discussed. Future work concerning the spatiotemporal variability of the hydraulic conductivity is still needed to draw any conclusions on this topic. The combination of various environmental tracers would help to differentiate between contributions from the surface and bottommost zones of the lakes.

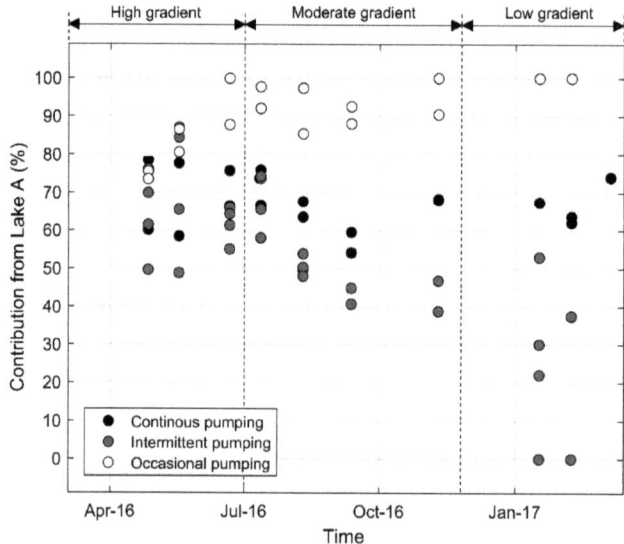

Figure 10. Estimated contribution from Lake A to the pumping wells according to the reference scenario. Lake A end-member is defined as a time-varying electrical conductivity (EC) signal, which was derived from the observed EC values at <1 m depth. Lake B end-member is a fixed EC value which corresponds to the mean of the wintertime and summertime depth-average value.

Contribution from Lake A at the occasionally pumping wells is typically >90%. However, in April and May, it was estimated that the former wells were receiving a relatively smaller contribution from Lake A (from 74% to 86%). This result possibly reflects that pore water with EC >500 µS/cm and originating from Lake A could have been stored during winter within the sandy bank and pumped only in April and May. Hence, mean residence time of the water in the sandy bank could reach months when the pumping wells are not active. Greater attenuation of the surface water temperature signal, at the occasionally pumping wells, also testifies to longer residence times. We thus highlighted the need for considering the variability of the residence time of water into the sediments when applying a time-variant mixing model.

In sum, contribution from Lake A is typically greater than Lake B throughout the year. However, the mixing ratios are temporally and spatially variable. Strong variability was found especially during the period of low hydraulic gradient at the intermittently and occasionally pumping wells. In such a context, the pumping regime seems to have a decisive impact on the mixing ratios and, ultimately, on water quality of the bank filtrate. Under high hydraulic gradient the mixing ratios tend to be more similar, regardless of the pumping regimes.

4.3.3. Sensitivity Analysis

A sensitivity analysis was conducted in order to investigate (1) the representativity of EC end-members values and (2) the uncertainties related to the EC measurements. Mixing ratios were therefore recalculated according to various scenarios where Lake A and Lake B end-member values varied from 458 µS/cm (i.e., minimal observed value) to 576 µS/cm (i.e., maximal observed value) and from 824 µS/cm to 924 µS/cm (i.e., reference scenario ±50 µS/cm), respectively. Also, a variation of ±40 µS/cm was applied to all the EC measurements at the pumping wells. Differences between the results of the scenarios were typically <10%, except for the ones concerning the Lake A end-member. When considering fixed EC values for the Lake A end-member, the estimation of the mixing ratios diverged up to 30% compared to the reference scenario. This result helped to quantify the importance of considering the temporal variability of the end-members to obtain reliable results when estimating mixing ratios. Despite the sensitivity of the model to Lake A end-member variability, general trends for mixing ratios were conserved for all the scenarios. This result was expected, as the mixing model was linear. Overall, the sensitivity analysis revealed that the relative estimations of mixing ratios were acceptable and that measurement errors were not likely to influence our conclusions.

The temporal resolution of the applied monitoring program did not allow for discussion of the short-term (i.e., hourly to daily) EC variability. Hence, it is not clear whether hourly variations in pumping rates could influence the observed EC values at the pumping wells.

4.4. Dominant Controls on the Origin of the Bank Filtrate

The time-variant binary mixing model highlights that the contribution of Lake A to the bank filtrate can vary from 0% to 100%. This section aims to understand the competing roles of anthropic and meteorological forcings on the origin of the bank filtrate. We define anthropic forcing as a process via which the origin of the bank filtrate at a given well is affected by its own pumping scheme and/or rate. Anthropic forcing can also occur indirectly, as drawdown of the water table in the vicinity of a given pumping well can influence the origin of water at less active adjacent pumping wells. Meteorological forcing is considered a natural process. Concerning our study site, the surface elevations of Lake A and Lake B showed seasonal variations, which are mainly controlled by meteorological conditions allowing or limiting surface water inputs into Lake A. Hence, we considered that the hydraulic gradient between Lake A and Lake B is meteorological forcing acting on the BF system.

Figure 11a,b shows EC against the one-month average pumping rate prior to the sampling date. Distinction between the pumping regimes (i.e., continuous, intermittent and occasional) is illustrated in a, while hydraulic gradients between Lake A and Lake B (i.e., high, moderate, low) are represented in b. Figure 11c is a schematic representation of the dominant forcing in relation to the different hydraulic contexts. First, in Figure 11a,b, we observed little variability in EC measurements if the pumping rate was >1000 m^3/day. In fact, most samples associated with high pumping rates (Figure 11a) showed EC values ranging from 583 µS/cm to 689 µS/cm, despite the variability of the hydraulic gradient (Figure 11b). Hence, for high pumping rates (i.e., >1000 m^3/day), it appears that anthropic forcing is dominant over the meteorological forcing (Figure 11c). However, two samples (see downward arrow in Figure 11c) showed lower EC while being operated at >1500 m^3/day. These exceptions were observed exclusively when the hydraulic gradient between Lake A and Lake B was maximal (i.e., in May and June). Under such hydraulic conditions, the anthropic forcing cannot counteract the meteorological forcing, resulting in an increase in the contribution from Lake A during springtime. Second, higher EC values (>750 µS/cm) are associated with the intermittent pumping regime, while low EC values (<600 µS/cm) are mostly related to the occasionally pumping regime (Figure 11a). For these two pumping regimes, meteorological forcing was clearly dominant over the anthropic one (Figure 11c). In fact, high EC values were strictly observed during the low hydraulic gradient period (Figure 11b). In short, this revealed that when a pumping rate of approximately 1000 m^3/day is applied continuously, the mixing ratios are less variable due to direct anthropic forcing. When wells are operated only intermittently or occasionally, indirect anthropic and/or meteorological forcings control the mixing between Lake A and Lake B waters.

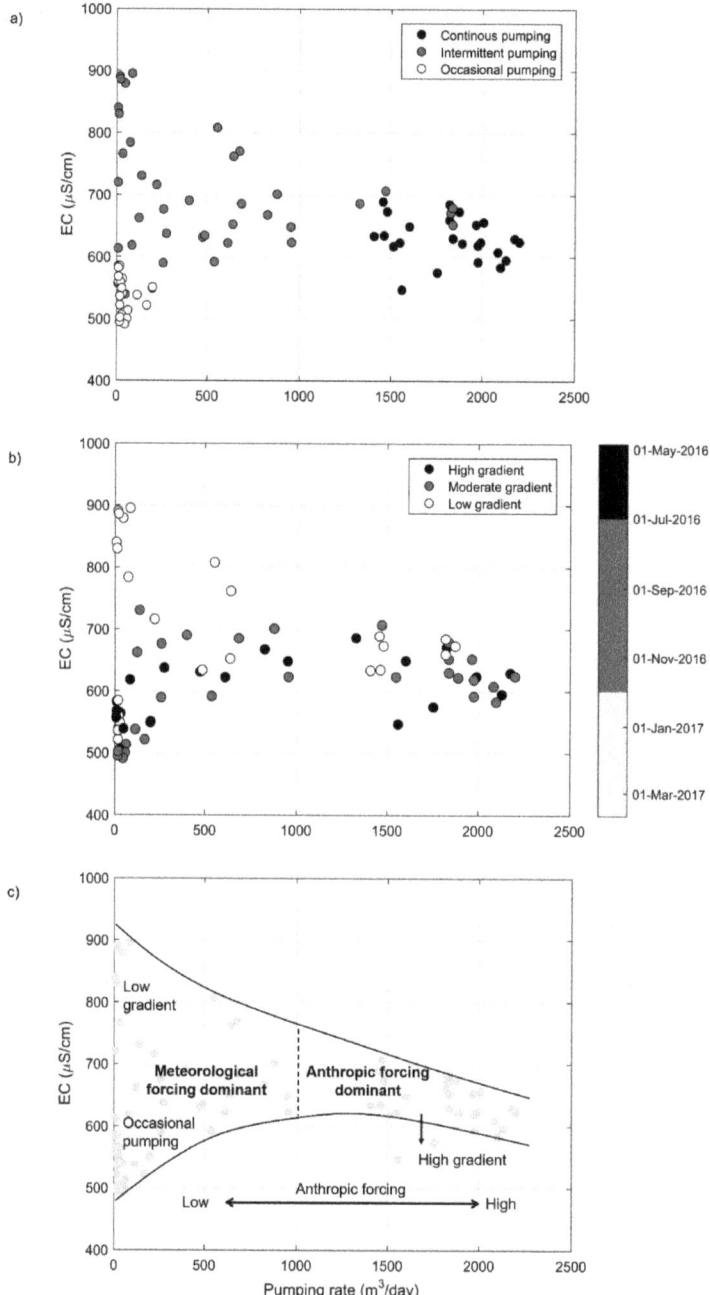

Figure 11. Relationship between electrical conductivity (EC) and the one-month average pumping rate prior to the sampling date, according to (**a**) the pumping regime and (**b**) the hydraulic gradient between Lake A and Lake B. A schematic representation of the dominant forcing is illustrated in (**c**), where solid and dashed lines represent the range of observed values and the delimitation between regimes where meteorological and anthropic forcings are dominant, respectively.

4.5. Implications for the Quality of the Bank Filtrate

In Section 4.2, we highlighted that geochemical analyses showed spatial variability in both total Fe and Mn concentrations at the pumping wells. This section aims to discuss the relationship between total Fe and Mn concentrations and the origin of water.

High total Fe concentrations were found at the occasionally pumping wells (i.e., at P5 and, to a lesser extent, at P4) and were associated with the highest contributions from Lake A (see Figure 10) and more reduced conditions (see Figure 5d). In comparison to the more anthropized section of the BF system, the residence times of the infiltrating water in the vicinity of the occasionally pumping wells are likely to be longer, because meteorological forcing alone is controlling groundwater flows (see Section 4.4). Since relatively low temperatures were observed at P4 and P5, it is also likely that higher viscosity, resulting in lower hydraulic conductivity, was responsible for longer residence times of the bank filtrate in the vicinity of these wells [48–50]. The longer residence times are potentially responsible for the high total Fe concentrations at P4 and P5. Evolution of redox conditions (from oxic to anoxic) is typically observed at BF systems [51] and can result in the dissolution of iron and/or manganese along flow paths [52]. However, as dissolved Fe concentrations are very low (i.e., generally <LOD), total Fe is controlled predominantly by the particulate fraction. Hence, it is more likely that the high rate and/or occasional pumping are causing the mobilization and resuspension of particulate Fe at P4 and P5. In fact, when activated for monthly sampling and monitoring, P4 and P5 typically operate at 150 m^3/h, while the other wells operate at lower rates (i.e., from 40 m^3/h to 125 m^3/h). Moreover, P5 was the only pumping well equipped with a 4 m long screened section (i.e., half of those of the other wells). The mean effective velocity of water entering P4 and P5 screens was from 2 to 4 times greater than at the other wells. The total Fe concentration at P4 and P5 could thus potentially be reduced by lowering the hourly mean pumping rates and operating on a daily basis. However, such engineered operational strategy would not help to lower the total Fe concentration to <0.2 mg/L (see Figure 12a).

The highest total Mn concentrations were concomitant with the highest fraction of Lake B water at the intermittently pumping wells. The presence of total Mn in the raw water could also be explained by the evolution of redox conditions along the flow path. Besides this, an elevated concentration in total Mn (1.06 mg/L) was measured at a 6 m depth in Lake B. As the latter was found to be geochemically stratified, relatively reduced conditions can develop in the epilimnion and promote the solubilization of Mn. Hence, it is also likely that Mn reaches the pumping wells by advective transport with water originating from the deeper zones of Lake B. Further investigation is needed to better understand the site-specific drivers of the Mn occurrence in the bank filtrate, since its mobility is controlled by numerous factors, such as travel times, temperature, pH, microbial activity, the extent of a clogging layer, and the degree of oxygen consumption [52]. Figure 12b illustrates the relationship between total Mn concentrations and the one-month average pumping rate prior to the sampling date. Total Mn concentrations decrease with higher pumping rates (for intermittently and continuously pumping wells). This suggests that pumping rate can be used as an operational tool to control the total Mn concentration in the pumped water.

In sum, high total Fe and Mn concentrations in the pumped water are governed by two distinct processes. Total Fe seems to originate from particulate iron mobilization and resuspension when effective velocities of water entering the screens of the pumping wells are high, whereas high total Mn concentrations seem to be associated with an increase in the contribution from the bottom of Lake B. Total Fe and Mn concentrations could potentially be regulated by lowering the mean effective velocity of water entering the screens and adjusting mixing ratios (i.e., by operating at adequate pumping rates).

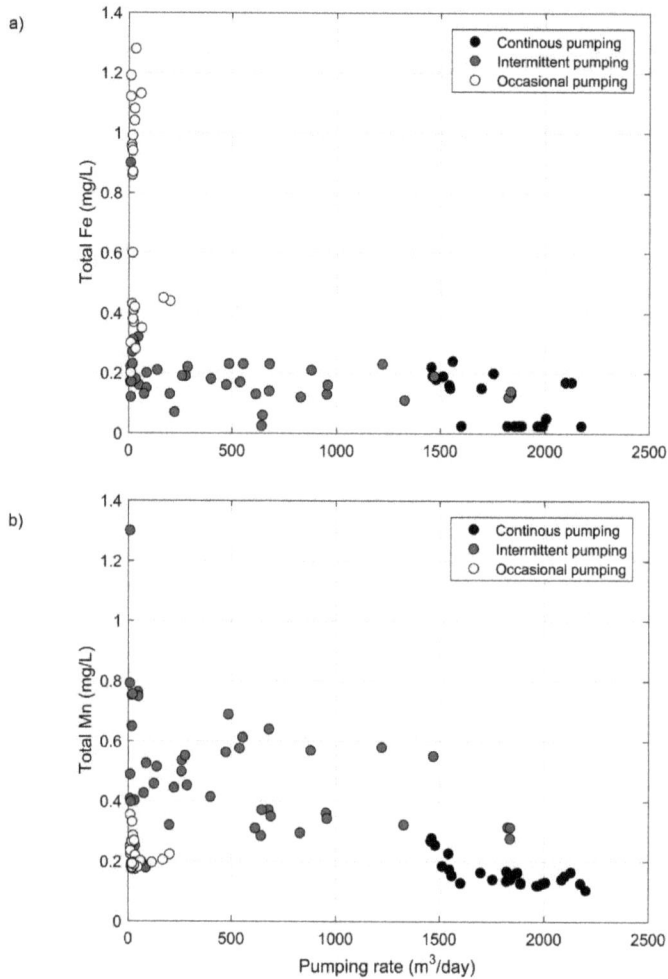

Figure 12. Relationship between (**a**) total Fe and (**b**) total Mn concentrations and the 1-month average pumping rate prior to the sampling date.

5. Conclusions

In this study, we demonstrated the controls of variable meteorological conditions and pumping schemes on the origin and quality of bank filtrate. Through a pumping rate analysis, the pumping schemes could be separated into three categories, namely the continuously, intermittently, and occasionally pumping wells. The continuously pumping wells (i.e., P3 and P6) supported 71% to 83% of the total pumping rate, except in summertime, when they contributed from 52% to 63%, of the total pumping rate as it increased from approximately 4000 m³/day to 7500 m³/day. An investigation of the geochemical facies of Lake A, Lake B, regional groundwater, and the bank filtrate revealed that the geochemistry of the pumped water is governed by mixing of Lake A and Lake B. Therefore, a two end-member mixing model was developed to estimate the contribution from both lakes to the pumping wells over a one-year period. To this end, EC measurements were used as a quantitative environmental tracer. A time-varying EC signal was considered for the Lake A end-member, whereas a fixed EC value was used to depict the Lake B end-member. This simple mixing model revealed the following:

- By considering the relative pumping rates and the estimated contributions from Lake A at each well, it was estimated that 62% of the annual pumped volume originates from Lake A;
- All the pumping wells typically receive >50% of water from Lake A, but the competition between anthropic (i.e., pumping regime) and meteorological forcings (i.e., relative water level of both lakes) leads to a large variability of the mixing ratios (i.e., from 0% to 100% of water originating from Lake A);
- When the meteorological forcing is high, the pumping regime has little influence over the origin of water and the mixing ratios are similar at all the pumping wells. When the meteorological forcing is low, the pumping regime is a decisive factor on the fraction of the contributing sources to the pumping wells;
- When a pumping rate of >1000 m^3/day is applied continuously, the mixing ratios are less variable due to direct anthropic forcing. When wells are operated only intermittently or occasionally and at a rate of <1000 m^3/day, indirect anthropic and/or meteorological forcings govern the mixing ratio between Lake A and Lake B waters;
- A sensitivity analysis revealed that the relative estimation of the mixing ratios was acceptable and that measurement errors were not likely to influence our calculations. It also helped to quantify the importance of considering the temporal variability of the lakes' end-members to obtain reliable results when estimating mixing ratios;
- The pumping regime influences total metals (i.e., Fe and Mn) concentrations in the raw abstracted waters. High Fe concentrations originate from particulate iron mobilization and resuspension when effective velocities of water entering the screens of the pumping wells are high, whereas high Mn concentrations are associated with an increase in the contribution from Lake B.

This study highlights how understanding the competition between anthropic and meteorological forcings can help to recommend guidelines for rapid decision-making regarding the quality of the pumped water. For instance, by identifying contexts for which the anthropic forcing is dominant, one can control the origin of the bank filtrate. Moreover, predicting periods under which the meteorological forcing is governing the flow patterns can help to adjust post-BF treatment in order to secure high quality of the distributed drinking water.

Author Contributions: Conceptualization, J.M.-D., P.B. and F.B.; Data curation, J.M.-D.; Formal analysis, J.M.-D.; Funding acquisition, P.B.; Investigation, J.M.-D., M.P., C.P., F.P.-B., M.M. and S.V.; Methodology, J.M.-D., P.B. and F.B.; Project administration, J.M.-D. and P.B.; Supervision, P.B., F.B., P.P. and B.B.; Visualization, J.M.-D.; Writing—original draft, J.M.-D.; Writing—review and editing, P.B., F.B., P.P. and B.B.

Funding: This research was funded by NSERC, grant numbers CRSNG-RDCPJ: 523095-17 and CRSNG-RGPIN-2016-06780.

Acknowledgments: The authors gratefully acknowledge the Town and M. Rybicki to allow access and water sampling on their property. The authors would also like to thank the students (A. Siméon, T. B. Quézel, T. Crouzal, D. Dufresne, J.-S. Grenier, just to name a few) who participated in the fieldwork. For the laboratory work, we would like to thank M. Leduc and J. Leroy from the Laboratoire de géochimie de Polytechnique Montréal.

Conflicts of Interest: The authors declare no conflict of interest. The funders had no role in the design of the study; in the collection, analyses, or interpretation of data; in the writing of the manuscript, or in the decision to publish the results.

References

1. Haas, R.; Opitz, R.; Grischek, T.; Otter, P. The AquaNES project: Coupling riverbank filtration and ultrafiltration in drinking water treatment. *Water* **2018**, *11*, 18. [CrossRef]
2. Ross, A.; Hasnain, S. Factors affecting the cost of managed aquifer recharge (MAR) schemes. *Sustain. Water Resour. Manag.* **2018**, *4*, 179–190. [CrossRef]

3. Gillefalk, M.; Massmann, G.; Nutzmann, G.; Hilt, S. Potential impacts of induced bank filtration on surface water quality: A conceptual framework for future research. *Water* **2018**, *10*, 1240. [CrossRef]
4. Hiscock, K.M.; Grischek, T. Attenuation of groundwater pollution by bank filtration. *J. Hydrol.* **2002**, *266*, 139–144. [CrossRef]
5. Gunkel, G.; Hoffmann, A. Bank filtration of rivers and lakes to improve the raw water quality for drinking water supply. In *Water Purification*; Gertsen, N., Sonderby, L., Eds.; Nova Science Publishers, Inc.: Hauppauge, NY, USA, 2009; pp. 137–169.
6. Ronghang, M.; Gupta, A.; Mehrotra, I.; Kumar, P.; Patwal, P.; Kumar, S.; Grischek, T.; Sandhu, C. Riverbank filtration: A case study of four sites in the hilly regions of Uttarakhand, India. *Sustain. Water Resour. Manag.* **2019**, *5*, 831–845. [CrossRef]
7. Dash, R.R.; Prakash, E.V.P.B.; Kumar, P.; Mehrotra, I.; Sandhu, C.; Grischek, T. River bank filtration in Haridwar, India: Removal of turbidity, organics and bacteria. *Hydrogeol. J.* **2010**, *18*, 973–983. [CrossRef]
8. Sahu, R.L.; Dash, R.R.; Pradhan, P.K.; Das, P. Effect of hydrogeological factors on removal of turbidity during river bank filtration: Laboratory and field studies. *Groundw. Sustain. Dev.* **2019**, *9*, 100229. [CrossRef]
9. Harvey, R.W.; Metge, D.W.; LeBlanc, D.R.; Underwood, J.; Aiken, G.R.; Butler, K.; McCobb, T.D.; Jasperse, J. Importance of the colmation layer in the transport and removal of Cyanobacteria, viruses, and dissolved organic carbon during natural lake-bank filtration. *J. Environ. Qual.* **2015**, *44*, 1413–1423. [CrossRef]
10. Romero, L.G.; Mondardo, R.I.; Sens, M.L.; Grischek, T. Removal of Cyanobacteria and cyanotoxins during lake bank filtration at Lagoa do Peri, Brazil. *Clean Technol. Environ. Policy* **2014**, *16*, 1133–1143. [CrossRef]
11. Otter, P.; Malakar, P.; Sandhu, C.; Grischek, T.; Sharma, S.K.; Kimothi, P.C.; Nuske, G.; Wagner, M.; Goldmaier, A.; Benz, F. Combination of river bank filtration and solar-driven electro-chlorination assuring safe drinking water supply for river bound communities in India. *Water* **2019**, *11*, 122. [CrossRef]
12. Massmann, G.; Greskowiak, J.; Dunnbier, U.; Zuehlke, S.; Knappe, A.; Pekdeger, A. The impact of variable temperatures on the redox conditions and the behaviour of pharmaceutical residues during artificial recharge. *J. Hydrol.* **2006**, *328*, 141–156. [CrossRef]
13. Hamann, E.; Stuyfzand, P.J.; Greskowiak, J.; Timmer, H.; Massmann, G. The fate of organic micropollutants during long-term/long-distance river bank filtration. *Sci. Total Environ.* **2016**, *545–546*, 629–640. [CrossRef] [PubMed]
14. Grunheid, S.; Amy, G.; Jekel, M. Removal of bulk dissolved organic carbon (DOC) and trace organic compounds by bank filtration and artificial recharge. *Water Res.* **2005**, *39*, 3219–3228. [CrossRef] [PubMed]
15. Nagy-Kovacs, Z.; Laszlo, B.; Fleit, E.; Czihat-Martonne, K.; Till, G.; Bornick, H.; Adomat, Y.; Grischek, T. Behavior of organic micropollutants during river bank filtration in Budapest, Hungary. *Water* **2018**, *10*, 1861. [CrossRef]
16. Dragon, K.; Gorski, J.; Kruc, R.; Drozdzynski, D.; Grischek, T. Removal of natural organic matter and organic micropollutants during riverbank filtration in Krajkowo, Poland. *Water* **2018**, *10*, 1457. [CrossRef]
17. Greskowiak, J.; Prommer, H.; Massmann, G.; Nutzmann, G. Modeling seasonal redox dynamics and the corresponding fate of the pharmaceutical residue phenazone during artificial recharge of groundwater. *Environ. Sci. Technol.* **2006**, *40*, 6615–6621. [CrossRef]
18. Burke, V.; Greskowiak, J.; Asmuss, T.; Bremermann, R.; Taute, T.; Massmann, G. Temperature dependent redox zonation and attenuation of wastewater-derived organic micropollutants in the hyporheic zone. *Sci. Total Environ.* **2014**, *482–483*, 53–61. [CrossRef]
19. Munz, M.; Oswald, S.E.; Schafferling, R.; Lensing, H.J. Temperature-dependent redox zonation, nitrate removal and attenuation of organic micropollutants during bank filtration. *Water Res.* **2019**, *162*, 225–235. [CrossRef]
20. Groeschke, M.; Frommen, T.; Winkler, A.; Schneider, M. Sewage-borne ammonium at a river bank filtration site in Central Delhi, India: Simplified flow and reactive transport modeling to support decision-making about water management strategies. *Geosciences* **2017**, *7*, 48. [CrossRef]
21. Ahmed, A.K.A.; Marhaba, T.F. Review on river bank filtration as an in situ water treatment process. *Clean Technol. Environ. Policy* **2017**, *19*, 349–359. [CrossRef]
22. Jiang, Y.; Zhang, J.J.; Zhu, Y.G.; Du, Q.Q.; Teng, Y.G.; Zhai, Y.Z. Design and optimization of a fully-penetrating riverbank filtration well scheme at a fully-penetrating river based on analytical methods. *Water* **2019**, *11*, 418. [CrossRef]

23. Dillon, P.; Stuyfzand, P.; Grischek, T.; Lluria, M.; Pyne, R.D.G.; Jain, R.C.; Bear, J.; Schwarz, J.; Wang, W.; Fernandez, E.; et al. Sixty years of global progress in managed aquifer recharge. *Hydrogeol. J.* **2019**, *27*, 1–30. [CrossRef]
24. Kvitsand, H.M.L.; Myrmel, M.; Fiksdal, L.; Østerhus, S.W. Evaluation of bank filtration as a pretreatment method for the provision of hygienically safe drinking water in Norway: Results from monitoring at two full-scale sites. *Hydrogeol. J.* **2017**, *25*, 1257–1269. [CrossRef]
25. Derx, J.; Blaschke, A.P.; Farnleitner, A.H.; Pang, L.; Blöschl, G.; Schijven, J.F. Effects of fluctuations in river water level on virus removal by bank filtration and aquifer passage—A scenario analysis. *J. Contam. Hydrol.* **2013**, *147*, 34–44. [CrossRef] [PubMed]
26. Shamrukh, M.; Abdel-Wahab, A. Riverbank filtration for sustainable water supply: Application to a large-scale facility on the Nile River. *Clean Technol. Environ. Policy* **2008**, *10*, 351–358. [CrossRef]
27. Zhu, Y.G.; Zhai, Y.Z.; Du, Q.Q.; Teng, Y.G.; Wang, J.S.; Yang, G. The impact of well drawdowns on the mixing process of river water and groundwater and water quality in a riverside well field, Northeast China. *Hydrol. Process.* **2019**, *33*, 945–961. [CrossRef]
28. Dillon, P.J.; Miller, M.; Fallowfield, H.; Hutson, J. The potential of riverbank filtration for drinking water supplies in relation to microsystin removal in brackish aquifers. *J. Hydrol.* **2002**, *266*, 209–221. [CrossRef]
29. Gupta, A.; Singh, H.; Ahmed, F.; Mehrotra, I.; Kumar, P.; Kumar, S.; Grischek, T.; Sandhu, C. Lake bank filtration in landslide debris: Irregular hydrology with effective filtration. *Sustain. Water Resour. Manag.* **2015**, *1*, 15–26. [CrossRef]
30. Hu, B.; Teng, Y.G.; Zhai, Y.Z.; Zuo, R.; Li, J.; Chen, H.Y. Riverbank filtration in China: A review and perspective. *J. Hydrol.* **2016**, *541*, 914–927. [CrossRef]
31. AGEOS. *Alimentation en eau Potable: Demande D'autorisation en Vertu de l'Article 31 du Règlement sur le Captage des Eaux Souterraines: Rapport D'expertise Hydrogéologique 2010-723*; AGEOS: Brossard, QC, Canada, 2010; Volumes 1–2.
32. Pazouki, P.; Prevost, M.; McQuaid, N.; Barbeau, B.; de Boutray, M.L.; Zamyadi, A.; Dorner, S. Breakthrough of cyanobacteria in bank filtration. *Water Res.* **2016**, *102*, 170–179. [CrossRef]
33. Grischek, T.; Bartak, R. Riverbed clogging and sustainability of riverbank filtration. *Water* **2016**, *8*, 604. [CrossRef]
34. Pholkern, K.; Srisuk, K.; Grischek, T.; Soares, M.; Schafer, S.; Archwichai, L.; Saraphirom, P.; Pavelic, P.; Wirojanagud, W. Riverbed clogging experiments at potential river bank filtration sites along the Ping River, Chiang Mai, Thailand. *Environ. Earth Sci.* **2015**, *73*, 7699–7709. [CrossRef]
35. AGEOS. *Alimentation en eau Potable: Suivi des Fluctuations Piézométriques de la Nappe et des Niveaux des Lacs: Période du 27 Avril 2012 au 17 Décembre 2015: Rapport Annuel 2015*; 29 février 2016; AGEOS: Brossard, QC, Canada, 2016; p. 42.
36. Gran, G. Determination of the equivalence point in potentiometric titrations. Part II. *Analyst* **1952**, *77*, 661–671. [CrossRef]
37. Sprenger, C. *Hydraulic Characterisation of Managed Aquifer Recharge Sites by Tracer Techniques*; Demeau: Berlin, Germany, 2016; p. 15.
38. Baudron, P.; Barbecot, F.; Gillon, M.; Aróstegui, J.L.G.; Travi, Y.; Leduc, C.; Castillo, F.G.; Martinez-Vicente, D. Assessing groundwater residence time in a highly anthropized unconfined aquifer using bomb peak 14C and reconstructed irrigation water 3H. *Radiocarbon* **2013**, *53*, 933–1006. [CrossRef]
39. Baudron, P.; Sprenger, C.; Lorenzen, G.; Ronghang, M. Hydrogeochemical and isotopic insights into mineralization processes and groundwater recharge from an intermittent monsoon channel to an overexploited aquifer in eastern Haryana (India). *Environ. Earth Sci.* **2016**, *75*, 434. [CrossRef]
40. Boving, T.B.; Choudri, B.S.; Cady, P.; Cording, A.; Patil, K.; Reddy, V. Hydraulic and hydrogeochemical characteristics of a riverbank filtration site in rural India. *Water Environ. Res.* **2014**, *86*, 636–648. [CrossRef]
41. Buzek, F.; Kadlecova, R.; Jackova, I.; Lnenickova, Z. Nitrate transport in the unsaturated zone: A case study of the riverbank filtration system Karany, Czech Republic. *Hydrol. Process.* **2012**, *26*, 640–651. [CrossRef]
42. Glorian, H.; Bornick, H.; Sandhu, C.; Grischek, T. Water quality monitoring in Northern India for an evaluation of the efficiency of bank filtration sites. *Water* **2018**, *10*, 1804. [CrossRef]
43. Lorenzen, G.; Sprenger, C.; Baudron, P.; Gupta, D.; Pekdeger, A. Origin and dynamics of groundwater salinity in the alluvial plains of western Delhi and adjacent territories of Haryana State, India. *Hydrol. Process.* **2012**, *26*, 2333–2345. [CrossRef]

44. Wett, B.; Jarosch, H.; Ingerle, K. Flood induced infiltration affecting a bank filtrate well at the River Enns, Austria. *J. Hydrol.* **2002**, *266*, 222–234. [CrossRef]
45. Government of Canada. Guidelines for Canadian Drinking Water Quality: Guideline Technical Document–Iron. Available online: https://www.canada.ca/en/health-canada/services/publications/healthy-living/guidelines-canadian-drinking-water-quality-guideline-technical-document-iron.html (accessed on 9 August 2019).
46. Government of Canada. Guidelines for Canadian Drinking Water Quality: Guideline Technical Document–Manganese. Available online: https://www.canada.ca/en/health-canada/services/publications/healthy-living/guidelines-canadian-drinking-water-quality-guideline-technical-document-manganese.html (accessed on 9 August 2019).
47. Arnoux, M.; Gibert-Brunet, E.; Barbecot, F.; Guillon, S.; Gibson, J.; Noret, A. Interactions between groundwater and seasonally ice-covered lakes: Using water stable isotopes and radon-222 multilayer mass balance models. *Hydrol. Process.* **2017**, *31*, 2566–2581. [CrossRef]
48. Des Tombe, B.F.; Bakker, M.; Schaars, F.; van der Made, K.J. Estimating travel time in bank filtration systems from a numerical model based on DTS measurements. *Ground Water* **2018**, *56*, 288–299. [CrossRef] [PubMed]
49. Glass, J.; Li, T.; Sprenger, C.; Stefan, C. Investigation of viscosity effects caused by seasonal temperature fluctuations during MAR. In Proceedings of the International Symposium on Managed Aquifer, Madrid, Spain, 20–24 May 2019.
50. Liu, S.D.; Zhou, Y.X.; Kamps, P.; Smits, F.; Olsthoorn, T. Effect of temperature variations on the travel time of infiltrating water in the Amsterdam Water Supply Dunes (The Netherlands). *Hydrogeol. J.* **2019**, *27*, 2199–2209. [CrossRef]
51. Henzler, A.F.; Greskowiak, J.; Massmann, G. Seasonality of temperatures and redox zonations during bank filtration—A modeling approach. *J. Hydrol.* **2016**, *535*, 282–292. [CrossRef]
52. Grischek, T.; Paufler, S. Prediction of iron release during riverbank filtration. *Water* **2017**, *9*, 317. [CrossRef]

© 2019 by the authors. Licensee MDPI, Basel, Switzerland. This article is an open access article distributed under the terms and conditions of the Creative Commons Attribution (CC BY) license (http://creativecommons.org/licenses/by/4.0/).

Article

Evaluating Bank-Filtration Occurrence in the Province of Quebec (Canada) with a GIS Approach

Marc Patenaude [1,2,3,*], Paul Baudron [1,2,3,4,*], Laurence Labelle [1,2,3] and Janie Masse-Dufresne [1,2,3]

1. Department of Civil, Geological and Mining Engineering, Polytechnique Montréal, Montreal, QC H3T 1J4, Canada; laurence.labelle@polymtl.ca (L.L.); janie.masse-dufresne@polymtl.ca (J.M.-D.)
2. Geotop Research Center, 201 ave. du Président-Kennedy, Montréal, QC H2X 3Y7, Canada
3. CentrEau Research Center, 1065 avenue de la Médecine, Québec, QC G1V 0A6, Canada
4. Institut de Recherche pour le Développement, UMR G-EAU, 34196 Montpellier, France
* Correspondence: marc.patenaude@polymtl.ca (M.P.); paul.baudron@polymtl.ca (P.B.)

Received: 20 November 2019; Accepted: 20 February 2020; Published: 1 March 2020

Abstract: Due to the abundance of surface water in the province of Quebec, Canada, it is suspected that many groundwater wells are pumping a mixture of groundwater and surface water via induced bank filtration (IBF). The regulatory framework in Quebec provides comprehensive guidelines for the development and monitoring of surface water and groundwater drinking water production systems. However, the regulations do not specifically address hybrid groundwater-surface water production systems such as IBF sites. More knowledge on the use of IBF in the province is needed to adjust the regulations with respect to the particularities of these systems. In order to provide a first evaluation of municipal wells potentially using IBF and the corresponding population served by these wells, a Geographic Information Science framework (GISc) was used to implement an IBF spatial database and calculate the distance from each well to the nearest surface water body. GISc is based on open source GIS programs and openly available data, to facilitate the reproducibility of the work. From this provincial scale approach, we show that nearly one million people are supplied by groundwater from municipal wells located <500 m from a surface water body, and half a million have a significant probability to be supplied by IBF wells. A more focused look at the watershed scale distribution of wells allows us to improve our interpretations by considering the aquifer type and other regional factors. This approach reveals strong spatial variability in the distribution of wells in proximity to surface water. Of the three selected regions, one has a high potential for IBF (Laurentides), one requires additional information do draw precise conclusions (Nicolet), and the third region (Vaudreuil-Soulanges) is unlikely to have widespread use of IBF. With this study, we demonstrate that extensive use of IBF is likely and that there is a need for improved understanding and management of these sites in order to properly protect the drinking water supply.

Keywords: managed aquifer recharge (MAR); induced bank filtration (IBF); geographic information science (GISc); geographic information systems (GIS); drinking water supply; guidelines

1. Introduction

Induced bank filtration (IBF) is a widely used method of managed aquifer recharge (MAR) [1]. In an IBF system, surface water is drawn through the banks and bed of a lake or river towards a pumping well by an induced hydraulic gradient. This results in a pumped mixture of both groundwater and infiltrated surface water. This process reduces the risk of intensive use [2] of the aquifer and improves water quality relative to surface water [3]. The drawback of this method, however, is that there is a higher risk of contamination from certain sources in comparison to standard groundwater exploitations [4–7], which makes it important to develop a specific regulatory framework adapted to this type of drinking water supply system.

In recent reviews of the international use of bank filtration [8–11], it is clear that IBF has been used worldwide and, in particular, in Europe for more than 100 years. In the last decades, the use of IBF also became relatively popular in the USA [3 and references therein]. More recently, few IBF systems have also been implemented in developing countries [3 and references therein]). The use of this technology is, however, completely overlooked in Canada and the province of Quebec in particular. As of this point in time, no inventory of IBF sites exists for the province, and the extent of its use throughout the province remains unknown. There is, however, a high probability that IBF is widely used. Indeed, Quebec is a province rich in both surface water and groundwater, with 22% of its surface covered by water either in the form of lakes, rivers, or wetlands [12]. It is estimated that Quebec contains 3% of the Earth's renewable freshwater resources [13]. In addition, Quebec was settled through its waterways, which has had long-term effects on its population distribution. The abundance of water and the distribution of the population mean that there is a high probability that pumping wells are located in close proximity to surface water. The process of unintentional IBF is thus likely to be occurring in many municipalities throughout the province.

In addition to the number of IBF sites throughout Quebec being unknown, the current regulations and guidelines in Quebec do not specifically address wells that pump a mixture of surface water and groundwater. In Quebec, the *Règlement sur le prélèvement des eaux et leur protection* (RPEP) provides comprehensive guidelines for protecting surface water and groundwater extractions from contamination [14]. This regulation provides a framework for monitoring contaminants that allows for appropriate intervention when groundwater resources contain surface microbiological contaminants, which ensures that the population is not likely to be affected by changes in water quality. Additionally, drinking water production systems that are considered Groundwater under the Direct Influence of Surface Water (GWUDI) are subjected to the same regulations as surface water, which are more stringent than for all groundwater wells. For instance, in such a case, the raw pumped water is tested weekly for microbiological parameters, whereas groundwater can be tested monthly [15]. Despite its name, the GWUDI classification does not aim at characterizing infiltration from surface water bodies. Rather, what is referred to as "surface water" within this classification system is in reference to any source of contamination from the surface (including septic tanks) that might provide recurrent microbial or viral contamination to a well. There is, therefore, no correlation with surface water bodies that are hydraulically connected to an aquifer unless said surface water bodies is considered a potential source of contamination or if there is persistent bacteriologic or virologic presence in the pumping well during initial characterization. Existing non-GWUDI IBF sites are therefore necessarily treated as standard groundwater extractions. This means that even if the groundwater wells are located within a few meters of a water body, and there is a hydraulic connection between them, there is a lack of protection guidelines for the nearby surface water.

Due to the lack of regulations specific to hybrid groundwater-surface water systems, many problems common to IBF sites may be overlooked or unanticipated during the planning and operation of these sites. IBF sites are sensitive to changes in surface water quality [6,16], changing redox conditions [17], and changes in the hydraulic conditions of the site [18]. Additionally, there are many undesirable chemical components and contaminants that can be found at pumping wells in IBF sites including Mn [19–21], Fe [22,23], NO^{-3-} [24], organic micropollutants [25–28], cyanobacteria [29–31], coliforms [32]. In order to have a more resilient water supply, the risk posed by these contaminants should be identified as early on in the development of the site as possible in order to minimize unforeseen costs and develop plans to reduce the risk. By identifying the potential contaminants at the sites related to changing hydraulic and chemical conditions, the risks associated with them would be reduced by allowing for strategic planning to avoid the conditions associated with poorer water quality. For example, well configuration and pumping schemes could be modified to increase transit times from surface water to pumping wells producing a more consistent water quality.

Geographic Information Systems (GIS) and Multi-Criteria Decision Analysis (MCDA) approaches are being developed for site suitability mapping for the development of MAR ([33] and references therein) and IBF [34,35]. As an example, Jamarillo Uribe [36] has studied the impact of stream morphology on bank filtration sites. Nonetheless, to our knowledge, this is the first study that uses a GISc approach to identify the number of existing IBF sites. The objective of this study is to provide a first overview of the potential extent of IBF in the province of Quebec. The framework is based on the concept of Geographic Information Science (GISc), described in [37] and [38]. GISc is based on the use of open-source GIS programs and openly available data. This concept facilitates reproducible research to other areas of study and data sets and could improve the transparency of research using GIS. First, we have processed and homogenized the sources of information from three different agencies. Secondly, we have carried out a pre-quantification of municipal wells with a higher likelihood of providing drinking water through IBF using easily available government data. Following this pre-selection, zones with varying characteristics are considered in greater detail, and the likelihood of IBF taking place in these areas is discussed.

2. Materials and Methods

2.1. Data Set

2.1.1. Well Data

The well data in the province of Quebec is contained in a variety of databases. All private wells require that various information is reported following drilling, such as coordinates, ownership, well design (i.e., type of tubing and depth of well), and stratigraphic sequence. This data is compiled in the *Système d'Information Hydrogéologique* (SIH) database [39], which has previously been described by Sterckx [40]. The SIH database consists of information largely provided by well-drilling companies. This can lead to inconsistencies in the geological descriptions and the precision of the coordinates. This database is, however, considered a reliable source of information concerning the depth of the contact between overburden and bedrock [41].

The database with the most extensive information is available through a series of studies entitled *"Programme d'Acquisition de Connaissances sur les Eaux Souterraines"* (PACES) [42]. These projects were initiated in 2008 and aimed at improving knowledge of groundwater resources in the southern regions of the province of Quebec in order to protect them and ensure their sustainability. These studies have led to a number of subsequent publications [43–46]. As of today, the PACES studies have been completed on a total of 13 regions throughout the province and led to the compilation of varied information (i.e., depth of well, depth of screen, depth of water table, type of aquifer, type of well and coordinates) on a total of roughly 180,000 wells. Supplemental information, including geochemistry, geology, and hydrogeological data (e.g., hydraulic conductivity), was also compiled for a small subset of wells (n = 15,162). The PACES database contains significant overlap with the SIH database, and therefore, some issues with the reliability of the coordinates and geological descriptions also affect this database.

The third database is comprised of municipal wells and surface water extraction points [47]. This database is typically used by decision-makers for public health and land-use planning. The version of the database used in this study was acquired in 2017 (i.e., prior to the most recent update in November 2018). The information contained in this database is centered on geographic coordinates, population served, and types of potabilization treatment. It contains information on 2116 individual wells and surface water extraction points. The geographic coordinates are generally more recently acquired and derived from official declaration documents, resulting in a better precision of the localization of the wells than the two other databases.

The municipal database formed the basis of this study with complementary information pulled from the other sources. The decision to focus on the municipal well database was made since (i) IBF requires sufficiently high pumping rates to induce a hydraulic gradient from the surface water to the well, (ii) the localization of the wells is precise and (iii) all these wells are supplying drinking water to the population. The PACES database was used to extract complementary information and draw some general conclusions about the distribution of wells within the province. Table 1 summarizes the data sets and the variables used in this study.

Table 1. Summary of data available in the databases.

Information	SIH	PACES *	Municipal *
Number of wells	~216,000	~180,000	~2000
Depth of well	X	X	
Depth of screen		X	
Geology	(X)	(X)	
Chemistry		(X)	
Population Served			X
Type of aquifer	(X)	X	
Type of well		X	X
Type of treatment			X
Coordinates	X	X	X

* Databases used for subsequent calculations; (X) not systematically compiled.

2.1.2. Surface Water Bodies and Other Data

Natural Resources Canada's (NRCAN) website contains a variety of vector data in the Canvec portion of the site [48]. The surface water files are subdivided into two categories, "watercourses" and "water bodies". Many of the features in the "watercourses" files are drainage ditches and ephemeral streams that are not likely to be supplying bank filtrate to wells throughout the year. The features in the "water bodies" files correspond to water bodies of larger size and more permanent nature, which are more likely to contain a sufficient volume of water to support a municipal water supply. Considering the above, it was decided to conduct our province-wide study with the "waterbodies" files only. The named rivers from the "watercourses" files were added to the regional studies. These files contain two types of data, i.e., polygons and polylines, respectively representing the water bodies and shorelines.

2.2. Description of the GISc Framework

In this study, we used a GISc framework in order to calculate the minimal distance between each municipal pumping well and the nearest surface water body (i.e., lake or river). Throughout this study, the distance to surface water is calculated with respect to the "waterbodies" files retrieved from the NRCAN website, as mentioned above. The work process has been carried out with the Quantum GIS program (QGIS) [49]. In this subsection, we list a number of steps that were taken in order to perform the spatial analysis.

2.2.1. Processing and Homogenization of Spatial Data Sets

- Removal of duplicate wells

The work process is initialized with spatial data debugging. The database of municipal wells contained a number of duplicated geometries, i.e., wells with identical coordinates, which caused issues when performing the calculations. The duplicate data most often resulted when wells were supplying water to multiple municipalities. In order to remedy this, duplicate geometries were automatically identified and removed from the spatial database. In cases where discrepancies were found in the attribute table, the correct points were identified and retained.

- Homogenization of the spatial reference system

Coordinates needed to be converted into a common projected coordinate system in order to do subsequent distances calculations. Spatial data were reprojected from the original coordinate system (i.e., NAD 83) to Lambert EPSG: 32,198. The Quebec Lambert was selected since it is suitable for use in Quebec (Canada) and for applications with an accuracy of <2 m [50].

- Conversion of geometries

First, the geometries were converted from multipart to single part. This step was necessary in order to ensure that the conversion to lines in the following step inserted all the necessary points. Without this step, only the existing nodes of the polyline file were converted to points, which led to an uneven and wide-ranging spread of points along the shorelines.

Second, waterbody files were converted from lines to points with 2 m spacing with the function *Convert lines to points* in the SAGA [51] toolbox. This function was run as a batch process for all of the 1:50,000 National Topographic System (NTS) zones individually.

2.2.2. Performing Distance Calculation

- Calculate the distance from each well to the nearest point

Using the smaller 1:50,000 zones allowed for the direct calculation from each well to the nearest point in each of the NTS zones. This calculation was completed using the *Distance to nearest hub* function. This algorithm identifies the nearest feature to each point and the Euclidean distance between all points. This, however, led to the creation of roughly 220 different files, each containing one distance value for each well. This avoided some problems that occurred when the calculation was done in the reverse order, but it created extremely large files that were not easy to work with in QGIS, especially for the larger PACES files.

- Calculate the minimum value for each well

The subsequent step was to merge all of these roughly 220 files and calculate the minimum value for each well. This was done in QGIS using the *merge* function, and the minimum selection was made using the *Select by expression* dialogue box and the following line of code HubDist = minimum (HubDist, Well ID). This could only be done in QGIS for the smaller municipal well files. For the larger PACES files, the merge and distance calculations were completed using R [52] programming language.

Once the distance was calculated, manipulation of the data was done manually using filters in QGIS, and R. Additional data was also joined to the well files either manually, or using common fields or spatial relationships using a variety of functions in QGIS.

This process was repeated for the regional studies with named rivers from the watercourse file and the water bodies used in the province-wide study.

3. Results and Discussion

There are two main conditions that will determine whether a well is pumping a mixture of surface water and groundwater. First, a hydraulic connection between the surface water and the aquifer is necessary. Second, assuming groundwater typically discharges into surface water bodies in the study region [53], it is necessary to pump a sufficient volume to reverse that natural gradient and draw surface water toward the pumping well. In order to conduct a study at the scale of the province, we used the municipal well database (description available in Section 2.1.1), which reports the geographical coordinates and the population served by each of the 2075 municipal wells and surface water extraction points. Since the type of aquifer is not reported within this database, the distance from the surface water bodies was deemed to be the most important criterion for estimating the number of IBF sites available for a province-wide scale. Then, we selected three areas with distinct characteristics in order to conduct regional analyses, including integrating the type of aquifer and other hydrogeological information. The results of both province and regional scale approaches are presented and discussed in the next subsections.

Bank filtration sites can be located at various distances from surface water bodies [32,54] from a few meters to greater than one kilometer. In this study, as a starting point, distances >500 m are considered less likely to be performing IBF, as wells were located <500 m from surface water in several studies on IBF [32,54,55]. Groundwater will discharge from aquifers towards rivers and lakes in most of Quebec for the majority of the year, especially during drier months [53]; therefore, wells must be able to reverse that natural gradient in order to perform IBF. In addition, most of the sites have been operational for at most a few decades, are not likely to have experienced intensive use [2], and therefore, large scale flow regime reversals, as seen in the Netherlands [54], are not likely. These combined factors make it unlikely that wells at a distance >500 m are performing IBF since reversal of a gradient over those distances would be more difficult. This choice for the cut-off could potentially omit some bank filtration sites from the province; however, this is considered to have minimal impact on the overall conclusions of this study.

3.1. Overview of the Province

3.1.1. Municipal Drinking Water Supply: Surface Water vs. Groundwater

Municipal drinking water sources can be classified into three categories, i.e., groundwater, surface water, and groundwater considered surface water, as shown in Figure 1. The first category (in red) includes all municipal drinking water sources which rely on one or multiple groundwater wells. Contrastingly, the second category (in blue) refers to municipal drinking water sources supplied by surface water. The third category (in green) corresponds to the few municipal drinking water sources relying on groundwater wells, which are documented as GWUDI according to the protocol detailed in the *Guide de conception des installations de production d'eau potable* described in Section 1 [15]. Of the 2075 municipal extraction points, 87% (n = 1799) are groundwater pumping wells, whereas 13% (n = 276) are directly extracting surface waters. This results in approximately 15% of the population of Quebec relying on groundwater resources for drinking water, representing roughly 1,260,000 citizens. A similar estimate (i.e., 20%) was also reported by the *Ministère de l'environnement et lutte contre les changements climatiques* [42]. The discrepancy between these two estimates is likely due to the proportion of the population supplied by private wells rather than a municipal distribution network.

In the more densely populated areas, surface water is the main water source for drinking water supply systems. This is likely a consequence of the greater drinking water demand in the cities, and the larger population's ability to support the more costly water treatment required for surface water. In smaller municipalities, groundwater sources are preferred as they are less costly to operate. In fact, in rural areas of the province, 90% of the population is served by groundwater sources [56]. As explained above, the important number of surface water bodies and the population's distribution results in a high likelihood of having groundwater wells near a surface water body, suggesting that many municipal drinking water systems may be benefiting from IBF processes.

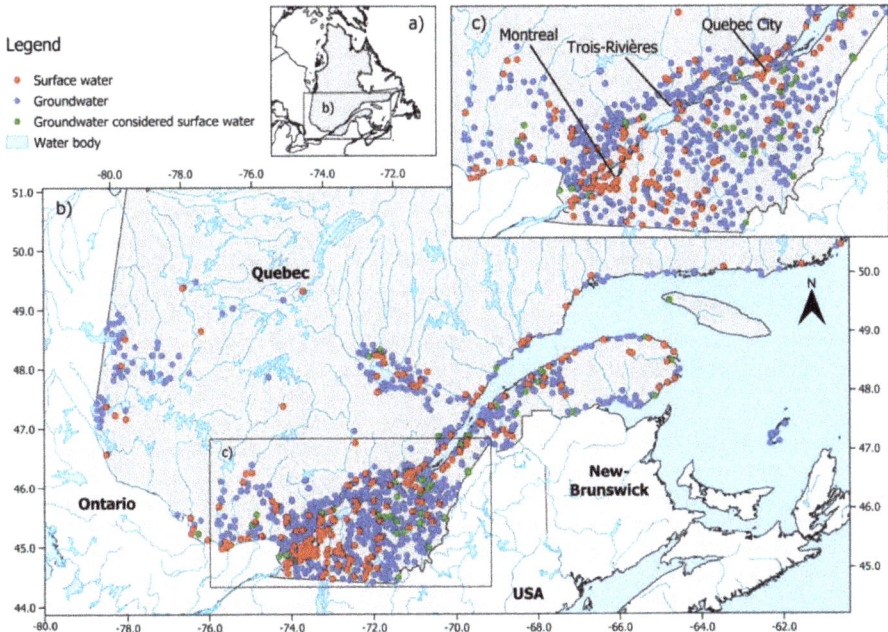

Figure 1. Spatial distribution of municipal drinking water sources. The water sources are classified into three categories, i.e., groundwater (in blue), surface water (in red), and groundwater considered surface water (in green). (**a**) overview of the study area; (**b**) view of the distribution of all wells throughout the province; (**c**) view of the most densely populated area of the province along the Saint-Lawrence River

3.1.2. Water Bodies Distribution around Wells

Among the municipal wells, almost all (97% of the cases; n = 1749) are located at <2000 m from a surface water body. As illustrated in Figure 2a, it is evident that the closest water body to most municipal wells is lakes (72%; n = 1262). This is likely due to the extremely high number of lakes in Quebec, which means that there is likely always a lake within a reasonable distance from any given point in the province, resulting in a geographically homogeneous distribution of municipal wells near lakes. The distribution of wells in close proximity to rivers (28%; n = 487) is less homogenously distributed throughout the province compared to those in proximity to lakes (see Figure 2). Many municipalities are located in close proximity to a river due to the settlement of the province through its waterways. Secondly, the area around rivers can often have more favorable properties for larger-scale water extraction due to the higher likelihood of containing sandy granular deposits compared to the more common glacial-marine deposits that cover the rest of the province [57].

Figure 2b also reveals the same trend of homogenous distribution of wells in proximity to lakes for each bin of 10 m width. In fact, 14% (n = 245) of the municipal wells are located at <200 m from a river. The overall distribution of wells shows a significant decrease in the density of wells at around the 120 m marks. Since the number of wells in proximity to lakes remains relatively constant in all 10 m bins, and the distribution of wells near rivers decreases around the 120 m marks, the overall trend in this graph of more wells in the first 120 m is controlled mostly by the greater number of wells in proximity to rivers.

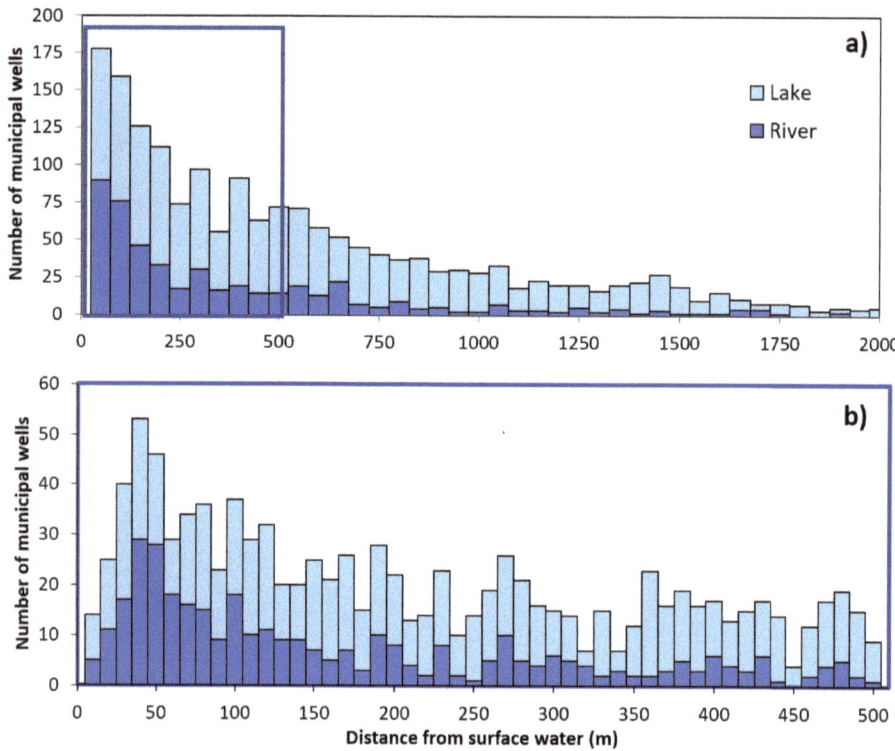

Figure 2. The distance of municipal wells from water surfaces for (**a**) 0 to 2000 m and (**b**) 0 to 500 m.

3.1.3. Insights into the Potential Population Supplied by IBF

Figure 3 illustrates the cumulative population served by municipal wells that are located at a distance of 0 m to 2000 m from a surface water body. It shows that approximately 1,200,000 people are served by those municipal wells. It also reveals that outside of the major cities, wells located at less than 500 m from a surface water body account for 74% of the population (n = 920,000) whose drinking water is supplied by groundwater from municipal wells. In fact, there is a rapid increase in the population served by wells within the first few hundred meters of a water body. Wells within 250 m and 100 m account for 57% (n = 720,000) and 34% (n = 410,000) of the population served by municipal wells respectively. Although many of these sites may not be using IBF due to various factors, these results reveal a significant probability that half of the groundwater-fed population connected to a municipal drinking water supply in Quebec, i.e., more than half a million citizens, might depend on IBF. This initial overview demonstrates the great opportunity of a better assessment of IBF occurrence in the province of Quebec in order to better protect our resources. This high proportion of close-to surface water wells is likely an indication that planners and developers are, in fact, selecting pumping sites in close proximity to surface water. Still, as aforementioned, the existing knowledge of IBF is not fully integrated into the planning process when developing and managing water abstraction plants.

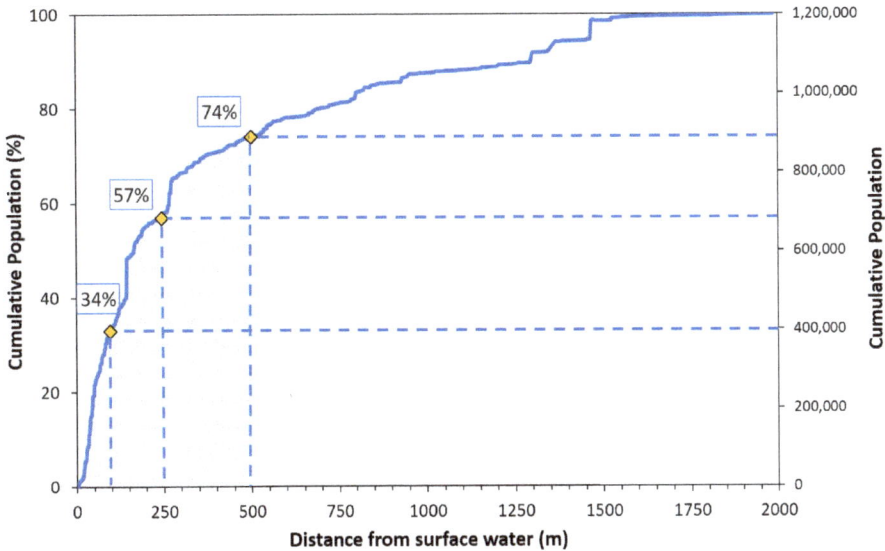

Figure 3. Cumulative population supplied by municipal groundwater wells with respect to the distance from the surface water body.

3.2. Insights on Selected Sub-Basins

In light of the results presented in the previous subsection, it was determined that making widespread generalizations about the province would not be straightforward. We opted to focus on a selection of areas with diverse population sizes, diverse geological settings, a variety of distances from surface water, a variety of surface water body types, and a variety of land cover types. By cross-checking the ID and localization of wells in the available databases, it was possible to manually assign the type of aquifer (fractured bedrock, unconfined granular, confined granular) to 101 municipal wells within the three areas, namely Laurentides (area #1), Nicolet (area #2) and Vaudreuil-Soulanges (area #3). These areas correspond to watersheds "du Nord", "Nicolet", and "Vaudreuil-Soulanges"respectively. It is important to note that the "Vaudreuil–Soulanges" watershed also extends into the neighboring province of Ontario to the West, but the investigation is limited to the portion within the Quebec border. The localization and the extent of each area are illustrated in Figure 4. The main geological contexts are also reported in this figure. The Grenville Province (area #1) is mainly composed of Archean autochthonous rocks dominated by highly metamorphosed gneissic complexes. The Appalachian Province (area #2) is composed of various types of strongly deformed rocks (i.e., sedimentary, volcanic, and ophiolitic rocks), whereas the St-Lawrence Platform (area #3) corresponds to sedimentary rocks. The criteria used to evaluate each region are the nature of quaternary deposits, the type of nearest water body in proximity, the population of the municipalities, the distance of wells, as well as certain qualitative factors that are mentioned in the PACES reports for each region [58–61].

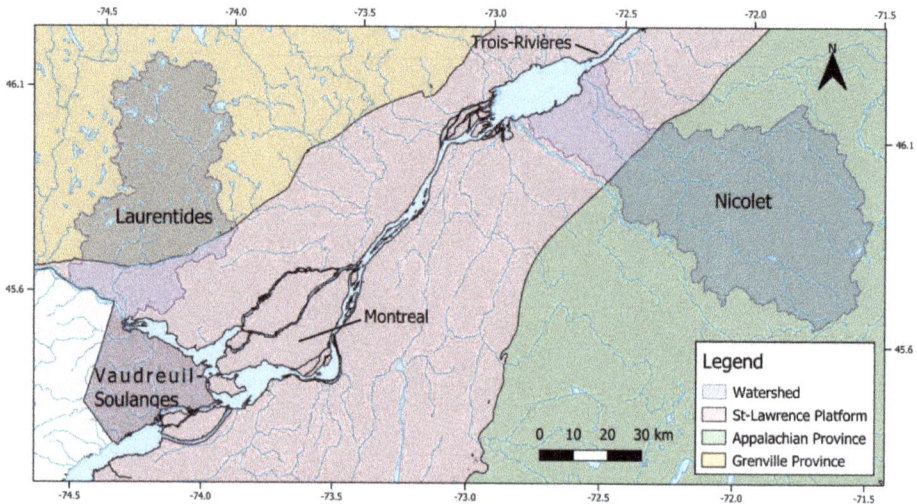

Figure 4. Location of the three selected watersheds (area#1: Laurentides, area #2 Nicolet, and area #3 Vaudreuil–Soulanges).

3.2.1. Area #1: Laurentides

The area of Laurentides has a mixture of forested areas and urban areas and is underlain by the hilly Grenville Province. It has not been covered by regional PACES studies at this time, and therefore, a detailed description of the surficial deposits underlying the region is somewhat dated. Portions of the area along the Saint Lawrence and Ottawa Rivers were described by Lajoie [58]. Bedrock in this area is overlain by till that varies in thickness from 7 to 12 m. Deposits of sand and gravel can also be found within this region, that were deposited by glacial rivers. Along existing rivers, including the Rivière-du-Nord, recent fluvial deposits composed mainly of loam are present [59]. Supplemental Materials containing the map of the quaternary geology and well distribution for each area is provided with this paper (Figure S1, Tabel S1).

In this area, 31 municipalities (of a total of 58) rely on groundwater to produce drinking water from a total of 136 municipal wells, and 91% (n = 124) are located <500 m from a surface water body. A total of 63 municipal wells (51%) are located at <500 m from a lake, the remaining part (49%; n = 61) being at <500 m from a river. As shown in Figure 5, a significant number (31%; n = 39) of municipal wells are found along the main river (i.e., Rivière-du-Nord). These wells are often serving municipalities with a population of >1000 people and are typically found in unconfined granular aquifers. There are three other rivers and a lake with wells in close proximity, namely, Red River with four wells serving two municipalities, Ottawa River with four wells serving three municipalities, Lac de Deux Montagnes with 12 wells serving one municipality, Achigan River with two wells serving one municipality. Further from the Rivière-du-Nord, the population generally decreases, and the wells are more frequently found in proximity to lakes and in fractured bedrock aquifers. The type of aquifer could be compiled for a total of 56 wells. Of these wells, 73% (n = 41) are found in an unconfined granular aquifer, with the majority located <100 m from a surface water body. Since the largest municipalities are more likely to be pumping sufficiently to induce a hydraulic gradient forcing the surface water to infiltrate the aquifer, the municipalities with the highest likelihood of performing IBF are the more populated ones in proximity to the Rivière-du-Nord. Meanwhile, wells located further from surface water and installed in fractured bedrock present a lower confidence level in the probability that IBF is taking place. It is important to note that the above-mentioned results include some wells outside of the selected watershed, as they shared a similar geological setting.

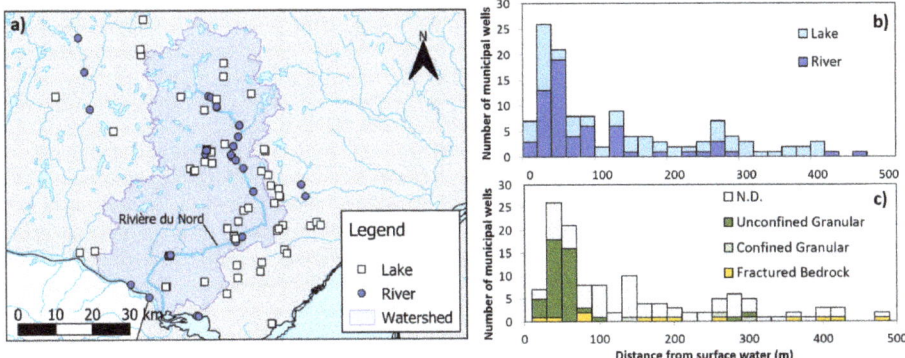

Figure 5. (a) The spatial distribution of municipal wells located at less than 500 m from lakes and rivers in area #1 and the distribution of wells according to (b) the type of surface water and (c) the type of aquifer.

3.2.2. Area #2: Nicolet

The Nicolet area straddles two geologic provinces, the Saint-Lawrence Platform in the lower altitude portion to the North and the Appalachian Orogen in the higher altitudes to the South. This area is principally covered by agricultural land. A total of 84 municipal wells are actively producing drinking water from groundwater resources to serve a total of 35 municipalities (of a total of 39). These municipalities are generally smaller than those in other regions, with many serving <500 people. As illustrated by Figure 6, in area #2, a high number of wells are in the vicinity of different rivers and tributaries, similar to area #1. In fact, a total of 48 municipal wells (57%) are in the 0–500 m range from a surface water body. Of these wells, most are located close to rivers (71%; n = 34), and the remaining 29% (n = 14) are located near lakes. However, the distance of these municipal wells varies more widely than in area #1. The Nicolet River and its tributaries are the rivers with the largest number of wells in close proximity. There are five rivers in the Nicolet river system with wells in close proximity. In addition, the Saint-Francois River has three wells in proximity serving two municipalities. Within this region, there are many more rivers whose quality could impact drinking water quality than in the other regions.

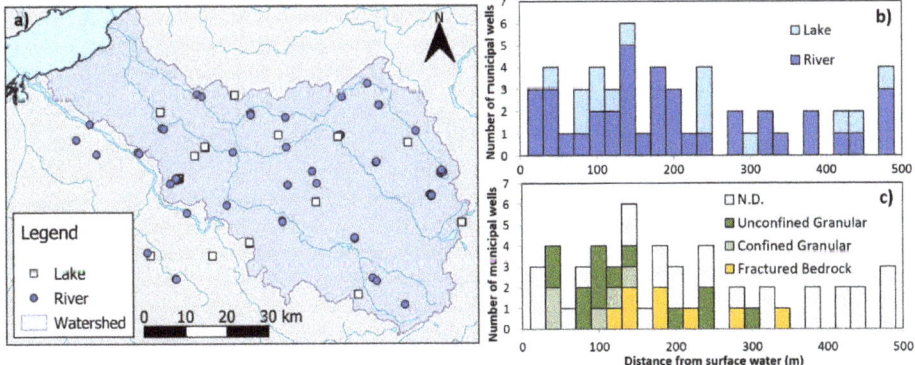

Figure 6. (a) The spatial distribution of municipal wells in the Nicolet area that are located at less than 500 m from lakes and rivers in area #2 and the distribution of wells according to (b) the type of surface water and (c) the type of aquifer.

The Nicolet region has a distinctly different sequence of quaternary deposits when compared to the other two zones. As reported in the PACES study [60], thick quaternary deposits exceeding 100 m in certain areas overlie the bedrock in certain parts of the region. A rough 20 km wide band along the Saint–Lawrence River is underlain by a significant thickness of marine clays, which can be partially or completely overlain by marine and lacustrine sands. The central portion of the basin located between elevations of 80 m and 120 m is dominated by aeolian and coastal sands, underlain by impermeable till around which peatlands can form. Thick glacial-fluvial sand and gravel deposits can lie directly on the bedrock in certain areas. Superficial deposit thicknesses at the municipal well locations calculated from the raster database that accompanies the PACES study [60] ranged from <5 m up to 30 m. Supplemental Materials containing the map of the quaternary geology and well distribution for each area is available (Figure S2, Table S1).

Geological information could be compiled for a total of 26 municipal wells located <500 m from a water body. In this region, 50% (n = 13) of the municipal wells procure water from unconfined granular aquifers. Of these wells, 77% (n = 10) are located <200 m from a surface water body. These results suggest that there are favorable aquifers for pumping near rivers, while the regions more distant from rivers are less favorable. This region, although it consists mostly of smaller municipalities, has a high probability of IBF taking place, especially in wells within the first few hundreds of meters from a river. More information relative to the pumping rates for these wells, combined with a geochemical and isotopic approach, would be needed to better estimate the potential use of IBF in this region.

3.2.3. Area #3: Vaudreuil–Soulanges

The area of Vaudreuil–Soulanges is located near Montreal and within the St-Lawrence Platform. The main types of land uses are agricultural and residential. This area, unlike the Nicolet area, is underlain solely by more recent quaternary deposits that were deposited uniquely during the last glacial cycle. The region is predominantly covered by marine clays (64% of the surface area). Only a small number of areas, along the Ottawa River and higher relief areas (i.e., till deposits, Mount Rigaud, "butte Saint-Lazare", and "butte de Hudson"), remain uncovered by these clays. Certain uncovered areas are composed of glacial-fluvial deposits that host productive granular aquifers. Examples of these deposits can be found along the Ottawa River in the northern portion of the zone and also in the Saint-Lazare region. Supplemental Materials containing the map of the quaternary geology and well distribution for each area is available (Figure S3, Table S1).

In this area, 36 wells are in operation and produce drinking water from groundwater resources for eight municipalities (of a total of 13) with populations typically >1000 people. As illustrated in Figure 7, the area contains a distribution of wells mostly in proximity to lakes. In fact, 66% (n = 24) of the municipal wells are located <500 m from a lake. Among these wells, 92% (n = 22) are located along a 6 km-long North-South transect near the region of Saint–Lazare. The PACES report highlights the presence of a thick and unconfined sandy deposit with a very productive aquifer in this zone [60]. Wells, in proximity to rivers, for the most part, are located along the periphery of this zone near the more major water bodies. Another zone of concentrated wells (20%; n = 10) is found in the region of Mount Rigaud, the majority of these wells are located in fractured rock or clay deposits [60], making the probability of IBF unlikely. It can also be seen in Figure 7 that there is a limited number of wells (33%; n = 16) located within the first 200 m of a water body, which sharply contrasts with the other two regions. In this region, wells are located in proximity to two rivers only. The Viviry River with four wells serving one municipality and the Saint–Lawrence River with four wells serving one municipality are the two most prominent ones with some smaller streams making up the difference. The type of aquifer could be compiled for 22 municipal wells in area #3. From these wells, the majority is found in confined granular aquifers (31%; n = 7) and fractured bedrock aquifers (40%; n = 9). The fractured bedrock aquifers are known to be confined [60], except in the regions with the highest concentrations of wells (i.e., St-Lazare, Hudson, and Rigaud).

Overall, the greater distance from surface water in this area seems to indicate that IBF in most of the municipalities in this area is not likely. We identified only two municipalities with a total of 3 wells in close proximity to surface water. However, these wells are in confined granular or fractured bedrock aquifers and, thus have limited potential for IBF. Moreover, there is a strong possibility that the permeable areas of the Saint–Lazare region are influenced by infiltrating water from the spring thaw and precipitations.

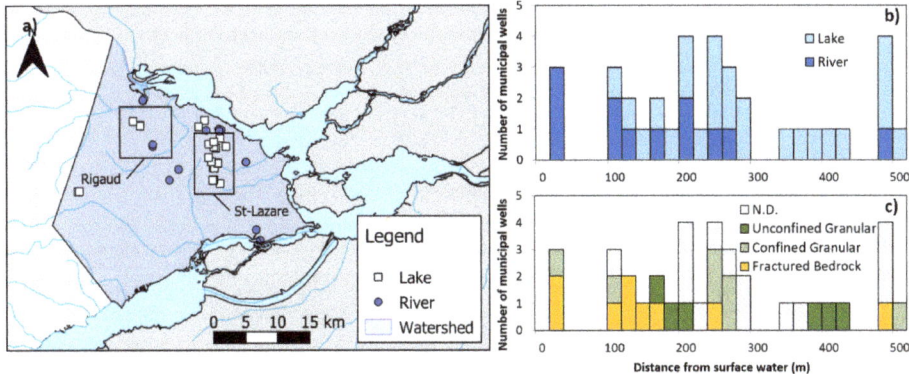

Figure 7. (a) The spatial distribution of municipal wells located at less than 500 m from lakes and rivers in area #3 and the distribution of wells according to (b) the type of surface water and (c) the type of aquifer.

4. Conclusions

This research has provided a reliable starting point for determining the impact of IBF on the population of Quebec. A simple process based on GISc was used by incorporating several open-source programs (QGIS, SAGA, and R) and openly available data. The initial use of distance from a surface water body as well as additional information extracted from the government database revealed that nearly one million people in the province of Quebec are supplied drinking water via wells in close proximity (i.e., <500 m) to a river or a lake. This first overview has also demonstrated that there is a high degree of regional variability with regard to the probability that IBF is being performed. One investigated area has the greatest potential use of IBF (area #1: Laurentides), a second one requires more information to draw precise conclusions (area #2: Nicolet), and the third one has the lowest probability of IBF (area #3: Vaudreuil–Soulanges).

There are a number of shortcomings that are evident in the current development, management, and understanding of hybrid groundwater-surface water extraction sites in the province of Quebec. The current regulations do not contain general insights into the protection or management of these sites. The existing GWUDI classification is not designed to assess risk at an IBF site as it does not expressly consider the contribution of surface water to a pumping well, and only under certain circumstances are IBF wells considered GWUDI. With mounting pressures on our water resources from climate change and population growth, we hope that this work demonstrates the need to better understand and improve the regulatory framework to specifically address hybrid groundwater-surface systems such as IBF in order anticipate problems common to IBF sites and ensure that our water resources are well protected and exploited sustainably.

It will be important to precisely develop guidelines for determining what municipalities should be targeted within future regulations. The policy changes would require the definition of a new category of wells for IBF sites that requires short, medium, and long-term assessment of risk related to the permanent or intermittent contribution of surface water bodies to wells.

In order to determine more precisely which wells are using IBF, a possible approach would be based on a dedicated sampling program for specific environmental tracers. These tracers would be used to quantify the spatio-temporal evolution of transit times (e.g., stable isotopes of water) and relative proportions of groundwater and surface water (e.g., electrical conductivity). A longer characterization period would allow for the minimization of unforeseen costs over the life of the water treatment plant.

Supplementary Materials: The following are available online at http://www.mdpi.com/2073-4441/12/3/662/s1, Figure S1: Map of Areas #1 (Laurentides) with quaternary geology and well distribution, Figure S2: Map of Areas #2 (Nicolet) with quaternary geology and well distribution, Figure S3: Map of Areas #3 (Vaudreuil Soulanges) with quaternary geology and well distribution, Table S1: Codes for the quarternary geology units in Figures S1–S3.

Author Contributions: Conceptualization, M.P. and P.B.; Data curation, M.P.; Funding acquisition, P.B.; Investigation, M.P.; Methodology, M.P. and P.B.; Project administration, P.B.; Resources, P.B.; Supervision, P.B.; Validation, L.L. and J.M.-D.; Visualization, L.L. and J.M.-D.; Writing—original draft, M.P.; Writing—review & editing, M.P., P.B., L.L. and J.M.-D. All authors have read and agreed to the published version of the manuscript.

Funding: This research was funded by FRQNT through grant number FQRNT 2018-NC-205703 and NSERC through grants number CRSNG—RDCPJ: 523095-17 and CRSNG-RGPIN-2016-06780.

Acknowledgments: We would like to thank the "Ministère de l'Environnement et Lutte contre les changements climatiques" for the data used in this study, Marc-André Bourgault for help during conceptualization, Guillaume Meyzonnat and Sylvain Gagné for discussions on various zones and insights into the reliability of data sources, Geneviève Boisjoly for help with the GIS calculations, and Gabriel Dion for the initial study that evolved into this one. Great thanks to Francisco Gomariz Castillo for his constructive review from a GIS expert point of view.

Conflicts of Interest: The authors declare no conflict of interest. The funders had no role in the design of the study; in the collection, analyses, or interpretation of data; in the writing of the manuscript, or in the decision to publish the results.

References

1. Umar, D.; Ramli, M.; Aris, A.; Sulaiman, W.; Kura, N.; Tukur, A. An overview assessment of the effectiveness and global popularity of some methods used in measuring riverbank filtration. *J. Hydrol.* **2017**, *550*, 497–515. [CrossRef]
2. Llamas, M.R.; Custodio, E. *Intensive Use of Groundwater: Challenges and Opportunities*; CRC Press: Boca Raton, FL, USA, 2002.
3. Gillefalk, M.; Massmann, G.; Nützmann, G.; Hilt, S. Potential Impacts of Induced Bank Filtration on Surface Water Quality: A Conceptual Framework for Future Research. *Water* **2018**, *10*, 1240. [CrossRef]
4. Hiscock, K.; Grischek, T. Attenuation of groundwater pollution by bank filtration. *J. Hydrol.* **2002**, *266*, 139–144. [CrossRef]
5. Tufenkji, N.; Ryan, J.; Elimelech, M. The Promise of Bank Filtration. *Environ. Sci. Technol.* **2002**, *36*, 422A–428A. [CrossRef]
6. Sprenger, C.; Lorenzen, G.; Hülshoff, I.; Grützmacher, G.; Ronghang, M.; Pekdeger, A. Vulnerability of bank filtration systems to climate change. *Sci. Total Environ.* **2011**, *409*, 655–663. [CrossRef]
7. Hellauer, K.; Karakurt, S.; Sperlich, A.; Burke, V.; Massmann, G.; Hübner, U.; Drewes, J. Establishing sequential managed aquifer recharge technology (SMART) for enhanced removal of trace organic chemicals: Experiences from field studies in Berlin, Germany. *J. Hydrol.* **2018**, *563*, 1161–1168. [CrossRef]
8. Stefan, C.; Ansems, N. Web-based global inventory of managed aquifer recharge applications. *Sustain. Water Resour. Manag.* **2018**, *4*, 153–162. [CrossRef]
9. Dillon, P.; Stuyfzand, P.; Grischek, T.; Lluria, M.; Pyne, R.; Jain, R.; Bear, J.; Schwarz, J.; Wang, W.; Fernandez, E.; et al. Sixty years of global progress in managed aquifer recharge. *Hydrogeol. J.* **2019**, *27*, 1–30. [CrossRef]
10. Sprenger, C.; Hartog, N.; Hernández, M.; Vilanova, E.; Grützmacher, G.; Scheibler, F.; Hannappel, S. Inventory of managed aquifer recharge sites in Europe: Historical development, current situation and perspectives. *Hydrogeol. J.* **2017**, *25*, 1909–1922. [CrossRef]
11. Stefan, C.; Ansems, N. Web-GIS of global inventory of managed aquifer recharge applications. In Proceedings of the 9th International Symposium on Managed Aquifer Recharge (ISMAR9), Mexico City, Mexico, 20–24 June 2016.
12. MDDELCC. *Rapport sur l'état de l'eau et des écosystèmes aquatiques au Québec*; Government of Quebec: Quebec City, QC, Canada, 2014.

13. Boyer, M. *Freshwater Exports for the Development of Quebec's Blue Gold*; Montreal Economic Institute: Montreal, QC, Canada, 2008.
14. Government of Quebec. *Water withdrawal and Protection Regulation-Environment Quality Act*; Publications Quebec: Québec, QC, Canada, 2019; Volume Q-2, p. 56.
15. Ministère de l'Environnement et Lutte contre les changements climatiques. *Guide de Conception des Installations de Production d'eau Potable*; Direction générale des politiques de l'eau, Government of Quebec: Quebec, QC, Canada, 2019; Volume 1, p. 297.
16. Stuyfzand, P.J. Hydrology and water quality aspects of rhine bank groundwater in The Netherlands. *J. Hydrol.* **1989**, *106*, 341–363. [CrossRef]
17. Massmann, G.; Dünnbier, U.; Heberer, T.; Taute, T. Behaviour and redox sensitivity of pharmaceutical residues during bank filtration-Investigation of residues of phenazone-type analgesics. *Chemosphere* **2008**, *71*, 1476–1485. [CrossRef] [PubMed]
18. Masse-Dufresne, J.; Baudron, P.; Barbecot, F.; Patenaude, M.; Pontoreau, C.; Proteau-Bédard, F.; Menou, M.; Pasquier, P.; Veuille, S.; Barbeau, B. Anthropic and Meteorological Controls on the Origin and Quality of Water at a Bank Filtration Site in Canada. *Water* **2019**, *11*, 2510. [CrossRef]
19. Paufler, S.; Grischek, T.; Benso, M.; Seidel, N.; Fischer, T. The Impact of River Discharge and Water Temperature on Manganese Release from the Riverbed during Riverbank Filtration: A Case Study from Dresden, Germany. *Water* **2018**, *10*, 1476. [CrossRef]
20. Paufler, S.; Grischek, T.; Bartak, R.; Ghodeif, K.; Wahaab, R.; Boernick, H. Riverbank filtration in Cairo, Egypt: Part II-detailed investigation of a new riverbank filtration site with a focus on manganese. *Environ. Earth Sci.* **2018**, *77*, 318. [CrossRef]
21. Paufler, S.; Grischek, T. Herkunft und Verhalten von Mangan bei der Uferfiltration. *Grundwasser* **2018**, *23*, 277–296. [CrossRef]
22. Grischek, T.; Paufler, S. Prediction of Iron Release during Riverbank Filtration. *Water* **2017**, *9*, 317. [CrossRef]
23. Romero-Esquivel, L.; Grischek, T.; Pizzolatti, B.; Mondardo, R.; Sens, M. Bank filtration in a coastal lake in South Brazil: Water quality, natural organic matter (NOM) and redox conditions study. *Clean Technol. Environ. Policy* **2017**, *19*, 2007–2020. [CrossRef]
24. Lee, H.; Koo, M.; Kim, Y. Impacts of Seasonal Pumping on Stream-Aquifer Interactions in Miryang, Korea. *Groundwater* **2017**, *55*, 906–916. [CrossRef]
25. van Driezum, I.; Derx, J.; Oudega, T.; Zessner, M.; Naus, F.; Saracevic, E.; Kirschner, A.; Sommer, R.; Farnleitner, A.; Blaschke, A. Spatiotemporal resolved sampling for the interpretation of micropollutant removal during riverbank filtration. *Sci. Total Environ.* **2019**, *649*, 212–223. [CrossRef]
26. Dragon, K.; Górski, J.; Kruć, R.; Drożdżyński, D.; Grischek, T. Removal of Natural Organic Matter and Organic Micropollutants during Riverbank Filtration in Krajkowo, Poland. *Water* **2018**, *10*, 1457. [CrossRef]
27. Trásy, B.; Kovács, J.; Hatvani, I.; Havril, T.; Németh, T.; Scharek, P.; Szabó, C. Assessment of the interaction between surface-and groundwater after the diversion of the inner delta of the River Danube (Hungary) using multivariate statistics. *Anthropocene* **2018**, *22*, 51–65. [CrossRef]
28. Moeck, C.; Radny, D.; Popp, A.; Brennwald, M.; Stoll, S.; Auckenthaler, A.; Berg, M.; Schirmer, M. Characterization of a managed aquifer recharge system using multiple tracers. *Sci. Total Environ.* **2017**, *609*, 701–714. [CrossRef] [PubMed]
29. Rose, A.; Fabbro, L.; Kinnear, S. Cyanobacteria breakthrough: Effects of Limnothrix redekei contamination in an artificial bank filtration on a regional water supply. *Harmful Algae* **2018**, *76*, 1–10. [CrossRef] [PubMed]
30. Pazouki, P.; Prevost, M.; McQuaid, N.; Barbeau, B.; de Boutray, M.L.; Zamyadi, A.; Dorner, S. Breakthrough of cyanobacteria in bank filtration. *Water Res.* **2016**, *102*, 170–179. [CrossRef]
31. Grützmacher, G.; Wessel, G.; Klitzke, S.; Chorus, I. Microcystin Elimination during Sediment Contact. *Environ. Sci. Technol.* **2010**, *44*, 657–662. [CrossRef]
32. Kumar, P.; Mehrotra, I.; Gupta, A.; Kumari, S. Riverbank Filtration: A sustainable process to attenuate contaminants during drinking water production. *J. Sustain. Dev. Energy* **2018**, *6*, 150–161. [CrossRef]
33. Sallwey, J.; Bonilla Valverde, J.; Vásquez López, F.; Junghanns, R.; Stefan, C. Suitability maps for managed aquifer recharge: A review of multi-criteria decision analysis studies. *Environ. Rev.* **2018**, *27*, 138–150. [CrossRef]
34. Wang, L.; Ye, X.; Du, X. Suitability evaluation of river bank filtration along the Second Songhua River, China. *Water* **2016**, *8*, 176. [CrossRef]

35. Lee, S.-I.; Lee, S.-S. Development of site suitability analysis system for riverbank filtration. *Water Sci. Eng.* **2010**, *3*, 85–94. [CrossRef]
36. Jaramillo Uribe, M. Evaluation of the Potential for Riverbank Filtration in Colombia. Ph.D. Thesis, Universidad Nacional de Colombia-Sede Medellín, Medellin, CO, USA, 2015.
37. Singleton, A.; Spielman, S.; Brunsdon, C. Establishing a framework for Open Geographic Information science. *Int. J. Geogr. Inf. Sci.* **2016**, *30*, 1507–1521. [CrossRef]
38. Malczewski, J.; Rinner, C. GIScience, Spatial Analysis, and Decision Support. In *Multicriteria Decision Analysis in Geographic Information Science*; Malczewski, J., Rinner, C., Eds.; Springer: Berlin/Heidelberg, Germany, 2015; pp. 3–21. [CrossRef]
39. MELCC. *Système d'information Hydrogéologique*; MELCC, Ed.; MELCC: Quebec, QC, Canada, 2018.
40. Sterckx, A. Étude des facteurs influençant le rendement des puits d'alimentation de particuliers qui exploitent le roc fracturé en Outaouais, Québec, Canada. Master's Thesis, Université Laval, Québec, QC, Canada, 2013.
41. INRS-ETE. Protocole pour la préparation des livrables: 15-Estimation de l'épaisseur des formations superficielles et 16-Topographie du roc. In *Programme d'acquisition de Connaissances sur les eaux Souterraines du Québec*; INRS-ETE: Quebec City, QC, Canada, 2012; Unpublished work.
42. Ministère de l'Environnement et Lutte contre les changements climatiques. *Programme d'acquisition de Connaissances sur les Eaux Souterraines (PACES)*; Government of Quebec: Quebec City, QC, Canada, 2009–2015.
43. Ghesquière, O.; Walter, J.; Chesnaux, R.; Rouleau, A. Scenarios of groundwater chemical evolution in a region of the Canadian Shield based on multivariate statistical analysis. *J. Hydrol. Reg. Stud.* **2015**, *4*, 246–266. [CrossRef]
44. Beaudry, C.; Lefebvre, R.; Rivard, C.; Cloutier, V. Conceptual model of regional groundwater flow based on hydrogeochemistry (Montérégie Est, Québec, Canada). *Can. Water Resour. J. Rev. Canadienne des Ressources Hydriques* **2018**, *43*, 152–172. [CrossRef]
45. Nadeau, S.; Rosa, E.; Cloutier, V. Stratigraphic sequence map for groundwater assessment and protection of unconsolidated aquifers: A case example in the Abitibi-Témiscamingue region, Québec, Canada. *Can. Water Resour. J. Rev. Canadienne des Ressources Hydriques* **2018**, *43*, 113–135. [CrossRef]
46. Gagné, S.; Larocque, M.; Pinti, D.; Saby, M.; Meyzonnat, G.; Méjean, P. Benefits and limitations of using isotope-derived groundwater travel times and major ion chemistry to validate a regional groundwater flow model: Example from the Centre-du-Québec region, Canada. *Can. Water Resour. J. Rev. Canadienne des Ressources Hydriques* **2018**, *43*, 195–213. [CrossRef]
47. MELCC. *Demande d'accès no. 2017-09-151: Sites de prelevement d'eau municipa ux et installations alimentées.xlsx*; MELCC, Ed.; Ministère de l'environnement et lutte contre les changements climatiques: Québec, QC, Canada, 2017.
48. NRCAN. *Canvec*; Canada, N.R., Ed.; Government of Canada: Ottawa, ON, Canada, 2016.
49. Team, Q.D. *QGIS Geographic Information System*; Open Source Geospatial Foundation: Beaverton, OR, USA, 2020.
50. Geomatic Solutions. Georepository-NAD83/Quebec Lambert. Available online: https://georepository.com/crs_32198/NAD83-Quebec-Lambert.html (accessed on 16 November 2019).
51. Conrad, O.; Bechtel, B.; Bock, M.; Dietrich, H.; Fischer, E.; Gerlitz, L.; Wehberg, J.; Wichmann, V.; Böhner, J. System for automated geoscientific analyses (SAGA) v. 2.1. 4. *Geosci. Model. Dev.* **2015**, *8*, 1991–2007. [CrossRef]
52. Team, R.C. *R: A Language and Environment for Statistical Computing*; R Foundation for Statistical Computing: Vienna, Austria, 2019.
53. Larocque, M.; Fortin, V.; Pharand, M.; Rivard, C. Groundwater contribution to river flows-using hydrograph separation, hydrological and hydrogeological models in a southern Quebec aquifer. *Hydrol. Earth Syst. Sci. Discuss.* **2010**, *2010*, 7809–7838. [CrossRef]
54. Stuyfzand, P.; Juhàsz-Holterman, M.; de Lange, W. Riverbank Filtration in the Netherlands: Well Fields, Clogging and Geochemical Reactions. In *Proceedings of Riverbank Filtration Hydrology*; Springer: Dordrecht, The Netherlands, 2006; pp. 119–153.
55. Grischek, T.; Schoenheinz, D.; Syhre, C.; Saupe, K. Bank filtration practise in the German Federal State of Saxony. Available online: https://www.researchgate.net/publication/265098211_Bank_filtration_practise_in_the_German_Federal_State_of_Saxony (accessed on 26 February 2020).

56. Rousseau, A.N.; Mailhot, A.; Slivitzky, M.; Villeneuve, J.-P.; Rodriguez, M.J.; Bourque, A. Usages et approvisionnement en eau dans le sud du Qubec Niveau des connaissances et axes de recherche privilgier dans une perspective de changements climatiques. *Can. Water Resour. J. Rev. Canadienne des Ressources Hydriques* **2004**, *29*, 121–134. [CrossRef]
57. Ministère de l'Environnement et Lutte contre les changements climatiques. Aires protégées au Québec: Les Provinces Naturelles-Niveau I du Cadre écologique de référence du Québec-Les Principaux Descripteurs des Provinces Naturelles. Available online: http://www.environnement.gouv.qc.ca/biodiversite/aires_protegees/provinces/partie3.htm (accessed on 16 November 2019).
58. Lajoie, P.G. *Les sols des Comtés d'Argenteuil, Deux-Montagnes et Terrebonne (Québec)*; Service de recherches, ministere de l'Agriculture de Quebec et le College Macdonald, Universite, McGill: Montreal, QC, Canada, 1960.
59. Institute de recherche et de développement en agroenvironnement. Études pédologiques. Available online: https://www.irda.qc.ca/fr/services/protection-ressources/sante-sols/information-sols/etudes-pedologiques/ (accessed on 20 November 2019).
60. Larocque, M.; Gagné, S.; Barnetche, D.; Meyzonnat, G.; Graveline, M.; Ouellet, M. *Projet de connaissance des eaux souterraines de la zone Nicolet et de la partie basse de la zone Saint-François*; Université du Québec à Montréal: Québec, QC, Canada, 2015.
61. Larocque, M.; Meyzonnat, G.; Barbecot, F.; Pinti, D.; Gagné, S.; Barnetche, D.; Ouellet, M.; Graveline, M. *Projet de connaissance des eaux souterraines de la zone de Vaudreuil-Soulanges*; Université du Québec à Montréal: Québec, QC, Canada, 2015.

© 2020 by the authors. Licensee MDPI, Basel, Switzerland. This article is an open access article distributed under the terms and conditions of the Creative Commons Attribution (CC BY) license (http://creativecommons.org/licenses/by/4.0/).

Article

Reactive Barriers for Renaturalization of Reclaimed Water during Soil Aquifer Treatment

Cristina Valhondo [1,2,*], Jesús Carrera [1,2], Lurdes Martínez-Landa [2,3], Jingjing Wang [1,2,3], Stefano Amalfitano [4], Caterina Levantesi [4] and M. Silvia Diaz-Cruz [1]

- [1] Department of Environmental Chemistry, Institute of Environmental Assessment and Water Research (IDAEA), Severo Ochoa Excellence Center, Spanish National Research Council (CSIC), Jordi Girona 18-24, 08034 Barcelona, Spain; jesus.carrera.ramirez@gmail.com (J.C.); jingjing.wang.xiang@gmail.com (J.W.); silvia.diaz@idaea.csic.es (M.S.D.-C.)
- [2] Hydrogeology Group (UPC-CSIC), Associate Unit, Jordi Girona, 08034 Barcelona, Spain; lurdesm.landa@gmail.com
- [3] Department of Civil and Environmental Engineering, Universitat Politecnica de Catalunya (UPC), Jordi Girona 1-3, 08034 Barcelona, Spain
- [4] Water Research Institute, National Research Council, Via Salaria Km 29, 10-00015 Roma, Italy; amalfitano@irsa.cnr.it (S.A.); levantesi@irsa.cnr.it (C.L.)
- * Correspondence: cvalhondo@gmail.com

Received: 20 November 2019; Accepted: 27 March 2020; Published: 2 April 2020

Abstract: Managed aquifer recharge (MAR) is known to increase available water quantity and to improve water quality. However, its implementation is hindered by the concern of polluting aquifers, which might lead to onerous treatment and regulatory requirements for the source water. These requirements might make MAR unsustainable both economically and energetically. To address these concerns, we tested reactive barriers laid at the bottom of infiltration basins to enhance water quality improvement during soil passage. The goal of the barriers was to (1) provide a range of sorption sites to favor the retention of chemical contaminants and pathogens; (2) favor the development of a sequence of redox states to promote the degradation of the most recalcitrant chemical contaminants; and (3) promote the growth of plants both to reduce clogging, and to supply organic carbon and sorption sites. We summarized our experience to show that the barriers did enhance the removal of organic pollutants of concern (e.g., pharmaceuticals and personal care products). However, the barriers did not increase the removal of pathogens beyond traditional MAR systems. We reviewed the literature to suggest improvements on the design of the system to improve pathogen attenuation and to address antibiotic resistance gene transfer.

Keywords: organic amendments; managed aquifer recharge; contaminants of emerging concern (CECs); pathogens; new water challenges

1. Introduction

Climate change and the expansion of urban areas is a major worldwide threat to sustainable and safe drinking water supplies [1]. Managed aquifer recharge (MAR) is a technique that allows groundwater-dependent ecosystems, including rivers, to be maintained, enhanced, and/or protected with limited consumption of energy and chemicals [2,3]. MAR systems based on water filtration during soil passage have been proven to retain suspended particles and colloids, including microorganisms [4], and to favor biodegradation of chemical contaminants, resulting in significant water quality improvement [5–7]. The processes affecting pathogen transport in these aquifers are retention and inactivation, and an extensive number of factors influence them [8]. However, periodic detection of pathogens in groundwater, some with severe human health impacts [9–13], has led to strict quality requirements that effectively impede the use of lesser quality water for MAR. For instance, rainfall fails to meet Spanish regulations for reuse (too low pH and too high suspended solids), which are the regulations adopted in practice for MAR [14]. This is paradoxical because potable water treatment during the 19th century simply consisted of sand filtering to remove pathogens and resulted in a life expectancy increase of some 20 years [15,16]. This paradox is well reflected in the ongoing debate about quality requirements for artificial recharge. Health protection authorities recommend strict controls on the water used for MAR but, at the same time, several major cities have shown that recharge using wastewater can be safe [17]. As a result, the European Commission's Joint Research Center (JRC) failed to reach a consensus on MAR water quality recommendations [18]. The situation is inadequate. Prudence demands regulation, while fear hinders the actual implementation of MAR, which impedes the restoration of ecosystem services of groundwater-dependent water bodies.

Overcoming resistance requires the addressing of not only old problems (e.g., water scarcity, recovery of groundwater-dependent water bodies), but also emerging concerns [19]. Among these, we include chemicals of emerging concern (CECs), antibiotic-resistant bacteria (ARBs), and antibiotic resistance genes (ARGs). The term CECs encompasses a wide range of substances, including pharmaceuticals, personal care products, and nano- and micro-plastics, among others, which are characterized by their continuous release into the environment and their potential to impact aquatic ecosystems and eventually human health [20]. Several studies have demonstrated that even after extensive treatment, such as advanced oxidation processes and reverse osmosis, some recalcitrant CECs are still detectable in reclaimed water [21–23]. Until the turn of the millennium, it was unknown that these chemicals presented a hazard to the environment, as they generally occur at trace levels, and pharmaceuticals in particular were always found at concentrations far below the therapeutic doses prescribed for humans [24,25]. However, studies carried out since then have provided evidence that even sub-therapeutic concentrations of certain pharmaceuticals affect microbes, plants, fishes, and insects [26–28]. Consequently, the concentrations of CECs measured in reclaimed water can be biologically relevant or can increase to such levels in the unavoidable co-occurrence with other chemicals that may increase their biological activity [29]. Under the certainty that the reclaimed waters still contain CEC residues, the use of these waters as source waters in MAR may pose a risk to human and environmental health.

Biodegradation and sorption appear to be the main processes involved in water quality improvement during MAR, especially regarding CECs' behavior [30–32]. The biomass and biodiversity of the microbial community is relevant for CEC degradation [33,34] Therefore, parameters controlling microbial community such as temperature, and the amount of organic substrate available and its quality (which controls the redox conditions), have a direct effect on biodegradation rates [4,35,36]. Changing parameters within an aquifer could lead to an increase in the microbial biodiversity, and a continuous source of organic substrate should allow the biomass to increase [37].

Sorption might be relevant as well since it retards contaminants [38], thus increasing the time available for the microbial community to degrade them. The discussion on whether retardation is favorable remains open. On the one hand, increasing the residence time would increase the ability of microorganisms to degrade them. On the other hand, some authors argue that contaminants are not biologically available when adsorbed, and therefore they are not potentially biodegradable [39,40].

To favor these two processes, we proposed adding a reactive barrier at the bottom of the infiltration basin in a MAR system. The barrier provides a reactive surface and diverse sorption sites, and adds organic carbon to yield a range of redox states. Ideally, this should allow diverse microbial communities to develop, thus increasing CECs' removal.

In this context, the goal of this paper was two-fold. First, we summarized what we have learned in two experiences of MAR using a reactive barrier [6,7,40,41]. Second, based on this adquired knowledgeand the results of others, we have discussed how to improve the system design and operation to enhance not only the removal of CECs but also the attenuation of pathogens, to minimize the transport of ARGs.

2. Materials and Methods

2.1. The Concept of the Reactive Barrier

We designed a reactive barrier to be installed at the bottom of soil aquifer treatment (SAT, the specific term for intermittent infiltration of reclaimed water) infiltration basins. The barrier consisted on an organic substrate able to release dissolved organic carbon (DOC) to the infiltrated water and to provide potential sorption surfaces. The purpose of the reactive barrier is to favor biodegradation by generating a redox zonation and enhanced adsorption for the widest possible range of CECs.

Figure 1 displays redox zonation during infiltration periods in a conventional SAT system and in a system with a reactive barrier. The source water should contain a labile DOC concentration higher than 6–9 mg/L in order to consume the oxygen and start to consume nitrate as the next electron acceptor. The implementation of the reactive barrier increases the concentration of DOC in the recharged water, so that available electron acceptors are consumed and redox zonation is developed further, reaching Fe- and Mn-reducing conditions, and hopefully SO_4-reducing conditions. Conditions should return to aerobic during drying periods in both cases.

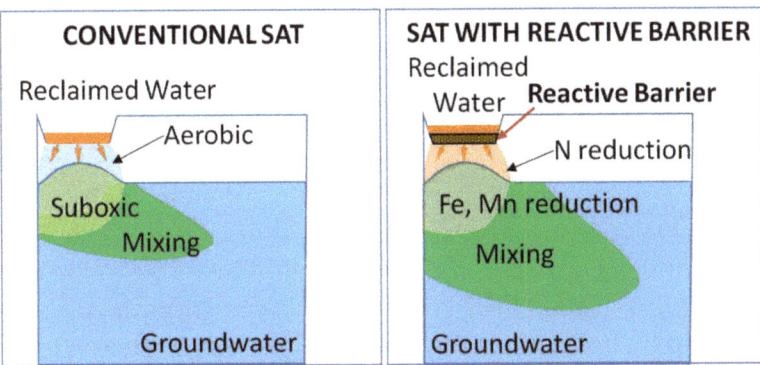

Figure 1. Schematic description of redox zonation during infiltration without and with a reactive barrier. The barrier adds dissolved organic carbon (DOC) to the recharge water, thus promoting highly reducing conditions. Ideally, the vadose zone becomes aerobic during drying periods in both cases.

This approach was tested at two sites and based on two organic substrates as organic carbon source, i.e., compost and woodchips. The first site was a pilot-scale MAR system located at Sant Vicenç dels Horts (close to Barcelona, Spain) where a reactive barrier based on compost was installed. The system operated with the barrier for four consecutive years. The second site consisted of six MAR systems with small variations in the configuration, located in Palamós (close to Gerona, Spain). One of the systems, the reference one, operated without a reactive barrier, four systems operated with reactive barriers based on compost, and the remaining system operated with a reactive barrier based on woodchips. To date, it appears that the implementation of these barriers has favored the infiltration capacity. The characteristics of each site and the performance of the tested reactive barriers were described in previous publications [6,7,41,42] and are summarized later.

Compost and woodchips were selected as organic substrates due to their capability to release DOC, their low cost, and the ease of their handling and transportation. The amount of DOC released by the organic substrate is expected to decrease with time (or with the volume of water infiltrated), but it may be compensated by the release from biomass growing in the basin, including plant roots. Still, after a period of operation, the barrier may have to be replaced.

2.2. Site Description

2.2.1. Sant Vicenç Dels Horts

The Sant Vicenç site is a complex of two basins (settlement and recharge, each ~5000 m^2) constructed at the side of the Llobregat River, some 15 km upstream of Barcelona (Figure 2A). The MAR system was constructed over the lower Llobregat valley sedimentary aquifer, formed mainly by gravels, sand, and a small fraction of clay [43]. The saturated aquifer thickness and vadose zone ranged from 12 to 14 m, and from 5 to 9 m, respectively, during the recharge experience (Figure 2C).

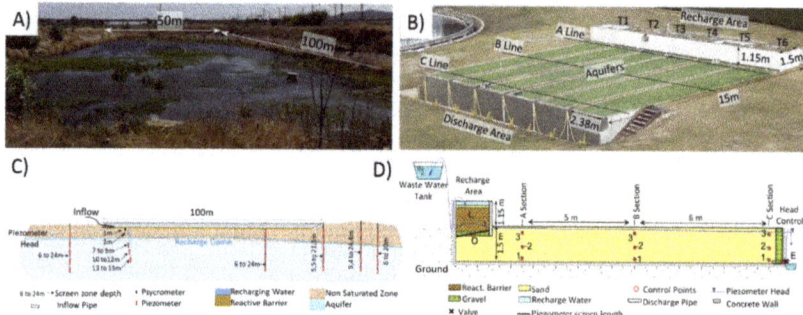

Figure 2. (**A**) Sant Vicenç dels Horts infiltration basin, and (**C**) the cross-section with the monitoring points and their screened sections; (**B**) six Palamós replicate MAR systems named T1 to T6, and (**D**) the cross-section of one generic replicate. Flow direction is from left to right in both cross-sections.

The MAR system was fed with the Llobregat River water, which is heavily impacted by wastewater treatment plants' (WWTPs') effluents [44]. The river water was diverted to the settlement basin, where it remained for 2 to 4 days. From there, water flowed to the recharge basin. Flow rate was measured hourly into the connecting pipe. The average infiltration rate was 1 m/d.

We installed a 65 cm thick reactive barrier on the bottom of the infiltration basin. This barrier consisted of vegetal compost and aquifer sediments in equal volumetric portions and a small quantity of clay and iron oxide. The role of the vegetable compost was to release degradable organic matter to the infiltrating water to favor changes in redox conditions underneath the basin, promoting microbiological diversity to enhance the removal of chemical contaminants [6,7,37,45], and to provide surfaces for neutral organic compound adsorption. Clay increased the sorption of cationic compounds and iron oxide facilitated the sorption of the anionic ones.

2.2.2. Palamós Site

The pilot MAR system was constructed in a municipal WWTP facility on the northeastern Spanish Mediterranean coast. This facility collects wastewater from several municipalities. The population served increases to include some 90,000 inhabitants during the summer, reaching the maximum treatment capacity of the plant. As a consequence, effluent water quality varies throughout the year.

We constructed six pilot recharge systems (15 × 15 m^2 excavated structures, divided into six 2.38 × 15 m channels; Figure 2B), to test the effect of the reactive barrier's composition and the role of plants on the fate of CECs and pathogens. The system was fed with the secondary treatment effluent of the WWTP, which infiltrated from the basin through the barrier and further flowed along the 15 m simulated aquifer, to finally discharge at the base of the 1.5 m thick aquifer. Indeed, the pilot MAR operated as a tertiary treatment (Figure 2D). In this case, two organic carbon sources were tested:

compost and wood chips. We assessed their performance by comparing the removal of more than 50 CEC and pathogen indicators to that of a reference system (infiltration without reactive barrier).

2.3. Analytical Methods

Pressure, temperature, and electrical conductivity were continuously recorded using conductivity, temperature and depth submersible dataloggers (CTD-Divers, Schlumberger water services, Delft, The Netherlands) in the source water and several monitoring points at San Vicenç dels Horts and Palamós sites (Figure 2C). Additionally, samples for chemical analysis were collected during several recharge events in both sites.

Target CECs were selected based on the frequency of their detection in the aquatic environment, and since the analytical methodology for each site was different, the final list of CECs analyzed in each case was defined according to the methodology requirements and the source water type (urban, hospital effluents, agricultural, or industrial).

At Sant Vicenç dels Horts, 51 CECs were analyzed in the collected samples following the method described by Nödler et al. [46] (Table S1). Briefly, the samples were allowed to settle overnight at 4 °C and the supernatant was recovered. A 500 mL aliquot of the supernatant was spiked with 10 µL of an internal standard solution and with 5 mL of a buffer solution before solid-phase extraction (SPE). The extraction and purification was performed using OASIS HLB (6 mL, 500 mg; Waters, Eschborn, Germany) cartridges. The analytes were eluted from the cartridges, and the extracts were evaporated with a stream of nitrogen and reconstituted with ammonium acetate solution before its transference to an auto-sampler LC-vial. The analyses were performed by high-performance liquid chromatography tandem-mass spectrometry (HPLC-MS/MS, Varian Inc., Palo Alto, CA, USA).

At the Palamós site, 58 CECs were determined by online solid-phase extraction coupled to high-performance liquid chromatography–tandem mass spectrometry (online-SPE-HPLC-MS/MS) in accordance with Gago-Ferrero et al. [47] (Table S2). In this method, water samples previously spiked with an isotopically labeled surrogate standard solution were isolated, pre-concentrated, and purified using an automated Symbiosis™ Pico online SPE-(Spark Holland; Emmen, the Netherlands). The online SPE of all samples, calibration standard solutions, and methodological blanks were performed by loading 5 mL of the water samples through PLRP-s cartridges. The trapped compounds were eluted from the cartridge to the HPLC column by the chromatographic mobile phase. The chromatographic separation was achieved with a HibarPurospher® STAR® HR R-18 ec. (50 mm × 2.0 mm, 5 µm) column from Merck using a mobile phase consisting of HPLC-grade water and acetonitrile, both with 0.1% formic acid for positive electrospray ionization, and with 5 mM ammonium acetate buffer (pH 6.8) for the negative ionization mode. MS/MS detection was performed on a 4000 Q TRAP™ MS/MS hybrid mass spectrometer from Applied Biosystems-Sciex (Foster City, CA, USA). Selected reaction monitoring (SRM) mode was applied for improved selectivity and sensitivity. Four identification points were considered, in compliance with the European Council Directive 2002/657/EC [48]

Additionally, a microbiological analysis was carried out at the Palamós site. Gram-positive and Gram-negative fecal bacteria indicator analysis was done via most probable number (MPN) detection tests following manufacturer's instructions (Colilert IDEXX, US). The log removal values were calculated considering the average of total coliforms and *Escherichia coli* concentrations in the source water, in the monitoring point immediately below the reactive barrier (O-points), and effluents (E-points) for each MAR system of Palamós site (Figure 2D). We analyzed also other substances, namely DOC, cations and anions, but they are out of the focus of the present study.

2.4. Assesing the Reactive Barrier Efficiency

We estimated first-order degradation rates (λ) and retardation coefficient (R) for 10 CECs at Sant Vicenç dels Horts site [41,49] and compared them with those reported in the literature from other experiments. While λs can be highly uncertain, they can be considered "relative measures for comparison" of results [50]. To this end, we measured pressure, electrical conductivity, and temperature at different distances from the infiltration basin. We also performed a pulse injection tracer test to obtain the residence time distribution of the recharged water at six monitoring points. The heads and breakthrough curves were used to estimate the flow and conservative transport parameters of the aquifer using a quasi-3D numerical model [49]. The model was built using the finite element code Transdens [51–53]. Secondly, CEC concentrations measured in the source water and nine monitoring points were used to estimate λ and R (in three and two defined subdomains, respectively) for 10 CECs [41,49]. The subdomains for λ estimation were BARR (reactive barrier), UZ (unsaturated zone), and AQF (aquifer). The subdomains for R estimation were BAR (reactive barrier) and UZ+AQF (unsaturated zone and aquifer). Pathogen indicators were not analyzed at the Sant Vicenç dels Horts site.

At the Palamós site, we compared the reduction in the concentration of 58 CECs, classified into four groups namely UV filters, paraben preservatives, pharmaceuticals, and total contaminants' load Table S2). Besides, the pathogen indicators (total coliforms and *E. coli*) were also measured along the reference system and compared to those obtained for the systems operating with the two reactive barriers to assess the efficiencies of these designs.

3. Results and Discussion

3.1. CECs Behavior

Figure 3 shows the comparison among the estimated degradation rates, λ (A), and retardation coefficients, R (B), at Sant Vicenç dels Horts and those reported from other studies carried out both in laboratory experiments and at field sites [54–62]. Considering the typical maximum duration for columns (1–2 years) and field experiments (10–20 years), the minimum value for λ was allowed to be 10^{-3}, and 10^{-4} d^{-1}, respectively, following the approach from Greskowiak et al. [63].

Estimated λs for the Sant Vicenç dels Horts site, operating with the reactive barrier, were similar to or higher than those reported in the literature. Indeed, the λs estimated for the reactive barrier subdomain (BARR) tended to be much larger than literature values, whereas the λs estimated for the aquifer domain (AQU) were comparable, suggesting the proper performance of the reactive barrier. Estimated Rs were also much higher than literature values for the barrier (BARR) and comparable for the aquifer (UZ + AQU). At this site, we did not have the opportunity to operate the system with and without reactive barrier simultaneously while keeping the remaining variables identical to assess the performance of the barrier. However, the proper assessment of the barriers' efficiency was performed at the Palamós site. There, we estimated removal efficiencies of the analyzed CECs for the reference system (T2) and the systems operating with the two reactive barriers (T4 and T5). For the sake of clarity, we compiled the information on the analyzed CECs grouped into four categories, as aforementioned i.e., UV filters (ΣUVF), paraben preservatives (ΣPBs), pharmaceuticals (ΣPhAC), and total load of the analyzed CECs (ΣTOTAL) (Table S2).

Figure 4 displays the removal efficiencies for the target contaminants (Table S2), estimated for the reference system (T2) and the systems operating with reactive barriers (T4 and T5) in two recharge periods, i.e., January and March 2018. Overall, significant removal, from 40% to 100%, was observed in all three systems, supporting the robustness of MAR in improving recharged water quality. Additionally, the systems operating with the reactive barrier performed equally well to or better than the reference system, indicating that the reactive barrier was successful at enhancing CEC removal. However, differences between the two barrier types' efficiencies were compound-dependent. This finding was in agreement with the results reported from a laboratory study carried out by Bertelkamp et al. [64], who investigated the sorption and biodegradation behavior of 14 CECs in soil columns under oxic conditions. They concluded that the presence of ethers and carbonyl groups increased biodegradability, whereas ring structures, amines, aliphatic ethers, and sulfides hindered degradation [64].

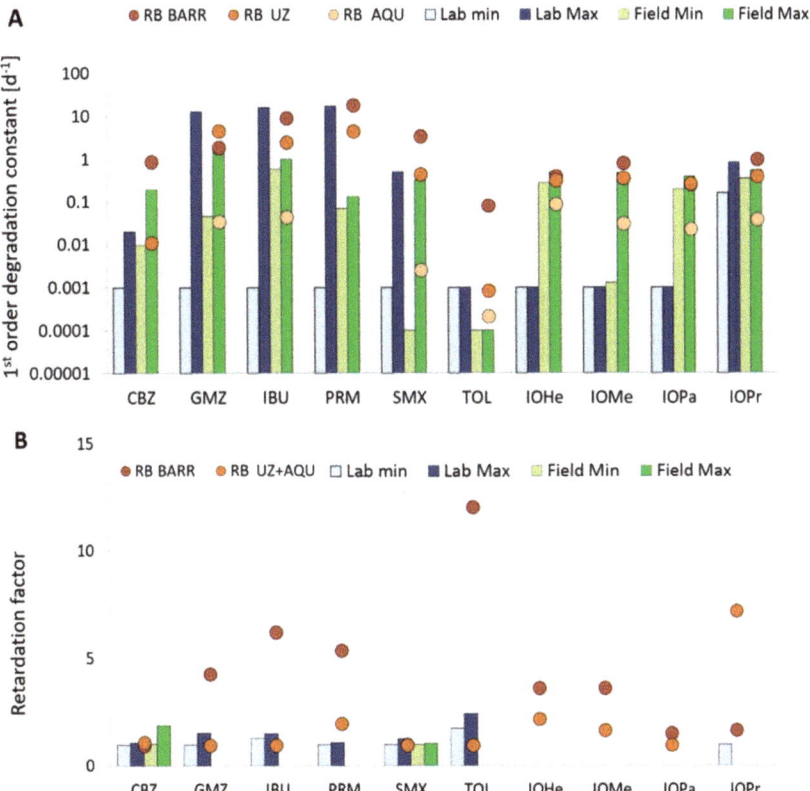

Figure 3. (**A**) First-order degradation rates (λ) (estimated for the three domains) and (**B**) retention factors (R) (estimated for two domains) at the Sant Vicenç dels Horts site for carbamazepine (CBZ), gemfibrozil (GMZ), ibuprofen (IBU), primidone (PRM), sulfamethoxazole (SMX), tolyltriazole (TOL), iohexol (IOHe), iomeprol (IOMe), iopamidole (IOPa), and iopromide (IOPr) and comparison with data from the literature obtained in field and laboratory column experiments.

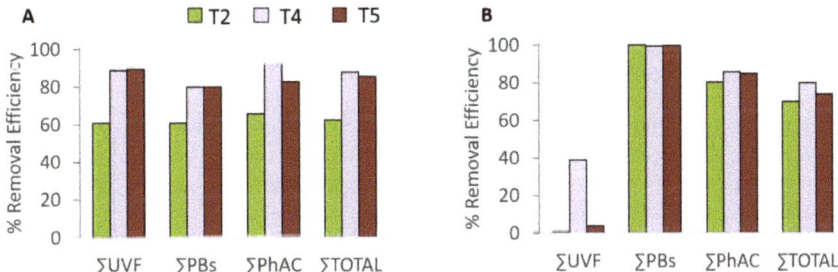

Figure 4. Removal efficiencies for organic UV filters (ΣUVF), paraben preservatives (ΣPB), pharmaceuticals (ΣPhAC), and total CECs (ΣTOTAL) estimated at the effluent of the reference system (T2), the system operating with the reactive barrier based on compost (T4), and the system operating with the barrier based on woodchips (T5) during (**A**) January 2018 and (**B**) March 2018 recharge episodes.

Our results demonstrate that a reactive barrier improved the removal of CECs, but other options are possible. Several authors have reported that oxic conditions and low biodegradable dissolved organic carbon (BDOC) favor CECs' degradation [57,65], whereas others suggest that anoxic conditions favor the degradation of a broader range of CECs [6,7,66–68]. Regnery et al. [69] proposed a method to reduce BDOC while boosting aerobic conditions during MAR by coupling two MAR systems, riverbank filtration followed by an aeration step prior to soil aquifer treatment. The goal was to reduce the BDOC during the riverbank filtration and to induce aerobic conditions with the aeration, providing oxic and low-BDOC conditions to the second MAR system [69,70]. In our experiences, we achieved reducing conditions during the recharge, assuming that aerobic conditions would be reached in the aquifer after recharged water mixed with native groundwater. This approach reduces the demands on the technique, since only one MAR system is needed. The issue is not settled, but it is clear that the optimal design of a MAR system is driven by several factors, including the water source, the hydrological characteristics of the aquifer, the geographical situation, and land availability, among others. Therefore, it is desirable to have a large set of MAR configuration options to select the most suitable for each particular case.

3.2. Pathogen Removal

The barriers' efficiency at reducing total coliforms and *E. coli* at Palamós during the recharge episode in March 2018 is shown in Figure 5. We compared their concentrations in the source water (INF) to those immediately below the vadose zone (O) and in the effluent (E) for the reference system (T2) and the systems operating with the reactive barriers based on compost and woodchips (T4 and T5, respectively). Figure 5 displays reductions between 2.5 and 5 log units for both, total coliforms and *Escherichia coli*. (*E. coli*). Similar results were observed during other recharge episodes. Overall, few or no differences among the systems were observed.

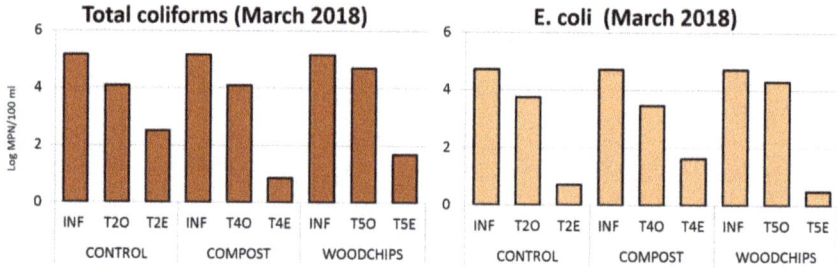

Figure 5. Concentrations (log MPN/100 mL) of total coliforms and *Escherichia coli* (*E. coli*)measured at the inflow (INF), immediately below the vadose zone (O-points), and effluent (E-points) of the reference system (T2, control) and the systems operating with reactive barriers based on compost (T4) and woodchips (T5) during the March 2018 recharge episode.

These results suggest that the barrier did not increase the attenuation of pathogens beyond traditional MAR systems, and, therefore, the system needs to be improved. Unfortunately, pathogen removal has traditionally been taken for granted in aquifers and, as a consequence, conceptual understanding is limited so far. Still, much research is available in the sand filtration literature, both rapid and slow sand filtration (more relevant for MAR) [71,72]. Materials other than sand have also been studied. Perez-Mercado et al. [73] explored the performance of biochar in reducing bacterial indicators from wastewater, with varying success.

While the extent of the supporting evidence is highly variable, the consensus is that pathogen fate is governed by two processes: retention and inactivation.

Retention refers to the immobilization of pathogens by straining or adsorption [8]. Straining refers to physical blocking of particles at small pores, and it is often assumed to occur at pores smaller than the bacteria size [74]. This contradicts what is known about colloidal straining, where filtration is largely caused by the lumping of particles, but it is hard to falsify. Adsorption refers to the retention of particles by electrostatic forces (actually, the set of mechanisms is much broader, ranging from diffusion into immobile water pores to the formation of surface complexes). Adsorption is usually explained using the double-, sometimes triple-layer theory.

Regardless of the actual mechanism (for quantification and upscaling purposes, identifying the retention mechanism is important), several factors affect pathogen retention, including size and accessible surface (in practice, they are hard to separate, since a small grain size leads to a high specific surface), hydraulic gradient (rather than velocity), pH (which controls the surface charge of both microorganisms and mineral surfaces), temperature, ionic strength (which affects surface potential, so that both increased and reduced retentions have been reported), and biofilm [75]. While a unified approach needs to be synthesized from the extensive literature on this topic, it is clear that a broad range of surface types and chemical states (ionic strength and pH appear to be the most controlling parameters) favor retention [76].

Inactivation is more difficult to ascertain. Pathogens tend to die outside the human body, which provides optimal conditions for their survival. The question, therefore, is how long it takes and whether it can be confirmed. Inactivation may occur in the liquid phase or, after adsorption, in the solid phase. Numerous factors contribute, including pH (in general, acidic conditions facilitate removal), temperature (survival has been observed to decrease with increasing temperature), and the presence of predators [77]. The conclusion from these studies is that most pathogens die off after a few weeks of residence in the soil. However, a few form spores or adopt spore-like forms that become inactive under unfavorable conditions but "resuscitate" when these conditions are favorable. For these, ascertaining elimination at the solid phase is critical [78].

It is precisely this variability in responses that underscores the need to observe a range of microorganisms. Given the impossibility to analyze all of them, it is common to use "indicator microorganisms" (bacteriophages, *E. coli*, *Cryptosporidium*, *Clostridium*).

The presence of metals favors inactivation. Urfer [79] showed that the addition of aluminum to (slow) sand filters enhanced the removal of bacteria. This is relevant because reclaimed water is often rich in aluminum (generally used for promoting flocculation during primary wastewater treatment). This aluminum will not precipitate as bauxite, but as gibbsite [$Al(OH)_3$] at neutral pH. Still, its trivalent state should favor retention and would explain why biofilm ageing increases pathogen retention. Park et al. [76] conducted column experiments with *Cryptosporidium parvum* oocysts (a frequent indicator microorganism) and found that retention was greatly enhanced by the presence of iron coatings on the sand medium and that suspended illite clay drastically enhanced oocyst deposition. Increasing ionic strength (up to a certain value) and decreasing pH also enhanced attachment efficiency.

Pathogenic behavior has been studied in numerous artificial recharge sites. Weiss et al. [80] calculated reductions from 2 to 4 log units for aerobic spores and 5.5 log units for total coliforms after travel distances between 24 and 177 m in three different riverbank filtration areas in the USA.

Bekele et al. [81] conducted a 39 month study on changes in water quality during infiltration through an unsaturated zone of 9 m and concluded that the elimination of microbial species was efficient: it detected adenovirus in only 6% and enteric viruses in 4% of the samples after 4.2 days.

Betancourt et al. [82] evaluated the elimination of enteric viruses in three artificial recharge facilities in the USA: one induced recharge (Colorado), an infiltration basin in Tucson, and another one in California. They concluded that enteric viruses group was below the bioanalytical method limit of detection after 5 days of transit time, and that residence time played a key role in the elimination of pathogens. The only infectious virus detected in the study was a reovirus. As it is difficult to associate reoviruses with a specific disease, they have not been paid much attention. However, they were found at higher concentrations than enteroviruses in treated and untreated wastewater, proving to be more resistant to UV disinfection than enteroviruses. Moreover, it appears that they also survive in water for long periods of time. According to these outcomes, reoviruses should be monitored in MAR system studies.

Elkayam et al. [83] studied several indicators at Shafdan (Israel), which recharges secondary effluent from a WWTP through 54 infiltration basins, covering an area of 270 ha in the coastal aquifer of Israel. The aquifer is formed of calcareous sandstones with intercalations of conglomerates, silts, and clay layers. Aquifer thickness is 180–200 m. The unsaturated area under the basins is 30–40 m deep. The wells are placed in two rings around the basins. The first ring extracts only recharged water from the rafts, whereas the second extracts also between 15% and 30% of water from the aquifer. The system has been operating on a large scale for more than 30 years. The study focused on indicators of bacteria, pathogenic viruses, coliphages, microbial source-tracking indicators (MST), and ARGs. The results showed a complete elimination of pathogenic viruses (enteroviruses, adenoviruses, noroviruses, parechoviruses), coliphages, and indicators of total and fecal bacteria and coliforms, fecal streptococcus, and bacteroides (MST) in the unsaturated area under the rafts. ARGs were detected in several wells, but they were also found in wells not impacted by the effluent, suggesting that these genes were related to the native microbial communities of the aquifer.

Beyond the actual mechanism controlling the fate of pathogens, it is important to take into account that (1) the soil is a living organism in itself, and (2) a broad range of pathogens, with different properties, may be present in treated wastewater or other source waters. The former implies that the soil system will be sensitive to external perturbations and its behavior will evolve in time. This, together with the specificities of every microorganism type, may explain the broad range of often contradictory sets of results reported in the literature.

The role of redox state has hardly been analyzed [84], which may reflect that most work is motivated by sand filters. Since we argue that a sequence of redox states improves the removal of CECs, it would also be desirable to understand the effect of redox state on the fate of pathogens. However, the indirect impact of redox variability becomes apparent. Bringing the water back into aerobic conditions favors the oxidation of iron (ferrous iron is mobile, while ferric iron tends to precipitate as goethite or its precursors). The positively charged surfaces of ferric oxides should retain pathogens that are mobile in other environments [84].

Thick unsaturated zones consistently lead to excellent removal rates of pathogens, and specifically viruses. While we attribute this success to air–water interfaces, which tend to retain colloids, other factors may be at play. The thickness of the unsaturated zone can only be controlled by pumping or, at the design stage, by site selection. Unfortunately, we did not have a thick unsaturated zone at our Palamós pilot site.

3.3. Antibiotic-Resistant Bacteria (ARB) and Antibiotic-Resistant Genes (ARGs)

Antibiotic resistance is a growing issue. Resistance is one of a number of consequences of the misuse and overuse of human and veterinary antibiotic drugs [85–88]. Antibiotics are substances that prevent the growth of bacteria. The term encompasses a wide range of pharmaceuticals with quite different physicochemical properties that are used in the treatment of bacterial infections, and also as prophylaxis for cattle and poultry. However, over the past decade, bacteria have been found to resist the drugs developed to suppress their growth and biological activity. In other words, bacterial strains are developing antibiotic resistance.

According to a recent report on the occurrence of antibiotic resistance in the USA [89], almost 3 million antibiotic-resistant infections are diagnosed every year in the USA, resulting in the deaths of more than 35,000 people. Globally, 10 million deaths per year caused by antibiotic-resistant bacterial infections are expected by 2050. Antibiotic-resistant bacteria, along with their resistance genes, are spread globally among foodstuffs, animals, plants, people, and the environment [90]. In this context, the World Health Organization (WHO) proposed a holistic action plan on microbial resistance involving humans and the environment [91], and the European Commission launched "One Health" as the European Action Plan to fight against antimicrobial resistance [92]. Additionally, the Watchlist for European Union monitoring defined in the Decisions 2018/840/EU (5 June 2018) included five antibiotics to be used to gather occurrence data to estimate the associated environmental risk of certain potentially hazardous compounds [93].

One of the reasons why ARGs cause great concern is because they are related to mobile genetic elements, and can therefore be easily transferred among microorganisms by horizontal gene transfer. This transfer can occur from bacteria, phages, free DNA, and dead cells to living cells [94].

WWTPs, in particular those treating urban sewage, have been recognized as one of the major receptors/sources of antibiotic resistance in the environment [88]. Even worse, ARGs may be enhanced in WWTPs [88]. The relationship between residual antibiotics in WWTPs and ARGs remains unclear. According to some studies, the presence of antibiotic residues during wastewater treatment may influence antibiotic resistance [87,95,96]. In contrast, other authors have pointed out that no correlation exists between the load of antibiotic residues in WWTPs and ARG abundance [97]. Nevertheless, a strong relationship between clinical and environmental antibiotic resistance has been reported. Temperature and humidity have been identified as two key factors controlling antibiotic resistance.

Since WWTPs are implicated as hotspots for the dissemination of antibiotic resistance into the environment and secondary effluents display higher relative abundance than the influents [98], it is important to assess whether ARGs could be reduced during soil aquifer treatment. Lack of a solid conceptual model for the fate of ARGs makes any proposal highly conjectural. However, some pieces of side evidence appear hopeful. The generation of anaerobic areas should help because the activity of microorganisms is reduced and the transfer of ARG is inhibited [99]. The presence of plants has proven efficient at removing ARGs in constructed wetlands [100]. The plant species selection deserves further research. In a recent study, the addition of biochar to soil resulted in notable changes in the microbial community, and these changes were different depending on the type of biochar used [101]. Changes in bacterial phylogenetic compositions can result in a change of ARGs. Therefore, the use of biochar as a component of reactive barriers might reduce ARGs. Nanotechnology may also be tested in MAR systems by including new nanomaterials in the infiltration pathway. To date, a few studies have pointed out the capability of selected nanomaterials to eliminate both ARB and ARGs [102].

3.4. Public Acceptance of MAR

A positive perception of MAR by the public is essential for its smooth implementation as a feasible and effective solution to increase water resources. The results of a public consultation about the use of treated wastewater in MAR operations were published recently [103]. Among the opposed respondents, the major concern was the lack of confidence in the wastewater treatment effectiveness before recharge. It is feared that if chemical and biological contaminants are recalcitrant to the treatments, recharge with WWTP effluents will lead to worsening groundwater quality. This kind of concern has led to broad negative public opinion on water reuse in general [104]. Negative perceptions may lead to failure [105–107].

In this regard, the implementation of operational strategies or designs to increase pathogen attenuation, chemical contaminant removal, and ARB and ARGs mitigation in MAR systems would facilitate public acceptance. If pilot studies are undertaken and results are successful, the MAR benefits supported by reliable scientific information should be stated and reasonably presented to the public. It is likely that the availability and understanding of this information will facilitate their positive perception and ultimately achieve public support.

Our concern, however, is that this may not suffice in times of "post-truth" and "fake news". Traditional approaches are needed: working with community organizations, promoting positive local media coverage of projects involving MAR, giving messages in clear non-technical language emphasizing its benefits and safety, offering public visits to the facilities, etc. Additional avenues of action that are specific to MAR include:

1. MAR (re)naturalizes water in that water quality improvement processes make it hard to distinguish from natural water;
2. Infiltration basins are beautiful, especially when covered with vegetation (Figure 2). This, together with the relatively large surface area of infiltration basins, suggests integrating them as part of landscape and territorial planning.

4. Conclusions and Current/Future Challenges in MAR

Our work supports the extensive literature body on water quality improvement during soil passage. Specifically, adding a reactive barrier improved the removal of chemical contaminants. Still, pathogen attenuation is significant (2–5 log units in our case), but was not particularly improved by the addition of the reactive barrier. Therefore, further improvement in the design of the reactive barriers and the operation of the system is needed.

We have discussed several options to enhance the degradation of recalcitrant chemical contaminants and the mitigation of pathogens. These include new compositions of the reactive barrier to broaden the types of sorption surfaces (biochar, zeolite, etc.), addition of metals to promote pathogen inactivation, implementation of thicker unsaturated zones to increase pathogen retention, and changes in the system operation to favor ferric oxide precipitation to create positively charged surfaces for further pathogen attenuation. We will test these approaches in the coming years at the Palamós pilot site and will assess the performance of the optimized system through the monitoring not only of CECs and pathogens, but also the development of antibiotic resistance, a serious emerging concern nowadays.

Public support must be achieved for the broad success of MAR. In the current context of climate change, where events of water scarcity and floods are occurring daily, improving water quality and increasing its quantity deserve determined action.

Supplementary Materials: Supplementary Materials are available online at http://www.mdpi.com/2073-4441/12/4/1012/s1.

Author Contributions: Conceptualization, C.V., J.C., L.M.-L., and M.S.D.-C.; Investigation and Methodology, C.V., L.M.-L., S.A., C.L., J.W. and M.S.D.-C.; Supervision, J.C. and M.S.D.-C. Writing—original draft, C.V., J.C. and M.S.D.-C.; Writing—review and editing, C.V., J.C., L.M.-L., S.A., C.L., J.W., and M.S.D.-C. All authors have read and agreed to the published version of the manuscript.

Funding: This research was funded by the Spanish Ministry of Science and Innovation CEX2018-000794-S), Water JPI (MARadentro-PCI2019-103603) and Catalan Water Agency (RESTORA-CA210/18/00040).

Acknowledgments: The authors acknowledge the Consorci Costa Brava for its constant support and the staff working at Palamós site for their unconditional help.

Conflicts of Interest: The authors declare no conflict of interest.

References

1. Wheeler, T.; Von Braun, J. Climate change impacts on global food security. *Science* **2013**, *341*, 508–513. [CrossRef] [PubMed]
2. Bouwer, H. Artificial recharge of groundwater: Hydrogeology and engineering. *Hydrogeol. J.* **2002**, *10*, 121–142. [CrossRef]
3. Dillon, P.; Stuyfzand, P.; Grischek, T.; Lluria, M.; Pyne, R.D.G.; Jain, R.C.; Bear, J.; Schwarz, J.; Wang, W.; Fernandez, E.; et al. Sixty years of global progress in managed aquifer recharge. *Hydrogeology* **2018**. [CrossRef]
4. Rauch-Williams, T.; Hoppe-Jones, C.; Drewes, J.E. The role of organic matter in the removal of emerging trace organic chemicals during managed aquifer recharge. *Water Res.* **2010**, *44*, 449–460. [CrossRef] [PubMed]
5. Patterson, B.M.; Shackleton, M.; Furness, A.J.; Bekele, E.; Pearce, J.; Linge, K.L.; Busetti, F.; Spadek, T.; Toze, S. Behaviour and fate of nine recycled water trace organics during managed aquifer recharge in an aerobic aquifer. *J. Contam. Hydrol.* **2011**, *122*, 53–62. [CrossRef] [PubMed]
6. Valhondo, C.; Carrera, J.; Ayora, C.; Barbieri, M.; Nödler, K.; Licha, T.; Huerta, M. Behavior of nine selected emerging trace organic contaminants in an artificial recharge system supplemented with a reactive barrier. *Environ. Sci. Pollut. Res* **2014**, 1–12. [CrossRef]
7. Valhondo, C.; Carrera, J.; Ayora, C.; Tubau, I.; Martinez-Landa, L.; Nödler, K.; Licha, T. Characterizing redox conditions and monitoring attenuation of selected pharmaceuticals during artificial recharge through a reactive layer. *Sci. Total Environ.* **2015**, *512*, 240–250. [CrossRef]
8. Stevik, T.K.; Aa, K.; Ausland, G.; Hanssen, J.F. Retention and removal of pathogenic bacteria in wastewater percolating through porous media: A review. *Water Res.* **2004**, *38*, 1355–1367. [CrossRef]

9. Hrudey, S.E.; Payment, P.; Huck, P.M.; Gillham, R.W.; Hrudey, E.J. A fatal waterborne disease epidemic in Walkerton, Ontario: Comparison with other waterborne outbreaks in the developed world. *Water Sci. Technol.* **2003**, *47*, 7–14. [CrossRef]
10. Craun, M.F.; Craun, G.F.; Calderon, R.L.; Beach, M.J. Waterborne outbreaks reported in the United States. *J. Water Health* **2006**, *4*, 19–30. [CrossRef]
11. Pedley, S.; Howard, G. The public health implications of microbiological contamination of groundwater. *Q. J. Eng. Geol.* **1997**, *30*, 179–188. [CrossRef]
12. Pandey, P.K.; Kass, P.H.; Soupir, M.L.; Biswas, S.; Singh, V.P. Contamination of water resources by pathogenic bacteria. *AMB Express* **2014**, *4*, 1–16. [CrossRef] [PubMed]
13. Marzouk, Y.; Goyal, S.M.; Gerba, C.P. Relationship of viruses and indicator bacteria in water and wastewater of Israel. *Water Res.* **1980**, *14*, 1585–1590. [CrossRef]
14. Boe, M.; Real, D. De 7 de diciembre, por el que se establece el régimen jurídico de la reutilización de las aguas depuradas. *boe 7 DICIEMBRE* **2007**, *1620*, 50639–50661.
15. Preston, S.H.; Van De Walle, E. *Urban French Mortality in the Nineteenth Century*; Taylor & Francis Ltd.: Abingdon, UK, 2016.
16. Preston, S.H.; Van De Walle, E. Urban French Mortality in the Nineteenth Century. *Popul. Stud.* **1978**, *32*, 275–297. [CrossRef]
17. Asano, T.; Cotruvo, J.A. Groundwater recharge with reclaimed municipal wastewater: Health and regulatory considerations. *Water Res.* **2004**, *38*, 1941–1951. [CrossRef] [PubMed]
18. Alcalde-Sanz, L.; Gawlik, B.M. Minimum Quality Requirements for Water Reuse in Agricultural Irrigation and Aquifer Recharge. Available online: https://publications.jrc.ec.europa.eu/repository/bitstream/JRC109291/jrc109291_online_08022018.pdf (accessed on 1 April 2020).
19. Dillon, P. Future management of aquifer recharge. *Hydrogeol. J.* **2015**, *13*, 313–316. [CrossRef]
20. EPA Contaminants of Emerging Concern including Pharmaceuticals and Personal Care Products. Available online: https://www.epa.gov/wqc/contaminants-emerging-concern-including-pharmaceuticals-andpersonal-care-products (accessed on 19 January 2019).
21. Kinney, C.A.; Furlong, E.T.; Werner, S.L.; Cahill, J.D. Presence and distribution of wastewater-derived pharmaceuticals in soil irrigated with reclaimed water. *Environ. Toxicol. Chem.* **2006**, *25*, 317–326. [CrossRef]
22. Urtiaga, A.M.; Pérez, G.; Ibáñez, R.; Ortiz, I. Removal of pharmaceuticals from a WWTP secondary effluent by ultrafiltration/reverse osmosis followed by electrochemical oxidation of the RO concentrate. *Desalination* **2013**, *331*, 26–34. [CrossRef]
23. Esplugas, S.; Bila, D.M.; Krause, L.G.T.; Dezotti, M. Ozonation and advanced oxidation technologies to remove endocrine disrupting chemicals (EDCs) and pharmaceuticals and personal care products (PPCPs) in water effluents. *J. Hazard. Mater.* **2007**, *149*, 631–642. [CrossRef]
24. Zimmermann, P.; Curtis, N. Antimicrobial Effects of Antipyretics. *Antimicrob. Agents Chemother.* **2017**, *61*, e02268–e02316. [CrossRef] [PubMed]
25. Sauvé, S.; Desrosiers, M. A review of what is an emerging contaminant. *Chem. Cent. J.* **2014**, *8*, 1–7. [CrossRef]
26. Wang, S.; Gunsch, C.K. Effects of selected pharmaceutically active compounds on the ammonia oxidizing bacterium Nitrosomonas europaea. *Chemosphere* **2011**, *82*, 565–572. [CrossRef] [PubMed]
27. Pennington, M.J.; Rothman, J.A.; Dudley, S.L.; Jones, M.B.; McFrederick, Q.S.; Gan, J.; Trumble, J.T. Contaminants of emerging concern affect Trichoplusia ni growth and development on artificial diets and a key host plant. *Proc. Natl. Acad. Sci. USA* **2017**, *114*, E9923–E9931. [CrossRef] [PubMed]
28. Pennington, M.J.; Rothman, J.A.; Jones, M.B.; McFrederick, Q.S.; Gan, J.; Trumble, J.T. Effects of contaminants of emerging concern on Myzus persicae (Sulzer, Hemiptera: Aphididae) biology and on their host plant Capsicum annuum. *Environ. Monit. Assess.* **2018**, *190*. [CrossRef] [PubMed]
29. Amariei, G.; Boltes, K.; Rosal, R.; Letón, P. Toxicological interactions of ibuprofen and triclosan on biological activity of activated sludge. *J. Hazard. Mater.* **2017**, *334*, 193–200. [CrossRef]
30. McMahon, P.B.; Chapelle, F.H. Microbial production of organic acids in aquitard sediments and its role in aquifer geochemistry. *Nature* **1991**, *349*, 233–235. [CrossRef]
31. Maeng, S.K.; Sharma, S.K.; Abel, C.D.T.; Magic-Knezev, A.; Amy, G.L. Role of biodegradation in the removal of pharmaceutically active compounds with different bulk organic matter characteristics through managed aquifer recharge: Batch and column studies. *Water Res.* **2011**, *45*, 4722–4736. [CrossRef]

32. Greskowiak, J.; Prommer, H.; Massmann, G.; Nützmann, G. Modeling seasonal redox dynamics and the corresponding fate of the pharmaceutical residue phenazone during artificial recharge of groundwater. *Environ. Sci. Technol.* **2006**, *40*, 6615–6621. [CrossRef]
33. Alidina, M.; Li, D.; Drewes, J.E. Investigating the role for adaptation of the microbial community to transform trace organic chemicals during managed aquifer recharge. *Water Res.* **2014**, *56*, 172–180. [CrossRef]
34. Li, D.; Sharp, J.O.; Saikaly, P.E.; Ali, S.; Alidina, M.; Alarawi, M.S.; Keller, S.; Hoppe-Jones, C.; Drewes, J.E. Dissolved Organic Carbon Influences Microbial Community Composition and Diversity in Managed Aquifer Recharge Systems. *Appl. Environ. Microbiol.* **2012**, *78*, 6819–6828. [CrossRef] [PubMed]
35. Alidina, M.; Shewchuk, J.; Drewes, J.E. Effect of temperature on removal of trace organic chemicals in managed aquifer recharge systems. *Chemosphere* **2015**, *122*, 23–31. [CrossRef] [PubMed]
36. Tran, N.H.; Urase, T.; Ngo, H.H.; Hu, J.; Ong, S.L. Insight into metabolic and cometabolic activities of autotrophic and heterotrophic microorganisms in the biodegradation of emerging trace organic contaminants. *Bioresour. Technol.* **2013**, *146*, 721–731. [CrossRef] [PubMed]
37. Li, D.; Alidina, M.; Ouf, M.; Sharp, J.O.; Saikaly, P.; Drewes, J.E. Microbial community evolution during simulated managed aquifer recharge in response to different biodegradable dissolved organic carbon (BDOC) concentrations. *Water Res.* **2013**, *47*, 2421–2430. [CrossRef] [PubMed]
38. Chefetz, B.; Mualem, T.; Ben-Ari, J. Sorption and mobility of pharmaceutical compounds in soil irrigated with reclaimed wastewater. *Chemosphere* **2008**, *73*, 1335–1343. [CrossRef] [PubMed]
39. Johnson, R.M.; Sims, J.T. Influence of surface and subsoil properties on herbigide sorption by Atlantic Coastal Plain soils. *Soil Sci.* **1993**, *155*, 339–348. [CrossRef]
40. Xu, J.; Wu, L.; Chen, W.; Chang, A.C. Adsorption and degradation of ketoprofen in soils. *J. Environ. Qual.* **2009**, *38*, 1177–1182. [CrossRef]
41. Valhondo, C.; Martinez-Landa, L.; Carrera, J.; Ayora, C.; Nödler, K.; Licha, T. Evaluation of EOC removal processes during artificial recharge through a reactive barrier. *Sci. Total Environ.* **2018**, *612*, 985–994. [CrossRef]
42. Valhondo, C.; Martínez-Landa, L.; Carrera, J.; Díaz-Cruz, S.M.; Amalfitano, S.; Levantesi, C. Six artificial recharge pilot replicates to gain insight into water quality enhancement processes. *Chemosphere* **2020**, *240*, 124826. [CrossRef]
43. Barbieri, M.; Carrera, J.; Sanchez-Vila, X.; Ayora, C.; Cama, J.; Köck-Schulmeyer, M.; de Alda, M.; Barceló, D.; Tobella Brunet, J.; Hernández García, M. Microcosm experiments to control anaerobic redox conditions when studying the fate of organic micropollutants in aquifer material. *J. Contam. Hydrol.* **2011**, *126*, 330–345. [CrossRef]
44. Köck-Schulmeyer, M.; Ginebreda, A.; Postigo, C.; López-Serna, R.; Pérez, S.; Brix, R.; Llorca, M.; de Alda, M.L.; Petrovic, M.; Munnì, A.; et al. Wastewater reuse in Mediterranean semi-arid areas: The impact of discharges of tertiary treated sewage on the load of polar micro pollutants in the Llobregat river (NE Spain). *Chemosphere* **2011**, *82*, 670–678. [CrossRef]
45. Alidina, M.; Li, D.; Ouf, M.; Drewes, J.E. Role of primary substrate composition and concentration on attenuation of trace organic chemicals in managed aquifer recharge systems. *J. Environ. Manag.* **2014**, *144*, 58–66. [CrossRef] [PubMed]
46. Nödler, K.; Licha, T.; Bester, K.; Sauter, M. Development of a multi-residue analytical method, based on liquid chromatography-tandem mass spectrometry, for the simultaneous determination of 46 micro-contaminants in aqueous samples. *J. Chromatogr. A* **2010**, *1217*, 6511–6521. [CrossRef] [PubMed]
47. Gago-Ferrero, P.; Mastroianni, N.; Díaz-Cruz, M.S.; Barceló, D. Fully automated determination of nine ultraviolet filters and transformation products in natural waters and wastewaters by on-line solid phase extraction-liquid chromatography-tandem mass spectrometry. *J. Chromatogr. A* **2013**, *1294*, 106–116. [CrossRef]
48. European Parliament and the Council of the European Union. In Proceedings of the 96/23/EC Commission Decision of Implementing Council Directive 96/23/EC Concerning the Performance of Analytical Methods and the Interpretation of Results (Notified under Document Number C (2002) 3044) (Text withEEA Relevance) (2002/657/EC); Off J Eur communities 8–36; European Parliament and the Council of the European Union, Brussels, Belgium, 12 August 2002. [CrossRef]

49. Valhondo, C.; Martinez-Landa, L.; Hidalgo, J.; Tubau, I.; De Pourcq, K.; Grau-Martinez, A.; Ayora, C. Tracer test Modeling for Local scale Residence Time Distribution characterization in an artificial recharge site. *Hydrol. Earth Syst. Sci.* **2016**, *20*, 4209–4221. [CrossRef]
50. Stuyfzand, P.J.; Segers, W.; van Rooijen, N. *Behavior of Pharmaceuticals and Other Emerging Pollutants in Various Artificial Recharge Systems in The Netherlands*; ISMAR: Arizona, AZ, USA; ACACIA: Arizona, AZ, USA, 2007; Volume 3, pp. 231–245.
51. Hidalgo, J.J.; Slooten, L.J.; Medina, A.; Carrera, J. A Newton-Raphson Based Code for Seawater Intrusion Modelling and Parameter Estimation. Available online: http://www.swim-site.nl/pdf/swim18_abstracts/Hidalgo.pdf (accessed on 1 April 2020).
52. Medina, A.; Carrera, J. Coupled estimation of flow and solute transport parameters. *Water Resour. Res.* **1996**, *32*, 3063–3076. [CrossRef]
53. Medina, A.; Carrera, J. Geostatistical inversion of coupled problems: Dealing with computational burden and different types of data. *J. Hydrol.* **2003**, *281*, 251–264. [CrossRef]
54. Bertelkamp, C.; Schoutteten, K.; Vanhaecke, L.; Bussche, J.V.; Callewaert, C.; Boon, N.; Singhal, N.; van der Hoek, J.P.; Verliefde, A.R.D. A laboratory-scale column study comparing organic micropollutant removal and microbial diversity for two soil types. *Sci. Total Environ.* **2015**, *536*, 632–638. [CrossRef]
55. Regnery, J.; Wing, A.D.; Alidina, M.; Drewes, J.E. Biotransformation of trace organic chemicals during groundwater recharge: How useful are first-order rate constants? *J. Contam. Hydrol.* **2015**, *179*, 65–75. [CrossRef]
56. Schaffer, M.; Kröger, K.F.; Nödler, K.; Ayora, C.; Carrera, J.; Hernández, M.; Licha, T. Influence of a compost layer on the attenuation of 28 selected organic micropollutants under realistic soil aquifer treatment conditions: Insights from a large scale column experiment. *Water Res.* **2015**, *74*, 110–121. [CrossRef]
57. Grünheid, S.; Amy, G.; Jekel, M. Removal of bulk dissolved organic carbon (DOC) and trace organic compounds by bank filtration and artificial recharge. *Water Res.* **2005**, *39*, 3219–3228. [CrossRef] [PubMed]
58. Henzler, A.F.; Greskowiak, J.; Massmann, G. Modeling the fate of organic micropollutants during river bank filtration (Berlin, Germany). *J. Contam. Hydrol.* **2014**, *156*, 78–92. [CrossRef] [PubMed]
59. Laws, B.V.; Dickenson, E.R.V.; Johnson, T.A.; Snyder, S.A.; Drewes, J.E. Attenuation of contaminants of emerging concern during surface-spreading aquifer recharge. *Sci. Total Environ.* **2011**, *409*, 1087–1094. [CrossRef] [PubMed]
60. Nham, H.T.T.; Greskowiak, J.; Nödler, K.; Rahman, M.A.; Spachos, T.; Rusteberg, B.; Massmann, G.; Sauter, M.; Licha, T. Modeling the transport behavior of 16 emerging organic contaminants during soil aquifer treatment. *Sci. Total Environ.* **2015**, *514*, 450–458. [CrossRef]
61. Ranieri, E.; Verlicchi, P.; Young, T.M. Paracetamol removal in subsurface flow constructed wetlands. *J. Hydrol.* **2011**, *404*, 130–135. [CrossRef]
62. Wiese, B.; Massmann, G.; Jekel, M.; Heberer, T.; Dünnbier, U.; Orlikowski, D.; Grützmacher, G. Removal kinetics of organic compounds and sum parameters under field conditions for managed aquifer recharge. *Water Res.* **2011**, *45*, 4939–4950. [CrossRef]
63. Greskowiak, J.; Hamann, E.; Burke, V.; Massmann, G. The uncertainty of biodegradation rate constants of emerging organic compounds in soil and groundwater: A compilation of literature values for 82 substances. *Water Res.* **2017**, *126*, 122–133. [CrossRef]
64. Bertelkamp, C.; Reungoat, J.; Cornelissen, E.R.; Singhal, N.; Reynisson, J.; Cabo, A.J.; van der Hoek, J.P.; Verliefde, A.R.D. Sorption and biodegradation of organic micropollutants during river bank filtration: A laboratory column study. *Water Res.* **2014**, *52*, 231–241. [CrossRef]
65. Massmann, G.; Dünnbier, U.; Heberer, T.; Taute, T. Behaviour and redox sensitivity of pharmaceutical residues during bank filtration—Investigation of residues of phenazone-type analgesics. *Chemosphere* **2008**, *71*, 1476–1485. [CrossRef]
66. Barbieri, M.; Licha, T.; Nödler, K.; Carrera, J.; Ayora, C.; Sanchez-Vila, X. Fate of β-blockers in aquifer material under nitrate reducing conditions: Batch experiments. *Chemosphere* **2012**, *89*, 1272–1277. [CrossRef]
67. Liu, Y.-S.; Ying, G.-G.; Shareef, A.; Kookana, R.S. Biodegradation of three selected benzotriazoles in aquifer materials under aerobic and anaerobic conditions. *J. Contam. Hydrol.* **2013**, *151*, 131–139. [CrossRef] [PubMed]

68. Maeng, S.K.; Sharma, S.K.; Lekkerkerker-Teunissen, K.; Amy, G.L. Occurrence and fate of bulk organic matter and pharmaceutically active compounds in managed aquifer recharge: A review. *Water Res.* **2011**, *45*, 3015–3033. [CrossRef] [PubMed]
69. Regnery, J.; Wing, A.D.; Kautz, J.; Drewes, J.E. Introducing sequential managed aquifer recharge technology (SMART) From laboratory to full-scale application. *Chemosphere* **2016**, *154*, 8–16. [CrossRef] [PubMed]
70. Hellauer, K.; Mergel, D.; Ruhl, A.S.; Filter, J.; Hübner, U.; Jekel, M.; Drewes, J.E. Advancing sequential managed aquifer recharge technology (SMART) using different intermediate oxidation processes. *Water (Switzerland)* **2017**, *9*, 221. [CrossRef]
71. Guchi, E. Review on Slow Sand Filtration in Removing Microbial Contamination and Particles from Drinking Water. *Am. J. Food Nutr.* **2015**, *3*, 47–55. [CrossRef]
72. Verma, S.; Daverey, A.; Sharma, A. Slow sand filtration for water and wastewater treatment—A review. *Environ. Technol. Rev.* **2017**, *6*, 47–58. [CrossRef]
73. Perez-Mercado, L.F.; Lalander, C.; Joel, A.; Ottoson, J.; Dalahmeh, S.; Vinnerås, B. Biochar filters as an on-farm treatment to reduce pathogens when irrigating with wastewater-polluted sources. *J. Environ. Manag.* **2019**, *248*, 109295. [CrossRef]
74. Matthess, G.; Pekdeger, A.; Schroeter, J. Persistence and transport of bacteria and viruses in groundwater—A conceptual evaluation. *J. Contam. Hydrol.* **1988**, *2*, 171–188. [CrossRef]
75. Kim, H.N.; Bradford, S.A.; Walker, S.L. Escherichia coli O157:H7 Transport in Saturated Porous Media: Role of Solution Chemistry and Surface Macromolecules. *Environ. Sci. Technol.* **2009**, *43*, 4340–4347. [CrossRef]
76. Park, Y.; Atwill, E.R.; Hou, L.; Packman, A.I.; Harter, T. Deposition of Cryptosporidium parvum Oocysts in porous media: A synthesis of attachment efficiencies measured under varying environmental conditions. *Environ. Sci. Technol.* **2012**, *46*, 9491–9500. [CrossRef]
77. Bichai, F.; Barbeau, B.; Dullemont, Y.; Hijnen, W. Role of predation by zooplankton in transport and fate of protozoan (oo)cysts in granular activated carbon filtration. *Water Res.* **2010**, *44*, 1072–1081. [CrossRef] [PubMed]
78. Sasidharan, S.; Bradford, S.A.; Šimunek, J.; Torkzaban, S. Minimizing virus transport in porous media by optimizing solid phase inactivation. *J. Environ. Qual.* **2018**, *47*, 1058–1067. [CrossRef] [PubMed]
79. Urfer, D. Use of bauxite for enhanced removal of bacteria in slow sand filters. *Water Sci. Technol. Water Supply* **2017**, *17*, 1007–1015. [CrossRef]
80. Weiss, W.J.; Bouwer, E.J.; Aboytes, R.; LeChevallier, M.W.; O'Melia, C.R.; Le, B.T.; Schwab, K.J. Riverbank filtration for control of microorganisms: Results from field monitoring. *Water Res.* **2005**, *39*, 1990–2001. [CrossRef] [PubMed]
81. Bekele, E.; Toze, S.; Patterson, B.; Higginson, S. Managed aquifer recharge of treated wastewater: Water quality changes resulting from infiltration through the vadose zone. *Water Res.* **2011**, *45*, 5764–5772. [CrossRef]
82. Betancourt, W.Q.; Kitajima, M.; Wing, A.D.; Regnery, J.; Drewes, J.E.; Pepper, I.L.; Gerba, C.P. Assessment of virus removal by managed aquifer recharge at three full-scale operations. *J. Environ. Sci. Heal. Part A Toxic Hazard. Subst. Environ. Eng.* **2014**, *49*, 1685–1692. [CrossRef]
83. Elkayam, R.; Aharoni, A.; Vaizel-Ohayon, D.; Katz, Y.; Negev, I.; Marano, R.B.; Cytryn, E.; Shtrasler, L.; Lev, O. Viral and Microbial Pathogens, Indicator Microorganisms, Microbial Source Tracking Indicators, and Antibiotic Resistance Genes in a Confined Managed Effluent Recharge System. *J. Environ. Eng.* **2018**, *144*, 05017011. [CrossRef]
84. Schijven, J.F.; Hassanizadeh, S.M. Removal of viruses by soil passage: Overview of modeling, processes, and parameters. *Crit. Rev. Environ. Sci. Technol.* **2000**, *30*, 49–127. [CrossRef]
85. Voigt, A.M.; Zacharias, N.; Timm, C.; Wasser, F.; Sib, E.; Skutlarek, D.; Parcina, M.; Schmithausen, R.M.; Schwartz, T.; Hembach, N.; et al. Association between antibiotic residues, antibiotic resistant bacteria and antibiotic resistance genes in anthropogenic wastewater—An evaluation of clinical influences. *Chemosphere* **2020**, *241*, 125032. [CrossRef]
86. Bengtsson-Palme, J.; Milakovic, M.; Švecová, H.; Ganjto, M.; Jonsson, V.; Grabic, R.; Udikovic-Kolic, N. Industrial wastewater treatment plant enriches antibiotic resistance genes and alters the structure of microbial communities. *Water Res.* **2019**, *162*, 437–445. [CrossRef]
87. Bengtsson-Palme, J.; Hammarén, R.; Pal, C.; Östman, M.; Björlenius, B.; Flach, C.F.; Fick, J.; Kristiansson, E.; Tysklind, M.; Larsson, D.G.J. Elucidating selection processes for antibiotic resistance in sewage treatment plants using metagenomics. *Sci. Total Environ.* **2016**. [CrossRef] [PubMed]

88. Berendonk, T.U.; Manaia, C.M.; Merlin, C.; Fatta-Kassinos, D.; Cytryn, E.; Walsh, F.; Bürgmann, H.; Sørum, H.; Norström, M.; Pons, M.N.; et al. Tackling antibiotic resistance: The environmental framework. *Nat. Rev. Microbiol.* **2015**, *13*, 310–317. [CrossRef] [PubMed]
89. Frieden, T. Antibiotic resistance threats in the United States. *Cent. Dis. Control Prev.* **2019**, 1–114. [CrossRef]
90. Larsson, D.G.J.; Andremont, A.; Bengtsson-Palme, J.; Brandt, K.K.; de Roda Husman, A.M.; Fagerstedt, P.; Fick, J.; Flach, C.F.; Gaze, W.H.; Kuroda, M.; et al. Critical knowledge gaps and research needs related to the environmental dimensions of antibiotic resistance. *Environ. Int.* **2018**, *117*, 132–138. [CrossRef] [PubMed]
91. World Health Organisation (WHO). *Global Action Plan on Antimicrobial Resistance*; WHO: Geneva, Switzerland, 2017; pp. 1–28.
92. European Commission. *A European One Health Action Plan against Antimicrobial Resistance*; Commun FROM Comm TO Counc Eur Parliam; European Commission: Brussels, Belgium, 2017.
93. EU Decision 2018/840. Commission Implementing Decision (EU) 2018/840 of 5 June 2018. *Off. J. Eur. Union.* **2018**, *141*, 9–12.
94. Sharma, V.K.; Johnson, N.; Cizmas, L.; McDonald, T.J.; Kim, H. A review of the influence of treatment strategies on antibiotic resistant bacteria and antibiotic resistance genes. *Chemosphere* **2016**. [CrossRef]
95. Gao, P.; Munir, M.; Xagoraraki, I. Correlation of tetracycline and sulfonamide antibiotics with corresponding resistance genes and resistant bacteria in a conventional municipal wastewater treatment plant. *Sci. Total Environ.* **2012**. [CrossRef]
96. Li, J.; Cheng, W.; Xu, L.; Strong, P.J.; Chen, H. Antibiotic-resistant genes and antibiotic-resistant bacteria in the effluent of urban residential areas, hospitals, and a municipal wastewater treatment plant system. *Environ. Sci. Pollut. Res.* **2015**, *22*, 4587–4596. [CrossRef]
97. Pärnänen, K.M.M.; Narciso-da-Rocha, C.; Kneis, D.; Berendonk, T.U.; Cacace, D.; Do, T.T.; Elpers, C.; Fatta-Kassinos, D.; Henriques, I.; Jaeger, T.; et al. Antibiotic resistance in European wastewater treatment plants mirrors the pattern of clinical antibiotic resistance prevalence. *Sci. Adv.* **2019**, *5*, eaau9124. [CrossRef]
98. Ju, F.; Beck, K.; Yin, X.; Maccagnan, A.; McArdell, C.S.; Singer, H.P.; Johnson, D.R.; Zhang, T.; Bürgmann, H. Wastewater treatment plant resistomes are shaped by bacterial composition, genetic exchange, and upregulated expression in the effluent microbiomes. *ISME J.* **2019**, *13*, 346–360. [CrossRef]
99. Du, J.; Geng, J.; Ren, H.; Ding, L.; Xu, K.; Zhang, Y. Variation of antibiotic resistance genes in municipal wastewater treatment plant with A2O-MBR system. *Environ. Sci. Pollut. Res.* **2015**, *22*, 3715–3726. [CrossRef] [PubMed]
100. Fang, H.; Zhang, Q.; Nie, X.; Chen, B.; Xiao, Y.; Zhou, Q.; Liao, W.; Liang, X. Occurrence and elimination of antibiotic resistance genes in a long-term operation integrated surface flow constructed wetland. *Chemosphere* **2017**. [CrossRef] [PubMed]
101. Cui, E.; Wu, Y.; Zuo, Y.; Chen, H. Effect of different biochars on antibiotic resistance genes and bacterial community during chicken manure composting. *Bioresour. Technol.* **2016**. [CrossRef] [PubMed]
102. Taylor, E.N.; Kummer, K.M.; Durmus, N.G.; Leuba, K.; Tarquinio, K.M.; Webster, T.J. Superparamagnetic iron oxide nanoparticles (SPION) for the treatment of antibiotic-resistant biofilms. *Small* **2012**, *8*, 3016–3027. [CrossRef]
103. Nijhawan, A.; Labhasetwar, P.; Jain, P.; Rahate, M. Public consultation on artificial aquifer recharge using treated municipal wastewater. *Resour. Conserv. Recycl.* **2013**. [CrossRef]
104. Smith, H.M.; Brouwer, S.; Jeffrey, P.; Frijns, J. Public responses to water reuse—Understanding the evidence. *J. Environ. Manag.* **2018**, *207*, 43–50. [CrossRef]
105. Lazarova, V.; Asano, T.; Bahri, A.; Anderson, J. *Milestones in Water Reuse: The Best Success Stories*; IWA Publishing: London, UK, 2013.
106. Hartley, T.W. Public perception and participation in water reuse. *Desalination* **2006**, *187*, 115–126. [CrossRef]
107. Hurlimann, A.; Dolnicar, S. When public opposition defeats alternative water projects—The case of Toowoomba Australia. *Water Res.* **2010**, *44*, 287–297. [CrossRef]

© 2020 by the authors. Licensee MDPI, Basel, Switzerland. This article is an open access article distributed under the terms and conditions of the Creative Commons Attribution (CC BY) license (http://creativecommons.org/licenses/by/4.0/).

Article

Potential for Managed Aquifer Recharge to Enhance Fish Habitat in a Regulated River

Robert W. Van Kirk [1,*], Bryce A. Contor [1], Christina N. Morrisett [2], Sarah E. Null [2] and Ashly S. Loibman [3]

1. Henry's Fork Foundation, P.O. Box 550, Ashton, ID 83420, USA; brycec@henrysfork.org
2. Department of Watershed Sciences, Utah State University, 5210 Old Main Hill, NR 210, Logan, UT 84322, USA; christina.morrisett@aggiemail.usu.edu (C.N.M.); sarah.null@usu.edu (S.E.N.)
3. Environmental Studies, Colgate University, 13 Oak Drive, Hamilton, NY 13346, USA; aloibman@colgate.edu
* Correspondence: rob@henrysfork.org; Tel.: +1-208-652-3567

Received: 20 November 2019; Accepted: 18 February 2020; Published: 1 March 2020

Abstract: Managed aquifer recharge (MAR) is typically used to enhance the agricultural water supply but may also be promising to maintain summer streamflows and temperatures for cold-water fish. An existing aquifer model, water temperature data, and analysis of water administration were used to assess potential benefits of MAR to cold-water fisheries in Idaho's Snake River. This highly-regulated river supports irrigated agriculture worth US $10 billion and recreational trout fisheries worth $100 million. The assessment focused on the Henry's Fork Snake River, which receives groundwater from recharge incidental to irrigation and from MAR operations 8 km from the river, addressing (1) the quantity and timing of MAR-produced streamflow response, (2) the mechanism through which MAR increases streamflow, (3) whether groundwater inputs decrease the local stream temperature, and (4) the legal and administrative hurdles to using MAR for cold-water fisheries conservation in Idaho. The model estimated a long-term 4%–7% increase in summertime streamflow from annual MAR similar to that conducted in 2019. Water temperature observations confirmed that recharge increased streamflow via aquifer discharge rather than reduction in river losses to the aquifer. In addition, groundwater seeps created summer thermal refugia. Measured summer stream temperature at seeps was within the optimal temperature range for brown trout, averaging 14.4 °C, whereas ambient stream temperature exceeded 19 °C, the stress threshold for brown trout. Implementing MAR for fisheries conservation is challenged by administrative water rules and regulations. Well-developed and trusted water rights and water-transaction systems in Idaho and other western states enable MAR. However, in Idaho, conservation groups are unable to engage directly in water transactions, hampering MAR for fisheries protection.

Keywords: climate adaptation; stream temperature; streamflow; Henry's Fork; fisheries; Snake River; Idaho; water rights

1. Introduction

In the Western USA, important aquatic ecosystems and recreational fisheries often occur in river basins with large irrigated agricultural diversions, resulting in conflicts between water for irrigation and environmental streamflow needs [1–3]. Climate change exacerbates these conflicts, as precipitation regimes shift from snowfall to rainfall and evaporative demand increases, leading to flashier streamflow in winter and spring and reduced baseflow through summer and fall [4,5]. Climate warming and reduced baseflow work in tandem to warm stream temperature and are expected to reduce habitat for cold-water ecosystems [6]. Increasing streamflow, particularly during summertime baseflow conditions, cools stream temperatures by increasing the assimilative heat capacity of rivers [7]. Management options to increase streamflow include re-operating reservoirs [8,9], reducing diversions through environmental water purchases [10], or conducting managed aquifer recharge (MAR) [11,12]. MAR is a promising strategy to enhance cold-water habitats while maintaining water resource benefits for people because excess water is intentionally recharged to raise aquifer levels, which increases baseflow, or may subsequently be pumped for irrigation. Groundwater additions to streams are particularly beneficial for cold-water fisheries because they create thermal refugia [13,14]. MAR is often recommended as a strategy to manage water for people and ecosystems with flashier runoff anticipated with climate change [12]. However, the potential of MAR to benefit cold-water ecosystems while maintaining irrigated agriculture requires (1) understanding the physical hydrology between the recharge site and the stream, (2) estimating temperature differences at groundwater seeps in the river and ambient temperatures, and (3) understanding administrative water rules to apply MAR to benefit cold-water habitat.

In streams that interact with local and regional aquifers, winter recharge enhances groundwater storage important for streamflow through the summer [14]. However, systems with shallow, unconfined aquifers are sensitive to climate variability [15] and may experience changes in the timing and magnitude of natural recharge [16], diminishing aquifer storage and groundwater-supported streamflow [16,17]. MAR can capture early rainfall or snowmelt and supplement late-summer return flows [18–20], by raising an aquifer's hydraulic head and creating groundwater seeps and shallow groundwater contributions that return to the stream in gaining reaches. Models have demonstrated that the lag time between MAR and return flow can delay the runoff peak [21–23], buffering a variable runoff regime [23] and potentially alleviating critical low-flow periods [24–26]. However, proportional contributions of MAR to streamflow depend, in part, on recharge site proximity [18,22,23].

Some studies have found that groundwater seeps and return flows mitigate the thermal effects of climate change on riverine habitats [27–29]. For example, measured water temperature at groundwater seeps have been 2–3 °C cooler than ambient river temperatures in the Pacific Northwest [21], and up to 4 °C cooler in Nevada's Walker River [12]. While shallow groundwater temperature is sensitive to long-term changes in air temperature, groundwater temperatures are less sensitive than surface water to changes in air temperature and are generally absent from heating by solar radiation [30–33]. Although studies note that MAR may increase summer baseflows [34], provide cool groundwater return flows to maintain cold-water fisheries during low-flow periods [26], and maintain aquatic ecosystems [18,35], field observations have yet to test these hypotheses. Furthermore, it is important to understand the extent and times that MAR can influence streamflow and stream temperature to maintain cold-water species in regulated rivers with climate change.

In the western USA, MAR must fit into the administrative rules of the Doctrine of Prior Appropriation, which allocates water for beneficial uses based on the seniority that water was first used. In most western states, the senior uses are mining and agriculture. Additionally, states must have well developed market and transfer mechanisms that provide administrative water for MAR within pre-existing allocation systems. However, western states that prioritize MAR, like Arizona, Colorado, California, and Idaho, each have different administration policies regarding which entities can implement MAR. Overall, implementation of MAR includes large-scale projects conducted by centralized public authorities, cities, and private companies in Arizona [36,37], smaller-scale projects implemented by landowners, local agencies, and counties in California [36,38], and a variety of MAR and recovery projects implemented by individual water right holders, local groundwater management districts, and cities in Colorado [37,39]. In Idaho, the state-run MAR program is primarily designed to increase discharge from the aquifer to the river for fulfillment of senior surface-water rights. Water transaction mechanisms in California and Colorado allow effective transfer of water to environmental uses, and conservation organizations can be direct participants in such transactions [40]. However, in Idaho, conservation groups cannot directly participate in transactions like water rental, which inhibits MAR for fisheries protection.

This study aims to understand the potential for MAR to benefit cold-water ecosystems while maintaining irrigated agriculture in Idaho's Henry's Fork Snake River. To do this, existing data and models were integrated and analyzed for a reconnaissance-level assessment. The following research questions were addressed: (1) What quantity and timing of streamflow response can MAR produce? (2) Does MAR increase streamflow by reducing channel loss to the aquifer or by increasing groundwater inflow to the stream? (3) Can groundwater inputs create local areas of decreased water temperature in the stream? (4) What legal and administrative hurdles exist for MAR to be used for cold-water fisheries conservation in Idaho? This research is instrumental to evaluate whether MAR is a water management strategy that has the potential to benefit managed fisheries. MAR is increasingly implemented throughout the western United States to meet human and environmental water objectives. There is a clear need to understand whether implementing MAR to benefit managed fisheries deserves additional effort and resources, and to identify current knowledge gaps when streamflow, stream temperature, and administrative water rules are considered for environmental MAR applications.

2. Materials and Methods

2.1. Study Area

The 93,000 km^2 upper Snake River basin in Idaho and Wyoming, USA, (Figure 1) is an ideal setting for assessing the potential of MAR to benefit cold-water ecosystems. The basin's water resources support an agricultural economy worth US $10 billion [41], as well as many ecologically important stream systems and recreational trout fisheries, which contribute US $100 million to local communities [42,43]. Mean annual surface water supply is 15,000 Mm^3 in the basin and 75% is withdrawn for irrigation. Irrigators also withdraw 1600 Mm^3 of groundwater from the Eastern Snake Plain Aquifer (ESPA; Figure 1) and 500 Mm^3 from tributary aquifers [41]. In a given year, over 10,000 km^2 of irrigated land produce hay, wheat, barley, potatoes, and dairy products for global companies such as Anheuser-Busch, General Mills, and Clif Bar. The ESPA is a highly transmissive, unconfined, regional aquifer hosted in sediments interbedded within fractured Quaternary basalts [44]. Water generally flows through the ESPA from northeast to southwest and discharges to the Snake River near American Falls Reservoir and in a 100 km reach immediately upstream of King Hill (Figure 1). Water levels in and discharge from the ESPA have been declining for 60 years due to a combination of decreased recharge incidental to surface water irrigation and increased groundwater pumping [41,44,45]. Declining aquifer levels have caused costly legal disputes, increased reliance on reservoir storage to meet irrigation demand, increased groundwater pumping costs, and decreased streamflow for fisheries.

As part of a comprehensive plan to increase storage in and discharge from the ESPA, Idaho has implemented publicly funded MAR, with an annual objective of 330 Mm^3 [41]. The primary management goal of the state's MAR program is to increase discharge from the aquifer to the river to fill senior surface water rights, rather than to store the water for future recovery via pumping. Increasing discharge over the long term requires increasing storage in the aquifer to maintain larger hydraulic gradients along connected river reaches. Higher storage volume, in turn, has ancillary benefits such as decreased pumping costs for groundwater users [41]. In addition, irrigation entities, cities, and private companies are using MAR on smaller scales to meet mitigation requirements of a 2015 legal settlement between senior surface water users and junior groundwater users. This settlement requires a specified reduction in groundwater pumping or mitigation with an equal amount of MAR. Concurrently, research describing benefits to aquatic systems from incidental and managed recharge in irrigated landscapes has motivated conservation organizations to consider MAR as a tool for maintaining and enhancing cold-water ecosystems in a changing climate [3,18,46].

An assessment was conducted in the Henry's Fork Snake River watershed (Figures 1 and 2), where the state of Idaho has recently invested US $1.5 million to expand and improve a MAR site known as Egin Lakes (Figure 1). The Henry's Fork and its tributaries have an annual surface-water supply of 3200 Mm^3 and irrigators withdraw ~1500 Mm^3 to apply on ~1000 km^2 of agricultural land; very little groundwater is used for irrigation in the watershed [47]. When natural streamflow is insufficient to meet irrigation demand—usually early July through early September—streamflow is augmented by draft of Island Park Reservoir, near the river's headwaters (Figure 1). Nearly all irrigation water is delivered through unlined canals constructed in the late 1800s. Thus, these canals have provided a large amount of incidental recharge to local and regional aquifers via seepage for over a century [48,49]. Historically, irrigation water was applied via flooding or furrow irrigation, but most application was converted to sprinklers in the 1980s and 1990s [49,50]. The lower one-third of the Henry's Fork, shown as the "modeled reach" in Figure 1, is hydraulically connected to local and regional aquifers. Previous research has shown that this reach gains water seasonally in response to locally increased water tables during irrigation season, but loses water to the regional ESPA during the winter [48,49,51,52]. The conversion to more efficient sprinkler application has reduced both total diversion and groundwater return flows to the river by around 250 Mm^3 per year since the late 1970s [49].

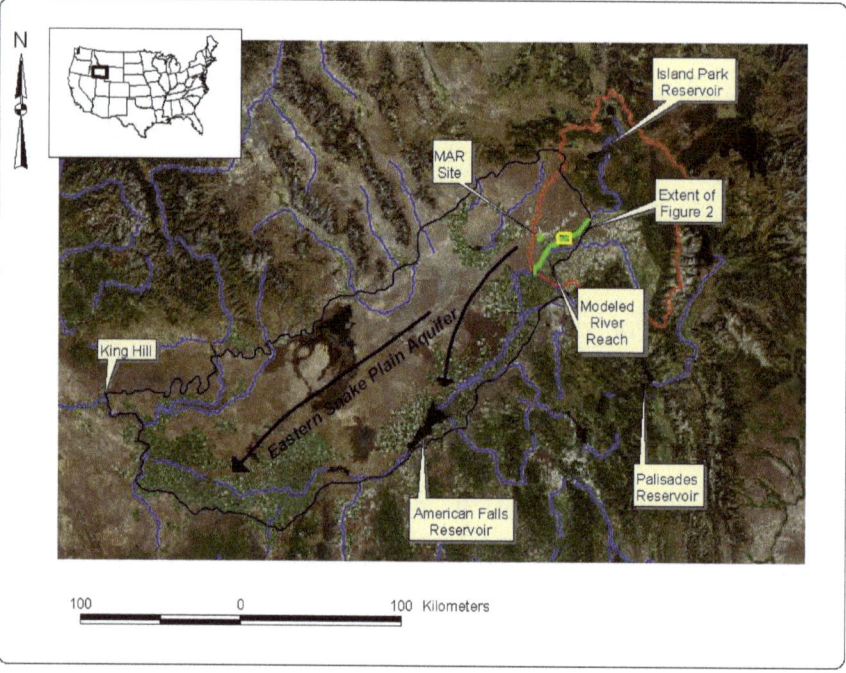

Figure 1. Upper Snake River basin, USA, showing the ESPA, modeled river reach of Henry's Fork (in green), and a nearby MAR site. The red polygon delineates the Henry's Fork watershed, and the yellow rectangle shows the area enlarged in Figure 2. Black arrows indicate primary groundwater flow paths on the ESPA. Data credit: Idaho Department of Water Resources.

The "field study reach" is the 12 river-km of the Henry's Fork immediately downstream of U.S. Geological Survey streamflow gage 13050500 at St. Anthony (Figure 2). This gage is the streamflow management point in the lower watershed, triggering additional releases from Island Park Reservoir when streamflow drops below a specified target at this gage [53]. However, four diversions in the field study reach downstream of the St. Anthony gage substantially reduce streamflow during the summer (Figure A1). The study reach supports an increasingly popular and economically valuable recreational sport fishery for wild brown trout (*Salmo trutta*) [54,55], which has an optimal summer temperature range of 12 °C to 19 °C. Habitat suitability for brown trout decreases as temperature increases above 19 °C to the lethal limit of 27 °C [56]. Over the summers of 2016, 2017, and 2018, daily mean water temperatures during July and August ranged from 16 °C to 20 °C at a water-quality monitoring station at the top of the field study reach and from 17 °C to 22 °C at a water-quality monitoring station at the bottom of the field study (Figures 2 and A2). Maximum instantaneous water temperature recorded at the lower station over this time period was 27.3 °C, and daily maxima frequently exceeded 22 °C. Due to high water temperatures, brown trout move either to local areas of groundwater input or out of the reach altogether during the summer [54].

Figure 2. Map of field study reach, showing locations of U.S. Geological Survey (USGS) streamflow gage at St. Anthony, temperature loggers deployed in 2010, water-quality monitoring locations, and stretch where springs were documented in 2019. Data credit: ESRI.

Summertime streamflow in the study reach could be increased by increasing draft of Island Park Reservoir 100 km upstream, but larger reservoir releases have numerous negative effects on other popular and economically important fisheries in the upper half of the watershed [3]. These include transport of suspended material out of the reservoir and resulting high turbidity during the peak fishing season, increased water temperatures downstream of the reservoir when it is drafted faster than thermal stratification can occur, and decreased trout survival during winter, when low outflow is required to refill the reservoir [57–59]. These effects do not propagate downstream to the study reach in the lower watershed. Thus, this study seeks to assess whether MAR has the potential to improve fisheries in the lower watershed without degrading those in the upper watershed. In particular, withdrawal of water for MAR at carefully identified times could increase groundwater inputs to the lower river during the summer, thereby increasing local trout habitat and water supply available for diversion there. In turn, increased summertime water supply in the lower river could limit reservoir draft, thereby simultaneously benefiting fisheries in the upper watershed.

2.2. Streamflow Response

We used an existing regional groundwater model to estimate response of streamflow in the Henry's Fork to MAR at the Egin Lakes site, located 8 km from the Henry's Fork (site location is shown in Figure 3, which also depicts the results). Modeling was done with the Idaho Department of Water Resources' Enhanced Snake Plain Aquifer Model Version 2.1 (ESPAM2.1) [52], a regional finite-difference flow model implemented in MODFLOW and configured with a single aquifer layer, monthly temporal resolution, and roughly 11,000 1.6-km grid cells. The model supports both steady-state and transient simulations. Although storage coefficients are typical of unconfined conditions, the transient rendition of the model uses time-constant aquifer transmissivity, making model results additive and scalable. The model was calibrated to 1980–2008 hydrologic conditions, using the first five of these as a burn-in period [52]. Calibration used over 43,000 aquifer water levels, 2000 river gain and loss estimates, and 2000 spring discharge measurements and was performed using PEST version 12.0, a nonlinear parameter estimation program [60]. The model was built specifically to estimate effects of aquifer pumping and recharge on river reaches and springs in hydraulic connection with the aquifer, so calibration optimized groundwater-surface water exchanges rather than hydraulic heads. The model does not simulate solute transport nor thermal changes in the aquifer or its discharge. In the model, the hydraulically connected section of the Henry's Fork is treated as a single, 75-km reach, referred to as the "modeled reach" (Figure 1), whereas our field study reach is only 12 km in the center of the ESPAM2.1 modeled reach (Figures 1 and 2). Monthly model calibration residuals for stream gains and losses in the modeled reach of the Henry's Fork were generally on the order of ± 25%, but monthly residuals as large in magnitude as −100% were observed early in the irrigation season. Over the period 1985–2008, the model underestimated cumulative river gain by around 10%. Thus, the model is suitable for our reconnaissance-level assessment when applied over long time periods.

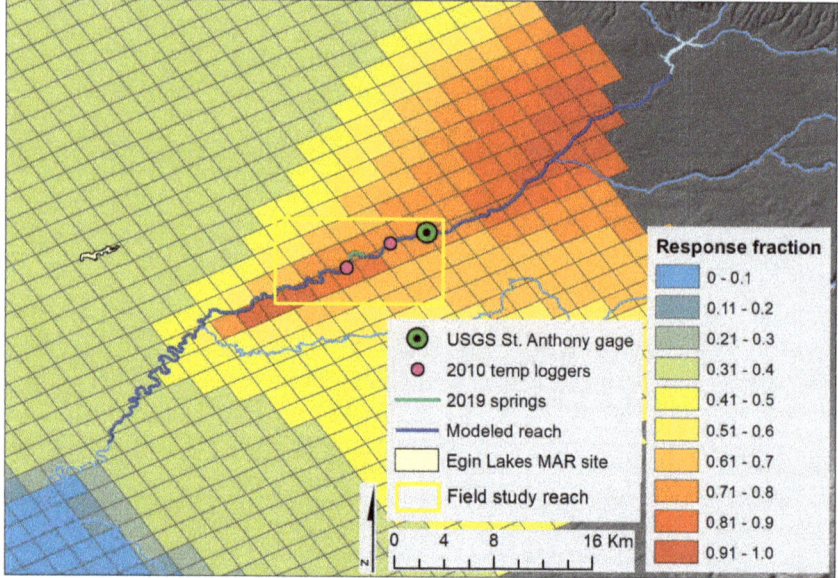

Figure 3. Steady-state discharge response in the modeled river reach to recharge conducted in a given model cell, as a fraction of total recharge volume. For example, a response fraction of 0.45 (white cells) indicates that 45% of the volume of water recharged in that cell will eventually contribute to streamflow in the modeled reach. The yellow rectangle indicates the field study reach.

The model was used in two ways. First, a steady-state simulation was used to calculate the fraction of total recharge in a given model cell that affects streamflow in the modeled reach of the Henry's Fork. Recharge was simulated in the model cells containing the Egin Lakes MAR site, as well as in other model cells in the vicinity of the lower Henry's Fork to assess whether developing MAR sites in other locations could increase streamflow response to MAR. Second, a 30-year transient simulation was used to estimate streamflow response in the modeled reach to a scenario in which 7.3 Mm^3 of water was withdrawn from the river and recharged at the Egin Lakes site in each of March,

April, and October and 25.6 Mm^3 of water was withdrawn from the river and recharged at Egin Lakes during November. This annual scenario was similar to operation of the Egin Lakes site during 2019 and was repeated in each of the 30 years of simulation. Model output was used to estimate net change to streamflow in the study reach by allocating total streamflow response for the modeled Henry's Fork reach proportionally to the field study reach and including the effect of diversion for MAR upstream of the study reach. The median 2000–2019 hydrograph for streamflow at the bottom of the study reach was used as a baseline condition, although the effect of modeled MAR was also assessed relative to streamflow in 2016, the driest year in the basin in over 40 years.

Recharge proximal to the lower Henry's Fork increases hydraulic gradients between the aquifer and the river, but if these gradients were initially negative (i.e., water flows from the river to the aquifer), a positive streamflow response from recharge would occur through decreased river losses rather than through increased river gains. Although the resulting increase in streamflow would be equivalent between the two mechanisms, the first mechanism would not provide the benefit of decreased water temperature during the summer. Thus, summertime water temperature was measured upstream and downstream of a reach known to be hydraulically connected with the underlying aquifer. These measurements were conducted in 2010, a decade after conversion from flood to sprinkler irrigation but six years prior to initiation of MAR at the expanded Egin Lakes site. Canal seepage, which has been roughly constant since 2000, was the only source of local groundwater recharge in 2010. Water temperature loggers were deployed from 1 June 2010 to 31 August 2010 at two locations in the upper half of our field study reach (Figure 2) and secured underneath overhanging riparian vegetation at ~40 cm water depth. The upstream logger was located immediately downstream of a reach through which the river flows over basalt bedrock and has little interaction with shallow groundwater. The other logger was placed 5 km downstream, in a reach where the river is well connected with shallow groundwater. Mean daily upstream temperature was subtracted from downstream temperature to create a time series of temperature differences. After accounting for serial autocorrelation with lag-3 autoregressive terms, two statistical models were fit to the time series—one with hypothesized zero mean and another with a non-zero fitted mean. Statistical significance of the fitted mean was assessed with the likelihood ratio test at a 0.05 level of significance (Appendix B).

2.3. Local Effects of Groundwater Inflow on Temperature

Whereas the 2010 temperature observations were made to assess the nature of summertime streamflow response to recharge solely from canal seepage, a separate field study in 2019 documented the locations and temperature of specific groundwater springs to investigate the potential for MAR to provide cool groundwater return flow to the river. In July 2019, locations of groundwater springs contributing water to the river channel were documented by walking a 1-km length of the right bank of the river, in the lower half of the 5 km reach studied in 2010 (Figure 2). A steep bluff approximately 5–10 m in vertical relief forms the boundary of the active floodplain on the right side of the river. Springs emerged from the face of the bluff and along its base, often between 1 and 50 m from the channel bank, and most spring outputs flowed into a secondary river channel. Each spring site was classified as (1) a single discharge point, where water originating from the bluff face created a separate channel that actively flowed into the river, or as (2) a "wall seep", where water emerged from continuously saturated sediments along the bluff face and contributed unchannelized flow to the river. At each spring site, a FLIR T450sc thermal infrared camera (FLIR Systems Inc., Wilsonville, Oregon, USA) was used to document differences between the surface temperatures of incoming groundwater springs and the river. The FLIR T450sc camera senses radiant stream surface temperatures in the 7.5 to 13 μm range, with an accuracy of ± 1 °C or 1% of the range of the reading [61]. To complement the imagery, instantaneous temperatures were measured with a handheld thermometer at three lateral locations: spring emergence, 0.6 m and 6 m into the river channel from where the spring entered the river. At wall seeps, lateral temperatures were measured at the upstream and downstream extent of each seep. In total, three temperature measurements were recorded at each of 20 spring sites. Temperature differences across the three lateral locations were analyzed with mixed-effects analysis of variance and Tukey's post-hoc test, treating spring site as a random effect. These tests were conducted at a 0.05 level of significance (Appendix B). The temperature analysis was not accompanied by assessments of whether physical habitat at these locations was otherwise suitable for or used by trout.

2.4. Water Administration

Potential streamflow and temperature benefits of MAR will not result in real changes in the river without sufficient availability of water for MAR at appropriate times. Thus, our assessment included analysis of physical and administrative availability of water for MAR in the upper Snake River basin within Idaho's prior appropriation system of water rights. This assessment relied on a formal review of Idaho's MAR program conducted for the Idaho Water Resource Board [62], to which two of the co-authors of this paper (RVK and CNM) contributed substantially. In addition, the state's water rights database, water-rights accounting manual [63], and water exchange procedures [64,65] were reviewed to identify opportunities for and limitations to conducting MAR for fisheries conservation purposes.

3. Results

3.1. Streamflow Response

The steady-state simulation using ESPAM2.1 predicted that 37% of the water volume delivered to the Egin Lakes MAR site will increase streamflow in the modeled reach of the Henry's Fork over the long term, and the balance will benefit other river reaches in the basin (Figure 3). If recharge were conducted closer to the river than the existing MAR site, the model predicted that >90% of recharge is realized as increased streamflow in the modeled reach. The modeled response fraction depended strongly on recharge location and decreased fairly rapidly with increasing distance between the river and recharge location (Figure 3). For example, 90% of water recharged in the red cells contributed to streamflow in the modeled reach, whereas less than 40% of the water recharged in the green cells contributed to streamflow in the modeled reach.

Transient simulation with ESPAM2.1 predicted that streamflow response to spring and fall recharge at Egin Lakes is relatively uniformly distributed over the year, with little month-to-month variability (Figure 4). Initial streamflow response to the spring-fall MAR scenario increased roughly linearly over time to reach 50% of its long-term value 6.5 years after first implementation of the annual MAR regime (Figure 4). Streamflow response increased more slowly after that, reaching ~90% of its long-term value 25 years after initial implementation of the annual MAR regime. Including the effects of water withdrawal from the river for delivery to the MAR site, the annual MAR scenario resulted in a 20%–25% decrease in streamflow during November and a 5%–10% decrease in streamflow during each of October, March, and April, relative to the current median hydrograph (Figure 5). Despite these decreases in streamflow due to withdrawal for MAR, median spring and fall streamflow still remained much higher than summertime lows, even after only five years of MAR. After 20 years of implementation of this annual MAR regime, median streamflow increased by 4%–7% during July and August of the median year (Figure 5), although increases during late June and early July were on the order of 10%–40% relative to streamflow in the dry year of 2016 (Figure A3).

Mean daily water temperature from 1 June 2010 to 31 August 2010 was 0.6 °C cooler at the downstream location influenced by groundwater inputs, and this difference was statistically significant ($\chi^2 = 5.3$, df = 1, $p = 0.02$). This indicates that during summer, streamflow response to seasonal aquifer recharge results from inflow of groundwater to the river not from reduced loss of water from the river to the aquifer. Since this result was observed when canal seepage was the only source of aquifer recharge in the vicinity of the field study reach, additional recharge from MAR will further increase flow of groundwater into the river in the field study reach.

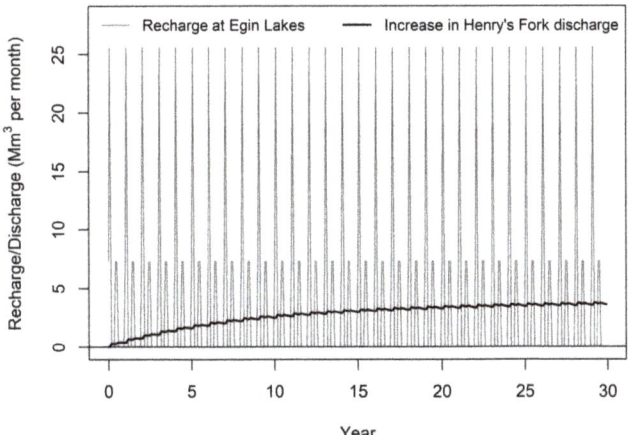

Figure 4. Recharge and discharge for annual MAR scenario of 7.3 Mm^3 per month in each of March, April, and October, and 25.6 Mm^3 per month in November, repeated every year for 30 years.

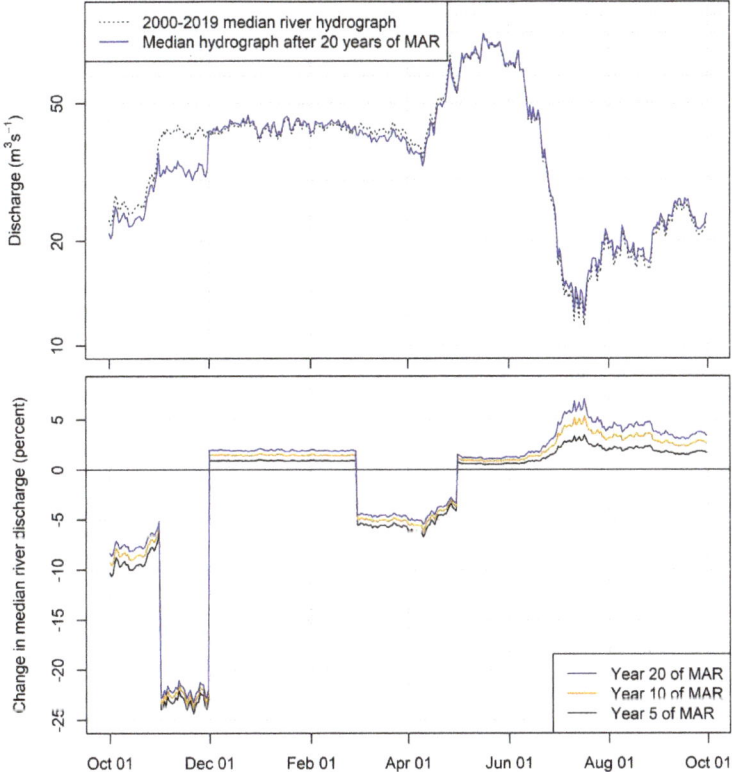

Figure 5. Net change in streamflow for annual scenario in which diversion for MAR from the study reach is 7.3 Mm^3 per month in each of March, April, and October, and 25.6 Mm^3 per month in November. Top panel shows median water-year hydrograph prior to and 20 years after initiation of annual MAR regime. Bottom panel shows percent change in streamflow 5, 10, and 20 years after initiation of annual MAR regime, respectively.

3.2. Local Effects of Groundwater Inflow on Temperature

In late July 2019, thermal imagery identified areas of cool water in the main river and its side channels near the points of spring inflow (Figure 6). Mean instantaneous water temperature at the 20 spring sites differed significantly across the three lateral locations: spring, 0.6 m from the streambank, and 6 m from the bank (F = 29.7, df_1 = 2, df_2 = 38, p < 0.001). All pairwise differences among the locations were significant (Tukey's Honest Significant Difference, adjusted p < 0.001). Mean water temperatures at the lateral locations, respectively, were 14.4 °C, 16.0 °C, and 18.3 °C (Figure 7). Ambient water temperatures during the time of the field observations ranged between daily minima of 18 °C and daily maxima of 23 °C at the top of the field study reach and between 18 °C and 25 °C at the downstream boundary of the study reach (Figure A2).

Figure 6. A side-by-side comparison of a visual image (**left**) and thermal infrared image (**right**) of the area where outflow from a groundwater spring located at 43°57′07.6″ N 111°43′26.7″ W entered a side channel of the river (flowing right to left). The photo was taken from a point 1.5 m from the margin of the side channel, looking toward the spring confluence. The spring emerged from the ground ~30 m from the confluence point. Temperature is indicated in °C.

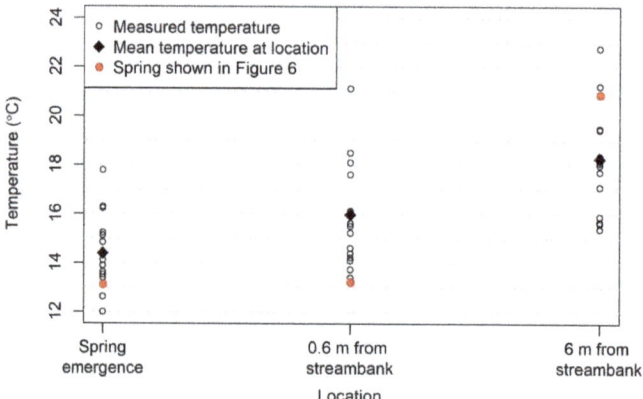

Figure 7. Water temperature at three locations measured at each of 20 distinct springs.

3.3. Water Administration

There are 84 decreed, permitted, or pending water rights for MAR in the upper Snake River basin, with a combined diversion rate of up to 725 m³/sec. However, every senior MAR right is very small, with combined diversion of only 0.59 m³/sec. The remaining 724.41 m³/sec are distributed among large water rights with priority dates of 1980 or later, in a prior-appropriation system where the majority of the irrigation rights have priority dates preceding 1910 and most reservoir storage priority dates precede 1940. Diversion for MAR allowed by the junior rights is available on an annual basis only in the winter and then only downstream of American Falls Reservoir (Figure 1). Considering only water rights specific to MAR, water is available for MAR at the Egin Lakes site in about half of all water years, usually between mid-May and early July. During irrigation season (1 April to 31 October), only water delivered to a designated, off-canal MAR site can be accounted as MAR, but during the winter, canal seepage also accounts as MAR, since canals would not customarily be delivering irrigation water then. Temporary transfers of senior water rights from irrigation or other uses to MAR occur through the state of Idaho's water supply bank, an administrative exchange bank rather than a physical storage bank such as those in Arizona and California.

A locally administered rental pool allows storage water held in Palisades Reservoir (Figure 1) to be rented for delivery to MAR sites anywhere in the basin, through administrative exchange or physical delivery, and these types of exchanges were used to provide the majority of water for MAR in the modeled scenario. Storage water rented in an administrative year (1 November to 31 October) must be delivered by December 1 and cannot be carried over any further into the subsequent year. Since the 2015 ESPA groundwater-surface water settlement was completed, new administrative rules have been enacted specifically to facilitate efficient but equitable water transactions for MAR. However, entities that do not hold water rights, including most conservation groups, cannot participate directly in water supply bank or rental pool transactions. Furthermore, the Idaho Water Resource Board is the only entity that can hold surface water rights for instream flow, regardless of whether those rights are permanent or temporary. Thus, even if conservation groups could participate in water transactions, there is no precedent for them to hold MAR rights specifically for environmental purposes.

3.4. Limitations

The importance of understanding the temporal response to water management actions is critical to address specific objectives. The one-month temporal resolution of ESPAM2.1 limits its ability to predict streamflow response during shorter time intervals that may be critical to trout survival. Furthermore, the model cannot distinguish between management actions that contribute groundwater and those that reduce streamflow losses to groundwater. Although our limited temperature observations suggest that summertime streamflow response to recharge near the field study reach is realized as groundwater inflow, the existing ESPAM2.1 model cannot predict whether MAR and other recharge strategies will change water temperature. The 1.6 km spatial resolution of the model also limits predictive use, especially in assessing response to MAR at hypothetical sites closer to the river, where response changes rapidly with distance away from the river. However, the greatest spatial-resolution limitation in ESPAM2.1 is the delineation of the river reach itself. The model cannot partition water across the 75 km lower Henry's Fork model reach to different locations, requiring simple proportional allocation to downscale model results, as was done here. Models with finer temporal and spatial resolution can be constructed [49], but calibrating higher-resolution models requires hydrologic observations made at the same scale, which are currently not available. A larger challenge to modeling groundwater flow in the Henry's Fork watershed is that the water budget cannot be closed using surface-water observations alone, whereas that for the larger ESPA can. This requires a priori assumptions about groundwater flux to calibrate parameters specifying boundary conditions. Groundwater flux is the most important model output in this case, so model output would essentially be pre determined by assumptions required to calibrate boundary parameters.

An alternative approach to constructing finer-scale models is to estimate local groundwater fluxes using fine-scale piezometer data and temperature mixing models, and use those fine-scale models to downscale the coarse aquifer-river interactions predicted by the regional model. Models can be verified by conducting measurements of streamflow gain and loss across short stream reaches. Continuous, high-resolution temperature data are inexpensive to collect and would contribute not only to temperature mixing models but also to identification of thermal refugia across the whole field study reach. Habitat surveys, observations of fish movement and habitat use, and water-quality analysis at and near areas of cooler water temperatures could then be used to determine whether reducing water temperature alone is sufficient to address factors limiting trout use of the study reach during summer.

4. Discussion

Overall, MAR can improve summer streamflow and stream temperatures for fish in localized areas of the basin. Thirty-seven percent of modeled Egin Lakes MAR returned to the study reach. Streamflow increased most (as a percentage change of baseflow) in the initial years following recharge, with 25–30 years needed to achieve steady-state response to an annually repeated MAR regime. Groundwater seeps confirmed that recharge was contributing to the river rather than merely reducing losses from the river to the aquifer. On average, stream temperature cooled 0.6 °C after traveling 5 km downstream during summer. July 2019 groundwater seep temperatures averaged 14.4 °C and were 16.0 °C about a half-meter from where groundwater seeps entered the river. These temperatures are in the suitable range for brown trout [56]. Average mid-summer ambient temperature was 20.1 °C for 2016 through 2018, which exceeds the optimal thermal brown trout threshold of 19 °C [61]. These findings suggest that MAR provides thermal refugia for managed fish species during summer. A review of administrative MAR rules for fisheries conservation provided more equivocal results. Some water is available for MAR in spring and fall of the wettest half of years, but substantial water for recharge is not consistently available, and canal capacity to the Egin Lakes MAR site limits recharge when water is available. Conservation organizations cannot participate directly in water transactions, but must partner with irrigators or water rights holders on MAR projects.

4.1. Physical and Administrative MAR Implications for Idaho and Henry's Fork

To maintain cool summer water temperatures at the reach scale for cold-water species, MAR volumes will likely need to offset declines in recharge that have occurred from improved irrigation efficiency, as sprinklers and center pivots have replace flood irrigation over the past few decades. Improved irrigation efficiency typically does not increase streamflow as more land is put into production or junior water rights come into priority [66,67]. Around 250 Mm^3 is needed annually to offset lost incidental recharge, which exceeds the physical capacity of the existing MAR site. This volume could be attained if recharge occurred year-round, so that winter canal seepage also contributes. Modeling showed that spring and fall recharge contributes to summer streamflow. Since only 37% of the total volume recharged at the Egin Lakes MAR site returns to Henry's Fork, the benefits to summer streamflow must be weighed against the negative effects of withdrawing larger amounts of water from the river at other times of year. These include negatively impacting aquatic habitat availability, species life history expressions (e.g., spawn timing), or fluvial geomorphic processes (e.g., floodplain maintenance). Developing MAR sites closer to the river could increase streamflow in the study reach per unit of water withdrawn.

Fine-scale field observations conducted in 2019 showed that at seep locations, groundwater inputs can cool ambient stream temperature by over 2 °C during summer, a difference that is biologically significant for trout. Even if cool groundwater inputs are not widespread across river reaches or the contributions are not large compared to river flow, springs may create local thermal refugia for fish, allowing greater survival throughout the summer than would otherwise occur given the same physical habitat availability and streamflow [68]. Further identification of groundwater inflows, their hydrogeologic properties, and their use by fish could inform water and fisheries management actions to enhance groundwater springs for fish populations. Understanding the connectivity of thermal refugia will also help managers understand where fish can become trapped or bottlenecks occur for movement [13,68].

However, it is also important to understand the quality of these groundwater return flows on these groundwater-dependent ecosystems. When conducted on working agricultural lands, MAR risks mobilizing nutrients and increasing chemical constituent loading to streams and riparian soils [20,69–71]. MAR can also facilitate groundwater contamination via crop-mediated aquifer contamination of fertilizers and pesticides [72]. Whereas aquifer recharge may introduce water quality concerns elsewhere, return flow from the ESPA is of high quality that is suitable for instream habitat uses [73]. This research is supported by monitoring conducted at Egin Lakes in 2019 by the Idaho Department of Environmental Quality that showed no increase in nitrogen or fecal coliform as a result of MAR (Aaron Dalling, Fremont-Madison Irrigation District, 2019, presentation to Henry's Fork Watershed Council, October 22). Thus, MAR operations in the Henry's Fork avoid such groundwater contamination and pollutant leaching by conducting recharge via canal seepage and infiltration at Egin Lakes.

While improved modeling and detailed field work can provide technical understanding of MAR benefits to summer habitat for trout, administrative and logistical hurdles must be overcome for implementation [74,75]. The most basic of these is the junior priority dates of water rights for MAR. In the larger context of prior appropriation and development of Idaho's water resources, MAR is a relatively new administratively-recognized beneficial use of water in a system in which water rights for agriculture and mining date back to the mid-19th century. The Idaho Water Resources Board implemented aggressive groundwater and conjunctive management policies and procedures in the 1980s and 1990s, including obtaining large-volume MAR rights with 1980 and 1998 priority dates. Idaho also has flexible and well-established water transaction mechanisms to facilitate transfer of senior water rights to MAR. In the Henry's Fork watershed, the state's MAR water rights are in priority in only half of all water years and then usually during irrigation season, when the canal system is already near capacity delivering irrigation water. Although new MAR infrastructure has been built throughout the basin since 2009, the majority of conveyance to MAR facilities occurs through the existing irrigation canal system. Additional canal capacity could alleviate this limitation during wet years but would go unused in the other half of years.

Because of summer canal capacity limitations, costs of new infrastructure, and the junior priority of MAR rights, storage water rental and other exchange mechanisms offer the greatest potential to increase MAR volumes. For example, in 2018 and 2019, reservoir storage water rented by groundwater users to meet mitigation requirements of their legal agreement with surface water users was not needed for direct delivery to the surface water users because of good water supply in those years. Instead, the rented water was assigned to the state for MAR, allowing recharge of an additional 84 Mm^3 in 2019 over what would have been available using the state's junior MAR rights alone (Wesley Hipke, Idaho Department of Water Resources, 2019, data distributed to stakeholders via email, December 6). The need to manage and administer groundwater and surface water conjunctively to meet the legal requirements of the ESPA settlement creates a pseudo-market to fund such exchanges. In 2017, 2018, and 2019, some of the water recharged at the Egin Lakes site was made available for MAR through water exchanges, including those described above, and delivered in spring and fall, outside of peak irrigation season.

Ideally, conservation groups could facilitate incentive-based irrigation reduction, rent the saved storage water for MAR, and keep that water in reservoirs throughout the summer, thus reducing negative effects of reservoir drawdown. The rented storage water could be diverted for MAR in the off-season, when natural streamflow is sufficient without reservoir releases. The separation of physical and administrative water that makes this possible is routine in Idaho under current administrative procedures. Canal capacity to deliver water for MAR is greater in the winter, when canals are not also delivering irrigation water, although the basin's sub-freezing temperatures present challenges in managing winter canal delivery and frozen soil can impede infiltration [76]. Canal seepage during the winter adds recharge that would not otherwise occur and is not simply an administrative replacement for historical seepage incidental to irrigation that occurs during the summer. Rental pool water is available in larger quantities and at lower prices during wet years, potentially allowing for more MAR during wet years, which would then contribute to streamflow in subsequent dry years. However, changes to water rental rules are needed to allow storage water rented in one administrative year to be diverted for MAR several months into the next administrative year [64]. In addition, conservation organizations cannot implement MAR projects, but instead must partner with irrigation entities or individual water rights holders. To fully capitalize on the high value anglers place on upper Snake

River fisheries, new administrative and transfer mechanisms are required to allow conservation organizations to participate directly in water transactions.

4.2. Physical and Administrative MAR Implications for other States in the Western USA

Most states in the western USA possess some of the physical and administrative features required for MAR to benefit cold-water ecosystems, but few have all of the requisite ingredients. Arizona's progressive and flexible administrative systems have been highly successful in facilitating recharge of Colorado River water, but the sole purpose of this MAR is to store the water for later recovery, not to enhance streamflow [36]. On the other hand, restrictive administrative rules in Colorado limit use of MAR in headwater alluvial aquifers [37], where snowmelt could be captured and recharged with the intent of enhancing streamflow later in the summer, and where Colorado's progressive water markets would allow conservation organizations to obtain this water for environmental uses [40]. California is now conducting "flood-MAR" in depleted aquifers using stormwater [38], but this water is junior to other rights and is available for MAR only when existing water rights and required environmental uses are fulfilled. Some regions of California are considering creating an environmental water account using MAR, although this idea is currently untested.

4.3. MAR as Climate Adaptation Strategy for Fisheries Conservation

Cold-water fish habitat is anticipated to decline substantially with climate change. In fact, brown trout are expected to lose 48% of their habitat in the interior western US by the 2080s [68]. These changes are driven by warmer stream temperature and increasing winter floods, and could have major repercussions for a local economy reliant on a US $100 million recreational fishery. Additionally, an increase in extreme climate events—i.e., a higher frequency of wet and dry years, with fewer 'normal' years is also expected [77]—which will alter instream conditions for biota [78]. Together, these changes suggest that climate adaptation strategies that provide mechanisms to reduce winter flooding, increase summer baseflow, cool summer stream temperature, and enhance thermal refugia are warranted. MAR for fisheries conservation is one such strategy. MAR is a promising climate-adaptive water management strategy because winter flows can be recharged to underlying aquifers to maintain baseflow and cold-water fish habitat throughout the year. However, junior water right holders will have considerable uncertainty from the increased inter-annual variability inherent with climate change [79]. Although this does not inherently reduce the utility of MAR for fisheries conservation in a warming climate, it does suggest that relaxing current administrative rules for greater flexibility to carry over reservoir rental water between years would improve the utility of MAR for fisheries conservation in a changing climate. Since many important recreational trout fisheries in the Western U.S. are located downstream of reservoirs, renting reservoir storage not used for irrigation in a given season and using it for MAR during the subsequent off-season is an innovative conservation mechanism that could have wide applicability.

5. Conclusions

MAR during spring and fall is expected to increase streamflow by around 5% during mid-summer, but only after 20 years of consistent MAR. Developing MAR sites closer to the river than the existing site would provide greater streamflow benefit per unit of water withdrawn and on shorter time frames. However, even relatively small increases in streamflow could have disproportionately greater benefits to trout, as streamflow response in our study area occurs in the form of increased groundwater inputs rather than decreased losses to the aquifer. July water temperatures were locally 2 °C cooler where groundwater flowed into the river. Because MAR is a new and junior water use in a priority system with irrigation rights dating back to the 19th century, MAR water rights are generally in priority only during late spring and early summer of years of above-average supply. However, administrative exchange that allows reservoir storage to be used for MAR can make water available during spring and fall, when MAR infrastructure capacity is greatest. Changes to current administrative rules could increase the effectiveness of such exchange mechanisms in providing water for MAR.

Author Contributions: Conceptualization: R.W.V.K.; methodology: R.W.V.K., B.A.C., C.N.M., and A.S.L.; software: B.A.C.; validation: R.W.V.K., B.A.C., and C.N.M.; formal analysis: R.W.V.K.; investigation: A.S.L., R.W.V.K., and C.N.M.; resources: R.W.V.K. and S.E.N.; data curation: R.W.V.K., B.A.C., and C.N.M.; writing—original draft preparation: R.W.V.K., B.A.C., C.N.M., and S.E.N.; writing—review and editing: B.A.C., C.N.M., S.E.N., R.W.V.K., and A.S.L.; visualization: C.N.M., B.A.C., and R.W.V.K.; supervision: R.W.V.K., C.N.M., and S.E.N.; project administration: R.W.V.K. and C.N.M.; funding acquisition: R.W.V.K. and C.N.M. All authors have read and agreed to the published version of the manuscript.

Funding: Collection of the 2010 water temperature data was funded by U.S. Department of Agriculture grant 2008-51130-19555 to R.V.K. C.N.M. received support from National Science Foundation grant no. 1633756. A.S.L. was supported during summer 2019 by an internship from Colgate University. Additional funding was provided by the Federal Highway Administration and Fremont County Idaho.

Acknowledgments: Matt Hively assisted with field data collection. Input from the academic editor and five anonymous reviewers greatly improved the manuscript.

Conflicts of Interest: The authors declare no conflict of interest. The funding sponsors had no role in the design of the study; in the collection, analyses, or interpretation of data; in the writing of the manuscript; or in the decision to publish the results.

Appendix A. Streamflow and Temperature Graphs

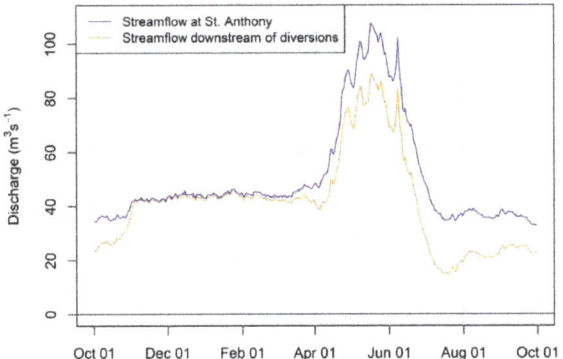

Figure A1. Streamflow at the St. Anthony gage (top of field study reach) and downstream of all diversions (roughly at the upstream extent of the springs identified in Figure 2).

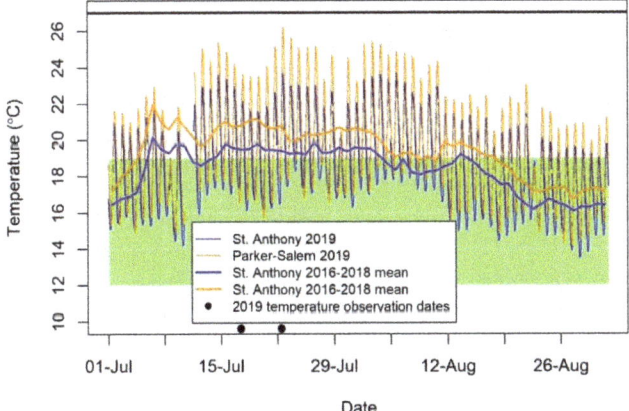

Figure A2. July–August water temperature at the St. Anthony water quality station (top of field study reach) and at the Parker-Salem water quality station (bottom of field study reach). Shaded polygon is optimal thermal range for brown trout. Horizontal line is lethal temperature limit for brown trout. The dates of the 2019 temperature observations are shown for reference.

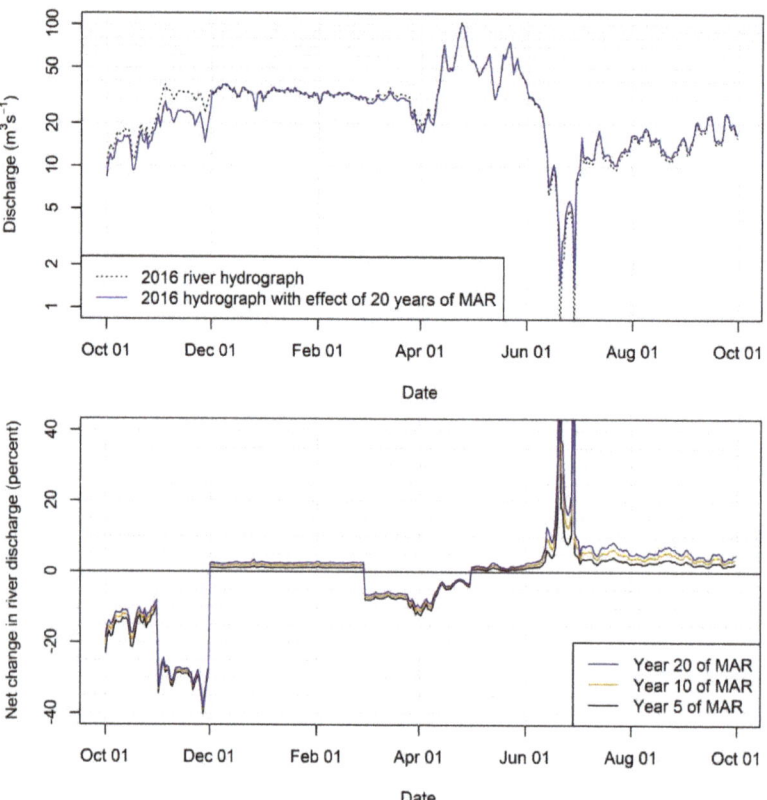

Figure A3. Net change in streamflow in a dry year for annual scenario in which diversion for MAR from the study reach is 7.3 Mm^3 per month in each of March, April, and October, and 25.6 Mm^3 per month in November. Top panel shows the 2016 water-year hydrograph and hypothetical effect of MAR implemented 20 years prior. Bottom panel shows percent change in observed 2016 streamflow 5, 10, and 20 years after initiation of annual MAR regime, respectively.

Appendix B. Statistical Methods

Appendix B.1. Time Series Analysis

Statistical hypothesis tests require independent observations for correct distribution of test statistics [80]. In time series such as daily water temperatures, observations are not independent of one another because of correlation between a given observation and the observations that precede it in the time series. This is referred to as serial or temporal autocorrelation. Its effect must be removed to obtain independence of observations before conducting hypothesis tests on time series. Autoregressive (AR) models are used to accomplish this [81]. The simplest AR model is the first-order model:

$$y_t = \mu + \phi_1(y_{t-1} - \mu) + \varepsilon, \tag{A1}$$

where y_t is the observation at time t, μ is the mean of the time series, ϕ_1 is the first-order autoregressive coefficient, and ε is random, independent, normally distributed error. The autocorrelation term $\phi_1(y_{t-1} - \mu)$ removes the dependence of y_t on y_{t-1}, allowing hypothesis tests to be conducted on the mean μ. In our case, serial autocorrelation was high enough that observations were correlated with those one, two, and three time steps prior. The resulting third-order model has the same form as

Equation (A1), but with three autocorrelation terms. Because observation y_t in our time series was the difference in temperature between the two locations, our null hypothesis was $\mu = 0$. If rejected, we infer $\mu \neq 0$. We used a significance level of 0.05, which is the default standard in statistical hypothesis testing. It represents the probability of having made an error in rejecting the null hypothesis, referred to as Type 1 error [80].

Appendix B.2. Tukey's Post-Hoc Test

Analysis of variance tests the single null hypothesis that all group means are equal. If the null hypothesis is rejected at a given significance level, the alternative hypothesis is simply that at least one group mean differs from at least one other. Additional tests must be done to assess which group mean(s) differ from which others. The probability of committing Type 1 error is compounded each time an additional test is performed. In our case, testing all possible differences between three group means requires three tests. If each is performed at a 0.05 significance level, the probability of committing at least one Type 1 error across the three tests is roughly 0.14. Tukey's post-hoc test is a method for conducting the tests for differences across all pairs of groups while maintaining the desired level of significance across the whole family of tests [80].

References

1. Slaughter, R.A.; Wiener, J.D. Water, adaptation, and property rights on the Snake and Klamath rivers. *JAWRA* **2007**, *43*, 308–321. [CrossRef]
2. Al-Chokhachy, R.; Sepulveda, A.J.; Ray, A.M.; Thoma, D.P.; Tercek, M.T. Evaluating species-specific changes in hydrologic regimes: An iterative approach for salmonids in the Greater Yellowstone Area (USA). *Rev. Fish Biol. Fish.* **2017**, *27*, 425–441. [CrossRef]
3. Van Kirk, R.; Hoffner, B.; Verbeten, A.; Yates, S. New approaches to providing instream flow for fisheries in the American West: Embracing prior appropriation and the marketplace. In *Multispecies and Watershed Approaches to Freshwater Fish Conservation*; Dauwalter, D.C., Birdsong, T.W., Garret, G.P., Eds.; American Fisheries Society: Bethesda, MD, USA, 2019; Symposium 91; pp. 515–564.
4. Ficklin, D.L.; Abatzoglou, J.T.; Robeson, S.M.; Null, S.E.; Knouft, J.H. Natural and managed watersheds show similar responses to recent climate change. *Proc. Natl. Acad. Sci. USA* **2018**, *115*, 8553–8557. [CrossRef] [PubMed]
5. Dai, A. Increasing drought under global warming in observations and models. *Nat. Clim. Chang.* **2013**, *3*, 52–58. [CrossRef]
6. Wenger, S.J.; Isaak, D.J.; Luce, C.H.; Neville, H.M.; Fausch, K.D.; Dunham, J.B.; Dauwalter, D.C.; Young, M.K.; Elsner, M.M.; Rieman, B.E.; et al. Flow regime, temperature, and biotic interactions drive differential declines of trout species under climate change. *Proc. Natl. Acad. Sci. USA* **2011**, *108*, 14175–14180. [CrossRef]
7. Poole, G.C.; Berman, C.H. An ecological perspective on in-stream temperature: Natural heat dynamics and mechanisms of human-caused thermal degradation. *Env. Manag.* **2001**, *27*, 787–802. [CrossRef]
8. Olden, J.D.; Naiman, R.J. Incorporating thermal regimes into environmental flows assessments: Modifying dam operations to restore freshwater ecosystem integrity. *Freshw. Biol.* **2010**, *55*, 86–107. [CrossRef]
9. Rheinheimer, D.E.; Null, S.E.; Lund, J.R. Optimizing selective withdrawal from reservoirs to manage downstream temperatures with climate warming. *J. Water Resour. Plann. Manag.* **2015**, *141*, 04014063. [CrossRef]
10. Elmore, L.R.; Null, S.E.; Mouzon, N.R. Effects of environmental water transfers on stream temperatures. *River Res. Appl.* **2016**, *32*. [CrossRef]
11. Scherberg, J.; Baker, T.; Selker, J.S.; Henry, R. Design of managed aquifer recharge for agricultural and ecological water supply assessed through numerical modeling. *Water Resour. Manag.* **2014**, *28*, 4971–4984. [CrossRef]
12. Ronayne, M.J.; Roudebush, J.A.; Stednick, J.D. Analysis of managed aquifer recharge for retiming streamflow in an alluvial river. *J. Hydrol.* **2017**, *544*, 373–382. [CrossRef]

13. Dzara, J.R.; Neilson, B.T.; Null, S.E. Quantifying thermal refugia connectivity by combining temperature modeling, distributed temperature sensing, and thermal infrared imaging. *Hydrol. Earth Syst. Sci.* **2019**, *23*, 2965–2982. [CrossRef]
14. Kløve, B.; Ala-Aho, P.; Bertrand, G.; Gurdak, J.J.; Kupfersberg, H.; Kværner, J.; Muotka, T.; Mykrä, H.; Preda, E.; Pekka, R.; et al. Climate change impacts on groundwater and dependent ecosystems. *J. Hydrol.* **2014**, *518*, 250–266. [CrossRef]
15. Sophocleous, M. Interaction between groundwater and surface water: The state of the science. *Hydrogeol. J.* **2002**, *10*, 52–67. [CrossRef]
16. Taylor, R.G.; Scanlon, B.; Döll, P.; Rodell, M.; van Beek, R.; Wada, Y.; Longuevergne, L.; Leblanc, M.; Famiglietti, J.S.; Edmunds, M.; et al. Ground water and climate change. *Nat. Clim. Chang.* **2013**, *3*, 322–329. [CrossRef]
17. Dams, J.; Salvadore, E.; Van Daele, T.; Ntegeka, V.; Willems, P.; Batelaan, O. Spatio-Temporal impact of climate change on the groundwater system. *Hydrol. Earth Syst. Sci. Discuss.* **2011**, *8*, 10195–10223. [CrossRef]
18. Kendy, E.; Bredehoeft, J.D. Transient effects of groundwater pumping and surface-water-irrigation returns on streamflow. *Water Resour. Res.* **2006**, *42*, W08145. [CrossRef]
19. Fernald, A.; Guldan, S.; Boykin, K.; Cibils, A.; Gonzales, M.; Hurd, B.; Lopez, S.; Ochoa, C.; Ortiz, M.; Rivera, J.; et al. Linked hydrologic and social systems that support resilience of traditional irrigation communities. *Hydrol. Earth Syst. Sci.* **2015**, *19*, 293–307. [CrossRef]
20. Niswonger, R.G.; Morway, E.D.; Triana, E.; Huntington, J.L. Managed aquifer recharge through off-season irrigation on agricultural regions. *Water Resour. Res.* **2017**, *53*, 5970–6992. [CrossRef]
21. Tague, C.; Grant, G.; Farrell, M.; Choat, J.; Jefferson, A. Deep groundwater mediates streamflow response to climate warming in the Oregon Cascades. *Clim. Chang.* **2008**, *86*, 189–210. [CrossRef]
22. Barber, M.E.; Hossain, A.; Covert, J.J.; Gregory, G.J. Augmentation of seasonal low stream flows by artificial recharge in the Spokane Valley-Rathdrum Prairie aquifer of Idaho and Washington, USA. *Hydrogeol. J.* **2009**, *17*, 1459–1470. [CrossRef]
23. Brunke, M.; Gonser, T.O.M. The ecological significance of exchange processes between rivers and groundwater. *Freshw. Biol.* **1997**, *37*, 1–33. [CrossRef]
24. Healy, R.W.; Cook, P.G. Using groundwater levels to estimate recharge. *Hydrogeol. J.* **2002**, *10*, 91–109. [CrossRef]
25. Palmer, M.A.; Lettenmaier, D.P.; Poff, N.L.; Postel, S.L.; Richter, B.; Warner, R. Climate change and river ecosystems: Protection and adaptation options. *Environ. Manag.* **2009**, *44*, 1053–1068. [CrossRef] [PubMed]
26. Fernald, A.G.; Cevik, S.Y.; Ochoa, C.G.; Tidwell, V.C.; King, J.P.; Guldan, S.J. River hydrograph retransmission functions of irrigated valley surface water-groundwater interactions. *J. Irrig. Drain. Eng. ASCE* **2010**, *136*, 823–835. [CrossRef]
27. Nichols, A.J.; Willis, A.D.; Jeffres, C.A.; Deas, M.L. Water temperature patterns below large groundwater springs: Management implications for Coho Salmon in the Shasta River, California. *River Res. Appl.* **2014**, *30*, 442–455. [CrossRef]
28. Snyder, C.D.; Hitt, N.P.; Young, J.A. Accounting for groundwater in stream fish thermal habitat responses to climate change. *Ecol. Appl.* **2015**, *25*, 1397–1419. [CrossRef]
29. Carlson, A.K.; Taylor, W.W.; Schlee, K.M.; Zorn, T.G.; Infante, D.M. Projected impacts of climate change on stream salmonids with implications for resilience-based management. *Ecol. Freshw. Fish* **2017**, *26*, 190–204. [CrossRef]
30. Cristea, N.C.; Burges, S.J. Use of thermal infrared imagery to complement monitoring and modeling of spatial stream temperatures. *J. Hydrol. Eng.* **2009**, *14*, 1080–1090. [CrossRef]
31. Taylor, C.A.; Stefan, H.G. Shallow groundwater temperature response to climate change and urbanization. *J. Hydrol.* **2009**, *375*, 601–612. [CrossRef]
32. Kurylyk, B.L.; MacQuarrie, K.T.B.; Voss, C.I. Climate change impacts on temperature and magnitude of groundwater discharge from shallow, unconfined aquifers. *Water Resour. Res.* **2014**, *50*, 3253–3274. [CrossRef]
33. Menberg, K.; Blum, P.; Kurylyk, B.L.; Bayer, P. Observed groundwater temperature response to recent climate change. *Hydrol. Earth Syst. Sci.* **2014**, *11*, 3637–3673. [CrossRef]
34. Scherberg, J.; Keller, J.; Patten, S.; Baker, T.; Milczarek, M. Modeling the impact of aquifer recharge, in-stream water savings, and canal lining on water resources in the Walla Walla Basin. *Sustain. Water Resour. Manag.* **2018**, *4*, 275–289. [CrossRef]

35. Larson, R.K.; Spinazola, J. Conjunctive management analyses for endangered species flow augmentation alternatives in the Snake River. In Proceedings of the Watershed Management and Operations Management 2000, Fort Collins, CO, USA, 20–24 June 2000; Flug, M., Frevert, D., Watkins, D.W., Jr., Eds.; American Society of Civil Engineers: Reston, VA, USA, 2000. [CrossRef]
36. Scanlon, B.R.; Reedy, R.C.; Faunt, C.C.; Pool, D.; Uhlman, K. Enhancing drought resilience with conjunctive use and managed aquifer recharge in California and Arizona. *Environ. Res. Lett.* **2016**, *11*, 035013. [CrossRef]
37. Mortimer, E. Managed aquifer recharge: An overview of laws affecting aquifer recharge in several western states. *Water Rep.* **2014**, *127*, 11–25.
38. California Department of Water Resources. *Flood-MAR: Using Flood Water for Managed Aquifer Recharge to Support Sustainable Water Resources*; White Paper: Sacramento, CA, USA, 2018.
39. Davis, D.; Li, Y.; Batzle, M. Time-Lapse gravity monitoring: A systematic 4D approach with application to aquifer storage and recovery. *Geophysics* **2014**, *73*, WA61–WA69. [CrossRef]
40. Szeptycki, L.; Forgle, J.; Hook, E.; Lorick, K.; Womble, P. *Environmental Water Rights Transfers: A Review of State Laws*; Water in the West; Stanford University: Stanford, CA, USA, 2015.
41. Idaho Water Resource Board. *Eastern Snake Plain Aquifer Comprehensive Aquifer Management Plan*; Idaho Water Resource Board: Boise, ID, USA, 2009.
42. Loomis, J. Use of survey data to estimate economic value and regional economic effects of fishery improvements. *N. Am. J. Fish Manag.* **2006**, *26*, 301–307. [CrossRef]
43. Grunder, S.A.; McArthur, T.J.; Clark, S.; Moore, V.K. *2003 Economic Survey Report*; Report IDFG 08-129; Idaho Department of Fish and Game: Boise, ID, USA, 2008.
44. Boggs, K.G.; Van Kirk, R.W.; Johnson, G.S.; Fairley, J.P.; Porter, P.S. Analytical solutions to the linearized Boussinesq equation for assessing the effects of recharge on aquifer discharge. *JAWRA* **2010**, *46*, 1116–1132. [CrossRef]
45. Johnson, G.S.; Sullivan, W.H.; Cosgrove, D.M.; Schmidt, R.D. Recharge of the snake river plain aquifer: Transitioning from incidental to managed. *JAWRA* **1999**, *35*, 123–131. [CrossRef]
46. Fernald, A.G.; Guldan, S.J. Surface water-groundwater interactions between irrigation ditches, alluvial aquifers, and streams. *Rev. Fish. Sci. Aquac.* **2006**, *14*, 79–89. [CrossRef]
47. U.S. Bureau of Reclamation. *Henrys Fork Basin Study Water Needs Assessment*; Technical Report PN-HFS-001; U.S. Bureau of Reclamation, Pacific Northwest Region: Boise, ID, USA, 2012.
48. Wytzes, J. Development of a Groundwater Model for the Henry's Fork and Rigby Fan Areas, Upper Snake River Basin, Idaho. Ph.D. Thesis, University of Idaho, Moscow, ID, USA, 1980.
49. Apple, B.D. Predicting Groundwater Effects Due to Changing Land Practices in the Intermountain West. Master's Thesis, Humboldt State University, Arcata, CA, USA, 2013.
50. Contor, B. *Delineation of Sprinkler and Gravity Application Systems. Eastern Snake Plain Aquifer Model Enhancement Project Scenario Document DDW-022*; Technical Report 04-005; Idaho Water Resources Research Institute, University of Idaho: Moscow, ID, USA, 2004.
51. Hortness, J.; Vidmar, P. *Seepage Study on the Henrys Fork and Snake River, Idaho.*; U.S. Geological Survey Idaho Water Science Center: Boise, ID, USA, 2003.
52. Idaho Department of Water Resources. *Enhanced Snake Plain Aquifer Model Version 2.1 Final Report*; Idaho Department of Water Resources: Boise, ID, USA, 2013.
53. Joint Committee. *Henry's Fork Drought Management Plan*; Fremont-Madison Irrigation District: St. Anthony, ID, USA, 2018.
54. Flinders, J.; Keen, D.; High, B.; Garren, D. *Fishery Management Annual Report, Upper Snake Region 2014*; Report IDFG 16-108; Idaho Department of Fish and Game: Boise, ID, USA, 2016.
55. Morrisett, C.; Van Kirk, R.; Loibman, A. *Lower Henry's Fork Hydrology and Habitat Assessment*; Progress Report Submitted to Meet Conditions of Ora Bridge Mitigation Agreement; Local Highway Technical Assistance Council: Boise, ID, USA, 2019.
56. Raleigh, R.F.; Zuckerman, L.D.; Nelson, P.C. *Habitat Suitability Index Models and Instream Flow Suitability Curves: Brown Trout*; U.S. Fish and Wildlife Service: Washington, DC, USA, 1984; 82(10.124).
57. Benjamin, L.; Van Kirk, R.W. Assessing instream flows and reservoir operations on an eastern Idaho river. *JAWRA* **1999**, *35*, 899–909. [CrossRef]

58. Laatsch, J.; Van Kirk, R.; Morrisett, C.; Manishin, K.; DeRito, J. Angler perception of fishing experience in a highly technical catch-and-release fishery: How closely does perception align with biological reality? In *Science, Politics, and Wild Trout Management: Who's Driving and Where are We Going? Proceedings of Wild Trout XII Symposium, West Yellowstone, MT, USA, 26–29 September 2017*; Carline, R.F., LoSapio, C., Eds.; pp. 47–53. Available online: https://www.wildtroutsymposium.com/proceedings-12.pdf (accessed on 21 February 2020).
59. McClaren, J.S.; Royer, T.V.; Van Kirk, R.W.; Muradian, M.L. Management and limnology interact to drive water temperature in a Middle Rockies river-reservoir. *JAWRA* **2019**, *55*, 1323–1334. [CrossRef]
60. Doherty, J. *PEST Model-Independent Parameter Estimation Users Manual*; Watermark Numerical Computing: Brisbane, Australia, 2004.
61. Beach, T.; Null, S.E.; Gray, C.A. An affordable method of thermal infrared remote sensing of wadeable rivers using a weather balloon. *J. Undergrad. Res.* **2016**, *7*, 26–31.
62. CH2M and Henry's Fork Foundation. *Eastern Snake Plain Aquifer (ESPA) Review of Comprehensive Managed Aquifer Recharge Program*; Final Report; Idaho Water Resource Board: Boise, ID, USA, 2016.
63. Olenichak, T. *Concepts, Practices, and Procedures Used to Distribute Water within Water District #1 Upper Snake River Basin Idaho*; Idaho Department of Water Resources Water District 01: Idaho Falls, ID, USA, 2015.
64. Idaho Department of Water Resources. *Water District 1 Rental Pool Rules*; Idaho Department of Water Resources: Idaho Falls, ID, USA, 2019.
65. Idaho Code Title 42, Chapter 17, Sections 1761–1766. Boise, Idaho, USA. Available online: https://legislature.idaho.gov/statutesrules/idstat/Title42/T42CH17/ (accessed on 17 November 2019).
66. Ward, F.A.; Pulido-Velazquez, M. Water conservation in irrigation can increase water use. *Proc. Natl. Acad. Sci. USA* **2008**, *105*, 18215–18220. [CrossRef] [PubMed]
67. Grafton, R.Q.; Williams, J.; Perry, C.J.; Molle, F.; Ringler, C.; Steduto, P.; Udall, B.; Wheeler, S.A.; Wang, Y.; Garrick, D.; et al. The paradox of irrigation efficiency. *Science* **2018**, *361*, 748–750. [CrossRef]
68. Fullerton, A.H.; Torgersen, C.E.; Lawler, J.J.; Steel, E.A.; Ebersole, J.L.; Lee, S.Y. Longitudinal thermal heterogeneity in rivers and refugia for coldwater species: Effects of scale and climate change. *Aquat. Sci.* **2018**, *80*, 1–15. [CrossRef] [PubMed]
69. Dillon, P.; Kumar, A.; Kookana, R.; Leijs, R.; Reed, D.; Parsons, S.; Ingerson, G. *Managed Aquifer Recharge—Risks to Groundwater Dependent Ecosystems—A review*; Water for a Healthy Country Flagship Report to Land & Water Australia; Commonwealth Scientific and Industrial Research Organisation: Canberra, Australia, 2009.
70. Scanlon, B.R.; Reedy, R.C.; Stonstrom, D.A.; Prudic, D.E.; Dennehy, K.F. Impact of land use and land cover change on groundwater recharge and quality in the southwestern US. *Glob. Chang. Biol.* **2005**, *11*, 1577–1593. [CrossRef]
71. Morway, E.D.; Gates, T.K.; Niswonger, R.G. Appraising options to reduce shallow groundwater tables and enhance flow conditions over regional scales in an irrigated alluvial aquifer system. *J. Hydrol.* **2013**, *495*, 216–237. [CrossRef]
72. Fienen, M.N.; Arshad, M. The international scale of the groundwater issue. In *Integrated Groundwater Management*; Jakeman, A.J., Barreteau, O., Hunt, R.J., Rinaudo, J.-D., Ross, A., Eds.; Springer Open: Basel, Switzerland, 2016; pp. 21–42.
73. Low, W.H. *Solute Distribution in Ground and Surface Water in the Snake River Basin, Idaho and Eastern Oregon*; United States Geologic Survey Hydrologic Atlas 696; United States Geological Survey: Reston, VA, USA, 2017. [CrossRef]
74. Langridge, R. Drought and groundwater: Legal hurdles to establishing groundwater drought reserves in California. *Environs* **2012**, *36*, 91–113.
75. Jacobs, K.L.; Holway, J.M. Managing for sustainability in an arid climate: Lessons learned from 20 years of groundwater management in Arizona, USA. *Hydrogeol. J.* **2004**, *12*, 52–65. [CrossRef]
76. Iwata, Y.; Hayashi, M.; Hirota, T. Comparison of snowmelt infiltration under different soil-freezing conditions influenced by snow cover. *Vadose Zone J.* **2008**, *7*, 79–86. [CrossRef]
77. Easterling, D.R.; Meehl, G.A.; Parmesan, C.; Changnon, S.A.; Karl, T.R.; Mearns, L.O. Climate extremes: Observations, modeling, and impacts. *Science* **2000**, *289*, 2068–2074. [CrossRef] [PubMed]
78. Rheinheimer, D.E.; Null, S.E.; Viers, J.H. Climate-Adaptive water year typing for instream flow requirements in California's Sierra Nevada. *J. Water Resour. Plann. Manag.* **2016**, *142*, 04016049. [CrossRef]
79. Null, S.E.; Prudencio, L. Climate change effects on water allocations with season dependent water rights. *Sci. Total Environ.* **2016**, *571*, 943–954. [CrossRef]

80. Sokal, R.R.; Rohlf, F.J. *Biometry*, 4th ed.; W.H. Freeman: New York, NY, USA, 2012.
81. Chatfield, C. *The Analysis of Time Series*, 4th ed.; Chapman and Hall: London, UK, 1989.

 © 2020 by the authors. Licensee MDPI, Basel, Switzerland. This article is an open access article distributed under the terms and conditions of the Creative Commons Attribution (CC BY) license (http://creativecommons.org/licenses/by/4.0/).

Article

Suitability Mapping for Managed Aquifer Recharge: Development of Web-Tools

Jana Sallwey [1,*], Robert Schlick [1], José Pablo Bonilla Valverde [2], Ralf Junghanns [1], Felipe Vásquez López [3] and Catalin Stefan [1]

1. Research Group INOWAS, Department of Hydro Sciences, Technische Universität Dresden, 01069 Dresden, Germany; robert.schlick@tu-dresden.de (R.S.); ralf.junghanns@tu-dresden.de (R.J.); catalin.stefan@tu-dresden.de (C.S.)
2. Instituto Costarricense de Acuedcutos y Alcantarillados, 10109 San Jose, Costa Rica; bonilla.jp@gmail.com
3. INERCO Consultoria, 110111 Bogota, Colombia; felvaslop@gmail.com
* Correspondence: Jana.sallwey@tu-dresden.de; Tel.: +49-351-46344146

Received: 12 August 2019; Accepted: 21 October 2019; Published: 28 October 2019

Abstract: Suitability maps for managed aquifer recharge (MAR) are increasingly used and hold the potential to be integrated into sustainable groundwater management plans. However, the quality of the maps strongly depends on the input data quality as well as the expertise of the decision-maker. The maps are commonly derived through GIS-based multi-criteria decision analysis (GIS-MCDA). To date, there is no common understanding of how suitability mapping should be conducted, as there is considerable variability concerning used GIS data and MCDA methodology. This study presents two web-tools that were conceptualized based on a review of GIS-MCDA studies in the context of MAR suitability mapping. The data retrieved from the review was compiled into a web-based query tool making the MAR- and MCDA-relevant information easily accessible. Based on the most commonly used MCDA practices in the assessed studies, we conceptualized and implemented a second web tool that comprises a simplified web GIS as well as supporting tools for weight assignment and standardization of the criteria. Both web tools will enable decision-makers to engage in MCDA for MAR mapping in a more structured and informed way. As the tools are open-source and web-based, they can facilitate the collaboration between multiple stakeholders and the easy sharing of results.

Keywords: managed aquifer recharge; web GIS; web tools; multi-criteria decision analysis; suitability mapping

1. Introduction

The application of managed aquifer recharge (MAR) is continuing to grow worldwide as a measure for sustainable groundwater management [1,2]. Before MAR schemes can be developed, comprehensive planning is required to ensure their long-term sustainability. While guidelines on the planning of MAR schemes exist [3–6], they mostly focus on their design and operation and put less focus on site selection. The selection of sites suitable for MAR is a critical step in the planning phase of a MAR project, as the location influences the recharge technique as well as the operational and maintenance parameters, such as the infiltration quantity and the recovery efficiency [7–10]. Site selection for MAR application is mostly conducted through field investigations. Suitability maps that show the potential of a foreseen area for the application of a certain MAR type can be generated as a preliminary step to field investigations. These maps are increasingly being used [11] and may fill a void in missing strategic MAR site planning. Their advantages for water management plans lie within the spatial display through maps [12], the quickness and simplicity of the analysis [13], the possibility to include projections of climate scenarios, population growth or land-use changes [14] as well as the assessment of different MAR techniques and their location [15].

While these maps are increasingly being used, there are no common guidelines on how the suitability mapping process should be conducted. The maps are generally generated by combining geoinformation of the surface and the subsurface with socio-economic criteria. This can be achieved by integrating multi-criteria decision analysis (MCDA) for solving spatial problems with GIS software [16]. A set of geospatial data must be chosen based on the study's objectives. The different GIS criteria are then weighted based on their importance for the study and combined into a suitability map. MCDA comprises a variety of methods for criteria weighting and combining [17]. A study showed that the GIS maps used and the methodologies applied for MCDA in the context of MAR, show a great variety [11]. The choice of GIS data and how important each dataset is seen for each study is dependent on the data availability and the local characteristics but also on the expert opinion and the problem statement. Finding common ground is near impossible, as these aspects are highly case-study dependent. However, the methodologies used for suitability mapping of MAR could potentially be synthesized. Rahman et al. [8] developed a GIS-based tool for MCDA site selection analysis and structured the methodology of site suitability mapping, making a first effort to standardize the GIS-MCDA methodology for MAR site selection.

This study continues the work of Rahman et al. [8] to structure and simplify the decision-making process for MAR suitability mapping. Our work is based on the knowledge generated from a previously published review on GIS-MCDA application for MAR suitability mapping [11]. Findings from the review were taken into consideration to design and implement the web tools presented in this paper. Two related tools were designed to help with the standardization of the MAR mapping process. All data collected from the review were implemented in a web-based query tool that makes the information easily accessible. From the review, the most frequently used methodologies for map generation were determined and included in a web GIS tool. This tool takes a systematic approach, engaging decision-makers in the MCDA process in a structured way. Links between the web GIS tool and the query tool support the decision-making process as they readily depict GIS criteria, criteria weighting, and MCDA methodologies.

While the previous work was dedicated to structuring the suitability mapping process, the present paper focuses on the development of user-friendly tools and their web-based implementation. Thereby, this work aids the decision-makers in undertaking a standardized mapping procedure and, thus, can help to increase the reliability of the method application for the generated maps. As the tools are web-based, they enable the collaboration of multiple stakeholders, thus, potentially improving the decision-making process as well as facilitating easy sharing of results. The web-based nature, as well as the open-access availability, thrive to enhance the usability of the tools, a particularity distinguishing them from existing desktop solutions.

2. Implications from Reviewing GIS-MCDA Studies

This section is based on an already published review on MAR suitability mapping [11] and focusses on analyzing the relevant parameters and methods for the comprehensive approach to GIS-based suitability mapping by Rahman et al. [8]. Their approach divides suitability mapping into four main steps. It follows the scheme of (a) problem definition, (b) screening of suitable areas (constraint mapping), (c) suitability mapping including the classification of thematic layers or criteria, standardization, weighting of the criteria, and layer overlaying by decision rule, and (d) sensitivity analysis.

Problem definition is the basis of choosing relevant GIS maps and weighting their importance for solving the problem statement. As this part of the approach is highly problem-specific, no general statement or implications could be formulated from the review.

Constraint mapping identifies the parts of the study area that are not suitable for the application of MAR or need to be excluded from the analysis as they are, for instance, natural reserves or private land. This is achieved through threshold values for the GIS criteria and by applying Boolean logic to clip respective areas from the final map. Half of the analyzed studies used constraint mapping as a tool to exclude unsuitable areas. As this methodology is widely used, options to constrain single GIS datasets or complete areas from the resulting map were envisaged for the web GIS tool.

Suitability mapping is the core of the MCDA process as it ranks the study area based on its suitability for the application of MAR. This step comprises the standardization of GIS maps, the assignment of weights to every map, and the combination of the weights and the standardized maps by decision rule. The most commonly used weight assignment methods are the rating method, the ranking method, the multi-influence factor (MIF), and the pairwise comparison. The rating and ranking methods are very simple methods comprising manual weight assignment on a predetermined scale [17]. MIF is a graphical weight assignment method where linkages between GIS datasets are drawn, and the weights are calculated based on the number and the importance of linkages between the criteria [18,19]. Pairwise comparison is the most used method for GIS-MCDA in the context of MAR. The weights are calculated through a matrix-based comparison of pairs of criteria [20]. The methods range from simple (rating method) to more complex (pairwise comparison). The advantage of the simple methods lies in the easiness of use, whereas the complex methods, such as pairwise comparison, offer a coefficient indicating the consistency of the decision-maker's choices. To account for the advantages of both the simple and more complex methods, the rating and ranking method, MIF and pairwise comparison were chosen to be incorporated into the web tool.

The decision rule states how the standardized datasets and their weights are combined to obtain the suitability map [17]. This integration can be based on threshold values (Boolean logic) or more elaborate integration rules, such as weighted linear combination (WLC). WLC comprises the summation of the weighted and standardized criteria and is the most commonly used decision rule. It has been developed further to its derivative analytical hierarchical process (AHP). AHP is the more structured approach that categorizes the GIS maps into hierarchical levels before aggregating und summing up the weighted criteria. It is used to solve more complex decision problems. The two most used methodologies, WLC and AHP, were chosen to be incorporated into the web tool.

To verify the obtained map, a sensitivity analysis should be conducted. It is used to display the effect of different standardization and weights on the final suitability map and indicates the robustness of the obtain suitability map [16]. While this is an important factor for strengthening the reliability of suitability maps, only 21% of the reviewed studies conducted this step.

All reviewed studies used desktop GIS software for their analysis. Some studies created their own tools for the analysis, e.g., Rahman et al. [8] developed the GIS-based Gabardine desktop decision support system. Other studies used tools available through standard software, e.g., an AHP tool has been implemented as an extension for ArcGIS [21], which has been applied by Anane et al. [22].

3. Materials and Methods

The developed tools are embedded into the INOWAS online platform (https://inowas.com). The INOWAS platform is an open-source collection of empirical, analytical, and numerical web-based models focusing on the planning, management, and optimization of MAR applications. The INOWAS platform and the tools are accessible with any state-of-the-art web browser. The platform works account-based, enabling the user to store and share their work. One key feature of the platform and its tools is the intuitive graphical user interface, guiding the users through the application of the tools.

The technical infrastructure of the platform is based on three components: the CLIENT, which is the user's internet device and browser, the SERVER, which is a standard Linux Server, and the WORKER, which is a Linux cluster with connection to the server. These components communicate with each other via a TCP/IP connection using standard protocols such as HTTP/HTTPS. The REST interface developed by the INOWAS group specifies the individual API calls and their function between the components.

For MAR suitability mapping, two tools were developed for the INOWAS platform. A query tool to filter information from the database that resulted from reviewing related studies and a web GIS system to obtain the suitability maps by following an integrated workflow.

The web-based query tool was designed to grant easy access to the information gathered from the GIS-MCDA review, namely data on different MAR related aspects as well as the MCDA methodology used in the studies. It is based on a pivot table approach. Decision-makers can sort, average, or sum up the database content by creating tables and graphs. Filters can be used to make specific queries, e.g., search studies for a certain MAR technique. The review tool is developed with ReactJs and is based on an open-source 3rd party project (https://github.com/nicolaskruchten/pivottable).

The INOWAS platform and web GIS system interfaces were created with ReactJS and Semantic UI (https://semantic-ui.com/) using some open-source 3rd party projects. Geodata is displayed in leaflet maps (https://leafletjs.com) and uses open street map layers as base layers (https://www.openstreetmap.org). Charts are displayed with ReCharts (http://recharts.org/en-US/), sliders with rc-slider (http://react-component.github.io/slider/) and network diagrams with visJS (http://visjs.org/). Raster calculations are done in JavaScript, using the mathJS library (http://mathjs.org). Users can save and share projects via a connection to the INOWAS backend server and their API entry points.

The full code of the INOWAS platform and the tools for MAR suitability mapping with version history is accessible through GitHub: https://github.com/inowas/inowas-dss-cra.

4. Results

All tools developed are available through https://inowas.com/tools/ using the "Start using now!" button. The tools are open-access and free of charge, but user registration is required. The tools can be accessed through the personal dashboard, which shows all available tools and stored projects. The projects can be shared with other users or can be made publicly available so that various users have access to the project and can edit it.

4.1. Database Query Tool

The database query tool is listed as tool T04 in the toolbox of the dashboard. Its basis is a database with information accumulated from the reviewed GIS-MCDA studies. The tool enables the user to research MAR specific information, such as the MAR type used, the water source, the objective of MAR application, or the location of the study. The main information stored is focused on MCDA related data, for example, the number and type of criteria used in the study, the weights assigned to the criteria, and the criteria standardization. Furthermore, information on weight assignment methods, decision rules, and the use of constraint mapping or sensitivity analysis by the authors of the study has been accumulated.

The different attributes from the database can be chosen to be displayed by dragging them into the column or the row fields of the tool, also enabling the display of combinations of attributes (Figure 1). Each attribute is equipped with a filter function where specific information queries can be chosen through class selection or conditional and numerical operators. The results can be visualized in different forms of tables and heat maps as well as be exported. They can be further modified through conditional and mathematical operators, including "Count", "Count Unique Values", "List Unique Values", and "Sum". The tool is supported by documentation explaining the underlying database and displaying three examples that help to get the user acquainted with the functionalities of the tool.

The database query tool can aid decision-makers at different steps along the MAR mapping procedure. During the problem statement step, it allows the decision-makers to investigate GIS criteria that are used often for certain MAR techniques or certain recharge water resources. During the weight assignment step, the importance that other decision-makers have given to a criterion can be assessed.

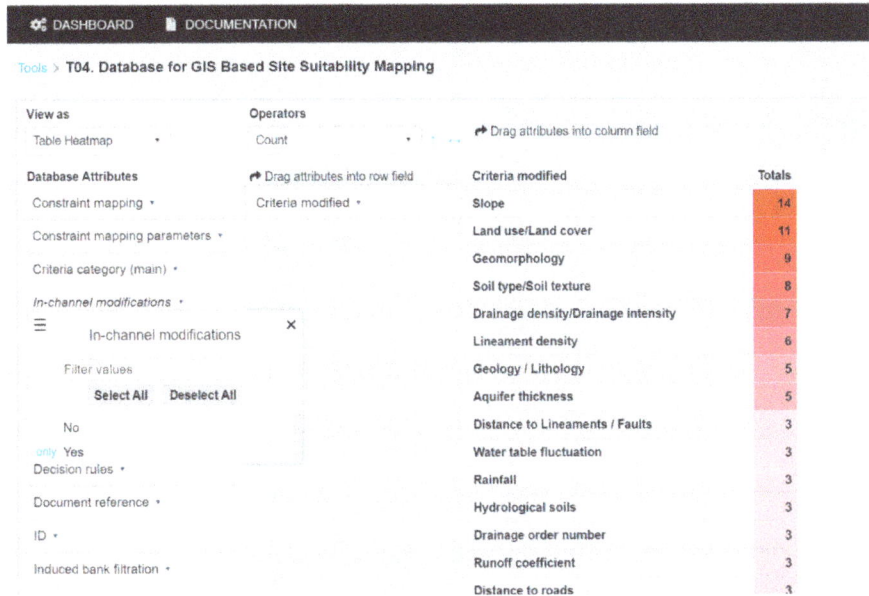

Figure 1. Interface of the database query tool showing the query for the most used criteria for the suitability mapping of in-channel modifications.

Figure 1 shows an example of the tool where the query was set to find the most used GIS criteria for suitability mapping of the in-channel method. Here, the filter was set to display the number of studies that used each criterion for this MAR method and to visualize them as a table heatmap. The results show that slope was the most used criterion, followed by land use and geomorphology. By adapting the attribute selection, it is also possible to show the 14 papers using the slope criterion. This enables the user to further study them regarding the use of the slope criterion, e.g., how it was standardized or classified in the studies.

4.2. Web GIS for Suitability Mapping

The web GIS is listed as tool T05 in the toolbox of the dashboard. The user is guided through the MCDA workflow with the help of a systematic approach indicated by the menu in the left column of the tool structure (Figure 2). It follows the workflow introduced by Rahman et al. [8] but excludes the steps of the problem statement and sensitivity analysis. The process starts with (1) the choice of GIS criteria, and continues with (2) the weights assignment, (3) the data upload, the constraint mapping and the reclassification, (4) the additional global constraint mapping, (5) the suitability mapping, and (6) the results visualization. The user is guided through the workflow by small green or orange circles that indicate whether the steps have been completed successfully. In case preceding information is required for the subsequent steps, the link in the navigation menu might be disabled until all requirements have been fulfilled.

A simplified example for MAR suitability mapping was prepared to depict the capabilities of the tool. It comprised suitability mapping for surface infiltration methods in southern Africa with four GIS criteria: geology, soil, land cover, and slope. A comprehensive example for MAR mapping in southern Africa was prepared in [23], including geoinformation on water sources and water demand. In this manuscript, the number of used criteria was reduced to account for better readability of the figures.

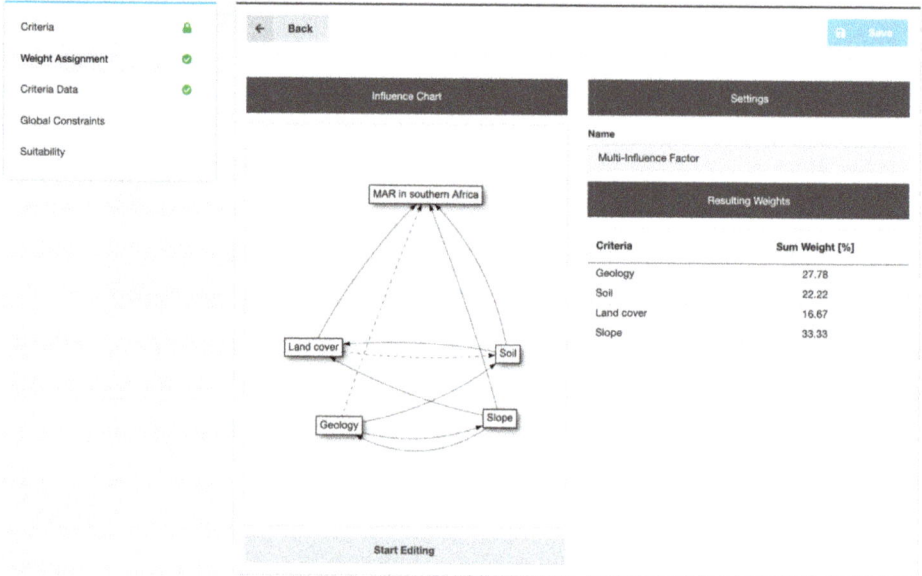

Figure 2. Interface of web GIS tool, with workflow display on the left and the weight assignment tool multi-influencing factor method. Here, the user can draw arrows between entities representing major or minor influences of a criterion upon another entity.

The problem definition step suggested by Rahman et al. [8] was not directly included in the tool approach. As this is a conceptual, case-specific step framing the project objectives and choosing criteria based on those objectives, it could not be incorporated into a tool. However, the database tool can deliver indications on the criteria choice based on the evaluation of previously conducted studies with similar objectives. In the case of surface infiltration methods, the four aforementioned GIS criteria are the most used criteria according to the database query tool.

Starting the tool, the user must choose whether to use the AHP or the WLC method as the decision rule, the latter being the default method of the tool. In both cases, all GIS criteria to be used for the suitability mapping need to be listed. Here, a link to tool T04 was included, enabling the user to analyze other MAR mapping studies regarding their choice of criteria. For each GIS criterion, the user needs to specify whether the criterion data is discrete or continuous and may set units to support their visualization. If AHP is activated, the main criteria classes need to be defined, and all GIS data need to be sorted by those main criteria classes. For this, a hierarchical tree is used and graphically presented. Weight assignments are done for the main criteria and ensuing for each branch with the respective sub-criteria. AHP is useful for complex problems with many GIS criteria. Dividing the procedure into separate branches simplifies the weight assignment step.

The weight assignment step allows performing any number of weight assignments through the integrated methods: ranking, rating, MIF, and pairwise comparison. This enables the user to try out different methods and compare the results, choosing the method and the subsequent weights most suitable for their project. All weighting methods are implemented in a visually appealing and user-friendly way.

For the MIF method, the user must draw connections between the different criteria and the project itself. Those connections represent the dependencies between the entities (Figure 2). The user can adapt the direction and strength of the connections. This information is then used to calculate the associated weights. In the example shown in Figure 2, the connections drawn between the four used criteria and the project itself correspond to weights that emphasize the importance of slope and geology.

For the ranking method, intuitive approaches, such as arrow buttons or the drag and drop method, help to order the criteria in a list, starting with the most important criterion and ending with the least important one. All ranks are summed up and then converted to weights in relation to the other criteria input values. The rating method is the most basic method of all, enabling the user to choose their own weight for each criterion. The fourth available weighting method is the pairwise comparison. On a predefined scale, the criteria are compared to each other by moving a slider towards one side, indicating the importance of a criterion compared to a second criterion (Figure 3). Based on these pairwise preferences, the criteria weights are calculated. The sliders indicate a preference of geology and slope over the other criteria, which is then depicted in the higher resulting weights. From the weighting choices, a consistency coefficient is calculated. The coefficient indicates the consistency of the user's preferences, and if it surpasses a threshold value, the users are asked to re-check their pairwise comparison choices. Again, a link to T04 is provided to the user, enabling the investigation of previously conducted studies regarding their criteria weighting choices.

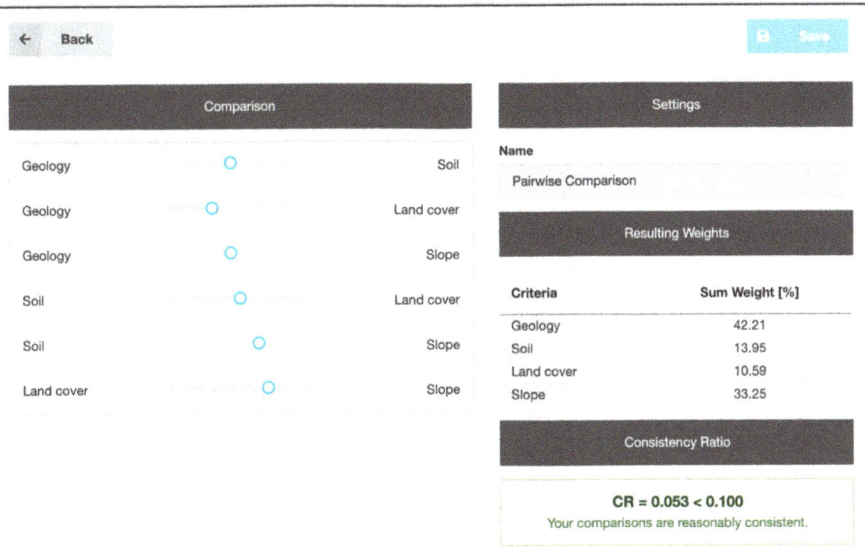

Figure 3. Interface of web GIS tool, with weight assignment tool pairwise comparison where user can set criteria preferences via sliders and indication of the robustness of the decision is given via consistency ratio.

During the third step, "Criteria data", data upload, constraining, and reclassification are performed for each GIS dataset. At the beginning, the final grid size of the project must be set. Then for each criterion, a GIS raster file must be uploaded. Uploaded raster files may be resampled through nearest-neighbor interpolation if the raster size differs from the final grid size. Constraints can be set in the second step, indicating the criterion classes that will not be used for the suitability mapping process. With discrete data, the automatically created classes can be disabled. For continuous data, ranges not to be considered in the final calculation can be defined by Boolean logic. Afterward, reclassification and normalization of the criterion data are performed. For continuous criteria, classes can be defined by indicating the minimum and maximum value and their respective reclassified value or by using a reclassification function. The reclassified values should be between zero and one. For discrete datasets, every existing criterion class is assigned a normalized value, with the possibility of several criterion classes forming one normalized class. All classes can be named and given a color for geo-visualization.

Finally, the results of the reclassification are displayed. It is possible to switch between the original, reclassified, constraint, and resulting data.

In the fourth, optional step, global constraints can be set by drawing polygons in a project area, which will then be disregarded from the final calculation of the suitability map.

For the fifth step, "Suitability mapping", the user must select one of the obtained weight assignment calculations. Then, the calculation is performed, combining all information on constraint mapping, criteria standardization, weighting, and decision rules. For the resulting map, the suitability classes can be redefined or left as default. Finally, the suitability map is displayed and can be downloaded as a text file for further processing in other GIS software (Figure 4). The map for southern Africa shows the geographic distribution of areas that, based on the criteria evaluated, are more suitability for the implementation of MAR schemes.

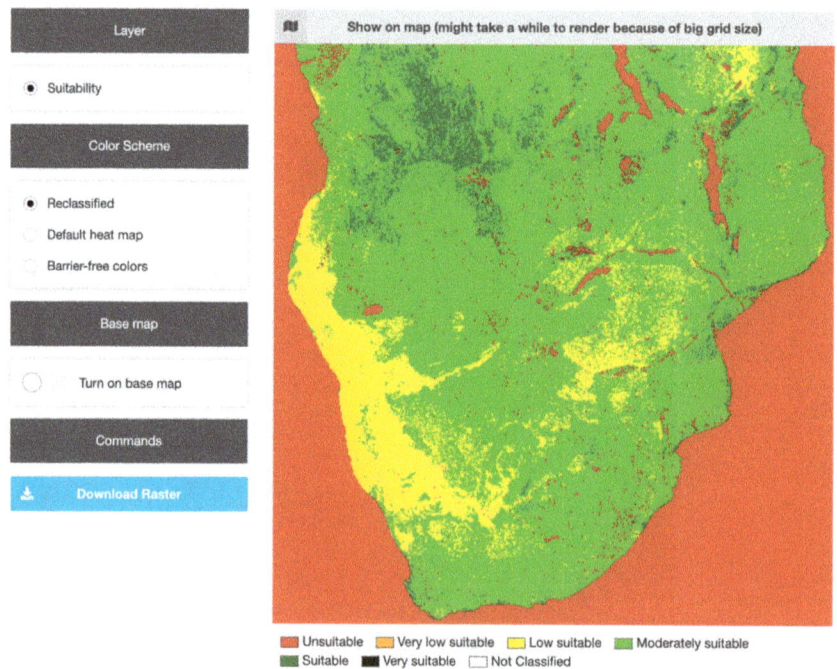

Figure 4. Interface of web-GIS tool, showing the final suitability map with redefined suitability classes.

The tool is supported by documentation explaining the tool functionalities as well as the underlying concepts and methodologies (https://inowas.com/tools/t05-gis-mcda/). One example case study is used to help to get the user acquainted with the functionalities of the tool. It is incorporated into a tutorial that provides the user with a step-by-step guide on how to generate a simple suitability map, also providing an example dataset (https://inowas.com/#tutorials, Tutorial 4).

To validate the correctness of the web-tool and the methodology incorporated, a case study was both prepared with ArcGIS and our web tool for comparison [23]. Minor differences occurred with a normalized root mean square error below 0.1. These differences were not attributed to the MCDA methodology but to the resampling during the data upload step. The nearest-neighbor algorithm was found to shift the raster by one half-pixel [24]. This caused a slight divergence between the map obtained from this tool and the map obtained through ArcGIS.

5. Discussion

This study indicates a trend in MCDA methodology applied for MAR suitability mapping, namely constraint mapping, suitability mapping by using pairwise comparison, and WLC or AHP, and less often a subsequent sensitivity analysis. Based on these findings, we designed an open-source web tool to guide the user through the MCDA process. We included several weight assignment methods so that next to pairwise comparison, the decision-maker can use rating and ranking method as well as MIF. While we kept those methods for their simplicity or advantage in visual decision-making, the combination of pairwise comparison with AHP must be highlighted as the methodology with the highest increase in usage [11] and the most benefits.

AHP offers a clear, systematic procedure that represents all aspects of the problem statement enforcing robust decision-making. In combination with pairwise comparison, the decision-maker must only give priorities considering two criteria at a time. Through a designated index, the methodology offers an indication of the consistency of the decision-maker's choices. The shortcomings of this method include the inability to include threshold values, no direct measure to assess the robustness of the criteria standardization values, and possible rank reverse issues [25]. Nevertheless, it can be asserted that, currently, AHP, in combination with pairwise comparison, is the state-of-the-art methodology for MAR suitability mapping. This can be further underlined by its frequent use in other environmental MCDA studies [25–27].

The established web tools cover all aspects of the suitability mapping process apart from the sensitivity analysis. A structured GIS-MCDA process should include a sensitivity analysis as it can help to generate more robust and reliable suitability maps. A future prospect of the web tool will be the integration of the sensitivity analysis process, which is relatively more complex and computation-intensive and, thus, was not included in the workflow yet. Furthermore, it is planned to improve the tool by increasing the possible maximum resolution of projects and by providing more flexibility for handling input raster data as well as supporting vector files.

The web-based implementation of the tools offers advantages over standard desktop GIS solutions. The user does not require advanced GIS-specific knowledge and does not need to install any GIS software unless pre-processing of the GIS data is necessary. Nevertheless, the platform provides an interface for data exchange with standard GIS software to allow for pre- or post-processing with conventional desktop-based software. The entire system works in a standard web browser, with no specific system requirements. The input data, as well as the resulting maps, are available online and can be easily shared among stakeholders allowing for collaboration on the mapping procedure as well as flexible sharing of the results. The web tools can be run in two operating modes (private and public), which offers a high degree of flexibility in data sharing as well as adequate privacy. The entire workflow is very transparent, with the possibility to revise and reverse steps. It provides a pre-defined workflow for inexperienced users while offering a comparison of several MCDA methods for more advanced users.

While MAR suitability mapping can be seen as a viable source for strategic water management, it is not a sufficient technology to point down to actual locations for MAR implementation. Maps can deliver an indication of areas of interest, but those need to be further assessed by numerical modeling [10,28] or on-site measurements of the local hydrology and hydrogeology [13,29,30]. For the numerical modeling and MAR scheme design and optimization, the INOWAS platform can be used as well, as it further offers numerical groundwater flow modeling tools as well as algorithms for MAR scheme optimization.

6. Conclusions

The developed web tools can help planners of MAR sites by increasing knowledge on MAR suitability mapping as well as by engaging in the MCDA processes in a structured and cooperative way. The tools are further envisioned to aid in capacity building measures as well as the education of water practitioners by accumulating knowledge on GIS-MCDA in the context of MAR and translating them into easy-to-use web tools. The clearly outlined process of map generation enforces standard methodology and can help to generate maps that are comparable due to a common methodological approach. Since MAR mapping has been increasingly used in recent years, the quality of the maps produced should be critically evaluated, and analysis and categorization of the methodologies used is one first step to improve the reliability of the maps.

While the tools can outline the map generation process, they cannot standardize one of the main sources of uncertainty—the datasets and respective weights assigned. Putting individual choices into perspective with similar studies retrieved from the database tool is a step towards decreasing the subjectivity of their weighting and standardization process. However, the problem statement and its specifics define the importance of a GIS dataset for each individual case study. Thus, the weights assigned to the criteria cannot be defined by rules. Furthermore, data availability and quality are major constraints in the mapping process. Thus, the tools at hand standardize and simplify the MAR suitability mapping process but cannot substitute the decision-maker's expertise in choosing relevant datasets and their importance for the specific study.

Author Contributions: J.S. and J.P.B.V. designed the overall study; J.S., F.V.L. and J.P.B.V. conducted the review, J.S., R.S. and F.V.L. devised the tools; R.S. and R.J. implemented the tools; J.S. and R.S. wrote the paper; C.S. contributed with discussion and advice.

Funding: This study was supported by the German Federal Ministry of Education and Research (BMBF), grant No. 01LN1311A (Junior Research Group "INOWAS").

Acknowledgments: We acknowledge the financial support of the graduate academy of TU Dresden as well as the Open Access Funding by the Publication Fund of the TU Dresden.

Conflicts of Interest: The authors declare no conflict of interest. The founding sponsors had no role in the design of the study; in the collection, analyses, or interpretation of data; in the writing of the manuscript, and in the decision to publish the results.

References

1. Stefan, C.; Ansems, N. Web-based global inventory of managed aquifer recharge applications. *Sustain. Water Resour. Manag.* **2018**, *4*, 153–162. [CrossRef]
2. Dillon, P.; Stuyfzand, P.; Grischek, T.; Lluria, M.; Pyne, R.D.G.; Jain, R.C.; Bear, J.; Schwarz, J.; Wang, W.; Fernandez, E.; et al. Sixty years of global progress in managed aquifer recharge. *Hydrogeol. J.* **2019**, *27*, 1–30. [CrossRef]
3. Government of India. *Manual on Artificial Recharge of Ground Water*; Government of India, Ministry of Water Resources, Central Ground Water Board: Faridabad, India, 2007.
4. NRMMC-EPHC-AHMC. *Australian Guidelines for Water Recycling. Managed Aquifer Recharge*; National Water Quality Management Strategy: Canberra, Australia, 2009.
5. Ministro de Obras Públicas. *Diagnóstico De Metodología Para La Presentación Y Análisis De Proyectos De Recarga Artificial De Acuíferos*. 2014. Available online: http://documentos.dga.cl/Resumen_Ejecutivo%20RA_v0.pdf (accessed on 27 October 2019).
6. Secretaría del Medio Ambiente y Recursos Naturale. *De medio ambiente y recursos naturales. In Norma Oficial Mexicana NOM-015-CONAGUA-2007, Infiltración Artificial De Agua A Los Acuíferos.-Características Y Especificaciones De Las Obras Y Del Agua*. 2009. Available online: http://www.conagua.gob.mx/CONAGUA07/Contenido/Documentos/NOM-015-CONAGUA2007.pdf (accessed on 27 October 2019).
7. Dillon, P. Future management of aquifer recharge. *Hydrogeol. J.* **2005**, *13*, 313–316. [CrossRef]
8. Rahman, M.A.; Rusteberg, B.; Gogu, R.C.; Lobo Ferreira, J.P.; Sauter, M. A new spatial multi-criteria decision support tool for site selection for implementation of managed aquifer recharge. *J. Environ. Manag.* **2012**, *99*, 61–75. [CrossRef]
9. Shankar, M.N.R.; Mohan, G. A GIS based hydrogeomorphic approach for identification of site-specific artificial-recharge techniques in the Deccan Volcanic Province. *J. Earth Syst. Sci.* **2005**, *114*, 505–514. [CrossRef]
10. Russo, T.A.; Fisher, A.T.; Lockwood, B.S. Assessment of Managed Aquifer Recharge Site Suitability Using a GIS and Modeling. *Groundwater* **2015**, *53*, 389–400. [CrossRef]

11. Sallwey, J.; Bonilla Valverde, J.P.; Vásquez López, F.; Junghanns, R.; Stefan, C. Suitability maps for managed aquifer recharge: A review of multi-criteria decision analysis studies. *Environ. Rev.* **2018**, *27*, 138–150. [CrossRef]
12. De Winnaar, G.; Jewitt, G.P.W.; Horan, M. A GIS-based approach for identifying potential runoff harvesting sites in the Thukela River basin, South Africa. *Phys. Chem. Earth Parts ABC* **2007**, *32*, 1058–1067. [CrossRef]
13. Selvam, S.; Dar, F.A.; Magesh, N.S.; Singaraja, C.; Venkatramanan, S.; Chung, S.Y. Application of remote sensing and GIS for delineating groundwater recharge potential zones of Kovilpatti Municipality, Tamil Nadu using IF technique. *Earth Sci. Inform.* **2016**, *9*, 137–150. [CrossRef]
14. Saidi, S.; Hosni, S.; Mannai, H.; Jelassi, F.; Bouri, S.; Anselme, B. GIS-based multi-criteria analysis and vulnerability method for the potential groundwater recharge delineation, case study of Manouba phreatic aquifer, NE Tunisia. *Environ. Earth Sci.* **2017**, *76*, 511. [CrossRef]
15. Raviraj, A.; Kuruppath, N.; Kannan, B. Identification of potential groundwater recharge zones using remote sensing and geographical information system in Amravathi Basin. In *Modelling Impact of Climatic Variability on Groundwater Dynamics*; Water Technology Centre, Tamil Nadu Agricultural University: Coimbatore, India, 2016; pp. 109–133, ISBN 978-93-83799-28-2.
16. Malczewski, J. *GIS and Multicriteria Decision Analysis*; John Wiley & Sons: New York, NY, USA, 1999.
17. Malczewski, J.; Rinner, C. *Multicriteria Decision Analysis in Geographic Information Science*; Springer: Berlin, Germany, 2015.
18. Shaban, A.; Khawlie, M.; Abdallah, C. Use of remote sensing and GIS to determine recharge potential zones: The case of Occidental Lebanon. *Hydrogeol. J.* **2005**, *14*, 433–443. [CrossRef]
19. Magesh, N.S.; Chandrasekar, N.; Soundranayagam, J.P. Delineation of groundwater potential zones in Theni district, Tamil Nadu, using remote sensing, GIS and MIF techniques. *Geosci. Front.* **2012**, *3*, 189–196. [CrossRef]
20. Saaty, T.L. *The Analytic Hierarchy Process*; McGraw-Hill: New York, NY, USA, 1980.
21. Marinoni, O. Implementation of the analytical hierarchy process with VBA in ArcGIS. *Comput. Geosci.* **2004**, *30*, 637–646. [CrossRef]
22. Anane, M.; Kallali, H.; Jellali, S.; Ouessar, M. Ranking suitable sites for Soil Aquifer Treatment in Jerba Island (Tunisia) using remote sensing, GIS and AHP-multicriteria decision analysis. *Int. J. Water* **2008**, *4*, 121–135. [CrossRef]
23. Schlick, R. *Visualizing the MAR Potential for Africa through GIS-MCDA and Web-Based Tool Design*; Technische Universität Dresden: Dresden, Germany, 2019.
24. Parker, J.A.; Kenyon, R.V.; Troxel, D.E. Comparison of Interpolating Methods for Image Resampling. *IEEE Trans. Med. Imaging* **1983**, *2*, 31–39. [CrossRef]
25. Cinelli, M.; Coles, S.R.; Kirwan, K. Analysis of the potentials of multi criteria decision analysis methods to conduct sustainability assessment. *Ecol. Indic.* **2014**, *46*, 138–148. [CrossRef]
26. Gounaridis, D.; Zaimes, G.N. GIS-based multicriteria decision analysis applied for environmental issues; the Greek experience. *Int. J. Appl. Environ. Sci.* **2012**, *7*, 307–321.
27. Esmail, B.A.; Geneletti, D. Multi-criteria decision analysis for nature conservation: A review of 20 years of applications. *Methods Ecol. Evol.* **2018**, *9*, 42–53. [CrossRef]
28. Brown, C.J.; Weiss, R.; Verrastro, R.; Schubert, S. Development of an Aquifer, Storage and Recovery (ASR) Site Selection Suitability Index in Support of the Comprehensive Everglades Restoration Project. *J. Environ. Hydrol.* **2005**, *13*, 1–13.
29. Jamali, I.A.; Mörtberg, U.; Olofsson, B.; Shafique, M. A Spatial Multi-Criteria Analysis Approach for Locating Suitable Sites for Construction of Subsurface Dams in Northern Pakistan. *Water Resour. Manag.* **2014**, *28*, 5157–5174. [CrossRef]
30. Satapathy, I.; Syed, T.H. Characterization of groundwater potential and artificial recharge sites in Bokaro District, Jharkhand (India), using remote sensing and GIS-based techniques. *Environ. Earth Sci.* **2015**, *74*, 4215–4232. [CrossRef]

© 2019 by the authors. Licensee MDPI, Basel, Switzerland. This article is an open access article distributed under the terms and conditions of the Creative Commons Attribution (CC BY) license (http://creativecommons.org/licenses/by/4.0/).

Article

Potential Benefits of Managed Aquifer Recharge MAR on the Island of Gotland, Sweden

Peter Dahlqvist [1],*, Karin Sjöstrand [2,3], Andreas Lindhe [3], Lars Rosén [3], Jakob Nisell [1], Eva Hellstrand [1] and Björn Holgersson [1]

1. Geological Survey of Sweden, Lund/Uppsala/Gothenburg, Sweden (SGU); Villavägen 18, 752 36 Uppsala, Sweden; Jakob.Nisell@sgu.se (J.N.); Eva.Hellstrand@sgu.se (E.H.); Bjorn.Holgersson@sgu.se (B.H.)
2. RISE Research Institutes of Sweden, 223 70 Lund, Sweden; karin.sjostrand@ri.se
3. Chalmers University of Technology, 412 96 Göteborg, Sweden; Andreas.Lindhe@chalmers.se (A.L.); Lars.Rosen@chalmers.se (L.R.)
* Correspondence: peter.dahlqvist@sgu.se; Tel.: +46-703-466-058

Received: 6 August 2019; Accepted: 16 October 2019; Published: 17 October 2019

Abstract: The Island of Gotland (3000 km^2), east of mainland Sweden, suffers from insufficient water availability each summer. Thin soils and lack of coherent reservoirs in the sedimentary bedrock lead to limited reservoir capacity. The feasibility of Managed Aquifer Recharge (MAR) is explored by identifying suitable areas and estimating their possible contribution to an increased water availability. MAR is compared to alternative water management measures, e.g., increased groundwater abstraction, in terms of costs and water availability potential. Results from GIS analyses of infiltration areas and groundwater storage, respectively proximity to surface water sources and surface water storage were classified into three categories of MAR suitability. An area of ca 7700 ha (2.5% of Gotland) was found to have good local conditions for MAR and an area of ca 22,700 ha (7.5% of Gotland) was found to have moderate local conditions for MAR. These results reveal the MAR potential on Gotland. The water supply potential of MAR in existing well fields was estimated to be about 35% of the forecasted drinking water supply and 7% of the total water demand gap in year 2045. It is similar in costs and water supply potential to increased surface water extraction.

Keywords: MAR; groundwater; mapping; Sweden; decision-support

1. Introduction

The Island of Gotland (3000 km^2), situated in the Baltic Sea 100 km from the mainland of Sweden (Figure 1a), suffers from insufficient water availability to supply the ever-increasing demand from society, especially during the tourist season (June–August, [1]). The annual precipitation on the island (ca 550 mm/year) is sufficient to cover a forecasted increase in water demand. However, intensive drainage of arable land, thin soil layers, and relatively impermeable rock lead to precipitation run-off and limited reservoir capacity in both surface water and groundwater reservoirs [2]. The already constrained water supply will be further aggravated in the future because the total water demand on the island is estimated to increase by 40% by 2045 [1]. The current water resources on the island will not meet this projected increase in demand. A high availability of water during the winter and a high demand for water in the summer makes MAR a suitable way to increase the water resources. Due to these factors, it is important to investigate the potential for MAR on Gotland.

The bedrock on Gotland consists mainly of Silurian limestone and marlstone, which represents the upper part of a 250–800-m-thick sequence with Palaeozoic sedimentary rocks overlying the crystalline basement [3]. The Quaternary overburden is generally thin (less than 2 m) and is largely composed of till and postglacial sand deposits. The relief is low, and the highest point is 82 m a.s.l. The main land uses are agricultural and forestry. The main aquifers on Gotland are situated within the bedrock where cracks, fractures and dissolution cavities store and transport the groundwater. Nevertheless, the soil layers have an important role to play because areas with soil, especially sand and gravel, act as infiltration and storage systems for the bedrock aquifer. The island may be considered one large aquifer, but with groundwater divides (created by relief) producing 7 (sub)aquifers, according to the European water framework directive. Saline groundwater is a problem because relict saltwater occurs under the entire island at a depth of 20–100 m b.s.l.

The total water demand on the island is estimated to increase by more than 40% by 2045, with increases of 30% in tourism, 20% in domestic demand, 20% in animal keeping, 15% in industry, and 100% in irrigation [1]. To enhance water resource security, and close the water supply and demand gap, several alternative water management measures are being examined. Managed Aquifer Recharge (MAR) is one of them. Today, the public water supply on Gotland relies on 14 well fields, two surface water catchments and a desalinisation plant. MAR is currently not used in any public water supply on Gotland, but may play an important role in the future if suitable areas can be found. Conversely, on the Swedish mainland, MAR has been in use for over 100 years, and accounts for approximately 20% of the public water supply [1,4]. MAR can be explained as the intended recharge into and storage of water in an aquifer [5]. It may be used to increase water security for uses including drinking water supply, irrigation, preventing saltwater intrusions, as well as providing environmental benefits [6]. MAR is widely distributed and applied on various scales around the globe, as well as in Europe [7]. The water source can be of varied origin, e.g., river water, seawater or sewer water. In some cases, there is a need for pre-treatment before groundwater recharge to minimize the risk of pollution or aquifer clogging [8]. The recharge can be made by spreading methods in areas with high infiltration capacity; by deep infiltration direct into the aquifers via wells; or as induced infiltration due to withdrawal [9]. Since the different MAR types are suitable for different conditions within hydrogeological settings (e.g., confined or unconfined aquifers), treatment opportunities, and land use, the selection of suitable recharge sites is crucial [10,11]. In this study, we focus on areas suitable for recharge through infiltration basins and natural conditions for storage. Conditions for well or induced infiltration are expected to be of minor importance because of the geomorphology, geology and hydrology of the island. Hence, the suitability of those MAR types is not investigated in this study.

To prioritize between alternative measures to improve water resources security (e.g., increased groundwater abstraction and desalination), useful decision support is needed. GIS-MCDA (Geographical Information System Multi Criteria Decision Analysis) [12] is a regularly applied method in MAR suitability assessment [13]. There are several possible criteria for mapping MAR suitability, the three most common being aquifer storage capacity, geomorphology and soil [13]. There are also concerns on limitations and discussion on the uncertainties of these GIS approach made visible by e.g., [14]. Although a GIS analysis will show where MAR might be successful, field work and numerical modelling will be important tools for increasing the success of MAR [15–19]. The economic assessment of MAR is an important question that has been studied previously [20–22]. The aim of this paper is to explore the feasibility of MAR on the island of Gotland by: identifying potential areas for MAR in proximity to a fresh water source and which are available for recharge; estimate the possible increase in groundwater recharge and groundwater extraction at existing wellfields; and compare MAR to other alternative measures in terms of costs and water availability potential.

2. Materials and Methods

2.1. Materials

Mapping of potential MAR locations on Gotland is based mainly on the existing data listed as follows: Intensive hydrogeological investigations from two campaigns of airborne transient electromagnetic surveys (2013–2015, SkyTEM resistivity measurements along flight lines covering 30% of the island, with 200 m spacing and geophysical soundings every 30 m) [23,24]; a site-specific overview in 2016 of groundwater catchments of the 14 existing well fields that showed that 30% of those well fields had a favorable geology and hydrology for spreading and induced infiltration-based MAR types [2]; 3D geological and hydrogeological models (Geoscene 3D by I-GIS) of the entire island (2015–2019, available online in 2020) based on resistivity models from the SkyTEM survey; existing geological information such as bedrock and soil maps (regional scale); seismic profiles and information from water wells; and national scale mapping (2018) of groundwater recharge and storage capacity [25] provide comprehensive additional data for assessing the potential of MAR on Gotland.

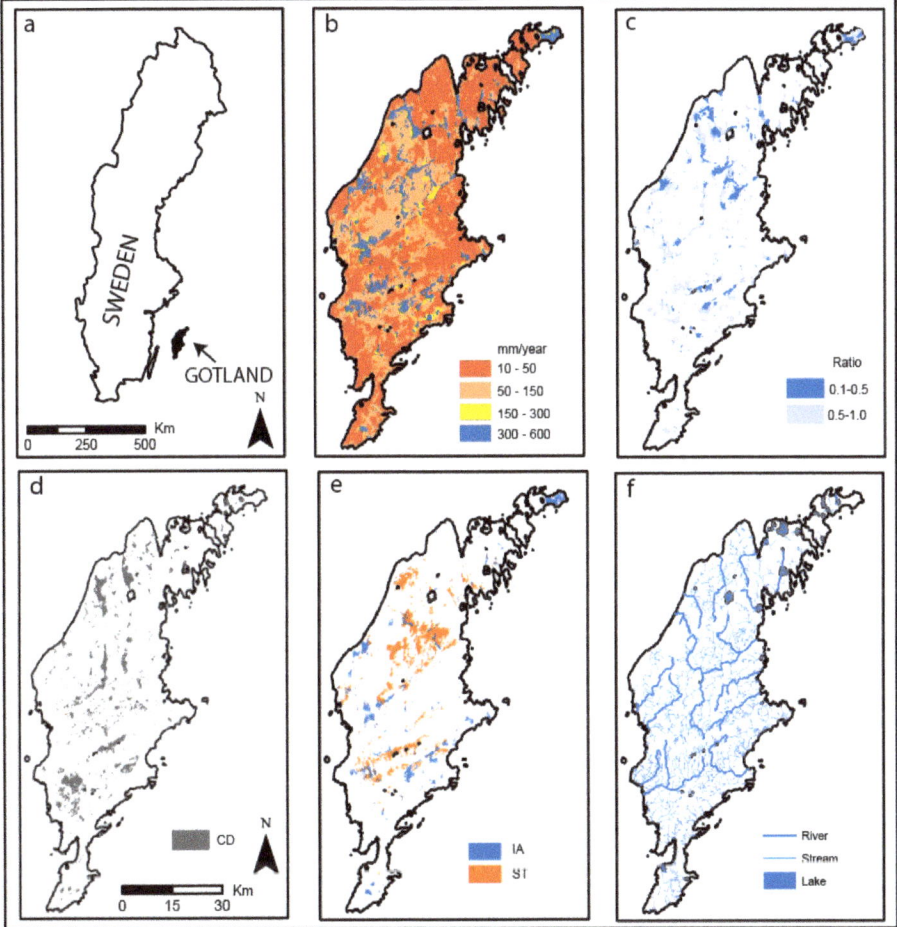

Figure 1. GIS maps and results from GIS analysis to identify areas with potential for MAR. (**a**) Map of Sweden and location of Island of Gotland. (**b**) Groundwater storage capacity (mm/year) [25]; (**c**) Ratio of groundwater recharge/storage capacity; only values below 1.0 are shown (GS); (**d**) Closed depressions (>1 ha) in the bedrock with no contact with the Baltic Sea (CD); (**e**) Areas from the lithological 3D model, blue: areas >4 m sand and/or gravel (IA) suitable for infiltration, red areas >4 m till and/or clay (ST) suitable for construction of surface storage dams; (**f**) Surface waters, i.e., raw water source (S).

2.2. Methods

A GIS-based (Boolean logic) approach was used to find suitable locations for MAR systems (in this study focusing on areas suitable for recharge through infiltration on the surface) on Gotland. No parameter weighting was included. Several mapping projects assessing suitability for MAR have been made globally [13]. GIS was used for analysis of 5 surface and subsurface datasets, which are presented below. Three of these (1–3) concern the MAR location, and two (4 and 5) further explore the MAR location to determine if there are local sources of water supply for the MAR site.

1. The aquifer storage capacity of the soil and bedrock were previously estimated by the Geological Survey of Sweden (SGU) on a national scale (Figure 1b) and should be used with caution on scales below 1:100,000 on Gotland. The groundwater storage capacity is based on assessments of porosity of soil and bedrock types, soil thickness, groundwater surface and possible drawdown caused by groundwater withdrawal [25]. A modeled annual groundwater recharge map for Sweden (national scale, coarse resolution) was made in 2006 [26]. Values of natural groundwater recharge on Gotland used in this analysis were either 200 or 260 mm/year, depending on location. A raster containing the ratio between groundwater recharge and groundwater storage capacity was created, and values below 1.0 indicate areas with potential to store more groundwater than the natural recharge, i.e., they might be suitable for MAR. This raster is abbreviated as GS and is shown in Figure 1c.

2. The geological 3D model for Gotland includes a bedrock surface. A GIS analysis identified depressions in the bedrock surface, assuming that these areas are generally favorable for storage of groundwater. Some of these areas coincide with lakes, whereas others are "hidden" depressions with little or no surface expression because they have been infilled by soils. A selection was made to show only closed depressions larger than 1 hectare and with no contact with the Baltic Sea. Water has a higher potential in these areas for storage without being lost as a shallow groundwater outflow through the permeable soil (Figure 1d). The resulting raster is named Closed Depressions (CD).

3. Through selection from the lithological 3D model of Gotland, areas with >4 m thickness of sand and/or gravel (Figure 1e) were identified. This geological environment is important on both local and regional scales because these high permeability deposits increase infiltration to the bedrock aquifer. The resulting raster is abbreviated as IA.

4. To assess the possibility of storing surface water in dams, areas with over 4 m thickness of till and or clay were selected from the lithological 3D model of Gotland (Figure 1e). In these areas the construction of sufficiently large storage dams will be a relatively easy and cheap operation since the construction material can be sourced on site. The resulting raster is abbreviated as ST (Surface water storage).

5. Because of arable land drainage, thin soil cover and relatively impermeable bedrock, most streams and rivers on Gotland have high flow rates during the winter (November–March; [1]). Even small streams can serve as good sources of water supply if the water can be stored (e.g., in man-made dams or wetlands) until the spring and summer. Lakes are rare on the island and mostly very shallow. Lakes and streams may both be regulated to increase the available source but that is not discussed in this paper. A GIS-based analysis of proximity to surface waters was made with a buffer of 0.2 km on smaller streams (sometimes intermittent) and 0.5 km for perennial rivers and lakes (Figure 1f, raster abbreviation S). This difference in distance reflects a variation in the estimation of cost effectiveness and can be further explored.

In addition to the above analyses, two further GIS analyses were completed with a combination (overlap with no priority weight) of data on favorable areas for infiltration (Figure 2a) and surface water source and storage (Figure 2b).

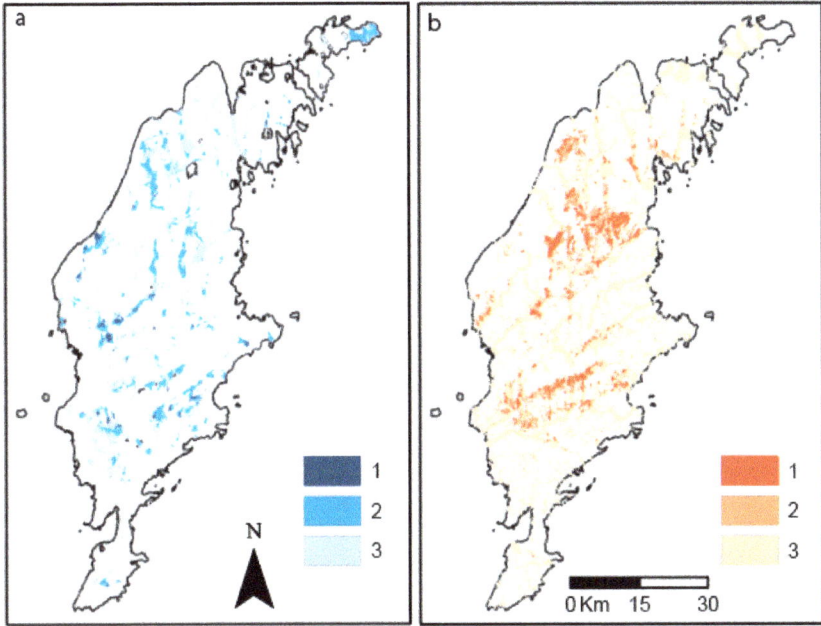

Figure 2. Maps of the combined GIS analysis. (**a**) Areas suitable as infiltration areas and for groundwater storage; (**b**) Areas close to a natural source (intermittent stream, perennial river, lake) and for construction of surface storage areas (dams). Data has been reclassified into classes; lower numbers more suitable.

To make estimations of possible increases in groundwater recharge and groundwater extraction through MAR, in relation to current well fields, favorable groundwater catchments were identified [5]. This was done using a GIS-aided and analytical approach. The values presented here for potential infiltration rate and increased withdrawal volumes are estimations with an inherent uncertainty. The estimates are based on access to surface waters of adequate size, presence and sufficient thickness of permeable sediments, the possibility of creating dams with local material, and current withdrawal capacity.

MAR was compared to other alternative measures (i.e., increased groundwater abstraction, enhanced water reuse for irrigation, increased surface water extraction, metered leak detection and desalination) in terms of annual water availability potential and economic viability. The measures were selected for inclusion in the analysis based on the outcome of a multidisciplinary stakeholder workshop, in which the participants were asked to identify measures with potential to improve the water resource security on the island. The comparative method was based on marginal abatement cost curves [27,28], including cost–benefit and cost-effectiveness analyses. The costs of the measures included investment costs, operating costs and cost savings. The measure costs were described by present values (PVs) [29], analyzed with a 3.5% discount rate over the 27-year time horizon from year 2019 to 2045 (corresponding to the current water plan period applied at Gotland) and based on 2018 prices. The NPVs were then expressed as equivalent annual costs (EACs) in SEK (Swedish krona) per year [30]. A theoretical maximum level of implementation was assumed for each measure category, except for desalination, which instead was based on estimates of one new desalination plant. The estimated cost and water availability input variables were based on a combination of literature data, expert judgements and GIS-based analyses. The unit cost of each measure was calculated as the ratio of the measure's EAC and annual water availability potential.

3. Results

3.1. Mapping of Suitable MAR and Source Areas

Results are described in three parts; areas where the geology is favorable for infiltration and/or groundwater storage (IA (Infiltration Areas) + GS (Groundwater Storage) + CD (Closed Depressions)); areas with proximity to a surface water source and/or suitable for surface water storage (S (Source) + ST Surface water storage)); and areas where these overlap (IA + GS + CD + S + ST).

3.1.1. Infiltration Areas and Areas for Groundwater Storage (IA + GS + CD)

Three data sets regarding the possibility for an area to be suitable for artificial groundwater recharge were combined into a raster set (Figure 2a). The resulting raster was divided into three classes: areas with three (class 1), two (class 2) or one (class 3) of the included favorable attributes. A raster with class 1 has a good potential for both infiltration and groundwater storage. Areas with class 2 or 3 are less suitable because one or two of the included raster sets (IA, GS, CD) is absent. Class 1 is present in 2068 ha (0.7% of Gotland), class 2 in 14,437 ha (4.8% of Gotland), and class 3 in 43,453 ha (14.4% of Gotland).

3.1.2. Source and Suitable Areas for Surface Water Storage (S + ST)

The source for artificial groundwater recharge in this investigation is natural surface waters from streams, rivers and lakes. To regulate and decide where there are favorable infiltration conditions there is also a need for a seasonal surface storage, in this investigation in the form of man-made dams. In this study, we do not discuss present land use and slope—two factors that might influence outcomes—but we regard them as having a minor influence (see Discussion) on Gotland. The combination of these sources with possibilities of constructing storage capacity is shown in Figure 2b. The resulting dataset is divided into three classes: (1) areas with both a nearby raw water source and good conditions for storage in surface dams (10,271 ha, 3.4% of Gotland); (2) areas with good conditions for dams, but more than 0.2 km (smaller intermittent streams) or 0.5 km (perennial rivers and lakes) from a source (10,345 ha, 3.5% of Gotland); and (3) areas with a source but the distance to a suitable storage area (dam) is more than 0.2 km (smaller intermittent streams) or 0.5 km (perennial rivers and lakes, 112,468 ha, 37% of Gotland).

3.1.3. Areas with Combination of Infiltration, Groundwater Storage, Source and Surface Water Storage (IA + GS + CD + S + ST)

To narrow down the selection of promising areas for MAR a raster set with different combinations of IA + GS + CD + S + ST is shown in Table 1 and Figure 3b. These are the best-adapted areas for MAR construction (recharge through infiltration) based on this study. There are a few stream catchment areas that appear more promising to work in. Please note that because the raster set with surface water source/storage is made with a buffer there will be areas with overlapping datasets (Figure 3a).

Class 1 (Table 1) is not present on Gotland in this analysis. The reason is in the construction of the analysis these conditions cannot exist in the same raster (ha scale). Class 2 is present in ca 7719 ha (2.5% of Gotland). Areas in this class have good conditions for successful MAR. Class 3 is present in 22,710 ha (7.6% of Gotland). In these areas, there are probably moderate conditions for successful MAR. Class 4 is present in 2765 ha (0.9% of Gotland). In these areas, there is no surface water source within the chosen distance.

Table 1. Synthesis from final step in GIS analyses. GS = Ratio >1.0 groundwater recharge/storage capacity; CD = Closed depression; IA = sand/gravel >4m; S = Raw water Source; ST = Storage. Location of mapped classes see Figure 3b.

Class	Groundwater	Surface Water		Area (ha)
1	IA + GS + CD	S + ST	Not present due to construction of analysis	0
2:1	IA + GS + CD	S	Good local conditions for MAR	980
1	IA + GS + CD	ST	Not present, due to construction of analysis	0
2:2	Two of IA, GS, CD	S + ST	Good local conditions for MAR	512
2:3	Two of IA, GS, CD	S	Good local conditions for MAR	7207
4:1	Two of IA, GS, CD	ST	No source within chosen distance	209
3:1	One of IA, GS, CD	S + ST	Probable local conditions for MAR	3579
3:2	One of IA, GS, CD	S	Probable local conditions for MAR	19,131
4:2	One of IA, GS, CD	ST	No source within chosen distance	2556

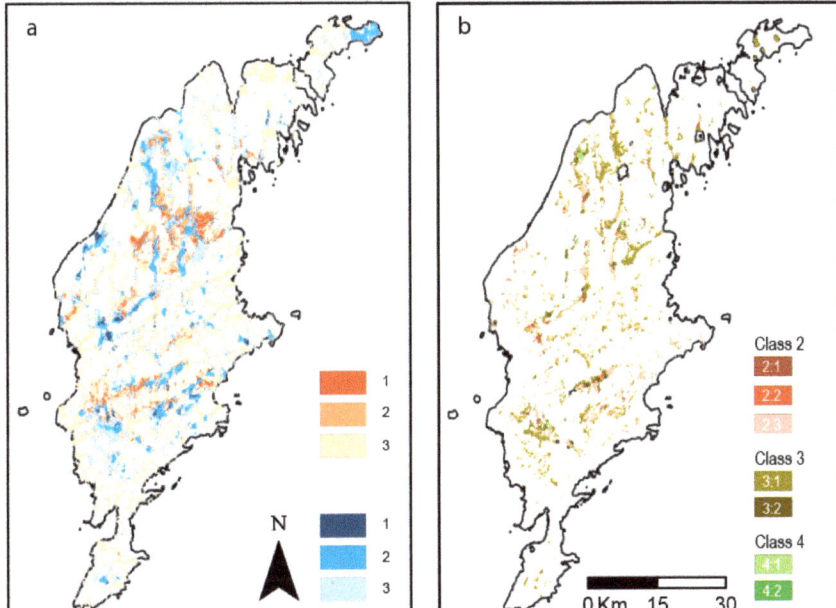

Figure 3. Maps from combined GIS analysis. (**a**) Combination of overlapping datasets from Figure 2a,b. Please note that since the raster set with surface water source/storage is made with a buffer there will be areas with overlapping datasets. (**b**) Map with the best areas for infiltration/groundwater storage and source/surface water storage. Classes are defined in the text and summarized in Table 1.

3.2. Estimation of Increased Groundwater Recharge and Groundwater Extraction at MAR Favorable Groundwater Catchments in Use Today

The presented values for potential infiltration rate and increased withdrawal volumes are estimations with considerable inherent uncertainty. Furthermore, there are large differences between the well fields. Site-specific conditions at existing abstraction areas differ due to, e.g., hydrogeology, number of wells, local water supply demand, quality, abstraction volume, etc. The abstraction volume in different well fields varies between 25 and 3300 m^3/day. There are also differences in estimation of infiltration capacity between 50–660 m^3/day (mostly based on local infiltration capacity) and the estimated increase in abstraction volume between 25–330 m^3/day (also includes the possibility of adding new wells). The percentage of predicted increased abstraction volume versus mean abstraction volume varies between 20–300% for estimated infiltration volume, and 10–150% of increased abstraction volume, in comparison to numbers for each well field, respectively.

3.3. Comparative Study of Alternative Measures

Figure 4 shows a marginal abatement cost curve for the analyzed measures, in which each measure is represented by a bar showing its unit cost (bar height) and annual water availability potential (bar width). As displayed, increased groundwater extraction and desalination had the largest potentials to improve water availability on Gotland. The water availability potential of desalination can be much larger, but the calculations here were based on assumptions of one new desalination plant. Increased groundwater extraction was associated with the lowest costs per cubic meter water provided, whereas desalination was associated with the highest unit costs. One reason for the high unit cost of desalination was the long pipelines needed to reach demand centers. In this comparative analysis, MAR was limited to groundwater recharge in the municipality's existing well fields. Hence, the water availability potential of MAR on Gotland may be significantly higher when not constraining the analysis to those areas. The unit cost of MAR in existing well fields on Gotland was in the same range as that for increased surface water extraction.

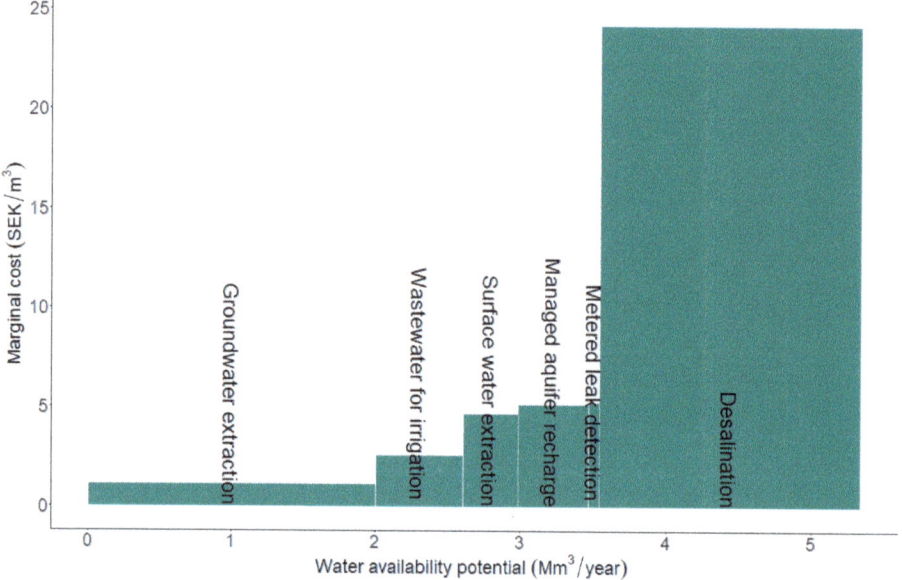

Figure 4. Marginal abatement cost curve for alternative measures to improve water resource security on the island of Gotland (100 SEK ≈ 11 USD).

4. Discussion

The presented data and analysis represent an early stage of mapping of MAR areas (focused on spreading methods in areas with high infiltration capacity) and estimates of potential and feasibility of this type of MAR on Gotland. Our project also includes mapping of good local conditions for a local source, e.g., water supply, which we believe further increases the utility of these study results for water management on the island. Mapping of suitable MAR areas with GIS is a widely used method [13]. There are uncertainties in both data and accuracy in the analysis, but within these limitations there is now a detailed picture on possible MAR and source areas which can be used by the municipality, farmers, and other stakeholders. Concerns regarding the limitations of these GIS analyses have been raised by, e.g., [14], who suggested that the use of sensitivity analyses of the factors used for MAR feasibility studies. In 2016, SGU made a first attempt to apply an overall assessment of the possibilities of MAR at existing abstraction areas [2]. Results from the present study can now improve the prior prioritization between MAR and alternative water measures for the island. The results should also be addressed on more than the water quantity. Increased groundwater recharge by MAR may influence the quality of the groundwater, e.g., by dilution effects, but also change the salinity levels in the bedrock aquifers. The municipality can analyze the results and compare with areas that are not used today but where the presented method shows good potential for MAR. There will be need for validation with numerical modelling and field tests to increase the strength and substance of the results, as shown in, e.g., [15–19].

This paper does not use any restrictions on unsuitable areas due to land use, a criterion that has been used in several analyses [13]. This can preferably be made by water management authorities, who are more suited to deciding between conflicting interests. Mainly because of the low relief of Gotland, the often-used parameter "slope" [13] is also not used here. A potentially much more important criterion is "closed depressions". This criterion uses the bedrock surface and the slope to calculate where there are bedrock depressions where water has more time to infiltrate and be stored as groundwater.

Even though there is a clear picture of the best areas from the perspective of infiltration/groundwater storage and source/dam, other areas can also take advantage of the presented data. Local conditions—for example, where the distance to the public water network is long—can make MAR solutions profitable in those more remote areas. The outcome of favorable areas (from 2000 ha (20 km^2) up to at most 43,000 ha (430 km^2)) in the analysis of infiltration areas, source areas and the combination of these should be compared to the area of the island (3000 km^2). The designated areas constitute only a small part of the island, and therefore care must be taken so that these are not destroyed by over-exploitation.

The degree of detail in the results is determined by the available data sets. SGU is working on an update on a few of the data sets, which will improve the certainty of the result. There are also several ongoing and future investigations associated with some of the data sets. For example, a few of the designated areas with a closed depression may be particularly suitable areas for groundwater dams. This is also the case with depressions that are not completely closed, not used in our investigation, and hence an interesting subject for future analysis for the viability of this MAR technique on Gotland. A groundwater dam is a man-made structure that obstructs the natural flow of groundwater and thereby can store larger quantities of water in the aquifer [31]. The results from this study are not validated by field studies. The data sets are; however, delivered to the local authorities for water resources and water management, and for analyses. Once the data sets are updated, the results may be integrated into the hydrogeological 3D model and calibration of parameters from field studies may improve future work with MAR on Gotland. The presented data sets may be tested in the newly developed online tools for suitability mapping, e.g., https://dss.inowas.com/tools [32,33]. The resulting suitability maps will be shared at the international MAR portal (https://apps.geodan.nl/igrac/ggis-viewer/viewer/globalmar/public/default).

The economic aspect of implementing MAR systems to improve potable and agricultural water supply has previously been investigated in different parts of the world (e.g., [21,22,34]). The associated capital costs are highly system specific, influenced by, e.g., hydrogeological, socioeconomic and legal factors [35]. As the economic analysis of MAR in this paper was based on rather small-scale complimentary infiltration of surface water at existing municipal well fields, no additional costs for new wells, treatment plants or pre-treatment were considered. The economic analysis was based on cost estimates associated with infiltration basins, raw water intake and new piping, resulting in a unit cost of approximately 5 SEK/m^3. This can be compared to cost estimates for MAR in Spain ranging between €0.08–0.58 per m^3 [20] (approximately 0.8–6 SEK/m^3).

The total water demand on Gotland is forecasted to increase by more than 40% by the year 2045 [1]. This will require water currently not available on the island. To make well-founded decisions on how to meet this forecasted demand and concurrently increase the preparedness for water scarcity situations, thorough decision support is needed. The presented data and analyses can be used to inform decision-making on measures to increase the amount of water that can be recharged on the island. Considering the entire island as a single groundwater aquifer would permit a holistic approach to groundwater management, and the water situation more robust. This is particularly important for the forecasted increase in public water demand but also for individuals, industry and farmers relying on private wells. By comparing MAR to alternative measures, in terms of costs and water availability potential, this paper also provides support in assessments of the measures' economic viability, usually an important decision criterion for municipalities, corporations and individuals alike.

5. Conclusions

This paper contributes results from analyses of possible MAR areas and their potential to increase water availability on the island of Gotland, Sweden. The method can be used to evaluate the MAR potential in similar areas with the same data sets. The results can be used on different scales, by authorities, the public water producer, farmers, industry and by people with private wells, for improved water resource security and further validated with field tests and more detailed models such as the afore mentioned hydrogeological model. Comprehensive field tests are probably the best for better understanding the problems in general. However, they are time consuming, and only a limited number of sites are available to accommodate the tests. In contrast, the GIS analysis allows us to explore and assess multiple sites with relative ease, yet the validity needs to be carefully checked. The main results are listed below.

- The best conditions for infiltration and groundwater storage occur in a total area of ca 2000 ha (0.7% of Gotland), second best in 14,400 ha (4.8% of Gotland), and third best in 43,000 ha (14% of Gotland).
- Areas with both proximity to a raw water source and conditions for storage in dams occur in a total area of ca 10,000 ha (3.3% of Gotland).
- An area of ca 7700 ha (2.5% of Gotland) has good local conditions for MAR and an area of ca 22,700 ha (7.5% of Gotland) has moderate local conditions for MAR.
- Decision support is provided by comparing MAR with other measures in a marginal abatement cost curve, contributing to informed prioritizations and decisions on water resource improvement on Gotland.
- MAR is not the alternative with the largest water availability potential, but it has significantly lower marginal costs compared, for example, with desalination, and the potential will increase if also considering new well fields and in preventing adverse consequences of increased abstraction.
- The water supply potential of MAR in existing well fields (public water supply) was estimated to be about 35% of the forecasted drinking water supply and 7% of the total water demand gap in year 2045. The total water supply potential of MAR on Gotland is much larger and is expected to exceed the demand.

Author Contributions: Conceptualization, P.D., K.S., A.L., E.H. and B.H.; Data curation, J.N.; Formal analysis, P.D., K.S., J.N. and E.H.; Funding acquisition, K.S.; Investigation, P.D., K.S., A.L. and B.H.; Methodology, P.D., K.S., A.L. and J.N.; Project administration, P.D.; Supervision, L.R.; Visualization, P.D. and K.S.; Writing—original draft, P.D. and K.S.; Writing—review & editing, P.D. and K.S.

Funding: The work was funded by the European Union's Horizon 2020 research and innovation program under the Marie Skłodowska-Curie grant agreement No 754412; Region Västra Götaland; Region Gotland; The Swedish Agency for Economic and Regional Growth; and the Swedish Research Council Formas contract no 942-2015-130.

Acknowledgments: The authors would like to thank Mikael Tiouls and Lars Westerlund at Region Gotland for contribution with local expertise in the comparative study.

Conflicts of Interest: The authors declare no conflict of interest.

References

1. Länsstyrelsen. *Regional Water Supply Plan of Gotland*; County Administrative Board of Gotland: Gotland, Sweden, 2018. (In Swedish)
2. Dahlqvist, P.; Thorsbrink, M.; Holgersson, B.; Nisell, J.; Maxe, L.; Gustafsson, M. *Wet Lands and Groundwater Recharge—Possibilities for Increased Capacity at Groundwater Catchments in Gotland*; SGU-Rapport 2017:01; Sveriges Geologiska Undersökning: Uppsala, Sweden, 2017; 73p. (In Swedish)
3. Erlström, M.; Persson, L.; Sivhed, U.; Wickström, L. *Description of the Bedrock on Gotland*; Sveriges Geologiska Undersökning K221; Sveriges Geologiska Undersökning: Uppsala, Sweden, 2009; 60p. (In Swedish)
4. Hansson, G. Artificial groundwater recharge—A method used in Swedish drinking water supply for 100 years. *VA Forsk* **2000**, *5*, 96–106, (In Swedish with English Summary).
5. Dillon, P. Future management of aquifer recharge. *Hydrogeol. J.* **2005**, *13*, 313–316. [CrossRef]
6. Stefan, C.; Ansems, N. Web-based global inventory of managed aquifer recharge applications. *Sustain. Water Resour. Manag.* **2017**, *4*, 153–162. [CrossRef]
7. Sprenger, C.; Hartog, N.; Hernández, M.; Vilanova, E.; Grützmacher, G.; Scheibler, F.; Hannappel, S. Inventory of managed aquifer recharge sites in Europe: Historical development, current situation and perspectives. *Hydrogeol. J.* **2017**, *6*, 1909–1922. [CrossRef]
8. National Water Quality Management Strategy. *Australian Guidelines for Water Recycling*; Water Quality Australia: Canberra, Australia, 2009; Volume 24, 251p.
9. IGRAC (International Groundwater Resource Assessment Centre). Global Inventory of Managed Aquifer Recharge (MAR) Schemes. 2015. Available online: https://ggis.un-igrac.org/ggis-viewer/viewer/globalmar/public/default (accessed on 22 June 2018).
10. Rahman, M.A.; Rusteberg, B.; Gogu, R.C.; Lobo Ferreira, J.P.; Sauter, M. A new spatial multi-criteria decision support tool for site selection for implementation of managed aquifer recharge. *J. Environ. Manag.* **2012**, *99*, 61–75. [CrossRef]
11. Russo, T.A.; Fisher, A.T.; Lockwood, B.S. Assessment of managed aquifer recharge site suitability using a GIS and modeling. *Groundwater* **2015**, *53*, 389–400. [CrossRef]
12. Malczewski, J. *GIS and Multicriteria Decision Analysis*; John Wiley & Sons: Hoboken, NJ, USA, 1999.
13. Sallwey, J.; Bonilla Valverde, J.P.; Vásquez López, F.; Junghanns, R.; Stefan, C. Suitability maps for managed aquifer recharge: A review of multi-criteria decision analysis studies. *Environ. Rev.* **2018**, *27*, 138–150. [CrossRef]
14. Maples, S.R.; Foglia, L.; Fogg, G.E.; Maxwell, R.M. Sensitivity of hydrologic and geologic parameters on recharge processes in a highly-heterogeneous, semi-confined aquifer system. *Hydrol. Earth Syst. Sci. Discuss.* **2019**. [CrossRef]
15. Asano, T.; Cotruvo, J.A. Groundwater recharge with reclaimed municipal wastewater: Health and regulatory considerations. *Water Res.* **2004**, *38*, 1941–1951. [CrossRef] [PubMed]
16. Bekele, E.; Pattersson, B.; Toze, S.; Furness, A.; Higginson, S.; Shackelton, M. Aquifer residence times for recycled water estimated using chemical tracers and the propagation of temperature signals at a managed aquifer recharge site in Australia. *Hydrogeol. J.* **2014**, *22*, 1383–1401. [CrossRef]
17. Greskowiak, J.; Prommer, H.; Massmann, G.; Johnston, C.D.; Nützmann, G.; Pekdeger, A. The impact of variably saturated conditions on hydrogeochemical changes during artificial recharge of groundwater. *Appl. Geochem.* **2005**, *20*, 1409–1426. [CrossRef]
18. Moeck, C.; Radny, D.; Auckenthaler, A.; Berg, M.; Hollender, J.; Schirmer, M. Estimating the spatial distribution of artificial groundwater recharge using multiple tracers. *Isot. Environ. Health Stud.* **2017**, *53*, 484–499. [CrossRef] [PubMed]
19. Franssen, H.J.H.; Kaiser, H.P.; Kuhlmann, U.; Bauser, G.; Stauffer, F.; Muller, R.; Kinzelbach, W. Operational real-time modelling with ensemble Kalman filter of variably saturated subsurface flow including stream-aquifer interaction and parameter updating. *Water Resour. Res.* **2011**, *47*. [CrossRef]
20. Fernández Escalante, E.; Calero Gil, R.; San Miguel Fraile, M.Á.; Sánchez Serrano, F. Economic assessment of opportunities for managed aquifer recharge techniques in Spain using an advanced geographic information system (GIS). *Water* **2014**, *6*, 2021–2040. [CrossRef]
21. Megdal, S.B.; Dillon, P. Policy and economics of managed aquifer recharge and water banking. *Water* **2015**, *7*, 592–598. [CrossRef]

22. Ross, A.; Hasnain, S. Factors affecting the cost of managed aquifer recharge (MAR) schemes. *Sustain. Water Resour. Manag.* **2018**, *4*, 179–190. [CrossRef]
23. Dahlqvist, P.; Triumf, C.-A.; Persson, L.; Bastani, M.; Erlström, M.; Jørgensen, F.; Thulin Olander, H.; Gustafsson, M.; Thorsbrink, M.; Schoning, K.; et al. *SkyTEM-Survey on Gotland*; Rapporter och Meddelanden 136; Sveriges Geologiska Undersökning: Uppsala, Sweden, 2015; 108p. (In Swedish)
24. Dahlqvist, P.; Triumf, C.-A.; Persson, L.; Bastani, M.; Erlström, M.; Schoning, K. *SkyTEM-Survey on Gotland, Part 2*; Rapporter och Meddelanden 140; Sveriges Geologiska Undersökning: Uppsala, Sweden, 2017; 135p. (In Swedish)
25. Geological Survey of Sweden. *Report of Government Mandate: Groundwater Recharge and Resources in Sweden*; SGU-Rapport 2017:09; Geological Survey of Sweden: Uppsala, Sweden, 2017; 45p. (In Swedish)
26. Rodhe, A.; Lindström, G.; Rosberg, J.; Pers, C. *Groundwater Recharge in Swedish Soils—A General Calculation with a Water Balance Model*; Institutionen för Geovetenskaper, Luft-och Vattenlära, Uppsala Universitet, Rapport, Serie A; Uppsala Universitet: Uppsala, Sweden, 2006; Volume 66, 35p. (In Swedish)
27. Addams, L.; Boccaletti, G.; Kerlin, M.; Stuchtey, M. *Charting Our Water Future—Economic Frameworks to Inform Decision-Making*; 2030 Water Resources Group: Washington, DC, USA, 2009.
28. Eory, V.; Pellerin, S.; Carmona Garcia, G.; Lehtonen, H.; Licite, I.; Mattila, H.; Lund-Sørensen, T.; Muldowney, J.; Popluga, D.; Strandmark, L.; et al. Marginal abatement cost curves for agricultural climate policy: State-of-the art, lessons learnt and future potential. *J. Clean. Prod.* **2018**, *182*, 705–716. [CrossRef]
29. Hastings, N.A.J. *Physical Asset Management: With an Introduction to ISO55000*; Springer International Publishing: Cham, Switzerland, 2015.
30. Brealey, R.A.; Myers, S.C.; Allen, F. *Principles of Corporate Finance*, 10th ed.; McGraw-Hill/Irwin: New York, NY, USA, 2010.
31. Onder, H.; Yilmaz, M. Underground dams. A tool of sustainable development and management of groundwater resources. *Eur. Water* **2005**, *11*, 35–45.
32. Sallwey, J.; Schlick, R.; Bonilla Valverde, J.P.; Junghanns, R.; Vásquez López, F.; Stefan, C. Suitability maps for managed aquifer recharge: Review and tool development. Abstract 33. In Proceedings of the 10th International Symposium on Managed Aquifer Recharge, Madrid, Spain, 20–24 May 2019.
33. Gorski, G.; van der Valk, M. Producing more interpretable maps of managed aquifer recharge suitability by visualizing sensitivity to subjective choices during mapmaking. Abstract 133. In Proceedings of the 10th International Symposium on Managed Aquifer Recharge, Madrid, Spain, 20–24 May 2019.
34. Gao, L.; Connor, J.; Dillon, P. The economics of groundwater replenishment for reliable urban water supply. *Water* **2014**, *6*, 1662–1670. [CrossRef]
35. Maliva, R. Economics of managed aquifer recharge. *Water* **2014**, *6*, 1257–1279. [CrossRef]

© 2019 by the authors. Licensee MDPI, Basel, Switzerland. This article is an open access article distributed under the terms and conditions of the Creative Commons Attribution (CC BY) license (http://creativecommons.org/licenses/by/4.0/).

Article

Managed Aquifer Recharge as a Strategic Storage and Urban Water Management Tool in Darwin, Northern Territory, Australia

Anthony Knapton [1], Declan Page [2,*], Joanne Vanderzalm [2], Dennis Gonzalez [2], Karen Barry [2], Andrew Taylor [2], Nerida Horner [3], Chris Chilcott [3] and Cuan Petheram [4]

1. CloudGMS Pty Ltd. 3 Wright Street, Edwardstown, Adelaide SA 5039, Australia
2. CSIRO Land and Water Waite Laboratories, Waite Rd, Urrbrae, Adelaide SA 5064, Australia
3. CSIRO Land and Water, Private Mail Bag 44, Winnellie, Darwin, NT 0822, Australia
4. CSIRO Land and Water, College Road, University of Tasmania, Sandy Bay, Hobart 7005, Australia
* Correspondence: declan.page@csiro.au; Tel.: +61-8-83038748

Received: 24 June 2019; Accepted: 5 September 2019; Published: 9 September 2019

Abstract: Population growth and increased irrigation demand have caused a decline in groundwater levels that limit water supply in the Darwin rural area. Managed Aquifer Recharge (MAR) is a practical solution that can be adopted to augment stressed groundwater systems and subsequently increase the security of water supply. Aquifer storage capacity is considered to be the primary constraint to MAR where unconfined dolostone aquifers rapidly recharge during the tropical, wet season and drain again in the dry season. As a result, there is a general understanding that aquifers of this nature recharge to full capacity each wet season. However, the aquifer storage capacity and the potential for niche opportunities for MAR to alleviate declining groundwater levels has not previously been examined. This paper uses the Darwin rural area's Proterozoic Koolpinyah Dolostone aquifer and the existing Koolpinyah Groundwater System to evaluate the prospects of MAR using both infiltration and injection techniques. Direct injection wells in an aquifer storage transfer and recovery (ASTR) scheme were favoured in this area, as injection wells occupy a smaller surface footprint than infiltration basins. This assessment suggested MAR during the early to mid-dry season could alleviate the impact of the dry season decline in groundwater levels in the Darwin rural area. The use of a larger aquifer storage and recovery (ASR) system (5,000,000 m^3/year) was also assessed as a potentially viable technical solution in the northern part of the aquifer where it is understood to be confined. The ASR scheme could potentially be scaleable to augment the urban water system and provide strategic long-term storage. Consideration must also be given not only to the strategic positioning of the ASR water bank, but also to the hydrogeology of the aquifers in which the systems would be developed. Not all locations or aquifer systems can successfully support a strategic storage ASR system. Scheme-scale feasibility assessment of an ASR water bank is required. The study reported here is an early phase of a series of investigations that would typically be required to demonstrate the viability of any proposal to apply MAR to increase the reliability of conjunctive groundwater and surface water supplies in stressed water resources systems. It focusses on assessing suitable storage areas in a lateritic aquifer.

Keywords: Managed Aquifer Recharge (MAR); aquifer storage and recovery (ASR); strategic storage; Northern Australia

1. Introduction

Urban potable water supply systems require a high degree of reliability and security. This can be challenging where rainfall is seasonal such as in the vicinity of the city of Darwin, Northern Territory,

Australia. Darwin experiences distinct wet and dry seasons, with 95% of rain falling in the wet-season months (November to April). The annual average rainfall is 1423 mm.

Darwin's reticulated water network has traditionally relied upon surface water reservoirs with a minor component (~15%) from groundwater [1]. Borefields used for urban water supply are in the peri-urban Darwin rural area and target the Koolpinyah Dolostone aquifer [2]. Reticulated water demand by urban and industrial users has produced an immediate system yield shortfall of approximately 5,000,000 m^3/year [1]. Locally, the water from this dolostone aquifer is also relied upon heavily for drinking and irrigation as the Darwin rural area is not connected to reticulated water supply. A consequence of this is that the residents of the Darwin rural area are particularly vulnerable to consecutive years of poor rainfall.

In 2016, the groundwater levels for the most part were low if not the lowest recorded for the past ten years for most of the Darwin rural area in the vicinity of the municipal borefield area (Figure 1). End of dry-season water levels can result in risk to the water supply for many groundwater users in this area and to nearby groundwater dependent ecosystems, such as Howard Springs. Figure 1b shows that the dry season groundwater levels approach the level at which Howard Springs reportedly ceases to flow [1,3]. MAR has been put forward as an option to help augment the stressed groundwater resource in the Darwin rural area. Previously, the potential for MAR in this area has been dismissed [4] without any technical assessment of viability using available data.

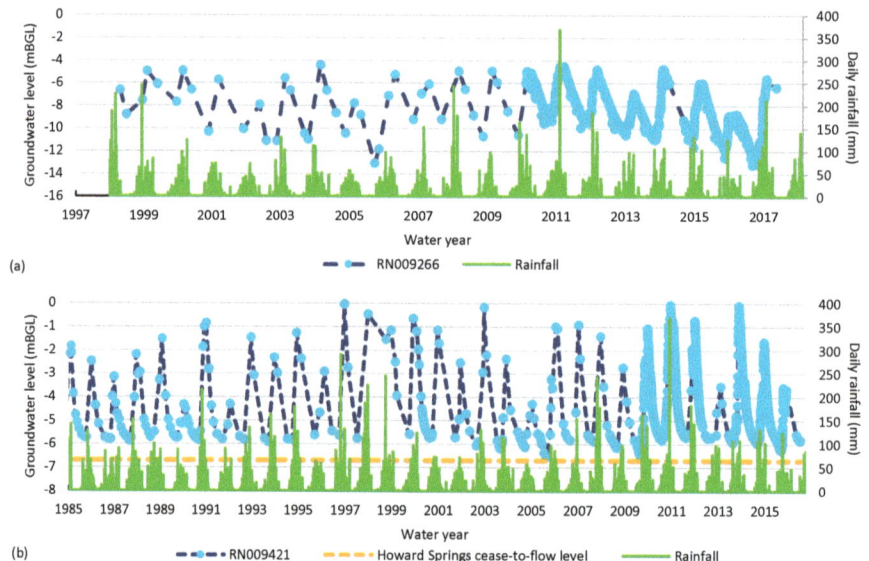

Figure 1. Hydrographs of bores in the Darwin rural area: (**a**) RN009266 at Middle Point and (**b**) RN009421 at Howard Springs.

This study investigates the potential for managed aquifer recharge (MAR) to: (i) reduce the risk of water stress for residents in the Darwin Rural Area; and (ii) provide addition storage capacity for reticulated supplies for urban and industrial use in Darwin. It examines how the options can interact to provide urban water supply security for the Darwin rural area and the potential for the development of a strategic storage for the City of Darwin. Specifically, it considers the potential for MAR into the most water stressed parts of the Koolpinyah Dolostone aquifer (e.g., MAR1-5 in Figure 2).

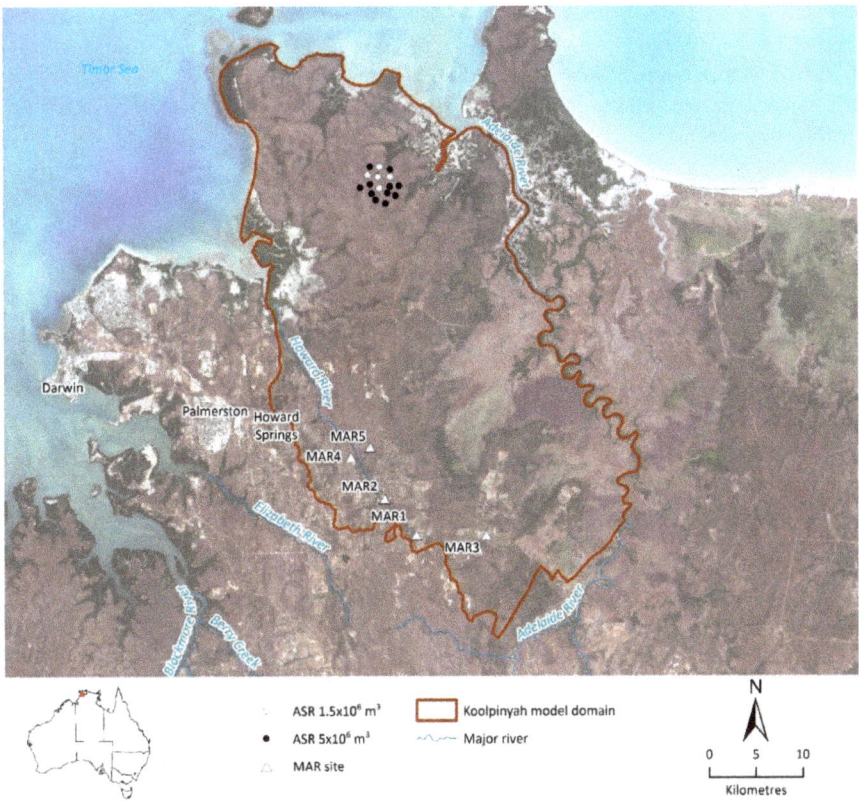

Figure 2. Location of modelling scenarios: Infiltration basin and well injection scenarios where aquifer is unconfined at MAR1-5; ASR Water bank scenarios where aquifer is confined in the northern section. White and black circles indicate a prospective location for a strategic ASR water bank (1.5 and 5 Mm^3/year). MAR scenarios are described in detail in Table 1.

Both the augmentation of the Darwin rural peri-urban system and a strategic ASR scheme are favoured in an aquifer that has generally high hydraulic conductivity, low specific yields, a suitably large volume of unsaturated sediments, natural boundaries that limit vertical and horizontal losses of the stored water, and low salinity of the native groundwater. For the ASR scheme to be successful, the hydraulic conductivity of the storage aquifer must be high to allow high rates of infiltration or injection over a relatively short wet season period as well as enabling high rates of extraction of the stored water to meet urban requirements.

The general understanding is that unconfined aquifers in the Darwin catchments rapidly recharge to full capacity during the wet season and drain again in the dry season [4], a common occurrence in unconfined, shallow aquifers. Aquifer storage capacity in the wet season, when there is access to a source of water for recharge, is considered to be the primary constraint to replenishing these unconfined aquifers via MAR [5]. However, the aquifer storage capacity of the Koolpinyah Dolostone and the potential for MAR have not previously been examined. This study addresses this knowledge gap by considering niche opportunities for MAR, by assessing the additional volume of water that could be recharged to alleviate the impact of current pumping for urban water supply, rural residential use and horticultural water supply. This study can be considered as a pre-feasibility assessment using available information and has the following objectives:

- Can MAR be used to reduce the groundwater stress (reduce the decline in groundwater levels during the dry season) in the Darwin rural area?
- Which MAR type is most effective (i.e., infiltration or injection)?
- Is there potential for well injection in the confined part of the aquifer for a strategic urban storage during the wet season?

Table 1. Howard East modelling scenarios.

Modelling Scenario	Description	Summary of Key Results
Base Case	Base case without MAR, simulated 1996–2014 Groundwater contours on 30 November 2009 (end of dry season) used for comparison to MAR scenarios Base case groundwater levels on 30 November 2009 were used to set trigger values for MAR scenarios	
MAR scenarios, detailed below:	MAR scenarios, simulated 1996–2014 Groundwater contours 30 November 2009 (end of dry season) compared to the base case	
SC1a infiltration mid-dry season	Infiltration basins in five stressed locations, infiltration rate of 0.015 m/d, recharge targets layer 1 in the model, recharge triggered by water level (MAR1, 2, 4 and 5~18 mAHD; MAR3~10 mAHD)	Infiltration rate limited by storage capacity of the aquifer in locations tested Model discretization too coarse to represent the cone of depression and limits the trigger to recharge in MAR2, 4 & 5 Evaluation focuses on MAR1 & 3, where recharge was triggered
SC1b injection mid-dry season	ASTR wells in five stressed locations, injection rate of 1370 m^3/d, recharge targets layer 3 in the model, recharge triggered by water level (MAR1, 2, 4 & 5~18 mAHD; MAR3~10 mAHD)	Recharge triggered mainly between August and December Model discretization too coarse to represent the cone of depression and limits the trigger to recharge in MAR2, 4 & 5 Evaluation focuses on MAR1 & 3, where recharge was triggered
SC1c injection wet season	ASR bores in the confined part of the aquifer, to the north of the stressed locations, recharge targets layer 3 in the model Five ASR bores, approximately 100 m deep were located around 20 km to the north of the stressed area, injection rate of 2,740 m^3/d/bore for 120 days Particle tracking with random walk (longitudinal dispersity (α_L) = 100, transverse dispersity (α_T) = 10, effective porosity (η_e) = 0.04) was used to evaluate migration of injected water away from the ASR bores	Injected water does not migrate far from point of injection End wet season mound ~+10 m End dry season drawdown ~−5 m Model grid is relatively coarse in this area, monitoring wells are limited
SC2a injection early dry season	ASTR wells in two stressed locations (MAR1 & 3), injection rate of 1,370 m^3/d, recharge targets layer 3 in the model, recharge applied at the start of the dry season for ~3 months, pumping from municipal bores in stressed area removed Constant head maintained from 1 April (or when heads declined between trigger) to 1 July, MAR1 constant head of 36 mAHD, MAR3 constant head of 20 m AHD	Recharge applied April to July in MAR1 & 3
SC2b injection wet season	ASR bores in the confined part of the aquifer, to the north of the stressed locations, recharge targets layer 3 in the model, pumping from municipal bores in stressed area removed 16 ASR bores, approximately 100 m deep were located around 20 km to the north of the stressed area, injection rate of 2740 m^3/d/bore for 120 days Particle tracking with random walk (α_L =100, α_T = 10, η_e = 0.04) was used to evaluate migration of injected water away from the ASR bores	Injected water does not migrate far from point of injection End wet season mound ~+20 m End dry season drawdown ~−10 m Model grid is relatively coarse in this area, monitoring wells are limited

Available surface water is assumed to be the source of water for recharge. The current Darwin Regional Water Supply Strategy does not address the vulnerability of water supply in the Darwin rural area [1]. Instead, it focuses on diversifying supply options to increase security and sustainability of supply to the Greater Darwin Region. Therefore, this paper is the first investigation of the potential to apply MAR to increase the reliability of conjunctive groundwater and surface water supplies in the rural area, a seasonally stressed water resource. However, the detailed technical, social and economic feasibility of specific MAR configurations is not addressed. The Australian MAR Guidelines [6] provide a comprehensive risk-based framework for assessment of scheme-scale technical feasibility, addressing human health and environmental risks along with operational issues such as clogging.

2. Methodology

2.1. Koolpinyah Groundwater System Model

The Northern Territory Government's Koolpinyah Groundwater System model was used to investigate the potential for MAR in the Koolpinyah Dolostone (i) to augment seasonally declining groundwater levels to prevent private water supply bores failing during the dry season and (ii) to provide a larger, potentially long-term strategic storage for urban water supply. The Koolpinyah Dolostone is largely unconfined in the stressed areas where augmentation is assessed, while to the north it is confined by mudstone.

The current Koolpinyah Groundwater System model was developed using the FEFLOW code. The model is fully described by CloudGMS [2]. The model domain covers approximately 1600 km^2, including the Howard River catchment and the western part of the Adelaide River catchment and includes the extent of the Koolpinyah Dolomite based on the interpretation of airborne magnetics and electromagnetic data [2]. It comprises three layers: layer 1 (the upper layer) represents the laterite aquifer, layer 2 represents the Cretaceous sediments within the weathered dolomite and layer 3 represents the fractured zone of the Koolpinyah Dolomite. The Koolpinyah Groundwater System model was calibrated on groundwater levels and dry season discharge measurements, over the period from 1980 to 2014, using the automated inversion software PEST [2]. Extraction volumes are uncertain, as rural residential water use is not metered.

The model is reported to be Category 2, with the capacity to achieve Category 3, according the Australian Groundwater Modelling Guidelines [7], due to a reasonably long observation data set over the areas with greatest stress and since projected pumping scenarios have stresses similar in magnitude.

There are five types of boundary conditions in the model:

- No-flow boundaries around the western and southern portions of the model domain and at the bottom of layer 3. No-flow boundaries to the south-west represent the presence of low permeability basement rocks that would impede radial flow and increase mounding. It is acknowledged that the proximity of the area of interest (stressed areas) to the no-flow boundary may impact on simulation results.
- Constant head (Dirichlet) boundary conditions representing spring discharge from the dolostone.
- Seepage surface boundary conditions at the ground surface.
- Transfer (Cauchy) boundary conditions representing the fluxes at the coast and Adelaide River.
- Well boundary conditions to represent groundwater withdrawal.

2.2. Modelling Scenarios for Peri-Urban Groundwater Augmentation and a Strategic Storage

The modelling scenarios to consider the potential for MAR in this rural area were simulated over the period 1996 to 2014 and focused on five stressed locations selected for recharge augmentation: MAR1 to MAR5 (Figure 2). Infiltration (SC1a) and injection (SC1b, SC2a) techniques were both assessed. At MAR1–MAR5, the Koolpinyah Dolostone is unconfined, and hydraulic heads were used to trigger and halt the enhanced recharge under these scenarios.

A key constraint for inter-seasonal recharge is securing a source of water during the dry season when the storage capacity becomes available in the aquifer. Therefore, the potential of a scheme to store water in the wet season, when it is likely to be available, was investigated. For this purpose, a strategic water bank formed with an ASR wellfield (SC1c, SC2b) was simulated to the north of the drawdown area (Figure 2) where the aquifer is understood to be confined (Figure 2). Each MAR scenario was compared to the base case (without MAR). Modelling scenarios are summarised in Table 1 and described in the following text.

In the stressed area, the first stage of modelling assessed the feasibility and type of MAR during the dry season (SC1a and SC1b, mid dry season). Hydraulic heads at a point in the centre of the MAR sites were used to trigger recharge augmentation, via using either infiltration basins or injection wells (Aquifer Storage Transfer Recovery (ASTR)). Trigger values were set equal to the observed end of dry season (30 November 2009) hydraulic head without MAR (MAR1, 2, 4 and 5~18 mAHD; MAR3~10 mAHD). A node spacing was allowed between recharge (MAR) and recovery locations (existing bores) to reduce interference.

The second stage of modelling considered the potential for MAR via injection (ASTR) with injection commencing at the start of the dry season for approximately three months until the start of July at MAR1 and MAR3 (SC2a, early dry season). A constant head was maintained from 1 April (or when heads declined below the trigger) to 1 July (MAR1 constant head of 36 mAHD applied, MAR3 constant head of 20 mAHD applied). Importantly, this scenario removed pumping from town water supply bores to assess the potential benefits to the stressed areas of using the strategic ASR water bank for town water supply.

In the assessment of the ASR water bank, the first stage of modelling considered 1,500,000 m^3/year of injection and recovery (SC1c) (5 ASR bores, 27.4 m^3/day per bore for 120 days). The ASR water bank scenario was developed further in the second stage of modelling (SC2b), with a 5,000,000 m^3/year water bank (16 SR bores, 27.4 m^3/day per bore for 120 days) replacing extraction for town water supply in the stressed area.

3. Results and Discussion

This first stage of modelling was used to determine if MAR recharge via infiltration or injection was more feasible for this area. The opportunity to create a strategic ASR water bank further to the north in the confined part of the aquifer was also investigated.

Both infiltration and injection modelling scenarios in the stressed areas were limited by the established model construction and scale. In areas where the cone of depression was not well represented, the node spacing between recharge and recovery locations resulted in limited instances where recharge was triggered in areas MAR2, 4 and 5. Recharge was triggered at MAR1 and MAR3 (Figure 3) and both techniques achieved a comparable gain in groundwater level when recharge was triggered (Figure 4).

The end of dry-season hydraulic head within a currently stressed area can clearly be increased by MAR (Figure 5, e.g., for injection). At MAR1, the impact on end of dry-season water table was up to 8 m when recharge was applied midway through the dry season and ~2 m when applied at the start of the wet season (Figure 4). The results for MAR3 indicate that infiltration basins and injection wells applying recharge midway through the dry season could cause the water table to rise by ~10 to 15 m at the end of the dry season (Figure 4). Injection at the start of the dry season resulted in an approximately 6 m increase over the base case in end of dry-season water table (Figure 4). However, further assessment is required of the potential for any rise in hydraulic head at the end of the dry season to protect against bores running dry remains.

Overall, the injection (ASTR) method was considered more prospective than infiltration due to fewer land use constraints. In the infiltration scenarios, an infiltration rate of 0.015 m/d was adopted to prevent excessing mounding, which resulted in heads approaching the surface and the invoked seepage face boundary conditions. The properties assigned to layer 1, representing the surficial laterite layer, could allow higher infiltrations fluxes; however, the lower permeability of layer 2 impedes vertical groundwater movement and results in excessive mounding. With a low infiltration rate (<0.02 m/d) a large area of land (>100 ha) would be required for infiltration basins. Land availability for infiltration basins is influenced by land use, which may be achievable in areas with larger block sizes, but areas to the west may not be as prospective due to basin area relative to block sizes.

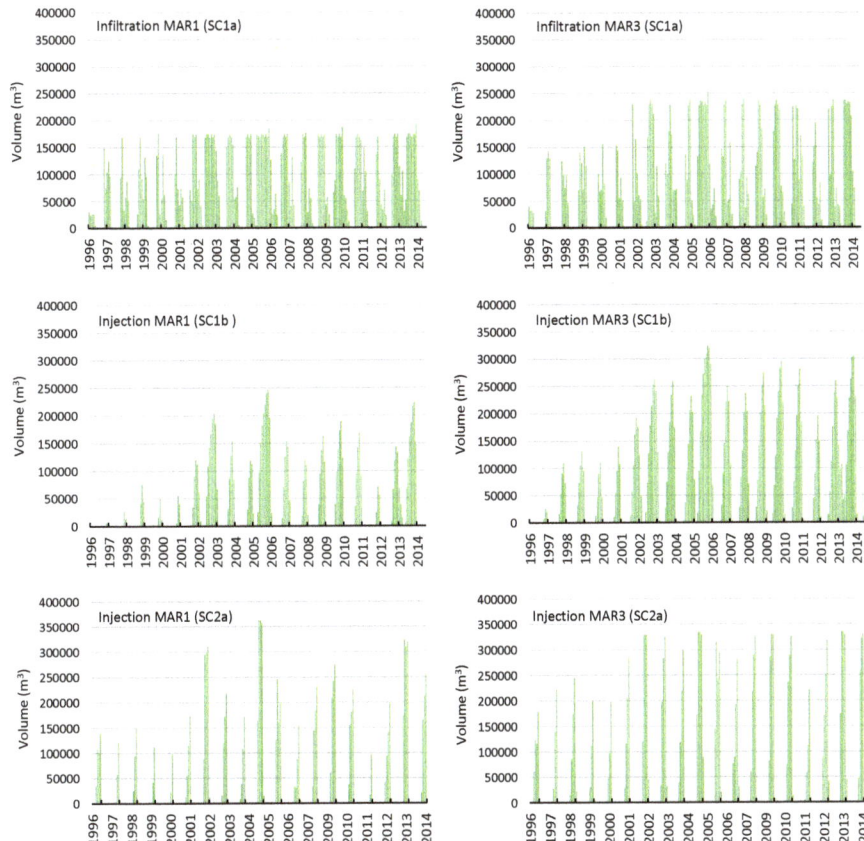

Figure 3. Comparison of monthly modelled recharge volume under infiltration (SC1a) or injection (SC1b, SC2a) type MAR scenarios: SC1a and SC1b recharge occurs during mid-dry season, recharge triggered by dry season minimum hydraulic head of MAR1~18 mAHD, MAR3~10 mAHD. SC2a recharge occurs early dry season, constant head maintained from 1 April to 1 July, MAR1 constant head of 36 mAHD, MAR3 constant head of 20 m AHD.

In the mid-dry season MAR scenarios, most of the augmented recharge was typically triggered between August and December each year. This interval coincides with the period of highest rural residential groundwater use (June to December) when surface water is in high demand for reticulated supply. Despite this competition for water, the volume of recharge required for the inter-seasonal MAR is small (i.e., 1–5,000,000 m³/year), and therefore is unlikely to have a significant impact on the volume of water in surface storage.

The potential for recharge at the start of the dry season was assessed in the second stage of modelling, through application of a constant head from the end of the wet season to the start of July (Table 1). Recharge was typically triggered for 3 months and the median recharge over the simulation period was 1,200,000 m³/year. The magnitude of the increase in hydraulic head at the end of the dry season was less than for the mid-dry season recharge scenario due to the time interval that recharge was triggered in, but presumably having an impact over a larger area. End of dry season hydraulic head declines suggest that sufficient storage remains available for wet-season recharge.

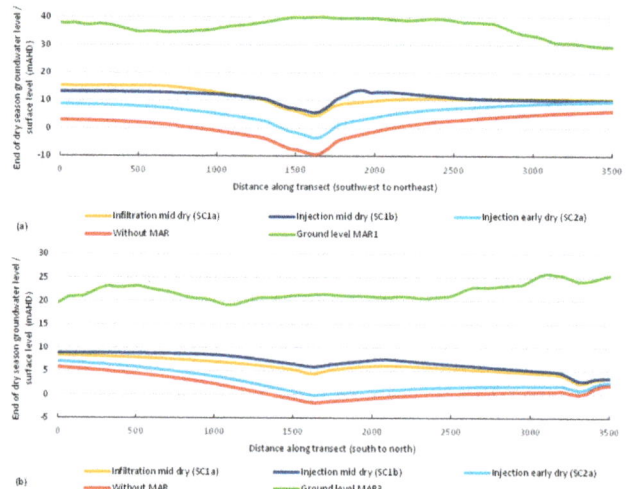

Figure 4. Example cross-section of hydraulic head response under infiltration (SC1a) or injection (SC1b, SC2a) type MAR scenarios at (**a**) MAR1 and (**b**) MAR3 for end of dry season groundwater level (30 November 2009). SC1a and SC1b Recharge occurs during the mid-dry season, recharge triggered by dry season minimum groundwater level of MAR1~18 mAHD, MAR3~10 mAHD. SC2a Recharge occurs in the early dry season, with constant groundwater level maintained from 1 April to 1 July, MAR1 constant head of 36 mAHD, MAR3 constant head of 20 mAHD. Location of the cross-section is shown on Figure 5.

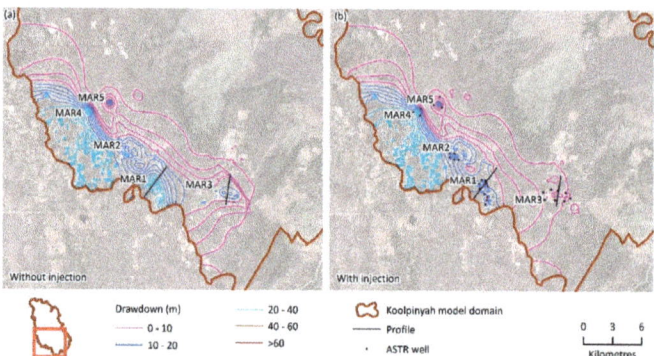

Figure 5. Hydraulic head response to an ASTR (injection well) MAR scheme (SC1b) triggered by declining hydraulic heads. Recharge is typically triggered between August and December. (**a**) The base case without MAR and (**b**) with the effect with MAR.

Maintaining hydraulic heads at the start of the dry season may be more practical and economical than waiting until later in the dry season when surface water resources are likely to be stressed and may not be available for MAR. Proximity to groundwater discharge locations would need to be assessed if this option was to be considered in more detail, to ensure the additional recharge is maintained within the aquifer. The scenarios considered indicate a reduction in the dry-season hydraulic head declines; however, the effectiveness of MAR in providing protection to individual groundwater users would need to take bore construction details into account. Infrastructure from the recharge source to recharge locations would also be required.

The ASR water bank scenario recharged a confined portion of the aquifer in the wet season and the model results suggest that the injected water remained localised (<2 km from the ASR bore) to the ASR borefield. With 1,500,000 m^3/year of injection and recovery, hydraulic head increased by ~10 m at the end of the wet season and declined by ~5 m at the end of the dry season, when compared to the base case without MAR. Increasing this to 5,000,000 m^3/year of injection and recovery, hydraulic head increased by ~20 m at the end of the wet season and declined by ~10 m at the end of the dry season, in relation to the base case.

The potential for development of a 5,000,000 m^3/year water bank to replace extraction for municipal supply in the stressed area of the Koolpinyah Dolostone was considered. The cessation of pumping from the municipal supply bores alone provided only localised benefits that did not alleviate groundwater declines in all of the five MAR locations assessed, due to multiple uses (irrigation, rural residential). The MAR water banking approach could serve as a longer-term option for urban water cycle resilience against reduced recharge during poor wet seasons. While the end user has not been defined, it is possible that the ASR water bank could contribute to Darwin urban water supply. Infrastructure from the recharge source to the ASR site and from the ASR site to the end user would be required. This approach is of broad international interest as it demonstrates that there is potential niche application of MAR in lateritic aquifers that have been previously dismissed as unsuitable for MAR. The study demonstrates that there may be opportunities to develop strategic urban water supplies in aquifers, though further investigations would be required to develop a specific scheme.

MAR scheme numerical modelling is typically undertaken at local (scheme) scale of a few kilometres, rather than the regional scale (10s to 100s of kilometres) of the Koolpinyah Groundwater System model. Due to the scale of the existing model, its discretization and the proximity of the stressed areas to the no-flow boundary, it is not possible to evaluate the hydraulic impact of individual MAR schemes. Nonetheless, this regional scale model was sufficient to undertake an entry-level assessment of MAR feasibility.

This pre-feasibility assessment suggests MAR may be beneficial in the Darwin rural area; however, scheme-scale feasibility assessment of an ASR water bank is required. The Australian MAR Guidelines [6] provide a comprehensive risk-based framework for assessment of scheme-scale technical feasibility, addressing human health and environmental risks along with operational issues such as clogging. Regardless of MAR technique, it is necessary to assess the risk of clogging due to recharge water quality and how this can be managed to ensure a sustainable operation.

This study is innovative, as the general area proposed for an ASR water bank has very little existing groundwater use or hydrogeological data. This is a common theme to many Mar investigations and at times may lead to pre-mature dismissal of MAR opportunities in aquifer such as these. Nevertheless, it will be essential to confirm the degree of confinement and the hydraulic response to MAR, the water quality within the storage zone and its potential impact on the utility of recovered water for urban supply and measures to minimize the impact of well clogging. In addition to MAR technical feasibility, it is essential to consider solutions which are supported by the community and are economically viable. Considerable investment in infrastructure would be required for MAR to alleviate stress across multiple areas in the Darwin rural area. However, a combined approach, with actions such as demand management and well maintenance, may prove to be lower cost than alternative water supply strategies.

4. Conclusions

Through this MAR modelling assessment, it was concluded that MAR (by a variety of means) is technically feasible to augment recharge and reduce the magnitude of groundwater decline in the Darwin rural area. Recharge via direct injection may be favoured due to having a smaller surface footprint, particularly in the more urbanised areas. The availability of a source of water for recharge during the dry season was considered a potential constraint for MAR. However, the volume of recharge required for the inter-seasonal MAR is small (i.e., 1-5,000,000 m3/year), and therefore is unlikely to have a significant impact on the volume of water in surface storage. Of particular interest internationally is that this study identified an opportunity for a strategic water bank of 5,000,000 m³/year using ASR wells to the north of the stressed area is another option to augment groundwater resources, storing wet-season excess surface water in the confined part of the aquifer for use when needed.

This pre-feasibility assessment suggests MAR may be beneficial in lateritic aquifers such as the Darwin rural area and that this assessment could be of broader interest in evaluating lateritic aquifers that have typically been deemed unsuitable for MAR.

Author Contributions: A.K. developed the groundwater model, applied the groundwater model to this investigation and contributed to paper writing. D.P. and J.V. conceived the MAR investigation and contributed to paper writing. D.G., K.B., N.H. contributed to data analysis and paper writing. A.T., C.C., C.P. provided oversight of the project and contributed to paper writing.

Funding: This research was funded by Australian Department of Agriculture grant number [22668] and is part of the CSIRO Northern Australia Water Resource Assessment, part of the National Water Infrastructure Development Fund: Water Resource Assessments.

Acknowledgments: This research was undertaken within the Northern Australia Water Resource Assessment as part of the Australian Government's Agricultural Competitiveness White Paper, the government's plan for stronger farmers and a stronger economy and was supported by the Department of Agriculture and Water Resources. The authors acknowledge the contributions of Des Yin Foo, Dale Cobban and Mardi Miles (Department of Environment and Natural Resources, NT); and David George, Shane Papworth, Trevor Durling (Power and Water Corporation). The authors gratefully acknowledge the thoughtful review comments of three anonymous reviewers and Peter Dillon, academic editor, in improving the quality of this paper.

Conflicts of Interest: The authors declare no conflict of interest.

References

1. Power and Water Corporation. *Darwin Region Water Supply Strategy*; Power and Water Corporation: Darwin, Australia, 2013.
2. CloudGMS. *Koolpinyah Groundwater System Groundwater Flow Model Update 1.0.*; Prepared for DLRM; CloudGMS: Adelaide, Australia, 2017.
3. Fell-Smith, S.A.; Sumner, J. *Technical Report—Koolpinyah Dolomite Aquifer Characteristics Project*; Department of Land Resource Management. Water Resources Branch: Darwin, Australia, 2011.
4. Cresswell, R.; Harrington, G.; Hodgen, M.; Li, L.; Petheram, C.; Buettikofer, H.; Davies, P. *Water Resource in Northern Australia*; CSIRO: Canberra, Australia, 2009.
5. Vanderzalm, J.; Page, D.; Gonzalez, D.; Barry, K.; Dillon, P.; Taylor, A.; Dawes, W.; Cui, T.; Knapton, A. *Assessment of Managed Aquifer Recharge (MAR) Opportunities in the Fitzroy, Darwin and Mitchell Catchments*; A Technical Report to the Australian Government from the CSIRO Northern Australia Water Resource Assessment, Part of the National Water Infrastructure Development Fund: Water Resource Assessments; CSIRO: Canberra, Australia, 2018.
6. NRMMC-EPHC-NHMRC. *Australian Guidelines for Water Recycling: Managed Aquifer Recharge*; National Water Quality Management Strategy Document No. 24; Natural Resource Management Ministerial Council, Environmental Protection and Heritage Council, National Health and Medical Research Council: Canberra, Australia, 2009.
7. Barnett, B.; Townley, L.R.; Post, V.; Evans, R.E.; Hunt, R.J.; Peeters, L.; Richardson, S.; Werner, A.D.; Knapton, A.; Boronkay, A. *Australian Groundwater Modelling Guidelines*; National Water Commission: Canberra, Australia, 2012.

© 2019 by the authors. Licensee MDPI, Basel, Switzerland. This article is an open access article distributed under the terms and conditions of the Creative Commons Attribution (CC BY) license (http://creativecommons.org/licenses/by/4.0/).

Article

Mapping Economic Feasibility of Managed Aquifer Recharge

Jean-Christophe Maréchal [1,2,*], Madjid Bouzit [1], Jean-Daniel Rinaudo [1,2], Fanny Moiroux [1,2], Jean-François Desprats [1,2] and Yvan Caballero [1,2]

1. Bureau de Recherches Géologiques et Minières (BRGM), 34000 Montpellier, France; m.bouzit@brgm.fr (M.B.); jd.rinaudo@brgm.fr (J.-D.R.); fanny.moiroux@outlook.fr (F.M.); jf.desprats@brgm.fr (J.-F.D.); y.caballero@brgm.fr (Y.C.)
2. University of Montpellier, 34000 Montpellier, France
* Correspondence: jc.marechal@brgm.fr; Tel.: +33-467-157-965

Received: 17 November 2019; Accepted: 18 February 2020; Published: 2 March 2020

Abstract: Managed aquifer recharge (MAR) constitutes a potential and promising solution to deal with several water management issues: water shortage, water level depletion, groundwater pollution, and saline water intrusion. Among others, the proper siting and cost evaluation of such a solution constitutes sources of uncertainty for the implementation of MAR schemes. In this study, we proposed a methodology for the assessment of the levelised cost of recharged water through an infiltration basin, including investment and operating costs. The method was implemented in a GIS-tool in order to build maps of levelised costs at the aquifer scale. The sensitivity analysis allows for the identification of the main natural characteristics (water quality and availability, etc.), technical (system life duration, recharge volume objective, etc.), and economic parameters (energy price, discount rate, etc.) that dominate the final cost estimate. The method was applied to a specific case study on an alluvial aquifer in Southern France. This new information on the economic feasibility of MAR scheme should be incorporated with more classical GIS-MCDA (relying on soil characteristics, aquifer storage capacity, land use, etc.) in order to properly site the system. Further information on financial and economic feedback from MAR implementation and research on the fate of recharged water are needed for a better benefits evaluation of this solution.

Keywords: infiltration basin; cost function; suitability map; groundwater; MAR; GIS-MCDA

1. Introduction

Groundwater is the world's largest freshwater resource. It provides an increasing quantity of water for irrigated agriculture and hence for global food security. If groundwater abstraction exceeds the natural recharge for a long period of time, intensive use and induced groundwater depletion occur. Depletion is widespread in large groundwater systems of the world [1]. Excessive extraction for irrigation where groundwater is slowly renewed is the main cause of the depletion, and climate change should exacerbate the problem in some regions where natural recharge is expected to decrease (the Mediterranean area for example). The effects of groundwater depletion are complex and dependent on the aquifer, and include (i) lowering of water tables leading to increased cost of pumping or drying up of wells [2]; (ii) reduced groundwater baseflow to streams, springs, and wetlands affecting ecosystems [2]; and (iii) land subsidence potentially damaging buildings and infrastructure [2]. Lowered water tables can lead to salinization by saltwater intrusion in coastal regions. Similarly, groundwater depletion can promote the spread of other types of pollution [2].

Managed aquifer recharge (MAR) constitutes very promising solutions for dealing with water shortage, water level depletion, groundwater pollution, and saline water intrusion [3]. It consists of infiltrating water abstracted from surface water resources (rivers, streams, lakes, etc.) through infiltration basins (indirect recharge) or injection wells (direct recharge [4]) in order to increase the natural groundwater recharge.

It is important to properly locate MAR systems according to the infiltration characteristics of the soil, the aquifer capacity to store water, the water resource location, and land use constraints. The site suitability assessment for MAR can be achieved by combining multi-criteria decision analysis (MCDA) for solving spatial problems with geographical information systems (GIS). For that purpose, many studies have been published in the literature describing methods for mapping the technical suitability of MAR solutions, most of them using GIS tools [5–7]. They combined the spatial analysis capacity of GIS with MCDA methodology that guides the decision making process—the resulting approach is called GIS-MCDA and has been recently implemented on web tools [8].

Despite many benefits and demonstrated advantages of the MAR, the growth of this solution has been much lower than expected due to the lack of a sound economic feasibility analysis. The performance and cost–benefit analysis of MAR scheme are key factors for the sustainability of this solution [9]. The costs of MAR schemes are influenced by a wide variety of hydrogeological, socio-economic, and legal and institutional factors [10].

A short review of GIS-MCDA for suitability mapping of MAR schemes shows that, except for a few studies [11], the cost is rarely included in such an analysis, maybe because of a lack of feedback on costs and financial data on MAR devices [10]. Despite the fact that several environmental variables (such as soil infiltration rate [12]) constitute a surrogate for economic evaluation, a specific economic analysis of MAR can bring about substantial information into a GIS-MCDA for MAR location [11].

In this study, we proposed to elaborate the cost function of recharge devices, taking into account capital and operating costs in order to compute the levelised cost of MAR. This cost function was spatially distributed in order to build a map of levelised cost in the study area. This method allowed for the identification of the part of a territory where the cost of MAR is expected to be lower compared with other regions, subject to the hydrogeological characteristics that affect aquifer storage capacity, the ability to recover water for high valued uses, and the environmental impacts of imposed changes to groundwater conditions.

2. Methodology

2.1. MAR Project Design

One MAR scheme can be divided into several engineering components (Figure 1): water abstraction system from the surface stream, water transfer pipe, pretreatment system, infiltration basin, and surface and groundwater monitoring.

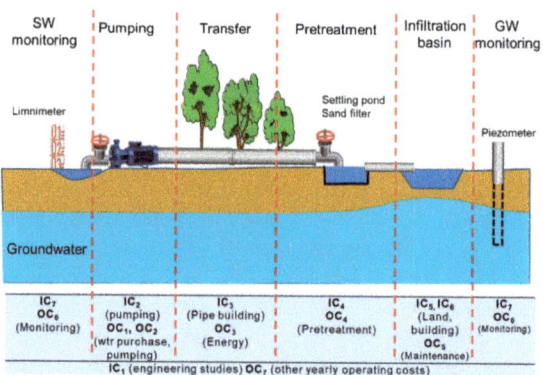

Figure 1. Main engineering components of a managed aquifer recharge (MAR) scheme with infiltration basin and associated investment (IC) and operating costs (OC).

Several parameters or natural characteristics constitute a set of values that characterize the MAR scheme project (Table 1). The size of an infiltration basin depends on the natural characteristics of the soil and the target water volume (Q) to be recharged. The soil infiltration rate (i) is a key parameter as the surface area (S_B) of the required infiltration basin is inversely proportional to this soil characteristic. The basin size is also linearly proportional to the target volume of recharge. The rate and the yearly duration of infiltration (N) are dependent on the water availability into the surface reservoir and provide the instantaneous flow rate (q) that are taken into account for calculating the diameter d_i of the pipes necessary for transferring the water from the abstraction place to the infiltration basin. Distance D and elevation difference Z between the surface water resource and the infiltration basin constitute the main characteristics to design the transfer infrastructure. The main parameters of such a project are listed in Table 1. These parameters are used for estimating the cost of the MAR scheme. Other site-specific parameters may also be incorporated in the analysis.

Table 1. Main parameters of an MAR scheme with infiltration basin project design.

Process	Parameter	Parameter	Unit	Comment
Water monitoring	No specific parameter	-	-	-
Water abstraction	Recharge rate	Q	m³/year	Annual recharge rate objective for the MAR
	Recharge duration per year	N	d/year	Yearly duration of the period during which water can be abstracted
	Flow rate	$q = \frac{Q}{N}$	m³/day m³/h	Daily/hourly flow rate for pipe diameter sizing
Water transfer	Distance	D	m	Between abstraction and recharge points
	Altitude difference	Z	m	
	Head losses	$H = -Z + 0.011D$	m	Assumption: linear head losses = 0.01 m/m of pipe
	Pipe diameter	$d_i = 22.9\, q^{0.4}$	mm	Hydraulic law
Water pretreatment	No specific parameter	-	-	-
Water infiltration	Soil infiltration rate	i	m/day	From in situ measurements
	Infiltration basin surface area	$S_B = \frac{q}{i}$	m²	
	Scheme surface area	$S_S = 1.1\, S$	m²	Assumption: 10% extra land necessary for neighbouring
	Basin depth	d	m	d between 1 and 3 m

2.2. Economical Approach-Cost Function

On the basis of the parameters/characteristics from Table 1, a cost function was built in order to compute the capital and operating costs of a MAR scheme using infiltration basins. These costs are described below according to the engineering component to which they are associated (Figure 1) and summarized in Table 2.

2.2.1. Investment Costs

The investment costs of a MAR project such as an infiltration basin cover seven main expenditure items:

- Cost of preliminary studies (IC_1): All preliminary characterization studies of the recharge site (e.g., geological and hydrogeological characterization, technico-economic study, impact study,

and preparation of the authorization file). In general, in "water" projects, this cost represents between 5% and 20% of the total investment cost depending on the size and complexity of the recharge project.
- Water abstraction cost (IC_2): cost of civil engineering works for the pumping of water out of the river/canal, as well as pumping equipment (in the case where gravity supply is not possible).
- Water transfer cost (IC_3): in most cases, it will be necessary to transfer the water to the recharge site. This investment item concerns the construction of water transfer infrastructure including the supply pipeline. Depending on distances (up to a few tens of kilometres) and volumes, this investment cost item can be significant in relation to the total investment.
- Cost of recharge water (pre)treatment units (IC_4): the quality of the recharge water must meet regulation standards for recharge authorization. At a minimum, intermediate settling and filtration basins (primary treatment) could be required to limit the clogging of the recharge structures. Additional treatment (secondary or tertiary treatment) may be required (especially in the case of direct recharge).
- Costs related to land acquisition (IC_5): the cost of purchasing land for the construction of infiltration basins, which may be significant depending on the location of the recharge site (rural or urbanized environment). It depends on the number and total surface area of the basins, which in turn will depend on the infiltration rate (i) and instantaneous flow rate (q) of the selected site.
- Cost of infiltration basins (IC_6): in general, this is the main investment item. These costs include the design (civil engineering) and construction of infiltration basins (injection wells in the case of direct recharge), as well as associated equipment.
- Other costs (IC_7): costs of monitoring equipment (e.g., construction of piezometers), and ancillary works (e.g., protection and development of the recharge site).

2.2.2. Operating Costs

Operating costs cover the operating and maintenance costs of the MAR device. These are annual and recurring costs, expressed in €/year. These expenses can also be grouped into seven main items:
- Water purchase cost (OC_1): if applicable, includes the purchase cost in the case of withdrawal from a water canal or network, as well as charges, levies, or other taxes.
- Maintenance cost of the water intake (OC_2): includes the maintenance of the recharge water pumping system in the river.
- Energy cost (OC_3): corresponds with the electricity consumption of the equipment and pumping system used to supply the recharge water to the recharge site (if not gravity-fed). It will depend on the flow rate and the price of energy.
- Pre-treatment operational cost (OC_4): the operational and maintenance costs of the infrastructure for pre-treatment of groundwater (excluding investment). They include, for example, the cost of maintaining and cleaning settling tanks, the cost of chlorination products, etc.
- Cost of maintenance and upkeep of infiltration basins (OC_5): includes the maintenance of the recharge device (e.g., cleaning of infiltration basins) and its surroundings.
- Monitoring cost (OC_6): all the costs related to the control and periodic monitoring of groundwater or recharge water quality (e.g., laboratory analysis cost) or the costs associated with checking the proper functioning of the device (essentially labour costs if an automated control system is not set up).
- Other annual expenses (OC_7): includes all financial expenses not mentioned above: administrative and personnel management expenses, financial expenses on investment and insurance loans, etc.

2.2.3. Levelised Cost

For a MAR scheme, the levelised costs can be defined as the constant level of cost each year to cover all the capital, operating, and maintenance expenses over the life of a MAR project divided by the annual volume of recharge (or infiltration). Levelised costs provide an effective means to compare the costs of water from alternative projects [13]. The levelised cost takes into account the duration life (T) of the MAR scheme and the depreciation (or discount) rate r, which is the rate at which the value of an asset is reduced each year.

The levelised cost is computed using equations described in Table 2, some of them from the cost Observatory of Rhône Mediterranean Corsica Water Agency (AERMC). The latest provided the levelised cost of pretreatment; therefore, for this cost only, we used assumptions on levelised cost. Finally, the levelised cost is computed using Equation No. 2 (Table 2), considering life duration of the MAR scheme and the discount rate.

Table 2. Summary of costs for a MAR scheme (recharge through infiltration basin option). α and β are fractional parameters in order to define specific costs as a fraction of other costs.

Process	Cost Description	Unit	Cost/Value	Comment
Other	IC_1: engineering studies	€	$IC_1 = \alpha_1 \sum_{i=2}^{7} IC_i$	α_1 ratio of engineering studies costs
	OC_7: other yearly costs	€/year	$OC_7 = \alpha_7 \sum_{i=1}^{6} OC_i$	α_7 ratio of yearly costs
Water abstraction	IC_2: pump installation	€	$IC_2 = 4520\,q + 180800$	q (l/s)
	OC_1: water cost	€/year	$OC_1 = SF + Pw\,Q$	SF: subscription fee (€/year) Pw: water price (€/m^3)
	OC_2: pump maintenance	€/year	$OC_2 = \alpha_2\,IC_2$	Assumption: portion of investment costs
Water transfer	IC_3: pipe building	€	$IC_3 = D(0.71\,di + 19.5)$	di: pipe diameter (mm)
	OC_3: lifting energy	€/year	$OC_3 = \frac{24\,N\,Pe\,Q\,H}{367\,\eta}$	Pe: electricity price (€/m^3) η: pump efficiency
Water treatment	IC_4: system building	€		30% of LC_4
	OC_4: system maintenance	€/year	$LC_4 = \beta_4$ (€/m^3)	70% of LC_4
Water infiltration	IC_5: land purchase	€/m^2	$IC_5 = LMV\,S_S$	Land market value (€/m^2)
	IC_6: basin building	€	$IC_6 = 2.28\,S_B d + 61100$	
	OC_5: basin maintenance	€/year	$OC_5 = \frac{(Pc+Ps)}{N_c} S_B H_C + 0.1\,Pm S_B$	Nc: Years between two dragging processes Hc: Sand height to be dragged (m) Pc: Sand dragging price (€/m^3) Ps: Sand price (€/m^3) Pm: Neighbours maintenance price (€/m^2/year)
Water monitoring	IC_7: monitoring equipment	€	$IC_7 = \beta_7$	Assumption
	$OC6$: yearly monitoring	€/year	$IC_6 = \beta_6$	Assumption
Total	Capital costs IC (CAPEX)	€	$IC = \sum_{i=1}^{7} IC_i$	Total of IC
	Operational costs OC (OPEX)	€/year	$OC = \sum_{i=1}^{7} OC_i$	Total of OC
	Operating life T	year	T	
	Discount rate r	Decimal	r	
	Capital recovery factor (CRF)	Decimal	$CRF = \frac{r(1+r)^T}{(1+r)^T - 1}$	No.1
	Levelised cost	€/m^3	$LC = \frac{CRF\,CC + OC}{Q}$	No.2

2.3. Costs Mapping Method

In the mapping approach, the study area is gridded according to the DEM resolution (50 × 50 m cell size). In each cell, a land value is determined according to local databases and the infiltration rate is deduced from permeability maps built for numerical modelling studies of the aquifer. For each cell (column i, raw j), shortest distances $D^n{}_{i,j}$ to surface resources (stream or lake) n are computed along with the corresponding elevation difference $Z^n{}_{i,j}$ between the abstraction cell and the infiltration cell (Figure 2). Using these data, one levelised cost is computed for each cell and n associated surface water resources. Then, for building the cost map, the minimum levelised cost is calculated among the n available resources according to:

$$LC_{i,j} = \min\left(LC^1_{i,j}, LC^2_{i,j} \ldots LC^n_{i,j}\right) \tag{1}$$

The minimum cost is therefore considered for each cell of the analysis domain.

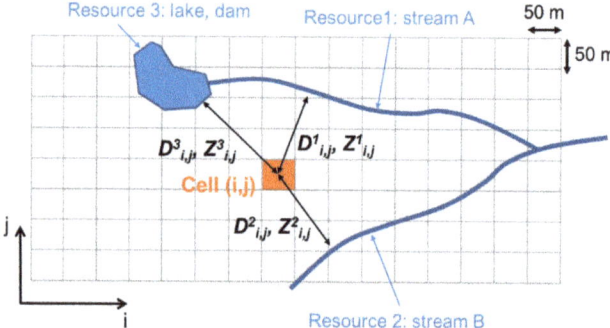

Figure 2. Cells gridding of the study area (example for a case with n = 3 available water resources: two streams and one lake). $D^n{}_{i,j}$ and $H^n{}_{i,j}$ are, respectively, minimum distance and elevation difference between the cell (i,j) and resource n.

3. Reference Case Study

3.1. Case Study Description

The reference case used was the Vistrenque and Costières Plain case study (VCP). This aquifer is located in Southern France between the Gardon River to the east and the Vidourle River to the west (Figure 3). The VCP area consists of a plain (Vistrenque) and a plateau at very low altitude (Costières) bordered to the north by the Nîmes garrigues and to the south by the Rhône plain and the Petite Camargue. Several types of unconsolidated rocks, among which alluviums largely dominate, constitute the aquifer. The VCP aquifer is unconfined on 84% of its surface area and confined on the rest.

Up until now, the aquifer is considered as being in a fragile hydraulic equilibrium, but the expected climate change impact on natural recharge and the increase of water abstraction induced by population concentration should lead to a potential water table decline, as already observed in the past during dry periods. Apart from water saving measures, MAR through infiltration basin using available surface water constitutes a possible alternative solution.

There are several surface water resources in the study area:

- The Vidourle River, to the west, is characterized by a very low baseflow and frequent flash floods. Water could be abstracted only during medium to high flow periods (excluding baseflow period).
- The Vistre River crosses the aquifer from northeast to southwest. Due to bad water quality, this river does not constitute a possible surface water resource for MAR.

- The Bas-Rhône and Languedoc regional water company (BRL) canal network conveys water from the Rhône River mainly for irrigation purposes. Its very well connected network of canals constitutes an efficient way of bringing surface water into the plains.

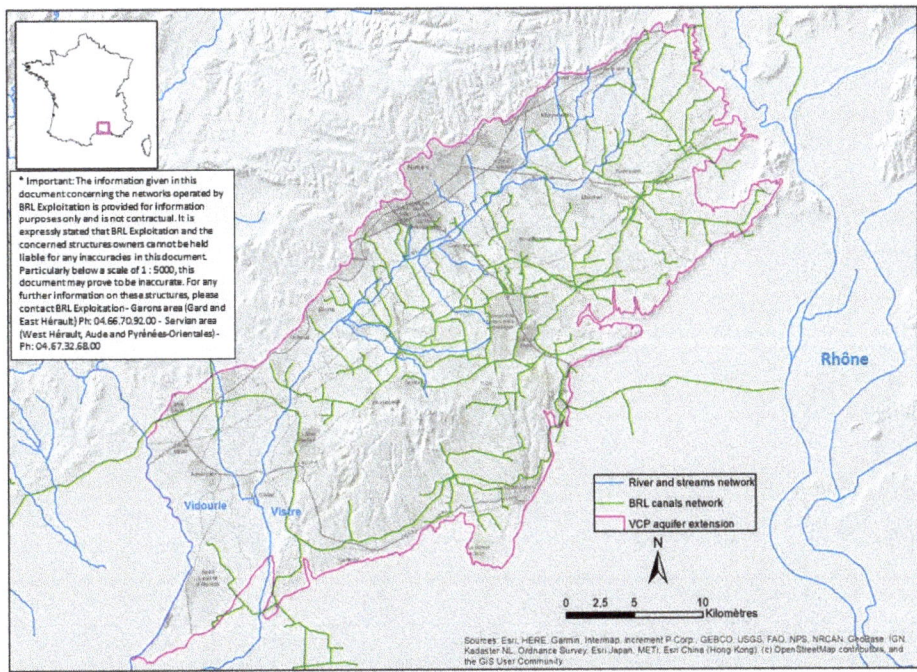

Figure 3. Map of the Vistrenque and Costières Plain (VCP) case study. Available surface water for MAR purposes is identified (in blue: rivers and streams; in green: Bas-Rhône and Languedoc regional water company (BRL) canal network).

3.2. MAR Design and Characteristics

The characteristics of the MAR scheme project on the VCP case study are summarized at Table 3. In that case study, the recharge rate objective was fixed at 1 Mm^3/year for 8 months (N = 243 days/year) because surface water (canals or rivers) was not available during 4 months/year in low stages. For the reference case, we assumed a distance and an elevation difference between the surface resource and the infiltration basin, respectively, of 1000 m and −10 m. We considered that the solution of free water (from the river) was preferable to the canal water which is costly. We assumed that a primary pretreatment was necessary to remove the silt and fine material present in the river water. Its levelised cost was LC_4 = 0.10 €/m^3. Other parameters (regarding basin dragging and maintenance) are listed in Table 3.

Table 3. Main characteristics of the MAR project on the VCP case study.

Process	Parameter	Value
Other	Engineering studies cost rate	$\alpha_1 = 0.10$
	Other yearly costs rate	$\alpha_7 = 0.10$
Water abstraction	Recharge rate	$Q = 10^6$ m^3/year
	Recharge duration per year	$N = 243$ days/year
	Flow rate	$q = 4115$ m^3/day $q = 171.5$ m^3/h
	Pump maintenance cost rate	$\alpha_2 = 0.10$
Water transfer	Distance	$D = 1000$ m
	Altitude difference	$Z = -10$ m
	Head losses	$H = 21$ m
	Pipe diameter	$di = 0.179$ m
	Pump efficiency	$\eta = 0.80$
Water treatment	Levelised cost	$\beta_4 = 0.10$ €/m^3
Water infiltration	Soil infiltration rate	$i = 1$ m/day
	Infiltration basin surface area	$S_B = 4115$ m^2
	System surface area	$S_S = 4527$ m^2
	Land market value	$LMV = 1$ €/m^2
	Basin depth	$d = 2.5$ m
	Duration between two dragging	$Nc = 5$ years
	Sand height to be dragged	$Hc = 0.30$ m
	Sand dragging and sand prices	$Pc = 3$ €/m^3 $Ps = 10$ €/m^3
Water monitoring	Investment cost	$\beta_7 = 20{,}000$ €
	Operating cost	$\beta_6 = 0.10$
Financial data	MAR scheme life duration	$T = 30$ years
	Local discount rate	$r = 0.04$

4. Results

4.1. Reference Case

The total investment cost obtained for the VCP case study was €1.7 million. The predominant cost items were the cost of implementing the pretreatment system (IC_4), as well as the cost (IC_2) of installing the water abstraction system (here, the reference example considered a river intake), corresponding to 53% and 23% of the total investment cost, respectively (Figure 4). The costs related to land purchase (IC_5) and monitoring costs (IC_7) represented a negligible part of the total investment costs. The investment cost of transferring water (IC_3) was low compared to IC_4 and IC_2. In cases where the distance D (1000 m in the reference example) between the surface water resource and the recharge area is higher, this cost item can take a significant part of the total investment cost.

The graph below summarizes the operating costs for the same reference case (Figure 4). The cleaning was fixed every $Nc = 5$ years, on a removal and replacement of $Hc = 30$ cm of gravel pack. The total operating cost obtained was nearly 130,000 €/year. The predominant cost item was the cost related to the pre-treatment of water (OC_4), corresponding to 54% of the total operating cost (Figure 4). The water transfer cost (energy, OC_3) was reasonably high in our reference example because an average altitude difference (−10 m) was chosen. If the water supply to the device was gravitational, OC_3 would have been negligible. In the case where water was drawn from canals or water networks, the water purchasing cost OC_1 became one of the main cost items. Although it was generally perceived as high by operators, the cost of maintaining the basin and its surroundings (OC_5) was found to be low compared to other cost items, such as OC_4.

In this reference case, both investment and operating costs were dominated by pretreatment cost (i.e., in this case an assumption of a pretreatment levelised cost of 0.10 €/m³).

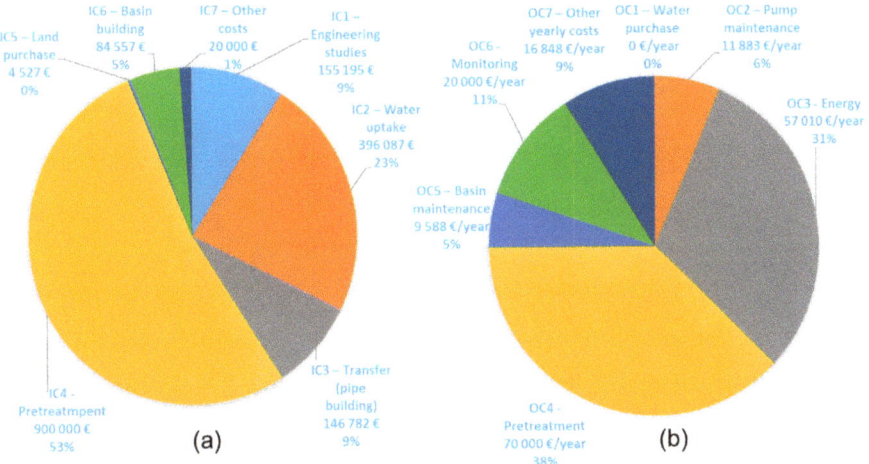

Figure 4. Partition of the various cost items for the reference case: (**a**) investment costs, IC (in €) and (**b**) operating costs, OC (in €/year).

4.2. Levelised Costs Mapping

The levelised cost map was obtained for the VCP case study applying Equation No. 2 (Table 2) using parameters from Table 3 and spatial variables distributed on the VCP maps (Figure 5a–d). The map of levelised cost is presented below (Figure 5e). The black areas correspond to the sectors excluded from the cost analysis (mask created from the unfavourable sectors from land use analysis), in order to isolate urbanized, artificialized areas, where it would be impractical and less interesting to install an MAR device (in addition to the difficulty of estimating land costs on these sectors).

The levelised cost LC ranged from 0.13 €/m³ (along the Vidourle) to nearly 0.55 €/m³ (on the contours of the entity; Figure 5). The average LC (average over the whole entity) was 0.29 €/m³. The most prominent criteria were:

- The purchase of water (OC_1) from BRL (high costs near the canals and raw water networks of the Gard);
- The levelised cost of water pre-treatment LC_4 (considered at 0.10 €/m³ for BRL resources and 0.05 €/m³ for Vidourle river);
- Distance to resource D (price decreased from surface water resources, linked to the increase in the cost of water transfer);
- The soil infiltration rate i (less infiltrating zones in orange, made up of Astian sands of the Costières, less permeable formation than the others, near Beauvoisin, Générac, Saint-Gilles, and Bellegarde);
- The difference in altitude Z (visible in particular in a flatter area in the commune of Cailar).

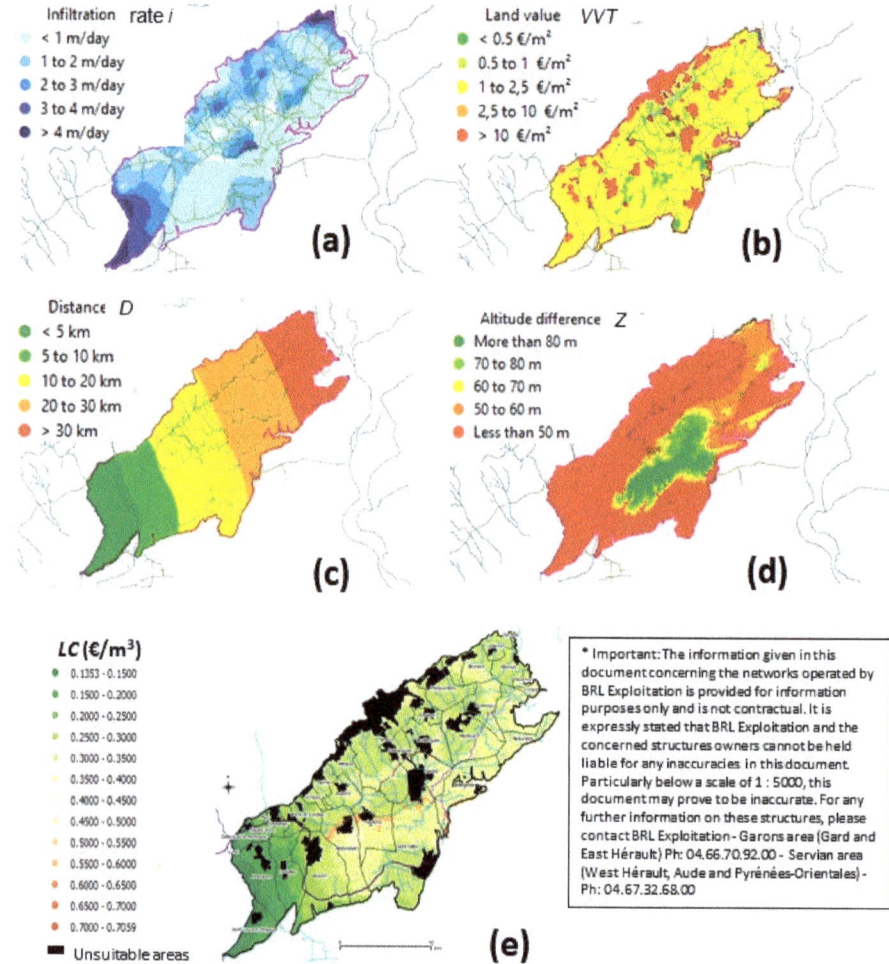

Figure 5. VCP case study: (**a**) map of soil infiltration rates i (m/day), (**b**) land market value LMV (€/m^2), (**c**) distance to one river D (Vidourle in this case), (**d**) elevation difference Z with closest river/stream, (**e**) minimum levelised cost LC map (in black: areas identified as unfavourable to infiltration basins due to land use constraints, most of the time corresponding to urban areas).

5. Discussion

5.1. Sensitivity Analysis

A systematic sensitivity analysis was performed to determine the effect of various parameters on the levelised cost of the MAR scheme. Sensitivity is defined as the rate of variation in one factor with respect to a variation in another factor. The normalized sensitivity is used to compare parameters and is defined as [14]:

$$S_{i,t} = \frac{\partial O}{\partial P_i / P_i} \qquad (2)$$

where $S_{i,t}$ is the normalized sensitivity of ith input parameter at time t, O is the output function of the system (i.e., the levelised cost in this case), and P_i is the ith input parameter of the system (in our case:

T, Q, r, D, etc.). The partial derivative of this equation can be approximated by a forward differencing formula as [15]:

$$\frac{\partial O}{\partial P_i} = \frac{O(P_i + \Delta P_i) - O(P_i)}{\Delta P_i} \quad (3)$$

The latest equation measures the influence that the fractional variation in a parameter, or its relative error, has on the output [15].

It must be specified that it is not a GIS-MCDA sensitivity analysis but a sensitivity analysis of the levelised cost of MAR scheme in the VCP case that has been computed for all the input parameters. They can be classified into three groups: (i) group 1 contains the highly sensitive parameters: life duration (T), recharge volume (Q), distance (D), and elevation difference (Z) between water uptake and infiltration basin, and water purchase and pretreatment costs; (ii) group 2 contains moderately sensitive parameters: yearly availability of water (N), soil infiltration rate (i), and discount rate (r); (iii) group 3 contains the lowly sensitive parameters: land market value (LMV), infiltration basin depth (p), duration between basin desilting (Nc), and thickness (Hc) of sands to scrab. The results for four of the most sensitive parameters are illustrated at Figure 6 with the following range of variation being explored: system life duration between 5 and 35 years, distance D between 0 and 5000 m, annual recharge volume between 0.01 and 5 Mm^3/year, and water treatment cost between 0 and 0.30 €/m^3. The cost sensitivity is positively linearly dependent on distance between abstraction and infiltration locations, and on water treatment cost, which tend to dominate the levelised cost beyond median values. The levelised cost appears to be highly dependent on annual recharge volume and life duration of the system, especially at low values of these parameters. This means, for example, that the levelised cost can be highly reduced by increasing the annual recharge volume up to 1 Mm^3/year ($\approx$0.2 of parameter range, x axis of graph at Figure 6). Regarding the system life duration, the minimum life duration should be above 20 years (>0.5 of parameter range).

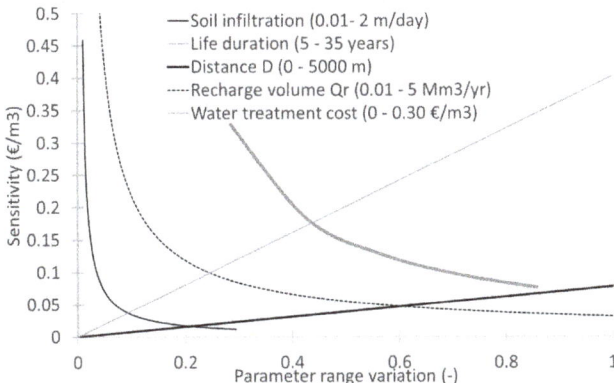

Figure 6. Levelised cost sensitivity analysis to water treatment cost, recharge water volume, MAR system life duration, and distance between the abstraction and recharge locations.

5.2. Approach Limitations and Outlines

This methodology relies on several assumptions regarding various costs and technical characteristics of the MAR scheme, which result in uncertainties in the final computed levelised cost screening. Therefore, the method should not be used as an accurate tool in a prefeasibility analysis but as a tool to compare several options concerning, for example, (i) the location of the MAR scheme, (ii) the water resource which will be used, or (iii) the required pretreatment processes. The tool can also be improved during the pre-project analysis as new data and information are collected in engineering studies.

The various criteria considered in the cost function are derived from various sources; economic and financial feedback; and, in some cases, are based exclusively on expert opinion, in the absence of information from the literature. It is therefore important to consider that the costs obtained at the end of this analysis are orders of magnitude, based on a certain number of assumptions. For this reason, the method should be used in a relative way, with the objective to site projects at locations with relatively low costs.

The cost assessment carried out in this study is similar to a cost–efficiency analysis (CEA), considering that all the volumes of water brought to the infiltration basin are recharged and stored in groundwater. No consideration was given to the capacity of the aquifer to store infiltrated water, nor the ability to retain it so that it can be recovered for high-valued uses. Part of the recharged water may enhance the discharge of groundwater to watercourses or other aquifers. Additional hydrogeological investigations are needed in order to evaluate the contribution of recharged water to intended economic and environmental benefits. The analysis carried out did not include the assessment of such benefits of MAR projects. This may require aquifer characterization and hydrodynamic modeling of the site.

6. Conclusions

This methodology was provided in order to approximate the levelised cost of an MAR scheme using an infiltration basin. Uncertainty in several input parameters and the lack of economic and financial feedback on MAR system costs introduce uncertainty into the calculated levelised cost.

The developed tool should be used as a way to identify the sensitivity of the cost for several input parameters, as well as guiding the sitting of the MAR device in a relative way. It should help in making decisions on the design of the MAR system. The method is general and can be applied in other contexts and other countries where information on MAR costs is available thanks to economic feedback.

Finally, this information on economical feasibility should be followed by more classical suitability analysis such as those relying on soil characteristics, aquifer storage capacity, and land use in order to properly site the MAR scheme. The levelised cost provides an effective means to compare the costs of MAR with alternative water projects.

Author Contributions: J.-C.M. and J.-D.R. had the original idea; M.B. and J.-C.M. conceived the methodology; F.M. and M.B. developed the tool; J.-F.D. developed the GIS application; F.M. applied the methods/tools and analysed the data; Y.C. contributed to materials and analysis tools; J.-C.M. wrote the paper. All authors have read and agreed to the published version of the manuscript.

Funding: This research was funded by Bureau de Recherches Géologiques et Minières (BRGM, French Geological Survey) and Agence de l'Eau Rhône-Méditerranée-Corse (Rhone Mediterranean Corsica Water Agency).

Acknowledgments: This study results from a scientific collaboration between Bureau de Recherches Géologiques et Minières (French Geological Survey) and Agence de l'Eau Rhône-Méditerranée-Corse (Rhone Mediterranean Corsica Water Agency).

Conflicts of Interest: The authors declare no conflict of interest.

Abbreviations

The following abbreviations are used in this manuscript:

MAR	managed aquifer recharge
MCDA	multi-criteria decision analysis
IC	investment cost
OC	operating cost
DEM	digital elevation model
GIS	geographic information system
VCP	Vistrenque and Costières Plain case study
AERMC	Rhone Mediterranean Corsica Water Agency
BRL	Bas-Rhône and Languedoc regional water company

References

1. Wada, Y.; Van Beek, L.P.H.; Van Kempen, C.M.; Reckman, J.W.T.M.; Vasak, S.; Bierkens, M.F.P. Global depletion of groundwater resources. *Geophys. Res. Lett.* **2010**, *37*. [CrossRef]
2. Aeschbach-Hertig, W.; Gleeson, T. Regional strategies for the accelerating global problem of groundwater depletion. *Nat. Geosci.* **2012**, *5*, 853–861. [CrossRef]
3. Dillon, P.; Stuyfzand, P.; Grischek, T.; Lluria, M.; Pyne, R.D.G.; Jain, R.C.; Bear, J.; Schwarz, J.; Wang, W.; Fernandez, E.; et al. Sixty years of global progress in managed aquifer recharge. *Hydrogeol. J.* **2019**, *27*, 1–30. [CrossRef]
4. David, R.; Pyne, G. *Groundwater Recharge and Wells*; Routledge: Boca Raton, FL, USA, 2017.
5. Stefan, C.; Ansems, N. Web-based global inventory of managed aquifer recharge applications. *Sustain. Water Resour. Manag.* **2018**, *4*, 153–162. [CrossRef]
6. Rahman, M.A.; Rusteberg, B.; Gogu, R.C.; Lobo Ferreira, J.P.; Sauter, M. A new spatial multi-criteria decision support tool for site selection for implementation of managed aquifer recharge. *J. Environ. Manag.* **2012**, *99*, 61–75. [CrossRef] [PubMed]
7. Sallwey, J.; Bonilla Valverde, J.P.; Vásquez López, F.; Junghanns, R.; Stefan, C. Suitability maps for managed aquifer recharge: A review of multi-criteria decision analysis studies. *Environ. Rev.* **2019**, *27*, 138–150. [CrossRef]
8. Sallwey, J.; Schlick, R.; Bonilla Valverde, J.P.; Junghanns, R.; Vásquez López, F.; Stefan, C. Suitability Mapping for Managed Aquifer Recharge: Development of Web-Tools. *Water* **2019**, *11*, 2254. [CrossRef]
9. Maliva, R.G. Economics of managed aquifer recharge. *Water* **2014**, *6*, 1257–1279. [CrossRef]
10. Ross, A.; Hasnain, S. Factors affecting the cost of managed aquifer recharge (MAR) schemes. *Sustain. Water Resour. Manag.* **2018**, *4*, 179–190. [CrossRef]
11. Escalante, E.F.; Gil, R.C.; Fraile, M.Á.S.M.; Serrano, F.S. Economic assessment of opportunities for Managed Aquifer recharge techniques in Spain using an advanced geographic information system (GIS). *Water* **2014**, *6*, 2021–2040. [CrossRef]
12. Dillon, P.; Arshad, M. Managed aquifer recharge in integrated water resource management. In *Integrated Groundwater Management: Concepts, Approaches and Challenges*; Springer International Publishing: Berlin/Heidelberg, Germany, 2016; pp. 435–452, ISBN 9783319235769.
13. Dillon, P.; Pavelic, P.; Page, D.; Beringen, H.; Ward, J. Managed Aquifer Recharge: An Introduction. 2009. Available online: http://hdl.handle.net/102.100.100/113803?index=1 (accessed on 20 February 2020).
14. Kabala, Z.J. Sensitiviby analysis of a pumping test on a well with wellbore storage and skin. *Adv. Water Resour.* **2001**, *24*, 483–504. [CrossRef]
15. Huang, Y.C.; Yeh, H. Der The use of sensitivity analysis in on-line aquifer parameter estimation. *J. Hydrol.* **2007**, *335*, 406–418. [CrossRef]

© 2020 by the authors. Licensee MDPI, Basel, Switzerland. This article is an open access article distributed under the terms and conditions of the Creative Commons Attribution (CC BY) license (http://creativecommons.org/licenses/by/4.0/).

Article

Dynamic Water Balance Modelling for Risk Assessment and Decision Support on MAR Potential in Botswana

Andreas Lindhe [1,*], Lars Rosén [1], Per-Olof Johansson [2] and Tommy Norberg [3]

1. Department of Architecture and Civil Engineering, Chalmers University of Technology, SE-412 96 Gothenburg, Sweden; lars.rosen@chalmers.se
2. Artesia Groundwater Consulting, Tunavägen 26, SE-186 41 Vallentuna, Sweden; per-olof.johansson@artesia.se
3. Mathematical Sciences, Chalmers University of Technology, SE-412 96 Gothenburg, Sweden; sten.tommy.norberg@gmail.com
* Correspondence: andreas.lindhe@chalmers.se; Tel.: +46-31-772-20-60

Received: 20 November 2019; Accepted: 3 March 2020; Published: 6 March 2020

Abstract: Botswana experiences a water stressed situation due to the climate and a continuously increasing water demand. Managed Aquifer Recharge (MAR) is considered, among other measures, to improve the situation. To evaluate the possibility for increased water supply security, a probabilistic and dynamic water supply security model was developed. Statistically generated time series of source water availability are used in combination with the dynamic storages in dams and aquifers, and the possible supply is compared with the demand to simulate the magnitude and probability of water supply shortages. The model simulates the system and possible mitigation measures from 2013 to 2035 (23 years), using one-month time steps. The original system is not able to meet the demand, and the estimated volumetric supply reliability in the year 2035 is 0.51. An additional surface water dam (now implemented) will increase the reliability to 0.88 but there will still be a significant water shortage problem. Implementing large-scale MAR can further improve the reliability to at least 0.95. System properties limiting the effect of MAR are identified using the model and show how to further improve the effect of MAR. The case study results illustrate the importance and benefit of using an integrated approach, including time-dependence and future scenarios, when evaluating the need and potential of MAR.

Keywords: water supply security model; risk assessment; decision support; dynamic; probabilistic; managed aquifer recharge; integrated water resource management

1. Introduction

Access to potable water is essential to human health and economic development. Water scarcity and drought are, however, major challenges on all continents [1] and must thus be managed to enable safe and secure access to clean water. Climate change, increased water demand and other factors will increase the problem of water scarcity, making this a key issue to reach the UN Sustainable Development Goals [2]. This calls for integrated water resources management, including measures to avoid water losses, the efficient use of water, the introduction of water saving technologies, as well as of water re-use and recycling.

Botswana is an example where the hydrological conditions and a continuously increasing water demand result in a water stressed situation. The arid to semi-arid climate provides a situation with low rainfall and high rates of potential evapotranspiration, resulting in low rates of surface runoff and low rates of natural groundwater recharge. To enable a reliable, safe, and sustainable water supply, it is widely accepted that a systems approach is needed, encompassing several water resources as well as technical and other mitigation measures. Managed Aquifer Recharge (MAR) to enhance groundwater recharge and store surface water (e.g., natural water, urban storm water, treated sewage, or desalinated water) [3,4] has, in several international studies, been pointed out as an important measure to mitigate water drought and scarcity situations [5–7]. MAR has also been identified as a possible measure to be implemented in Botswana [8,9]. However, to properly evaluate if MAR is a suitable option and the benefits it may provide, a holistic system analysis considering both current and future conditions is needed. Not only the potential storage and other aquifer specific properties should be determined, but also the reliability and access to water over time must be analysed in detail, considering risks and variations in water demand and supply. Hence, a risk-based approach is needed to provide decision support on water supply security and the effect of MAR. Water security is defined in different ways in the literature [10–12] but typically includes the water quantity aspect, which is the focus of this paper.

Different methods for selecting suitable sites for MAR exist and are typically based on GIS [13–15]. When designing the final MAR scheme and determining operational strategies etc., a detailed groundwater model is commonly used [16]. No generically accepted method exists, however, for assessing the overall effect of MAR on the water supply security. Several examples exist where system dynamic modelling [17] has been used to analyse the conjunctive use of surface water and groundwater. This approach is common when evaluating policy options [18,19] but examples including aquifer storage and recover exist [20]. Another type of modelling approach was applied by Clark et al. [21] to analyse the reliability of water supply from stormwater harvesting and MAR. A model combining stormwater hydrology with subsurface storage and demand was used and repeated simulations were performed to estimate the volumetric reliability. Gao et al. [22] analysed the reliability of an urban water supply using Monte Carlo analysis to model variations in surface water availability. Aquifer injection and banking was analysed with the aim to identify the most cost-effective way to meet specified criteria for supply reliability. These examples show the importance of creating models that include the entire system to be analysed and to compare the possible supply and water demand so that mitigation measures' effect on water security can be estimated.

An integrated and holistic approach is important to avoid overlooking interactions between subsystems and events, and to minimise the sub-optimisation of mitigation measures [23]. The importance of considering the entire supply system when assessing drinking water risks is emphasised by, for example, the World Health Organization (WHO) as part of a framework including Water Safety Plans (WSPs) [24]. A WSP is typically focused on water quality but the integrated approach is equally applicable to water supply security.

In this paper, a Water Supply Security Model (WSSM) is presented and applied to assess risks in the main drinking water supply in Botswana and to evaluate and provide decision support on the potential effects of MAR. The overall aim of the work was to enable a holistic assessment of the potential for using large-scale MAR to improve the water supply security in Botswana. The specific objectives were to (i) develop a comprehensive and dynamic water balance model, (ii) simulate and show the predicted water shortage over time, and (iii) model the potential of alternative mitigation measures and identify limiting factors.

2. Study Area

Botswana is located in south-central Africa and occupies an area of approximately 582,000 km^2. The total population of the nation is a little over 2 million, making Botswana one of the most sparsely populated countries in the world. The country is predominantly flat with some parts having a slightly rolling landscape. Botswana is dominated by the Kalahari Desert, covering about 70% of the total area. The north-east part of Botswana has an annual precipitation of about 600 mm, whereas the drier south-west receives, on average, only 200 mm per year. Due to the arid to semi-arid climate, potential evapotranspiration rates exceed the total rainfall at all times of the year [8]. There are no perennial streams originating in Botswana. It is estimated that the mean annual rates of surface runoff do not exceed 50 mm anywhere except in small steep rocky catchments. The annual recharge to aquifers from rainfall reaches a maximum of about 40 mm in small areas in the Chobe District in the north. For most of the Kalahari region, the natural groundwater recharge is less than 1 mm/year [8]. The largest groundwater resources are in the Kalahari sediments (including the Okavango Delta), the Ntane Sandstone, the Ecca Sandstones, and the Damaran and Ghanzi rock formations.

In eastern and southern Botswana, with its relatively high density of population and substantial water demand, several surface water dams have been constructed to collect and store ephemeral river flow. The largest dams are the Shashe, Dikgatlhong, Letsibogo, and Gaborone dams. The storages of the dams are very variable due to the highly seasonal, occasional, and variable river flows. In addition, the need to store water for drought periods and the flat topography in most areas result in large losses of water to evaporation from these dams. The surface water dams in eastern Botswana have been connected through a nearly 400-km long pipeline transfer system denoted the North–South Carrier (NSC), providing possibilities to transfer water to urban centres.

The NSC water supply system is the focus of this study and the included demand centres, surface water dams, aquifers and additional components are shown in Figure 1. The study was performed in the year 2013 and the descriptions of the system, planned measures, etc., are thus based on the situation at that time. The Dikgatlhong Dam was being constructed when the study was performed and thus included in one of the modelled scenarios to represent the coming system structure. In addition to the surface water dams, a few groundwater wellfields are connected or are planned to be connected to demand centres supplied with water from the NSC, e.g., Palla Road, Chepete, Masama, Makhujwane, Malotwane, and Palapye Wellfields.

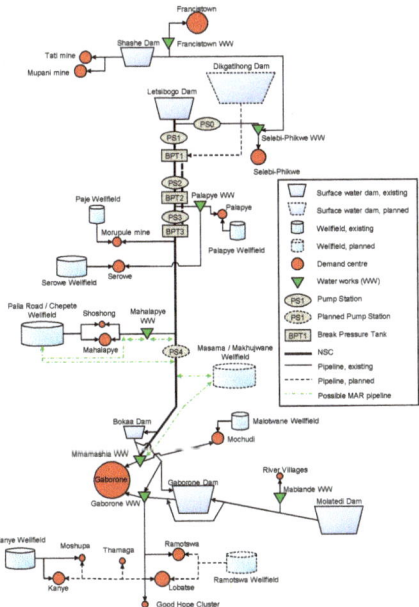

Figure 1. Schematic illustration of the water supply system linked to the North–South Carrier (NSC), according to the situation when the study was performed in the year 2013.

Due to the highly variable storage in surface water dams, the groundwater aquifers have a potential to support the NSC demand centres during drought periods. Because of the very limited natural recharge to these aquifers, their long-term sustainable capacity could be improved by managed recharge with surface water. Managed recharge (injection) with collected and treated surface water from dams may also reduce the total loss of water to evaporation. The focus of the case study is thus to evaluate the possibility and effects on the water supply security of MAR scenarios including the Palla Road/Chepete and Masama/Makhujwane Wellfields.

3. Materials and Methods

3.1. Water Supply Security Model (WSSM)

The WSSM is a dynamic water balance model where statistically generated time series of the availability of source water are used, together with dynamic storages in dams and aquifers, as well as water demands, to simulate the magnitude and probability of water supply shortages. Models have been previously developed for the water supply in Botswana but they have not considered MAR scenarios [8]. The WSSM is developed as a spreadsheet model in Excel since one of the goals of this study was to provide an easily accessible model that can be run without expert knowledge. To enable statistical analysis considering uncertainties in input data and results, an add-in software (Oracle ®Crystal Ball) is used to run Monte Carlo simulations.

The WSSM simulates the NSC system and connected components from 2013 to 2035 (23 years). The period is selected to match water demand forecast in the National Water Master Plan Review [8]. The simulations are performed with a time step of one month and for each month the demand, the available storage in dams and aquifers, as well as treatment capacities, water losses, etc., are considered.

The schematic illustration in Figure 2 shows the parameters considered in the model and the link to the model components. Based on historical data on inflow to the dams (see Section 3.2), a set of possible time series are generated and used to sample from when running the model. The generated time series consider the correlation between the dams and each generated data set includes all five dams. The annual inflow data is transformed into monthly data based on the closest historical annual inflow and the monthly distribution that year. Since the dams are spatially correlated, the historical data for the Gaborone Dam is used when transforming the simulated annual data for the Bokaa and Gaborone Dams. In the same way, the historical data for the Dikgatlhong Dam is used for the Letsibogo, Shashe, and Dikgatlhong dams.

For each dam, water balance calculations are performed for each month, considering initial storage, inflow, abstraction, evaporation, seepage, spill over, and additional parameters presented in Figure 2. The input data is based on [8] and information from personnel at the Water Utility Corporation (WUC) and the Department of Water Affairs (DWA). The evaporation is calculated based on the area–storage relationship and data from previous studies in the area [8,25]. Key inputs for modelling the dams in the WSSM are presented in Table 1.

Table 1. Storage properties and environmental flows for the dams included in the Water Supply Security Model (WSSM).

Dam Property	Gaborone	Bokaa	Letsibogo	Dikgatlhong	Shashe
Maximum storage (Mm^3)	140.59	18.20	108.00	397.60	75.05
Minimum operational storage (% of max storage)	15	3	5	4	15
Initial storage, 1 Jan 2013 (Mm^3)	49.65	7.08	31.00	198.80	59.62
Seepage (L/s)	0.0	0.0	0.0	30.0	0.8
Environmental flow, Feb; Nov	0; 0 *	0; 0 *	2.49; 0.49 **	5; 5 *	0; 0 *
Loss due to sedimentation (Mm^3/year)	0.22	0.04	0.35	0.70	0.61

* %; ** Mm^3, provided inflow > 0.

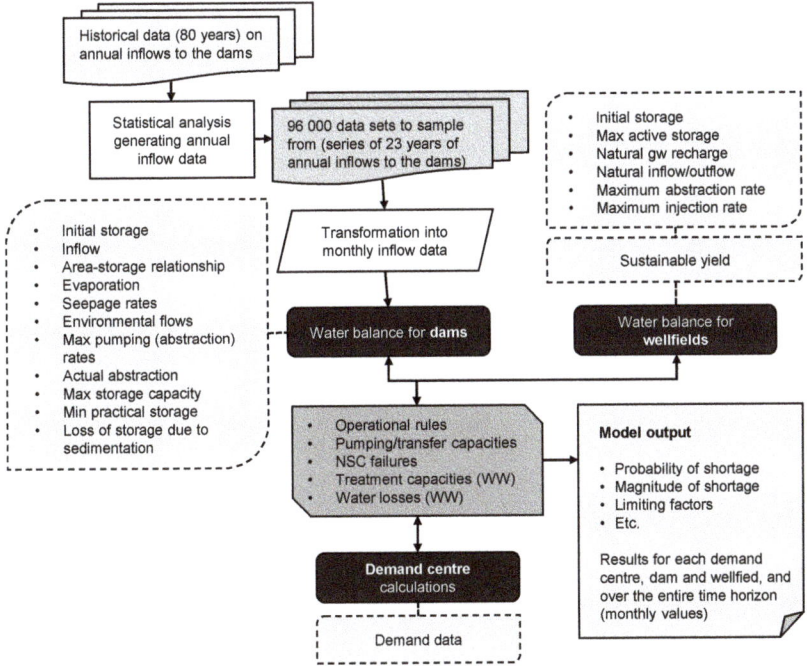

Figure 2. Overview of the parameters considered in the model and the link between them.

The water balance calculations for MAR wellfields are performed considering initial storage, natural recharge, inflow, injection, outflow, and abstraction. Necessary input data for the aquifers are based on [26–29], and the key figures used in the scenarios modelled here (see Section 3.3) are presented in Table 2. The wellfields are recharged by injecting water if the maximum storage is not reached and provided that water is available in the dams and is at a capacity to abstract, treat, transfer, and inject the water. A critical dam storage level (20%) was defined by the WUC and DWA and is used as an operational rule in the model stating that water for injection may only be abstracted when the dam storage is above this level. The abstraction of water from the MAR wellfields starts when the water demand cannot be met by the supply of treated surface water and the non-MAR wellfields. The abstraction is only limited by the abstraction rate and capacity to treat and distribute the water.

Table 2. Input data on Managed Aquifer Recharge (MAR) wellfields to the WSSM.

Wellfield Property	Palla Road/Chepete	Masama/Makhujuwane
Maximum active storage	42.8	40.0
Initial storage (Mm^3)	40.5	40.0
Natural groundwater recharge (m^3/d)	12,383	3379
Groundwater inflow (m^3/d)	0	0
Groundwater outflow (m^3/d)	7600–12,000 (function of active storage)	0–3600 (function of active storage)
Maximum abstraction and injection rates (m^3/d)	21,342	37,734

For the wellfields not considered relevant for MAR, an estimated sustainable yield is used as a maximum abstraction rate to not cause groundwater mining. The sustainable yield is estimated on an annual basis and defined in the model as a monthly maximum abstraction that may not be exceeded.

The total sustainable yield for the non-MAR wellfields is 7.2 Mm^3, and 11.7 Mm^3 when also including the Masama/Makhujuwane and Palla Road/Chepete Wellfields as non-MAR wellfields.

The water demand forecast from [8] is used to determine how much water must be abstracted from the dams and wellfields. The estimated change in water demand is based on a population forecast and an increase in specific water demand, including assumptions of changes from standpipe to yard and yard to house connections. The assumed industrial, commercial, and institutional annual growth rate is 3%. Unaccounted-for water, including technical losses and non-technical losses (unmetered consumption and illegal connections), is also considered in the forecast. Since the reported water consumption in 2012 was 4 Mm^3/year lower than the forecast for the same year, the original demand forecast was reduced by this volume. The total annual demand for the demand centres included in the WSSM is 81 Mm^3 in 2013 and 148 Mm^3 in 2035.

In addition to the capacities and other parameters presented above, a set of operational rules are used to determine, for example, when different sources are used and to what demand centres and the extent to which water is supplied. In all calculation steps, available abstraction rates, treatment capacities, water losses during treatment etc., are considered. The model is thus not used to optimize the supply from different sources but to estimate the performance based on the actual operational rules used to manage the system. The schematic illustration in Figure 1 shows how the different components of the system are connected and how water can be transferred. The actual supplied amount of water is compared with the demand for each demand centre and possible shortage etc., is calculated in each time step.

3.2. Dam Inflow Time Series

The time series of monthly inflows to the five dams (Gaborone, Bokaa, Letsibogo, Dikgatlhong and Shashe) are available based on measurements and hydrological modelling for the 80-year period of 1925 to 2004 [8] (vol. 11). The dams are grouped based on their spatial correlation and the annual inflows are presented in Figure 3. An analysis of the annual inflows is made to generate 96,000 future annual inflow time series (23 years) that are used to sample from when running the WSSM. This five-dimensional time series is modelled as a first-order stationary Gaussian Auto-Regressive, AR(1) sequence:

$$y_t - \mu = \Phi(y_{t-1} - \mu) + \varepsilon_t \quad (1)$$

where the column vector y_t is the annual inflow and $t = -79, -78, \ldots, 0$ ($t = 0$ corresponds to year 2004). The column vectors μ and ε_t denote the long-time yearly mean and white noise, respectively, the latter with covariance matrix Σ. To carry out a standard least squares (LS) estimation, the model is rewritten as follows:

$$y_t = \Phi y_{t-1} + b + \varepsilon_t \quad (2)$$

where $b = (I - \Phi)\mu$ (I denotes the identity). The model parameters Φ, b, and Σ are estimated by the method of least squares. Φ is a 5 by 5 coefficient matrix and y, b and ε_t are 5-dimensional column vectors. Also estimated is the spatial covariance matrix:

$$\gamma = E(y_t - \mu)(y_t - \mu)' \quad (3)$$

where the prime ′ denotes transpose. Future dam inflow values $y_1, y_2, \ldots, y_{23}$ are then repeatedly simulated from the estimated model, taking the uncertainty of the LS estimates into account.

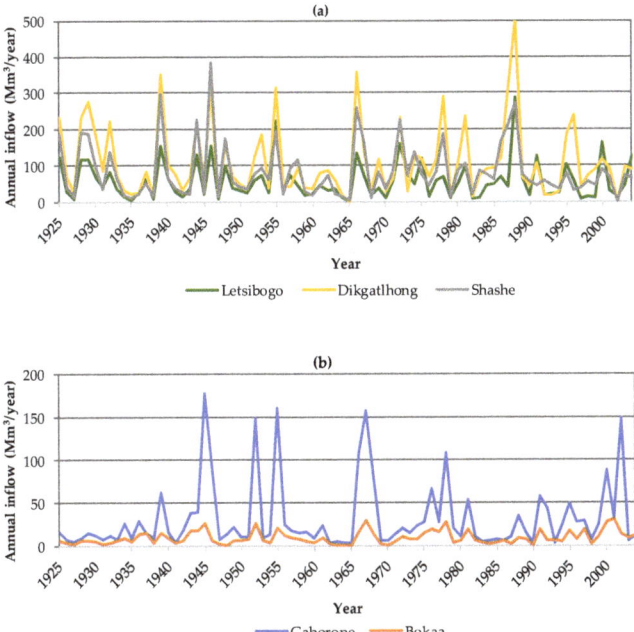

Figure 3. Historic inflow time series (80 years, 1925–2004) for the dam sites: (**a**) Letsibogo, Dikgatlhong and Shashe; (**b**) Gaborone and Bokaa [8].

3.3. Scenarios

The system included in the analysis and implemented in the WSSM includes: 6 surface water dams, 8 wellfields, 7 water works, and 18 demand centres. In addition to five dams previously presented, the Molatedi Dam in South Africa is also included in the model. A constant supply (80% of the maximum agreed transfer) from the dam is assumed since no historical time series are available. The same assumption has previously been used when evaluating the supply system [8]. The selection of wellfields for MAR scenarios was based on [9,26–33] and workshops including representatives from the DWA, WUC, and the authors. Several scenarios including MAR and non-MAR wellfields are possible, but we here focus on the three scenarios listed below as a basis for evaluating the potential of using large scale MAR in Botswana. The system structure in 2013 (Scenario A), i.e., when the study was performed, is used as a reference to illustrate the need and potential effects of mitigation measures (Scenarios B and C) on the water supply security. As mentioned above, the Dikgatlhong Dam was under construction when the study was performed and is now in operation. Hence, the dam is included in Scenario B and combined with MAR in Scenario C. The purpose is not to evaluate MAR as an alternative measure to the Dikgatlhong Dam but to see how MAR can further improve the system. To facilitate a relative comparison of the mitigation measures, i.e., the Dikgatlhong Dam and the MAR wellfields, they are included from the start of the simulated period (i.e., year 1).

A. Current system (as in year 2013)
B. Current system + Dikgatlhong Dam (implemented after the study)
C. Current system + Dikgatlhong Dam + Palla Road/Chepete Wellfields MAR + Masama/Makhujwane Wellfields MAR

4. Results

The results from the WSSM show that the supply system in Scenario A is clearly insufficient to meet the water demand within the simulated period of 23 years. Water shortage is likely to be a problem early in the simulated period and is expected in around 70% (mean value) of the months for most demand centres. Given a month with water shortage, the deficit varies between c. 20–60% of the demand. The water supply security can be assessed based on the volumetric reliability, i.e., the volume of water supplied divided by the demand in a given year. The results are presented in Figure 4 and show that the reliability is dramatically reduced for Scenario A over the simulated period. There is no reliability target level defined in Botswana but at the end of the simulated period the level is only 0.51 (mean value). As a comparison, case studies in Australia [21,22] have applied a 0.995 volumetric reliability target for potable supplies and a 0.95 target level for non-potable use.

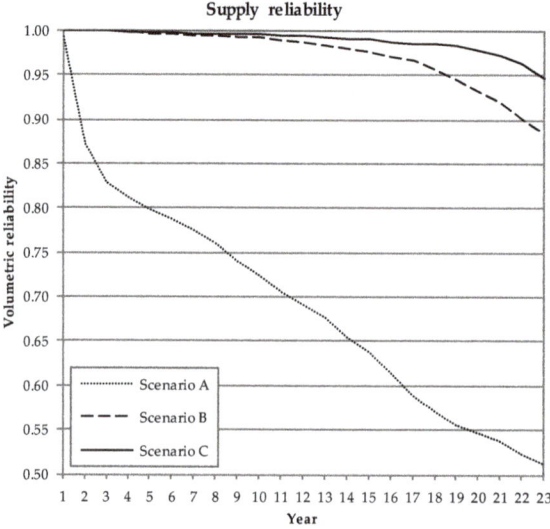

Figure 4. Annual volumetric reliability (mean value) of the supply to Great Gaborone over the simulated period (23 years).

In Figure 5, the expected (mean) probability of annual water shortage of different magnitudes is presented for Gaborone and the demand centres connected to the capital city, here referred to as Great Gaborone. The results are similar for most of the demand centres, and, for evaluating the potential effect of MAR, we focus on Great Gaborone. The total demand for Great Gaborone will increase over the analysed period from 44.8 (year 2013) to 82.4 Mm^3/year (year 2035). The water demand for Great Gaborone constitutes 62% of the total demand in the NSC system.

The connection of the Dikgatlhong Dam to the NSC and the related system upgrades (Scenario B) will have a large positive effect on the supply security, see Figure 6, and reduce the expected total water shortage (summed over the 23 years) by approximately 90%. The supply reliability will increase (Figure 4) and be >0.99 for approximately 10 years. However, in 2035, the reliability is estimated to be 0.88 and the results thus show that there still will be a significant risk for water shortage for Great Gaborone during the late part of the simulation period.

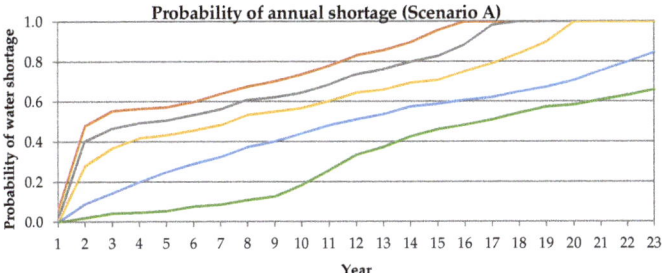

Figure 5. Expected probability of annual water shortages of different magnitudes in Great Gaborone for each year in Scenario A.

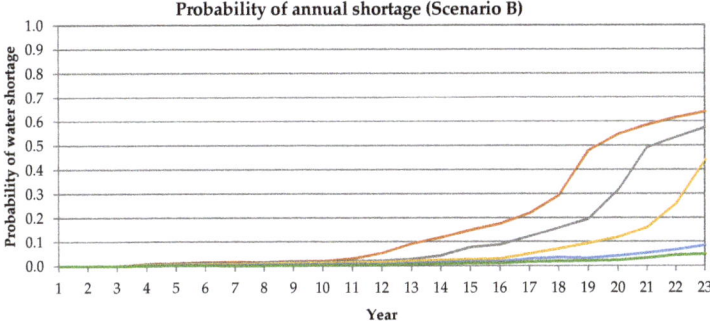

Figure 6. Expected probability of annual water shortages of different magnitudes in Great Gaborone for each year in Scenario B, including the Dikgatlhong Dam.

In Figure 7, the results are presented for Scenario C, i.e., including the MAR wellfields. The probability of shortage is further reduced compared to Scenario B. For example, the probability of having a 10% (8 Mm^3) water shortage in Great Gaborone in 2035 is reduced from 40% (Scenario B) to 10%. For Scenario C, the supply reliability is >0.99 in approximately 15 years and is estimated to be 0.95 in 2035. The effects of implementing MAR are, however, limited due to the capacity of different system components. As an example, Figure 8 shows what limiting the injection of water from the dams at the Palla Road Wellfield. Injection is expected to be needed in 41% of the months. In 34% of these months, injection up to the full storage or maximum injection rate is obtained. However, the dam storage (abstraction of water for injection only allowed if dam storage is >20%), the abstraction rate from dams, and the capacity of the treatment plants (only treated water is injected) are limiting the injection in 45%, 17%, and 4% of the cases, respectively. If the most critical technical system properties causing the limitations are eliminated, the positive effects of MAR in Scenario C further increase. For example, the supply reliability in 2035 will increase to 0.97 and the probability of an annual water deficit of 2 Mm^3 (2.5% of the demand) will reduce from 32% to 10%. This effect can be obtained without any substantial risk of mining the wellfields. The maximum active storage for the Masama/Makhujwane and Palla Road/Chepete Wellfields are 40 and 42.8 Mm^3, respectively (Table 2). The probability of having full storage at the end of the simulated period is 0.8 for both wellfields in Scenario C when the key limiting factors have been eliminated. If the limitations are included, the probability of full storage is 0.2 and 0.5, respectively.

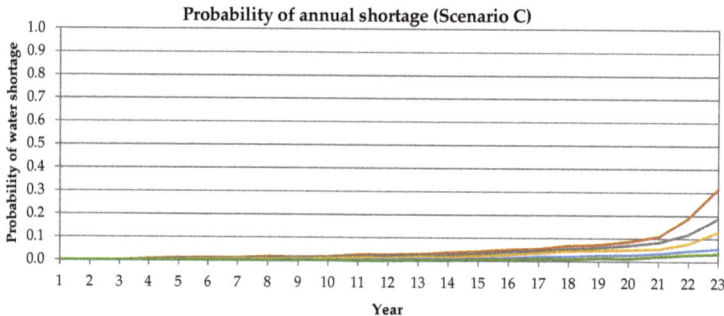

Figure 7. Expected probability of annual water shortages of different magnitudes in Great Gaborone for each year in Scenario C, including Masama/Makhujwane and Palla Road/Chepete, MAR.

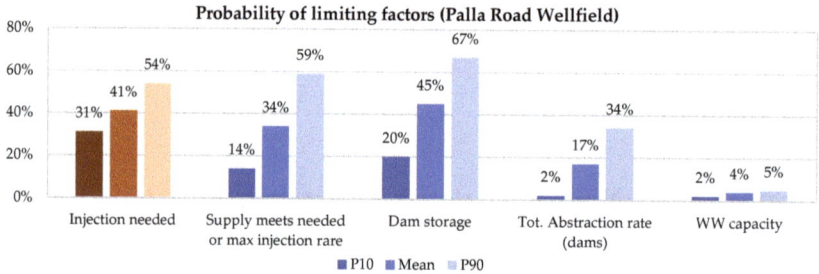

Figure 8. Factors limiting the injection at Palla Road Wellfield, Scenario C.

Provided that no water is abstracted and the maximum injection can be applied, it would take 5–6 years to recharge the MAR wellfield from 0 Mm^3 to full storage. If dependent on natural groundwater recharge only, the wellfields will only be recharged up to approximate half of the maximum storage after the simulated 23 years.

5. Discussion and Conclusions.

The performed case study addresses the possible effects of implementing MAR in the NSC system in Botswana. It is concluded that the NSC system including the Dikgatlhong Dam (Scenario B) but without MAR is not likely to be able to provide a safe water supply over the entire time period. When the water demand increases, the reliability of the system is reduced. Implementing MAR at the Palla Road/Chepete and Masama/Makhujwane Wellfields (Scenario C) will further improve the system, although not eliminate the risk of future water shortage. However, implementation of MAR may be of great importance in managing the water supply situation in eastern Botswana.

By including the Dikgatlhong Dam in the system (Scenario B), an additional water source is added and the potential volume of water that may be accessible is increased. Due to the spatial correlation between the dams, periods of no or limited inflow may affect several dams at the same time and thus cause a severe water shortage. By including the MAR wellfields, we add components that are dependent on the surface water dams to be recharged but the water stored in the aquifers can be used independently of the dams. The latter part is one of the key reasons for the increased reliability in Scenario C. The results show that over time there is enough water in the dams to be able to recharge the MAR wellfields. However, the possibility to inject water is partly limited in Scenario C due to both the capacity of the system components and access to the surface water. If these limitations are reduced or eliminated, the positive effect of implementing MAR will further increase.

The non-MAR wellfields are operated using the estimated sustainable yields as a maximum monthly abstraction rate. If this criterion would have been defined on an annual basis and allowed a varying abstraction over the year, it is possible that some smaller shortage events could have been avoided. Another criterion for sustainable yield could be implemented into the WSSM but would, however, not have a significant effect on the more severe shortage events. By allowing the same abstraction rates for the Palla Road/Chepete and Masama/Makhujwane Wellfields, as in Scenario C, but without MAR, the system would improve but only for a limited time. Due to the limited natural groundwater recharge, the aquifers would gradually be emptied. MAR is thus needed in order to provide for a long-term sustainable solution.

The results from the modelled scenarios do not show any cases where the implementation of MAR reduces the supply system security in Botswana. This could be the case if, for example, there are large losses of water or the abstraction rate is too low compared to the surface water dams.

The developed WSSM enables a thorough analysis and evaluation of the original system as well as the effect of both MAR implementation and other system changes. A key advantage of the model is the ability to not only model the possible enhanced groundwater recharge over time but to compare the possible supply and demand for all system components. The predicted water shortage over time in combination with dam and groundwater levels provides a comprehensive picture of the system performance. This makes it possible to evaluate system reliability with consideration to the entire system and compare MAR scenarios with other possible measures to avoid the sub-optimisation of risk mitigation measures.

Furthermore, system properties limiting the effects of MAR can be identified as shown in the case study. This result can guide further analysis and improvements to enhance the effect of implementing MAR. Hence, the WSSM can provide results to support decisions on both larger system changes and minor upgrading to improve water supply security.

The main conclusions of this study are:

- The developed WSSM is a comprehensive and dynamic water balance model that enables a thorough analysis of recharge and abstraction from MAR wellfields as well as the overall system reliability.
- The case study application demonstrates the practical applicability of the model and shows that useful decision support is provided.
- Future development of the WSSM approach will include consideration to climate change in dam inflow data and other relevant system parameters.
- The WSSM provides added value by enabling an integrated approach, including time-dependence and future scenarios, when evaluating the need and potential of implementing MAR.

Author Contributions: The study was initiated by all authors. L.R. was the project leader and P.-O.J. analysed and estimated input data related to aquifers and water demand. A.L. created the water supply security model with support from the other authors. T.N. analysed and modelled time series of inflows to surface water dams. A.L. performed the calculations and was the main author of the paper. All authors have read and agreed to the published version of the manuscript.

Funding: This research was funded by the Swedish International Development Cooperation Agency (Sida) and the Botswana Department of Water Affairs (DWA).

Acknowledgments: The authors gratefully acknowledge the contributions of the DWA and the Water Utility Corporation (WUC) in Botswana.

Conflicts of Interest: The authors declare no conflict of interest.

References

1. UN WATER–Water Scarcity. Available online: http://www.unwater.org/water-facts/scarcity/ (accessed on 26 February 2019).
2. UN Sustainable Development Goals. Available online: https://www.un.org/sustainabledevelopment/ (accessed on 26 February 2019).

3. Dillon, P.; Arshad, M. Managed aquifer recharge in integrated water resource management. In *Integrated Groundwater Management: Concepts, Approaches and Challenges*; Jakeman, A.J., Barreteau, O., Hunt, R.J., Rinaudo, J.-D., Ross, A., Eds.; Springer International Publishing: Cham, Switzerland, 2016; pp. 435–452.
4. Maliva, R.; Missimer, T. Overview and impacts on arid land water resources. In *Arid Lands Water Evaluation and Management*; Springer Berlin Heidelberg: Berlin/Heidelberg, Germany, 2012; pp. 975–999.
5. Murray, R.; Tredoux, G.; Ravenscroft, P.; Botha, F. *Artificial Recharge Strategy*; Version 1.3; Department of Water Affairs and Forestry & Water Research Commission: Pretoria, South Africa, June 2007.
6. Murray, E.C.; Tredoux, G. *Artificial Recharge: A Technology for Sustainable Water Resource Development*; Report No 842/1/98; Water Research Commission: Pretoria, South Africa, 1998.
7. Dillon, P.; Stuyfzand, P.; Grischek, T.; Lluria, M.; Pyne, R.D.G.; Jain, R.C.; Bear, J.; Schwarz, J.; Wang, W.; Fernandez, E.; et al. Sixty years of global progress in managed aquifer recharge. *Hydrogeol. J.* **2019**, *27*, 1–30. [CrossRef]
8. Department of Water Affairs. *National Water Master Plan. Review*; Ministry of Minerals, Energy & Water Resources, Government of Botswana: Gaborone, Botswana, 2006.
9. Groundwater Africa. *Managed Aquifer Recharge (MAR): Support to the Department of Water Affairs, Botswana, Final report*; Groundwater Africa: Somerset West, South Africa, 2012.
10. Grey, D.; Sadoff, C.W. Sink or Swim? Water security for growth and development. *Water Policy* **2007**, *9*, 545–571. [CrossRef]
11. Cook, C.; Bakker, K. Water security: Debating an emerging paradigm. *Glob. Environ. Chang.* **2012**, *22*, 94–102. [CrossRef]
12. Gerlak, A.K.; House-Peters, L.; Varady, R.G.; Albrecht, T.; Zúñiga-Terán, A.; de Grenade, R.R.; Cook, C.; Scott, C.A. Water security: A review of place-based research. *Environ. Sci. Policy* **2018**, *82*, 79–89. [CrossRef]
13. Rahman, M.A.; Rusteberg, B.; Gogu, R.C.; Lobo Ferreira, J.P.; Sauter, M. A new spatial multi-criteria decision support tool for site selection for implementation of managed aquifer recharge. *J. Environ. Manag.* **2012**, *99*, 61–75. [CrossRef] [PubMed]
14. Russo, T.A.; Fisher, A.T.; Lockwood, B.S. Assessment of managed aquifer recharge site suitability using a GIS and modeling. *Groundwater* **2015**, *53*, 389–400. [CrossRef] [PubMed]
15. Sallwey, J.; Schlick, R.; Bonilla Valverde, J.P.; Junghanns, R.; Vásquez López, F.; Stefan, C. Suitability Mapping for Managed Aquifer Recharge: Development of Web-Tools. *Water* **2019**, *11*, 2254. [CrossRef]
16. Ringleb, J.; Sallwey, J.; Stefan, C. Assessment of Managed Aquifer Recharge through Modeling—A Review. *Water* **2016**, *8*, 579. [CrossRef]
17. Sterman, J.D. *Business Dynamics: System Thinking and Modelling for Complex World*; McGraw-Hill: New York, NY, USA, 2000.
18. Qaiser, K.; Ahmad, S.; Johnson, W.; Batista, J.R. Evaluating water conservation and reuse policies using a dynamic water balance model. *Environ. Manag.* **2013**, *51*, 449–458. [CrossRef] [PubMed]
19. Wang, K.; Davies, E.G.R.; Liu, J. Integrated water resources management and modeling: A case study of Bow river basin, Canada. *J. Clean. Prod.* **2019**, *240*, 118242. [CrossRef]
20. Niazi, A.; Prasher, S.O.; Adamowski, J.; Gleeson, T. A system dynamics model to conserve arid region water resources through aquifer storage and recovery in Conjunction with a Dam. *Water* **2014**, *6*, 2300–2321. [CrossRef]
21. Clark, R.; Gonzalez, D.; Dillon, P.; Charles, S.; Cresswell, D.; Naumann, B. Reliability of water supply from stormwater harvesting and managed aquifer recharge with a brackish aquifer in an urbanising catchment and changing climate. *Environ. Model. Softw.* **2015**, *72*, 117–125. [CrossRef]
22. Gao, L.; Connor, J.D.; Dillon, P. The economics of groundwater replenishment for reliable urban water supply. *Water* **2014**, *6*, 1662–1670. [CrossRef]
23. Lindhe, A.; Rosén, L.; Norberg, T.; Bergstedt, O. Fault tree analysis for integrated and probabilistic risk analysis of drinking water systems. *Water Res.* **2009**, *43*, 1641–1653. [CrossRef] [PubMed]
24. WHO. *Guidelines for Drinking-Water Quality*; Fourth edition incorporating first addendum; World Health Organization: Geneva, Switzerland, 2017.
25. SMEC. *Study of Open Water Evaporation in Botswana*; Final Report; SMEC for DWA: Melbourne, Australia, 1987.
26. Wellfield Consulting Services & CIC Energy. *Mmamabula Energy Project. Bankable Feasibility Study*; Kudumatse Groundwater Resources Report; Wellfield Consulting Services/CIC Energy: Gaborone, Botswana, 2007.

27. Geo World. *Masama Groundwater Resources Evaluation Project–Groundwater modelling*; Final Report; Geo World: Gaborone, Botswana, 2009.
28. Water Resources Consultants. *Post-Auditing of the Palla Road Groundwater Model (Palla Road & Chepete Wellfields)*; Draft Modelling Report; Water Resources Consultants: Gaborone, Botswana, 2012.
29. Water Resources Consultants. *Post-Auditing of the Palla Road Groundwater Model (Palla Road & Chepete Wellfields)*; Final Modelling Report & Wellfield Report; Water Resources Consultants: Gaborone, Botswana, 2013.
30. Geotechnical Consulting Services. *Review of Monitoring Performed by DWA and DGS—Assessment of Water Resources and Improvement of Techniques*; Geotechnical Consulting Services: Gaborone, Botswana, 2000.
31. Geotechnical Consulting Services. *Kanye Emergency Works Water Supply Project*; Final Report, Appendix 3 Groundwater Modelling; Geotechnical Consulting Services: Gaborone, Botswana, 2006.
32. Water Surveys Botswana. *Costing of Botswana Water Supply and Demand Clusters Investigation*; Water Surveys Botswana: Gaborone, Botswana, 2008.
33. Water Surveys Botswana. *Gaotlhobogwe Wellfield Borehole Replacement Emergency Project*; Final Report, Volume 1 Main Report & Volume 3 Modelling; Water Surveys Botswana: Gaborone, Botswana, 2010.

© 2020 by the authors. Licensee MDPI, Basel, Switzerland. This article is an open access article distributed under the terms and conditions of the Creative Commons Attribution (CC BY) license (http://creativecommons.org/licenses/by/4.0/).

Article

Managed Aquifer Recharge in Africa: Taking Stock and Looking Forward

Girma Y Ebrahim [1,*], Jonathan F. Lautze [2] and Karen G. Villholth [2]

1   International Water Management Institute, P.O. Box 5689, Addis Ababa, Ethiopia
2   International Water Management Institute, Private Bag X813, Silverton, Pretoria 0127, South Africa; J.Lautze@cgiar.org (J.F.L.); K.Villholth@cgiar.org (K.G.V.)
*   Correspondence: g.ebrahim@cgiar.org

Received: 20 November 2019; Accepted: 8 June 2020; Published: 27 June 2020

Abstract: Climatic variability and change result in unreliable and uncertain water availability and contribute to water insecurity in Africa, particularly in arid and semi-arid areas and where water storage infrastructure is limited. Managed aquifer recharge (MAR), which comprises purposeful recharge and storage of surface runoff and treated wastewater in aquifers, serves various purposes, of which a prominent one is to provide a means to mitigate adverse impact of climate variability. Despite clear scope for this technology in Africa, the prevalence and range of MAR experiences in Africa have not been extensively examined. The objective of this article is provide an overview of MAR progress in Africa and to inform the potential for future use of this approach in the continent. Information on MAR from 52 cases in Africa listed in the Global MAR Portal and collated from relevant literature was analyzed. Cases were classified according to 13 key characteristics including objective of the MAR project, technology applied, biophysical conditions, and technical and management challenges. Results of the review indicate that: (i) the extent of MAR practice in Africa is relatively limited, (ii) the main objective of MAR in Africa is to secure and augment water supply and balance variability in supply and demand, (iii) the surface spreading/infiltration method is the most common MAR method, (iv) surface water is the main water source for MAR, and (v) the total annual recharge volume is about 158 Mm^3/year. MAR schemes exist in both urban and rural Africa, which exemplify the advancement of MAR implementation as well as its out scaling potential. Further, MAR schemes are most commonly found in areas of high inter-annual variability in water availability. If properly planned, implemented, managed, maintained and adapted to local conditions, MAR has large potential in securing water and increasing resilience in Africa. Ultimately, realizing the full potential of MAR in Africa will require undertaking hydrogeological and hydrological studies to determine feasibility of MAR, especially in geographic regions of high inter-annual climate variability and growing water demand. This, supported by increased research to gauge success of existing MAR projects and to address challenges, would help with future siting, design and implementation of MAR in Africa.

Keywords: managed aquifer recharge; water security; climate change; Africa

1. Introduction

Challenges posed by climate variability and change in Africa are widely recognized [1]. Sadoff et al. [2] highlight how rainfall variability disrupts productivity of rain-fed agriculture, contributes to disasters associated with floods and droughts, and stalls economic growth. An African Ministers' Council on Water (AMCOW) report [3] state that recurrent droughts in sub-Saharan Africa are a dominant climate risk that compromises livelihoods, water and food security and exerts a major negative effect on Gross Domestic Product (GDP) growth in one third of the continent's countries.

Groundwater currently contributes water supply to increasing populations in Africa, while also presenting a resource with significant potential for further development for multiple uses in parts of the continent [4]. Particularly in arid and rural areas, groundwater plays a key role in enhancing resilience. However, with increasing climate variability, frequency of extreme events, and population growth, water development in Africa needs to go forward in ways that consider all water sources in conjunction to enhance sustainable, reliable, climate-smart and equitable water availability and access [5].

Managed aquifer recharge (MAR) is a water management option that provides a means of intentionally recharging and storing water underground for subsequent recovery and beneficial use. It provides an important buffer against the impacts of climate variability and change, especially when additional water is recharged during wet or flooding periods for subsequent abstraction during dry or drought periods [6]. MAR has been identified to hold particular potential in arid and semi-arid areas where the control and storage of increasingly irregular surface runoff is challenging [7]. According to Tuinhof et al. [8], MAR is a significant adaptation option for coping with climate change and hydrological variability. Importantly, MAR provides water storage that is generally better protected against loss from evaporation [9,10]. MAR may take advantage of sources other than surface water runoff, e.g., by using treated wastewater, which often provides a more reliable source, while requiring stricter control of water quality [11].

Despite the large scope for MAR in Africa, there has been scant effort to take stock of MAR implementation and experience in the continent and to assess further potential with the exception of project report by Ebrahim et al. [12]. Dillon et al. [13] provided an overview of MAR in the Southern Africa, while Murray [14] provided an introductory guide to MAR in the Southern African Development Community (SADC) for the Groundwater Management Institute (SADC-GMI) and assessed the MAR case in the city of Windhoek, Namibia [15]. Similarly, Bugan et al. [16] and Jovanovic et al. [17] examined the Atlantis MAR scheme in South Africa [18]. Previous regional syntheses have also been undertaken in Europe [19] and Latin America and the Caribbean [20]. Recently, the research group 'Innovative Web-Based Decision Support System for Water Sustainability under a Changing Climate' (INOWAS) and the International Groundwater Resources Assessment Centre (IGRAC) produced the first global inventory of MAR sites. It contains key information and parameters related to implementation and bio-physical conditions of about 1200 MAR schemes in 62 countries worldwide [21]. No work has specifically synthesized past MAR experience in Africa.

As such, the objective of this article is to compile and synthesize experience on MAR across Africa from documented schemes in order to inform the feasibility and modalities of future MAR implementation in the continent.

2. Methods

2.1. Case Collation

Collation of cases for this article is primarily based on a review of the Global MAR Portal [22]. Case information in the portal generally includes country, site name, coordinates, MAR type, MAR key objective, operation start year, source of water, and final use. The portal contains 44 cases from Africa. Furthermore, a literature search was undertaken to expand the database with additional cases, as well as obtain Supplementary Information on existing cases. Eight additional cases were identified through this literature search. Hence, in total 52 cases were reviewed. All the cases were originally compiled from sources including: technical governmental documents (35%), peer reviewed publications (28%), conference presentations and proceedings (22%) and Master of Science theses (15%). The compiled set of MAR cases likely does not reflect a complete inventory of MAR practice in Africa, as documentation is sometimes contained in less easily accessible technical governmental reports (e.g., Wipplinger [23]) or in non-English language (e.g., French, Portuguese, and Arabic), which were not considered in this analysis (experts from non-English speaking countries were consulted to inquire about additional cases). For example, Gijsbertsen and Groen [24] reported that in the Kitui District in Kenya, more than 500 sand dams were constructed since 1994 by collaboration of a Non-Governmental Organization (NGO) called SASOL (Sahelian Solution Foundation) and local communities. However, only eight sand dam cases in Kenya are found in the Global MAR Portal and included in this article. Still, the database applied represents the current best available dataset. It is assumed that the cases included in the analysis are representative of the types of existing MAR in Africa. Both full scale operation and pilot cases were considered (Annex S1 of the Electronic Supplementary Material (ESM) specify which schemes belong to full scale and pilot scale). The focus of this review is on single MAR schemes with individual documentation, disregarding landscape type approaches to enhancing recharge.

2.2. Case Classification

Classification of MAR cases for this study was based on various scheme characteristics (Table 1) and derived from the classification in the Global MAR Portal [22] and additional sources. In total, 13 parameters were used for classification. Some parameters, related to the biophysical and environmental conditions at the MAR sites, were determined from separate sources, e.g., rainfall, inter-annual variability of water availability, soil type, and geology (Table 1). Other factors, related to socio-economics, costs (capital and operational) and detailed water quality aspects, are important, but scarcity of information on these parameters precluded their inclusion in the analytical framework.

Table 1. Managed aquifer recharge (MAR) classification framework.

Parameter	Explanation	Source
Location	Country and geographic coordinates	[13,22,25–30]
Start year	Year in which MAR scheme was put in operation	[13,22,25–30]
Main MAR type	Surface spreading/infiltration; open well, shaft and borehole injection; in-channel modification; rainwater and runoff harvesting and induced bank filtration	[13,22,25–30]
Source of water	River water, treated wastewater, multiple sources (e.g., river water and treated wastewater, stormwater runoff and treated wastewater), groundwater, and rooftop runoff	[13,22,25–30]
Key MAR objective	(a) to secure and augment water supply (balance variation in water supply and demand) (b) to improve water quality (c) to ameliorate groundwater level decline (d) to prevent seawater intrusion (e) to secure water supply in drought and emergency situations (f) to both augment water supply and prevent seawater intrusion (g) to enhance environmental flow	Reviewed cases
Recharge volume	Volume of water recharged through MAR per year	Reviewed cases
Final use (sector)	Domestic water supply, agricultural use, industrial use, environmental use	[13,22,25–30]
Challenges	Challenges related to technical, bio-physical, and managerial issues e.g., site selection and design, operation and maintenance	Reviewed cases
Rainfall	Long term average annual rainfall in the area of the MAR site	[31]
Inter-annual variability of water availability	Calculated as the standard deviation of available "blue" water divided by the mean. The available "blue" water approximates naturalized river discharge. It is an estimate of surface water availability after deducting water consumed upstream.	[32,33]
Location in transboundary aquifer	Whether the MAR scheme is located in an identified transboundary aquifer or not	[34]
Geology	Geology determined from global geology map and classified into two major rock types: sedimentary and hard rocks	[35]

Information on MAR schemes in Table 1 was mostly obtained from the Global MAR Portal, and missing information was obtained from the literature. Likewise, recharge volume, key MAR objective and challenges in MAR implementation were obtained through additional literature review. The key MAR objective classification adopted in this study are more specific than the three main MAR objective classes of the Global MAR portal; (i.e., maximizing natural storage, maximizing natural storage and physical aquifer management, and water quality improvement). Challenges reported associated with MAR applications are broad and may include technical, biophysical, managerial, socio-economic, regulatory, institutional issues, availability of MAR water, water quality and degradation, etc. These were broadly divided into challenges in site selection and design on the one hand, and challenges in operation and maintenance on the other.

Four biophysical parameters were considered. First, long-term average annual rainfall was determined for the 52 cases. Data were extracted from the Climate Hazards Group Infrared Precipitation with Stations version 2 (CHIRPS) [31]. CHIRPS combines $0.05° \times 0.05°$ resolution satellite imagery with in-situ station measurements. CHIRPS data are available from 1981–present (ftp://ftp.chg.ucsb.edu/pub/org/chg/products/CHIRPS-2.0).

Second, inter-annual variability of water availability were determined using geospatial data from Aqueduct Global Maps 2.1 [33]. Inter-annual variability of water availability is calculated as the standard deviation of the annual available "blue" water divided by the mean of annual available "blue" water (1950–2010). Water availability here is defined as the surface water available after accounting for sanctioned diversions upstream. Hence, "blue" water hereafter is referred to as surface water.

A third biophysical parameter considered was whether the aquifer or aquifer system for a particular MAR scheme is transboundary, i.e., whether the aquifer traverses a national border, as defined from present delineation or not. A map of the transboundary aquifers of the world [34] was used to classify the aquifers. MAR implementation in a transboundary aquifer might be a coincidence rather than planned.

Finally, the fourth biophysical parameter, surficial geology of the MAR sites, was classified using the Global Lithological Map (GLiM) developed by Hartmann and Moosdorf [35], which has an average resolution of 1:3,750,000 and consists of 16 lithological classes. These were lumped into two major rock types for the purpose of this study: sedimentary and hard rock (The geology represents the general surficial geology of the MAR sites. Whether this is also representing the actual geologic formation used to store MAR recharge, which could be the case for many sites, has not been assessed).

It should be noted that additional MAR site information, such as soil infiltration rates, unsaturated zone thickness, and aquifer hydraulic conductivities/transmissivity, were obtainable for a few cases from the literature review. However, these classifications are not included in the present study due to limited data. Detailed data sets of parameters in Table 1 for the MAR cases are provided in Annex S1 (MAR scheme information) and Annex S2 (MAR site information) of the ESM.

3. Results

The MAR cases are concentrated in nine countries in Africa. The vast majority of African countries have not implemented MAR. MAR is practiced in greatest abundance in South Africa with 17 reported cases, followed by Tunisia with 11 cases, Kenya with eight cases, and Algeria with five cases (Figure 1). Five additional countries also present evidence of MAR implementation (Egypt, Ethiopia, Morocco, Namibia and Nigeria), with up to four cases per country. It appears that MAR is more often practiced in Africa's wealthier, yet drier countries or where a certain MAR technology has gained popularity (like sand dams in Kenya, see below).

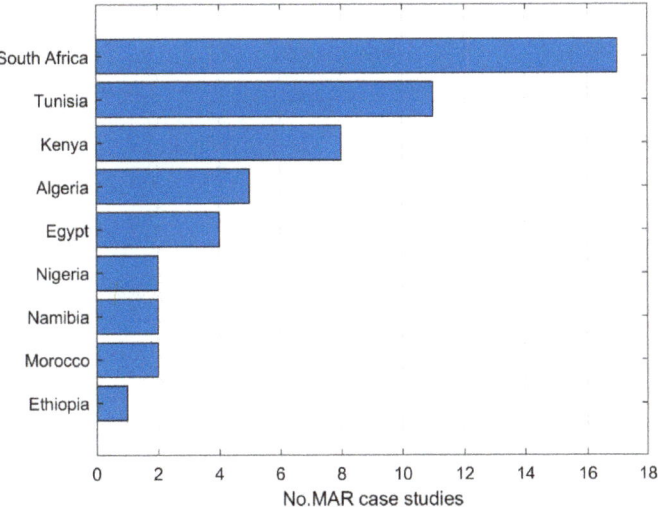

Figure 1. Number of MAR cases in Africa per country (n = 52).

The main MAR type is the surface spreading/infiltration method. The most common MAR type is the surface spreading/infiltration method, the second most common is open well, shaft and borehole injection, and the third is in-channel modification (Figure 2). In-channel-modification using sand dams is the most practiced MAR type in Kenya, while in Tunisia, it is the spreading/infiltration method, and in South Africa, the open well, shaft and borehole injection method (Table 2). The Sidfa Riverbank Filtration case in Egypt [36] is the only reported induced river bank filtration case in Africa. There are no reported cases of rainwater and runoff harvesting.

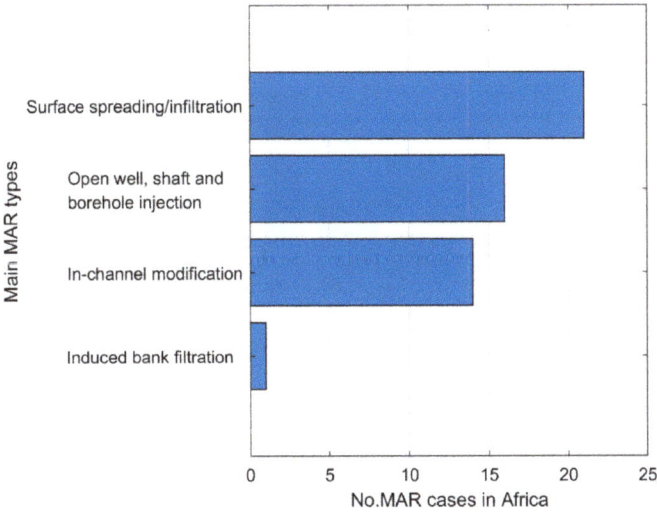

Figure 2. Number of MAR cases in Africa per main MAR type (n = 52).

Table 2. Number of main MAR types per country.

Country	Surface Spreading/Infiltration	Open Well, Shaft and Borehole Injection	In-Channel Modification	Induced Bank Filtration
Algeria	5	-	-	-
Egypt	3	-	-	1
Ethiopia	-	-	1	-
Kenya	-	-	8	-
Morocco	1	-	1	-
Namibia	1	1	-	-
Nigeria	1	1	-	-
South Africa	4	12	1	-
Tunisia	6	2	3	-
Total	21	16	14	1

Implementation of MAR started in the 1960s and has increased over time. The first MAR projects in Africa were launched in 1965, in Soukra, Tunisia [29] and Polokwane, South Africa [13], two followed in the 1970s, and three in the 1980s. The pace of MAR implementation increased substantially in the 1990s with seven new projects undertaken, 16 more projects in the 2000s, and seven projects in the 2010s (Figure 3). One particular example of MAR proliferation is sand dam implementation in Kenya in the 2000s. The apparently illogical decline in the number of MAR cases in the 2010s could be due to lag in reporting.

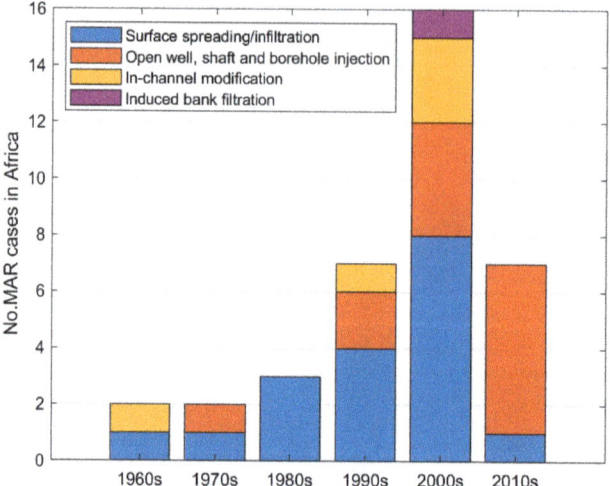

Figure 3. Historical development of MAR cases in Africa where starting date is known. For each decade, the number of new projects is given (n = 37).

The main source of water for the MAR schemes is river water. Thirty three out of the 52 cases use river water as a source and 11 cases use treated wastewater (Figure 4). Some schemes use several sources of water. The Atlantis scheme [18] uses both urban stormwater runoff and treated wastewater. The Windhoek scheme [15] uses river water and treated wastewater. Among the 33 cases that use river water, 19 cases use an ephemeral river as their water source, while eight cases rely on perennial rivers. Three of the eight cases that rely on perennial rivers are found in Egypt, using Nile River water. The Toushka and El Bustan area, with two MAR schemes rely on Nile flood water for their source [37], while the Sidfa river bank filtration project relies on perennial flow of the Nile River [36]. Treated wastewater is used in countries with limited water resources, such as Atlantis, South Africa [18], Windhoek, Namibia [38], and Korba Cap Bon, Souil Wadi, Nabelul, Soukra, Nabeul-Hammamet, El Hajeb-Sidi Abid, Boumerdes and Mahdia-Ksour Essef, Tunisia [29]. The five MAR cases, which rely on groundwater as source of water are at: Williston, South Africa [39], Sishen Mine (Khai Appel), South Africa [25], Elandsfontein, South Africa [25], Kolomela South Africa [13,26], and Koraro-01, subsurface dam, Ethiopia [40]. The Williston case in South Africa [39], uses groundwater abstracted from one aquifer to recharge another aquifer compartmentalized by dykes while the other three cases in South Africa use dewatering water from mines.

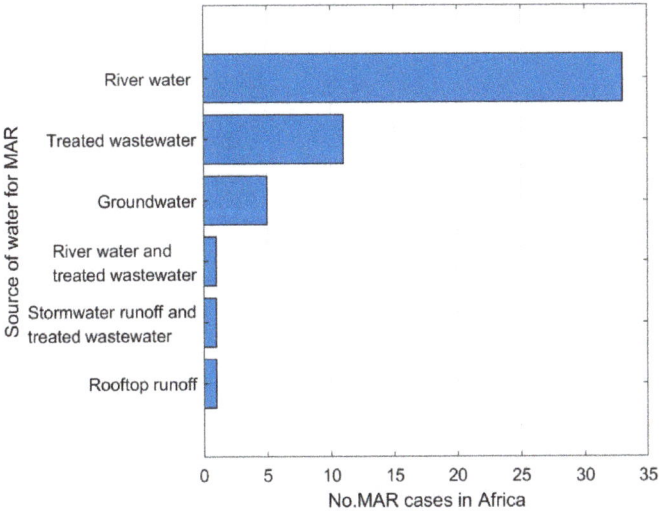

Figure 4. Number of MAR cases in Africa per water source (n = 52).

The key objective of the MAR schemes is to secure and augment water supply. Out of the 52 cases, 23 have a key objective focused exclusively on securing and augmenting water supply (Figure 5). Nine cases have key objectives focused on water quality improvement. Six cases have ameliorating groundwater level decline as key objective, while four cases have preventing seawater intrusion as key objective.

Examples of securing and augmenting water supply include purposes of increasing water availability to meet rural domestic demand during dry periods (e.g., sand dams in Kenya), of meeting summer peak demand (e.g., Prince Albert [41], and Plettenberg [42], South Africa). Examples of securing water supply in drought and emergency situations include Calvinia, South Africa [43] and Windhoek, Namibia [43]. Augmenting water supply and preventing seawater intrusion refers to storing water to meet domestic demand while simultaneously preventing seawater intrusion in the coastal aquifer. An example is the Atlantis scheme, South Africa [18], which separates the domestic and industrial wastewater and stormwater runoff based on salinity, recharges the low salinity water into the inland aquifer used for domestic supply, and recharges the more brackish water into the coastal aquifer to prevent seawater intrusion. Other examples of cases with a preventative seawater intrusion objective include El Khairat [44] and Teboulba [30] aquifers, Tunisia. Finally, the objective of improving water quality refers to water quality enhancement during infiltration (SAT) process (e.g., Ben Sergao, Morocco [45]). One case with an exclusive key objective of enhancing environmental flow is Elandsfontein, South Africa [25].

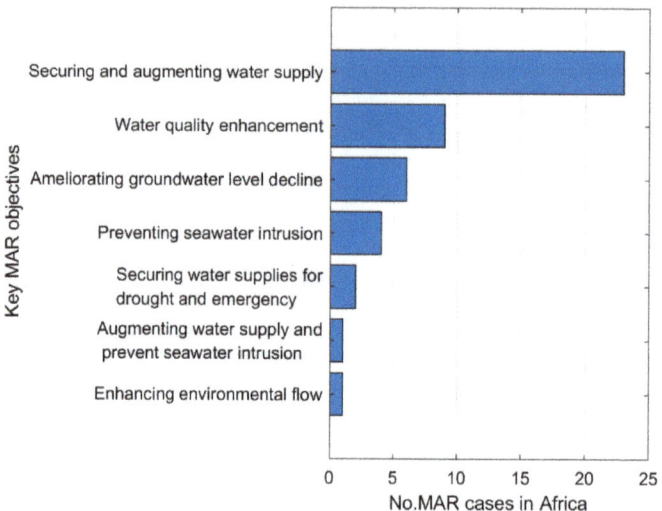

Figure 5. Number of MAR cases in Africa per MAR key objective where information is available (n = 46).

Rainfall across the MAR schemes varies widely. The average annual rainfall across the MAR sites ranges from 1–2225 mm/year (Figure 6a). The average annual rainfall across the MAR sites is 461 mm/year. The minimum and maximum average annual rainfall occurs in MAR sites in Toushka [37], Egypt and Michael Okpara University of Agriculture, Nigeria [46], respectively. The number of cases per five average annual rainfall interval classes is shown in Figure 7. While just over 65% of cases are located in the low annual rainfall class (<500 mm/year), concentration of MAR cases in low rainfall geographies is not dominant.

MAR schemes are concentrated in regions of above-average levels of inter-annual variability in water availability. Eighty three percent of the MAR cases are located in regions of above average inter-annual variability in water availability (Figure 6b). Twenty one of the MAR cases are located in regions with 'medium to high' inter-annual variability, and another 17 cases in regions with 'high' variability (Figure 8). Five cases are located in 'extremely high' variability regions. Saaipoort [25,27,28], Sishen Mine (Khai Appel) [25], Smouskolk (Vanwyksvlei) [25], Williston [39], South Africa and Windhoek, Namibia [43] are the five cases in the extremely high inter-annual variability class. Conversely, only nine cases are located in collective regions of 'low' or 'medium to low' variability.

MAR cases are concentrated in highly populated regions. In total, 31 cases (60%) are located in regions with population density greater than 100 inhabitants per square kilometer (ESM, Figure S1). Nonetheless, MAR experience is also found in less populated areas of e.g., Egypt, Namibia, South Africa and Tunisia. Cases with low population density could be explained by primary use for rural and irrigation use (Teboulba, Tunisia [30], El Hajeb-Sidi Abid, Tunisia [29]), or because water from MAR is subsequently transferred to population centers (Omaruru Delta (OMDEL), Namibia [47], Williston, South Africa [39]).

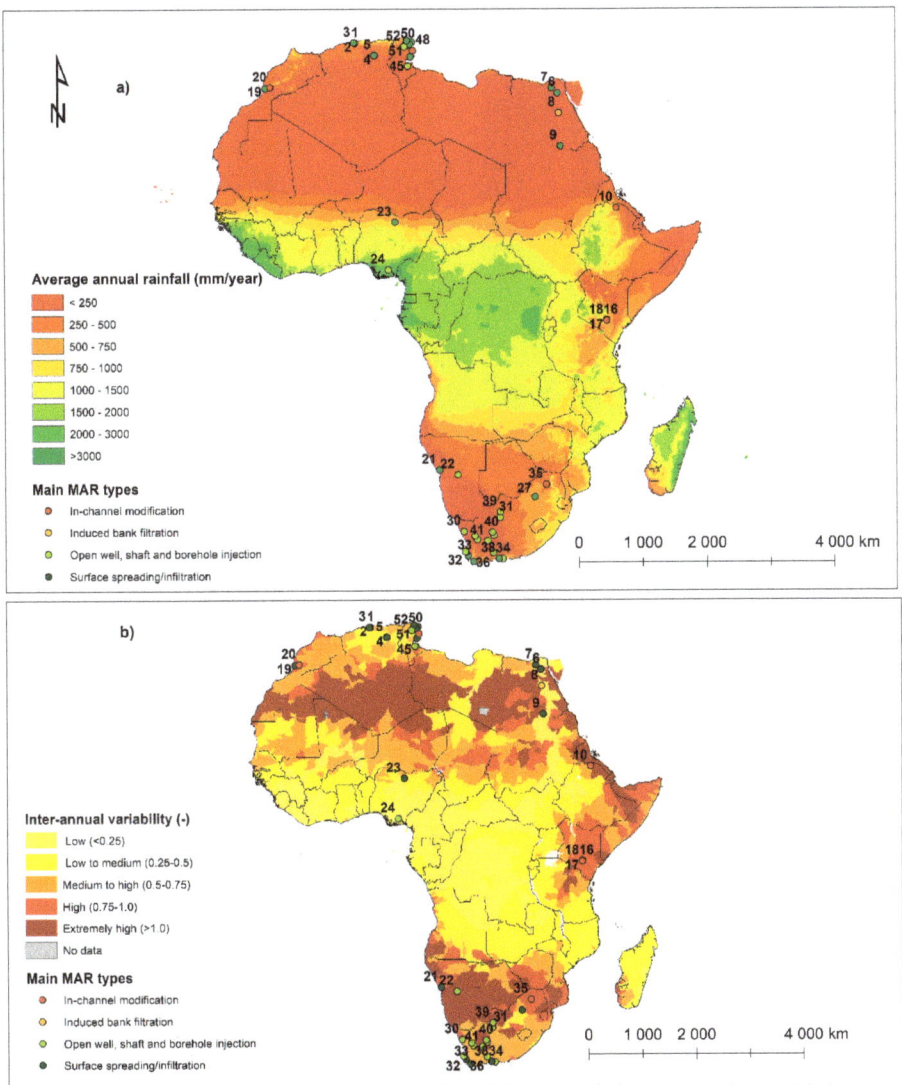

Figure 6. Location of MAR schemes in Africa overlaid on (**a**) average annual rainfall from Climate Hazards Group Infrared Precipitation with Stations (CHIRPS) [31] and (**b**) inter-annual variability in available surface water from Aqueduct Global Maps 2.1 [33]. MAR case labels corresponds to MAR cases numbering in Annex S1 and Annex S2 of the ESM.

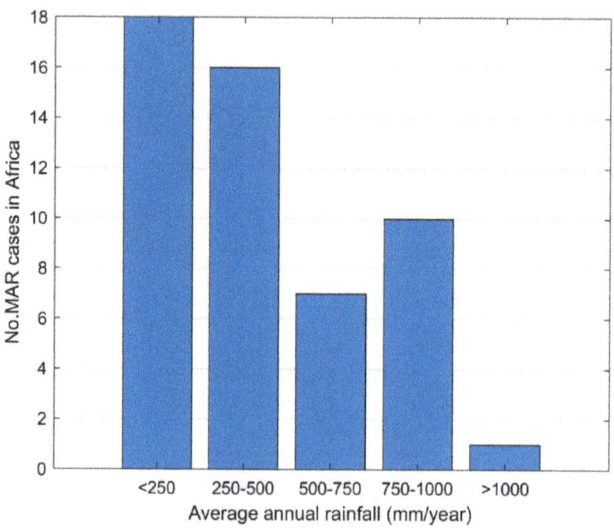

Figure 7. Number of MAR cases in Africa per average annual rainfall class (n = 52).

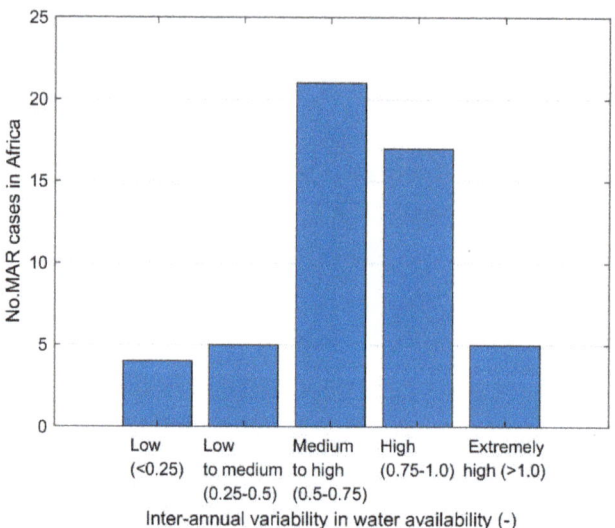

Figure 8. Number of MAR cases in Africa per inter-annual variability in surface water availability class (n = 52).

Few MAR schemes are located in transboundary aquifers. Six cases are found to be located in a transboundary aquifer. Four cases in Egypt are located in the Nubian Sandstone Aquifer system shared among Chad, Egypt, Libya and Sudan, one case in Ethiopia is located in the Mereb Aquifer shared between Ethiopia and Eritrea, and one case in Nigeria is located in the Lake Chad Aquifer shared among six countries: Algeria, Cameroon, Central Africa Republic, Chad, Niger and Nigeria. None of the MAR projects in these transboundary aquifers is believed to have significant transboundary concerns or impacts, as they are not located close to the national borders, and the magnitude of enhanced recharge is relatively small compared to the natural storage of the transboundary aquifers.

The primary geological environment of the MAR cases is sedimentary aquifers. The majority of the MAR sites is situated in sedimentary aquifers (n = 42) and the rest (n = 10) is located in hard rock aquifers. This is expected, as sedimentary formations (e.g., carbonate, gravel, and sand) with various degree of consolidation or cementing are usually targeted for MAR [48]. MAR sites situated in sedimentary rocks are found in Egypt, Ethiopia, Morocco, South Africa and Tunisia. Igneous and metamorphic rocks are often loosely referred as hard rock due to their poor drill-ability [49] and have low primary porosity and most of their porosity comes from secondary porosity (cracks and fractures). Examples of MAR in hard rock are found in Kenya, Nigeria, Namibia and South Africa. The Windhoek case, Namibia [15] is a good example of a large-scale municipal use MAR scheme in fractured rock aquifer setting. Although, half of the sand dams (n = 4) in Kenya are located in hard (metamorphic) rock, the weathering product of the same (sand and gravel) produce eroded sediments, which constitute the actual aquifer used for MAR.

Sectoral use of the MAR schemes is predominantly domestic. The final use of the recovered MAR water across all schemes is presented in Figure 9. The dominant use of MAR is to support domestic water supply (n = 23), followed by agricultural use (n = 17), while only one scheme is solely used for industry (Eland Platinum Mine, South Africa [50]). Multiple purpose schemes have multiple uses of the MAR water (typically agricultural and domestic use, e.g., Loeriesfontein, South Africa [39,51], and domestic and environmental use, e.g., Atlantis, South Africa [18,52]). Despite classifying sand dams into domestic water supply, it is likely that the water is used for multiple purposes, including, livestock and garden agriculture [24]. The Elandsfontein, South Africa [25] is the only case with dominant use of MAR water for environmental flows.

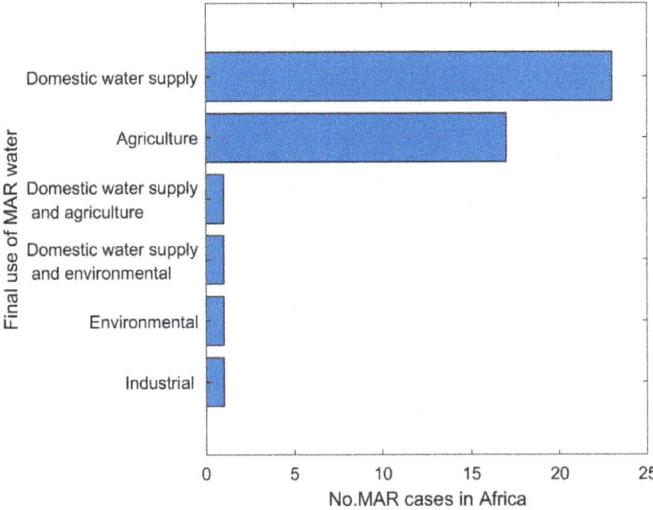

Figure 9. Number of MAR cases in Africa per final water use of stored water where information is available (n = 44).

Recharge volume per MAR scheme varies widely. Recharge volume estimates (compiled as recharge volume per year) are available for 23 MAR cases in seven of the nine countries with recorded MAR schemes (Table 3). The recharge volume ranges from 0.001–100 Mm^3/year, as estimated mostly from reported single daily recharge rates. The highest recharge volume was recorded for the Souss-Massa case, Morocco [53,54] and the lowest recharge volume was for the Koraro-01 subsurface dam (included in the in-channel modification MAR type), Ethiopia [40]. The high recharge volumes (Souss-Massa, Morocco [53,54]; Sidfa Riverbank Filtration, Egypt [36]; Omaruru Delta, Namibia [47]; and Atlantis, South Africa [18]) give an indication of the potential that MAR provides in terms of securing water supply (most notably in the dry season). Even the smallest recharge schemes, like Koraro-01, Ethiopia [40], provide a secured domestic water supply for rural communities.

Recharge volume per country is also quite variable (Table 4). The total annual recharge volume for Africa from full scale MAR is 158 Mm^3/year. To compare, the value for Latin America and the Caribbean is 340 Mm^3/year [20]. The Souss-Massa MAR scheme [53,54], which contributes the largest share of MAR recharge in Africa, makes use of Aoulouz and Imi El Kheng dams located in the upstream part of the Souss-Massa River basin to detain flood water and regulate release of water into the streambed downstream to match infiltration capacity of the streambed. The Souss-Massa River basin is one of the major River basins of Morocco located in the southwestern part of the country. The River basin is drained by two main rivers, the Souss River and the Massa River. The Souss River originates from the High Atlas Mountains in the northern part of the basin. The river is seasonal with occasional high floods between October and February. It is regulated by four big dams. The Massa River drains the Souss-Massa from the southern side of the River basin. Both rivers drain into the Atlantic Ocean. The Aoulouz and the Imi El Kheng dams, two big dams located in the upstream part of the Souss River, are used for MAR mainly due to their favorable geologic conditions and good water quality [55]. The storage capacity of the Aoulouz and the Imi El Kheng dams is 103 and 12 Mm^3, respectively. Based on monthly dam release data of 1991–2004, Bouchaou [53] estimate an annual recharge of 100 Mm^3 (86% of total annual release), which is consistent with other studies [56]. The annual recharge volume is estimated using piezometric water level fluctuations before and after dam release [54].

Table 3. Recharge volume per MAR case where information is available (n = 23).

MAR Case	Country	Recharge Volume (Mm^3/Year)		Reference
		Full Scale	Pilot Scale	
Oued Biskra	Algeria	1.7		[57]
Sidfa Riverbank Filtration	Egypt	22.0		[36]
Koraro-01	Ethiopia	0.001		[40]
Souss-Massa	Morocco	100		[53]
Omaruru Delta	Namibia	9.0		[47]
Windhoek	Namibia	2.8		[13]
Atlantis [1]	South Africa	4.2		[13,18]
Calvinia	South Africa		0.8	[43]
Elandsfontein [2]	South Africa	5.6		[25]
Hermanus	South Africa		1.5	[58]
Kharkams	South Africa	0.005		[47]
Kolomela	South Africa	0.7		[13]
Plettenberg Bay	South Africa		0.3	[42]
Polokwane	South Africa	4.5		[43]
Sedgefield	South Africa		0.5	[58]
Smouskolk (Vanwyksvlei)	South Africa	1.1		[25]
Williston	South Africa	0.1		[39]
El Hajeb-Sidi Abid	Tunisia	0.1		[29]
El Khairat	Tunisia	3.3		[29]
Jeffara Plain (Oum Zessar)	Tunisia	1.2		[57]
Khalidia	Tunisia	0.8		[59]
Korba, Cap Bon	Tunisia	0.5		[29]
Teboulba	Tunisia	0.2		[30]
Total		157.8	3.1	

[1] Recharge volume for the Atlantis scheme, South Africa is around 2.7 Mm^3/year, but an additional 1.5 Mm^3/year more brackish salinity water is recharged in coastal aquifers for seawater control. [2] Annual recharge volume is obtained from Fanus Fourie, DWS (Department of Water and Sanitation), and South Africa (personal communication).

Table 4. Recharge volume per country and recharge as percentage of groundwater use, arranged by recharge volume, where information is available (n = 23).

Country	Recharge Volume (Mm^3/Year)		Groundwater Use (Mm^3/Year) [3]	Recharge Volume as Percentage of Groundwater Use (%)
	Full Scale	Pilot Scale		
Morocco	100		3060	3.3
Egypt	22.0		7780	0.3
South Africa	16.2	3.1	3140	0.5
Namibia	11.8		150	7.9
Tunisia	6.1		2020	0.3
Algeria	1.7		2870	0.1
Ethiopia	0.001		1490	0.0
Kenya	-		620	
Nigeria	-		3440	
Total	157.8		24,570	0.6

[3] Groundwater use for 2010 from Margat and van der Gun [7].

Recharge volume as percentage of groundwater use per country is also variable. Recharge volume as percentage of groundwater use per country is calculated using recharge volume for operational MAR schemes and groundwater use data from 2010, as reported by Margat and van der Gun [7] (Table 4). Recharge volume as percentage of groundwater use is highest in Namibia (7.9%) followed by Morocco (3.3%) and South Africa (0.5%).

Recharge volume as percentage of groundwater use is the least for Africa compared to other continents. The total groundwater use in 2010 for countries in Africa reported by Margat and van der Gun [7] is 41,640 Mm^3/year. Based on this, MAR as percentage of groundwater use in Africa is 0.4%, which is close to the estimate for South America (0.5%) but low compared to other continents, e.g., Middle East (9.4%), Oceania (8.3%), Europe (6.3%), North America (2.3%) and Asia (1.8%) [13] (See also Table S4 of the ESM).

Summary of MAR cases per country. In Morocco, both surface spreading/infiltration and in-channel modification are practiced in equal share. The main objective of MAR is water quality improvement as well as ameliorating groundwater level decline. River water (Souss-Massa [53,54]) and treated wastewater (Ben Sergao [45]) are used as a main water sources for MAR in Morocco, and the final use of recovered MAR water is for agriculture. In Egypt, the main MAR type is surface spreading/infiltration, the main objective of MAR is to secure and augment water supply, the main water source is river water, and the final use of MAR water is domestic water supply. In South Africa, the main MAR type is open well, shaft and borehole injection, the main objective of MAR is to secure and augment water supply, the main water source for MAR is river water, and the final use of recovered MAR water is domestic water supply. In Namibia, both surface spreading/infiltration (Omaruru Delta [47]) and borehole injection (Windhoek [38,43]) are practiced. The main objective of MAR is to secure water supplies in drought and emergency situations as well as ameliorating groundwater level decline. River water and treated wastewater are used as a main water sources for MAR, and the final use of recovered MAR water is for domestic water supply.

In Tunisia, the main MAR type is surface spreading/infiltration method, the main objective of MAR is water quality improvement, the main source of water for MAR is treated wastewater, and the final use of MAR water is for agriculture. In Algeria, the main MAR type is surface spreading/infiltration methods, the main source of water is river water and the final use of recovered MAR water is for agriculture. In Ethiopia, the only analyzed case is a subsurface dam constructed near the village of Koraro [40]. The Koraro-01 subsurface dam supplies domestic water. In Kenya, the main MAR type is in-channel modification (sand dams), the main objective of MAR is to secure and augment water supply, the main water source is river water, and the final use of MAR water is domestic water supply.

In Nigeria, surface spreading/infiltration and open well, shaft and borehole injection are practiced equally prevalently. The main objective of MAR is to ameliorate declining groundwater levels, and the main source of water is river water.

4. Discussion and Conclusions

This article has compiled information on 52 well-reported MAR cases in Africa in order to develop a baseline in understanding of the breadth and status of MAR in Africa. A classification framework was developed and applied to the set of MAR cases, in order to make a first-order assessment based on existing data. Despite the relatively small sample (n = 52), the analysis has generated six significant findings.

First, MAR implementation is not extensive in Africa and is concentrated in nine of the continent's 54 countries. The extent of MAR implementation in Africa remains low compared to other continents [21]. However, the volume reported here (158 Mm^3/year) significantly exceeds the only documented summary of MAR in Africa, which was for Southern Africa, presented by Dillon et al. [13] (10 Mm^3/year for 2015). MAR as fraction of groundwater use in the Southern African region is now estimated at 0.6%, which is three times the reported value by Dillon, et al. [13]. This increase is partially explained by new cases coming into operation since 2015 (e.g., Elandsfontein [25]) or additional cases included in this study (e.g., Omaruru Delta [47]).

The number of MAR cases in Asia, Europe, North America, Oceania and South America from the Global MAR Portal is 281, 282, 308, 95 and 113, respectively [22]. Still, there are cases in Southern and Northern Africa that have more than 30 years of MAR practice (e.g., Atlantis, South Africa [18], Teboulba, Tunisia [30]). Overall, the extent of MAR practice in Africa is low considering the high variability in surface water availability. Major factors that limit wider application of MAR in Africa may include lack of: (1) awareness of and experience with MAR, (2) financial resources, (3) human and institutional capacity, (4) enabling policy frameworks, (5) sufficient and sufficiently well-functioning demonstration sites, (6) understanding of aquifer hydrogeology and geochemical properties.

Second, the key objective of the MAR schemes is to secure and augment water supply and balance inter-seasonal variability in supply and demand. This is consistent with the study in Latin America and the Caribbean by Valverde et al. [20]. For example, the Atlantis scheme, South Africa [18] contributes approximately 25–30% of the water supply of Atlantis Town and has recently been proposed for an expansion to support future drought resilience of Cape Town [60]. The Calvinia case, South Africa [43] is reported to have the potential to provide two to three months of the water supply to Calvinia Town. According to Murray [14], when developed in the next phase, the Windhoek case, Namibia [38,43] is expected to provide a drought buffer as the main water resource for the city for up to three years. As such, Africa presents an initial set of cases and experiences, on which to further develop and upscale the benefits of MAR in the continent.

Third, the main MAR type is the surface spreading/infiltration method. This is different from other continents, e.g., in Europe (induced bank filtration is dominant) [19], and Latin America and the Caribbean (in-channel modification) [20]. In Africa, while sand dams are practiced widely in Kenya, open well, shaft and borehole injection methods are relatively more common in Southern Africa.

Fourth, and linking to point two, a central bio-physical factor that appears to explain MAR location is variability in water availability. The majority (83%) of the MAR sites in Africa are located in geographic regions with above average inter-annual variability in water availability. This is expected as one of the main purposes of MAR is to buffer short-term, seasonal and across-year-variability in water supply. Population density, representing water demand (Figure S1, ESM) and technical and financial capacity are additional factors that may help explain MAR uptake in Africa. Surface spreading/infiltration methods and open well, shaft and borehole injection methods are mostly used for larger urban supply systems, whereas sand dams and subsurface dams for smaller rural systems. As such, there is an apparent dichotomy between MAR systems in Africa governed by demand setting, investment environment, water sources to be applied for MAR (with increasing use of treated wastewater in urban settings) and the management schemes required to address the various issues of financial feasibility and technical maintenance. It is important to note that wastewater reuse should only be encouraged if groundwater quality protection measures are demonstrated effective, which requires guidelines, monitoring and analysis capability. Water from MAR in rural areas also tend to have more variety in use requirements compared with municipal systems, which cater mostly to domestic and industrial uses.

Fifth, while not exhaustive, our study pinpointed some general challenges with MAR in Africa (ESM, Table S4). In total, 13 cases (28%) were reported to have a challenge related to site selection and design or to operation and maintenance. Shallow unsaturated zone thickness was identified as a problem in two surface spreading/infiltration MAR cases (Atlantis case, South Africa [18]) and Polokwane, South Africa [43]. The unsaturated zone of one of the basins of the Atlantis case is reported to be fully saturated during the rainy season, which affects the storage and water quality improvement capacity of the system [18]. In the Polokwane case, South Africa [43], due to insufficient unsaturated zone thickness, there is a concern that some bacteria, viruses or parasites could survive and contaminate groundwater.

Site selection problems were identified as main reasons for underperformance or failure of some sand dams in Kenya [24,61,62]. Site selection may affect rate of sand deposition behind the dam, hence, the time to reach full storage capacity. As an example, three sand dams located in the upstream part of a catchment were reported to have very low infiltration capacity and storage due to the existence of a silt layer in the accumulated sediment, and it took nine years for the sand dam to reach its full storage capacity [24]. Similarly, siting sand dams in areas with a high depth to impermeable bedrock was found to be a problem giving rise to excessive leakage of stored water underneath the dam and scouring of dam foundation [24]. Site selection was a problem in the Abu Rawash case, Egypt [63], where the infiltration basin was sited in clay soil with low infiltration rate.

Infrastructure design problems were reported in the Kharkams well injection case, South Africa [43] and in the in-channel modification Koraro-01 case, Ethiopia [40]. Filter design was an issue in the Kharkams injection MAR scheme, where sand was entering the injection borehole, resulting in a clogging problem [43], whereas the problem at Koraro-01 was the infrastructural design that resulted in breaching of above ground situated subsurface dam crest due to flooding [40]. For some infiltration schemes, i.e., Atlantis, South Africa [18], the Omaruru Delta, Namibia [47] and Abu Rawash, Egypt [63], clogging of infiltration basins or recovery wells, created problems. For example, in Atlantis, South Africa, elevated iron and sulfate in the groundwater caused biological iron-related clogging in recovery wells [18].

Mixing of recharge water with poor inherent water quality in the aquifer was identified as a concern at two MAR sites, Calvinia [43] and Eland Platinum Mine [50,64], both in South Africa. The Calvinia site uses underground natural breccia pipes as storage of MAR water due to their excellent hydraulic confinement and because of the poor surrounding water quality. According to Murray and Tredoux [43], the breccia pipes used in Calvinia are about 100 m in diameter and are very poorly connected to the Karoo sequence aquifer in which they are located. Since the surrounding aquifer already contains high pH, fluoride and arsenic that exceed drinking water standards, there is a concern that the quality of abstracted water could be compromised due to mixing of recharged water with inherently contaminated aquifer water [43]. Use of abandoned mining and quarry sites may reduce costs associated with infrastructure of MAR storage. However, risk of mixing with poor quality water from mining activities may be a concern. For the Eland Platinum Mine, South Africa [50,64], high nitrate and electrical conductivity levels were a concern. Risk of heavy metal contamination from the mining activity is also possible, but data are not available to determine the level of risk. Some of the measures for addressing the above challenges are presented in Table S5 of the ESM. Existing guidelines developed in Australia and elsewhere provide useful guidance for planning and implementation, monitoring and evaluation of MAR (e.g., [39,65–67]). By following approaches presented in these guidelines, problems can be anticipated and counteracted. Re-siting or re-designing projects, and pro-actively investigating how the schemes can best be monitored and managed are key to effective operation and sustainable outcomes.

Before concluding this discussion it is worth noting that there are other types of aquifer recharge enhancement in Africa that extend beyond those examined in this article. These include cases of road water harvesting [68,69]; spate irrigation [70], and rainwater harvesting [71–73]. Road water harvesting is a type of water harvesting technique where concentrated runoff generated by or along roads is collected using adjacent ponds or channeled to farm land [69,74], which assists road drainage as well as crops, and may incidentally enhance recharge. Spate irrigation is an irrigation technique, by which seasonal river water is diverted and spread to a larger riparian surface areas, primarily to enhance soil moisture for growing crops, and this would also incidentally enhance recharge.

5. Looking Forward

This paper is the first to systematically examine reported MAR schemes over the African continent and review experiences. While only 52 cases are reported, they provide about 158 Mm^3/year of additional water storage to enhance water supply and/or water quality. This volume represents 0.4% of continental groundwater use, but plays a critical role in locally supplying and securing water for large populations. Examples of technical challenges were revealed that could be a deterrent to future African investment in MAR, unless these are systematically addressed. Observations from MAR practices around the world suggest that guidelines and policies can help to prevent or overcome such problems. South Africa has advanced significantly in this regard with a national map showing the prospects for MAR [51], a manual on how to develop MAR projects [75], and well-documented case studies [58]. Wider adoption of this approach would well serve other African nations.

Furthermore, impact assessment, including pre-project baseline assessment and continuous monitoring are warranted for significant MAR schemes to reveal their performance and long-term outcomes with respect to water security and resilience. Maintenance and financial, institutional and socio-economic aspects are also key elements to sustainability, which needs increased attention. Hence, strong feasibility analysis, performance and impact assessment of existing and new MAR projects should be a primary focus going forward.

Supplementary Materials: The following are available online at http://www.mdpi.com/2073-4441/12/7/1844/s1. References [76–86] are cited in the supplementary materials. An Electronic Supplementary Material (ESM) summarizing MAR cases information. Figure S1: Location of 52 MAR cases in Africa overlaid on population density map of 2015, Table S1: MAR cases per key objectives per country (arranged by total number of MAR cases), Table S2: MAR case studies per source of recharge water per country, Table S3: MAR cases per final use of recovered water per country, Table S4: Recharge volume, absolute and as percentage of groundwater use in different continents or regions, ordered by MAR in 2015, Table S5: Challenges and possible solutions in MAR implementation, Annex S1: MAR cases in Africa (MAR scheme information), Annex S2: MAR cases in Africa (MAR site information).

Author Contributions: G.Y.E. led the development of this work including analysis of cases and write-up and prepared the original draft. J.F.L. contributed in conceptual direction and to draft preparation and in the review. K.G.V. contributed to the conceptual framework and review. All authors have read and agreed to the published version of the manuscript.

Funding: This study was undertaken in the context of and funded by the Potential Role of the Transboundary Ramotswa Aquifer-2 project, funded by the United States Agency for International Development (USAID) under the terms of Award No. AID-674-IO-17-00003, the CGIAR Research Program on Water, Land and Ecosystems (WLE) supporting the Groundwater Solutions Initiatives for Policy and Practices (GRIPP), and the Conjunctive Management across Borders in SADC project, also funded by USAID.

Acknowledgments: The authors would like to thank Fanus Fourie (Department of Water and Sanitation, South Africa), Habib Chaieb, (Directeur chez DHER-CRDA de Ben Arous—MARHP El Mourouj, Ben Arous Governorate, Tunisia), and Lhoussaine Bouchaou (Department of Earth Sciences, University Ibn Zohr Agadir, Morocco), and the International Groundwater Resources Assessment Center (IGRAC) for providing support in terms of case study information, and the two anonymous reviewers and the editor for providing constructive comments, which greatly improved the manuscript.

Conflicts of Interest: The authors declare no conflict of interest.

References

1. IPCC. *Climate Change 2007—The Physical Science Basis: Working Group I Contribution to the Fourth Assessment Report of the IPCC*; Solomon, S., Qin, D., Manning, M., Averyt, K., Marquis, M., Eds.; Cambridge University Press: Cambridge, UK; New York, NY, USA, 2007; Volume 4.
2. Sadoff, C.W.; Hall, J.W.; Grey, D.; Aerts, J.C.J.H.; Ait-Kadi, M.; Brown, C.; Cox, A.; Dadson, S.; Garrick, D.; Kelman, J.; et al. *Securing Water, Sustaining Growth: Report of the GWP/OECD Task Force on Water Security and Sustainable Growth*; University of Oxford: Oxford, UK, 2015; p. 180.
3. African Ministers Council on Water. *Water Security and Climate Resilient Development, Strategic Framework*; African Ministers Council on Water: Abuja, Nigeria, 2012.
4. Cobbing, J.; Hiller, B. Waking a sleeping giant: Realizing the potential of groundwater in Sub-Saharan Africa. *World Dev.* **2019**, *122*, 597–613. [CrossRef]
5. Lautze, J.; Holmatov, B.; Saruchera, D.; Villholth, K.G. Conjunctive management of surface and groundwater in transboundary watercourses: A first assessment. *Water Policy* **2018**, *20*, 1–20. [CrossRef]
6. Megdal, S.B.; Dillon, P. Policy and Economics of Managed Aquifer Recharge and Water Banking. *Water* **2015**, *7*, 592–598. [CrossRef]
7. Margat, J.; van der Gun, J. *Groundwater around the World: A Geographic Synopsis*; CRC Press: Boca Raton, FL, USA, 2013.
8. Tuinhof, A.; Foster, S.; van Steenbergen, F.; Talbi, A.; Wishart, M. *Appropriate Groundwater Management Policy for Sub-Saharan Africa: In Face of Demographic Pressure and Climatic Variability*; World Bank: Washington, DC, USA, 2011.
9. Bouwer, H. Integrated water management: Emerging issues and challenges. *Agric. Water Manag.* **2000**, *45*, 217–228. [CrossRef]
10. Bouwer, H. Artificial recharge of groundwater: Hydrogeology and engineering. *Hydrogeol. J.* **2002**, *10*, 121–142. [CrossRef]
11. Asano, T.; Cotruvo, J.A. Groundwater recharge with reclaimed municipal wastewater: Health and regulatory considerations. *Water Res.* **2004**, *38*, 1941–1951. [CrossRef]
12. Ebrahim, G.Y.; Fathi, S.; Lautze, J.; Ansems, N.; Villholth, K.G.; Nijste, G.-J.; Magombeyi, M.; Mndaweni, S.; Kenabatho, P.; Moehadu, M.; et al. *Review of Managed Aquifer Recharge (MAR) Experience in Africa and MAR Suitability Mapping for Ramotswa Transboundary Aquifer Area*; International Water Management Institute: Pretoria, South Africa, 2017.
13. Dillon, P.; Stuyfzand, P.; Grischek, T.; Lluria, M.; Pyne, R.D.G.; Jain, R.C.; Bear, J.; Schwarz, J.; Wang, W.; Fernandez, E.; et al. Sixty years of global progress in managed aquifer recharge. An overview of Managed Aquifer Recharge in Southern Africa. *Hydrogeol. J.* **2019**, *27*, 1–30. [CrossRef]
14. Murray, R. *Managed Aquifer Recharge, an Introductory Guide for the SADC Groundwater Management Institute Including the Windhoek Case Study*; Groundwater Africa: Cape Town, South Africa, 2017.
15. Murray, R.; Louw, D.; van der Merwe, B.; Peters, I. Windhoek, Namibia: From conceptualising to operating and expanding a MAR scheme in a fractured quartzite aquifer for the city's water security. *Sustain. Water Resour. Manag.* **2018**, *4*, 217–223. [CrossRef]
16. Bugan, R.D.; Jovanovic, N.; Israel, S.; Tredoux, G.; Genthe, B.; Steyn, M.; Allpass, D.; Bishop, R.; Marinus, V. Four decades of water recycling in Atlantis (Western Cape, South Africa): Past, present and future. *Water SA* **2016**, *42*, 577–594. [CrossRef]
17. Jovanovic, N.; Bugan, R.D.; Tredoux, G.; Israel, S.; Bishop, R.; Marinus, V. Hydrogeological modelling of the Atlantis aquifer for management support to the Atlantis Water Supply Scheme. *Water SA* **2017**, *43*, 122–138. [CrossRef]
18. DWA. *The Atlantis Water Resource Management Scheme: 30 Years of Artificial Groundwater Recharge*; PRSA 000/001/1169/10-Activity 17 (AR5.1); Department of Water Affairs: Pretoria, South Africa, 2010.
19. Sprenger, C.; Hartog, N.; Hernández, M.; Vilanova, E.; Grützmacher, G.; Scheibler, F.; Hannappel, S. Inventory of managed aquifer recharge sites in Europe: Historical development, current situation and perspectives. *Hydrogeol. J.* **2017**, *25*, 1909–1922. [CrossRef]
20. Valverde, J.P.B.; Stefan, C.; Nava, A.P.; da Silva, E.B.; Vivar, H.L.P. Inventory of managed aquifer recharge schemes in Latin America and the Caribbean. *Sustain. Water Resour. Manag.* **2018**, *4*, 163–178. [CrossRef]

21. Stefan, C.; Ansems, N. Web-based global inventory of managed aquifer recharge applications. *Sustain. Water Resour. Manag.* **2018**, *4*, 153–162. [CrossRef]
22. IGRAC. International Groundwater Resources Assessment Centre (IGRAC) Global MAR Portal. Available online: https://www.un-igrac.org/ggis/mar-portal (accessed on 6 May 2020).
23. Wipplinger, O. *The Storage of Water in Sand; an Investigation of the Properties of Natural and Artificial Sand Reservoirs and of Methods of Developing such Reservoirs*; South West Africa Administration Water Affairs Branch: Windhoek, Namibia, 1958.
24. Gijsbertsen, C.; Groen, J. A Study to Up-Scaling of the Principle and Sediment (Transport) Processes behind, Sand Storage Dams, Kitui District, Kenya. Master's Thesis, Vrije Universiteit, Amsterdam, The Netherlands, 2007.
25. Fourie, F.; Vermaak, N.; Hohne, D.; Murray, R. Authorising MAR projects within the context of South Africa's National Water Act. In Proceedings of the 9th International Symposium on Managed Aquifer Recharge (ISMAR9), Mexico City, Mexico, 20–24 June 2016.
26. Murray, R.; Baker, K. *Kolomela Mine: Results of the Injection Tests in Groenwaterspruit for Mine Water Disposal, Groundwater Africa, South Africa*; Groundwater Africa: Cape Town, South Africa, 2012.
27. Esterhuyse, C.J. *Carnarvon: Report on Artificial Recharge Investigation at Saaipoort, Northern Cape Province Report Prepared by SRK Consulting (South Africa) (Pty) Ltd. for Aurecon South Africa (Pty) Ltd.*; SRK Consulting: Cape Town, South Africa, 2019.
28. Hohne, D. Drought in the Northern Cape Karoo: Lessons learned and options for future use. In Proceedings of the 15th Biennial Ground Water Division Conference, Stellenbosch, South Africa, 14–18 October 2017.
29. Chaieb, H. Tunisian Experience in Artificial Recharge using Treated Waste Water. In Proceedings of the Regional Conference on Sustainable Integrated Wastewater Treatment and Reuse, Sharm El Sheikh, Egypt, 1–2 December 2014.
30. Bouri, S.; Dhia, H.B. A thirty-year artificial recharge experiment in a coastal aquifer in an arid zone: The Teboulba aquifer system (Tunisian Sahel). *C. R. Geosci.* **2010**, *342*, 60–74. [CrossRef]
31. Funk, C.C.; Peterson, P.J.; Landsfeld, M.F.; Pedreros, D.H.; Verdin, J.P.; Rowland, J.D.; Romero, B.E.; Husak, G.J.; Michaelsen, J.C.; Verdin, A.P. A quasi-global precipitation time series for drought monitoring. *US Geol. Surv. Data Ser.* **2014**, *832*, 1–12.
32. World Resources Institute. Aqueduct Global Maps 2.1 Data. Available online: https://www.wri.org/resources/data-sets/aqueduct-global-maps-21-data (accessed on 7 June 2018).
33. Gassert, F.; Luck, M.; Landis, M.; Reig, P.; Shiao, T. *Aqueduct Global Maps 2.1: Constructing Decision-Relevant Global Water Risk Indicators*; World Resources Institution: Washington, DC, USA, 2014.
34. IGRAC; UNESCO-IHP. Transboundary Aquifers of the World. 2015. Available online: http://ihp-wins.unesco.org/layers/geonode:tba_map2015 (accessed on 5 September 2018).
35. Hartmann, J.; Moosdorf, N. The new global lithological map database GLiM: A representation of rock properties at the Earth surface. *Geochem. Geophys. Geosyst.* **2012**, *13*. [CrossRef]
36. Shamrukh, M.; Abdel-Wahab, A. Riverbank filtration for sustainable water supply: Application to a large-scale facility on the Nile River. *Clean Technol. Environ. Policy* **2008**, *10*, 351–358. [CrossRef]
37. El Arabi, N. Environmental management of groundwater in Egypt via artificial recharge extending the practice to soil aquifer treatment (SAT). *Int. J. Environ. Sustain.* **2012**, *1*, 66–82. [CrossRef]
38. Tredoux, G.; Van Der Merwe, B.; Peters, I. Artificial recharge of the Windhoek aquifer, Namibia: Water quality considerations. *Boletín Geológico Min.* **2009**, *120*, 269–278.
39. DWA. *Water Banking: A Practical Guide to Using Artificial Groundwater Recharge*; Department of Water Affairs: Pretoria, South Africa, 2010.
40. Mohn, R.; Abdurahman, M.A.; Gardemann, J.; Mathys, W.; Nedaw, D.; Kruse, B.; Herbst, C. Subsurface Micro-Reservoirs for Rural Water Supply in the Ethiopian Highlands, v. Available online: https://www.hb.fh-muenster.de/opus/fhms/volltexte/2012/780/ (accessed on 26 October 2017).
41. Murray, R. Prince Albert Municipality. In *Groundwater Management and Artificial Recharge Feasibility Study*; Groundwater Africa: Cape Town, South Africa, 2007.
42. Murray, R. *Bitou Municipality Groundwater Management and Artificial Recharge Feasibility Study*; Groundwater Africa: Cape Town, South Africa, 2007.
43. Murray, E.; Tredoux, G. *Pilot Artificial Recharge Schemes: Testing Sustainable Water Resource Development in Fractured Aquifers*; Water Research Commission: Pretoria, South Africa, 2002.

44. Zammouri, M.; Brini, N.; Horriche, F.J. Assessment of Artificial Recharge Efficiency against Groundwater Stress in the El Khairat Aquifer. In *Groundwater and Global Change in the Western Mediterranean Area*; Springer: Berlin/Heidelberg, Germany, 2018; pp. 305–312.
45. Bennani, A.; Lary, J.; Nrhira, A.; Razouki, L.; Bize, J.; Nivault, N. Wastewater treatment of greater Agadir (Morocco): An original solution for protecting the bay of Agadir by using the dune sands. *Water Sci. Technol.* **1992**, *25*, 239–245. [CrossRef]
46. Igboekwe, M.U.; Ruth, A. Groundwater recharge through infiltration process: A case study of Umudike, Southeastern Nigeria. *J. Water Resour. Prot.* **2011**, *3*, 295. [CrossRef]
47. Murray, R. Artificial Recharge Lecture Notes. In *Artificial Recharge: The Intentional Banking and Treating of Water in Aquifers*; Department of Water Affairs and Forestry: Pretoria, South Africa, 2008.
48. Dillon, P.J.; Pavelic, P. *Guidelines on the Quality of Stormwater and Treated Wastewater for Injection into Aquifers for Storage and Reuse*; Urban Water Research Association of Australia: Melbourne, Australia, 1996.
49. Prathapar, S.; Sharma, B.R.; Aggarwal, P.K. *Hydro, Hydrogeological Constraints to Managed Aquifer Recharge in the Indo Gangetic Plains*; International Water Management Institute: Colombo, Sirlanka, 2012.
50. Botha, F. Can Man-Made Aquifers Provide a Solution for a Thirsty Mining Industry? In Proceedings of the International Symposium on Managed Aquifer Recharge (ISMAR 7), Abu Dhabi, UAE, 9–13 October 2009.
51. DWA. *Potential Artificial Recharge Areas in South Africa*; PRSA 000/00/11609/1—Activity 14 (AR04); Department of Water Affairs: Pretoria, South Africa, 2009.
52. Tredoux, G.; Genthe, B.; Steyn, M.; Engelbrecht, J.; Wilsenach, J.; Jovanovic, N. An assessment of the Atlantis artificial recharge water supply scheme (Western Cape, South Africa). *WIT Trans. Ecol. Environ.* **2009**, *127*, 403–413.
53. Bouchaou, L. MAR techniques and examples successfully applied in light of climate change adaptation: Case study in Morocco. In Proceedings of the Conference Presentation—Theme Opportunities for Managed Aquifer Recharge, Amman, Jordan, 11–14 December 2012.
54. Bouragba, L. Etude de la recharge artificielle des nappes en zone semi-aride (Application au bassin du Souss-Maroc). In *Study of Artificial Recharge of Groundwater in Semi-Arid Areas (Application to the Souss-Maroc basin)*, U.F.R Sciences & Techniques; Université de Franche-Comté École Doctorale Homme, Environnement: Santé, France, 2011.
55. Choukr-Allah, R.; Ragab, R.; Bouchaou, L.; Barceló, D. *The Souss-Massa River Basin, Morocco*; Springer: Berlin/Heidelberg, Germany, 2017.
56. Al-Turbak, A.S. Effectiveness of recharge from a surface reservoir to an underlying unconfined aquifer. In Proceedings of the Vienna Symposium of the Hydrology of Natural and Manmade Lakes, Vienna, Austria, 11–24 August 1991; pp. 191–196.
57. Ghiglieri, G. WADIS-MAR Project: Water harvesting and Agricultural technique in Dry lands: An integrated and Sustainable model in MAghreb Regions. In Proceedings of the WADIS-MAR: Achievements, Activities, WADIS-MAR final Conference International Workshop on Sustainable water Resources Management in Arid and Semi-Arid Regions, Sassari, Italy, 16 June 2016.
58. DWA. *Potential Artificial Recharge Schemes: Planning for Implementation*; Department of Water Affairs: Pretoria, South Africa, 2010.
59. Mhamdi, R.; Heilweil, V. *A Quantitative Evaluation of the Impacts of Artificial Recharge to the Mornag Aquifer System of Northern Tunisia*; Acacia Publications: Phoenix, AZ, USA, 2007; pp. 484–493.
60. DWS. *Water Outlook 2018 Report, Revision 25—Updated 20 May 2018 Produced by Department of Water and Sanitation (DWS) City of Cape Town*; Department of Water and Sanitation: Cape Town, South Africa, 2018.
61. De Trincheria, J.; Nissen-Petersen, E.; Walter, L.F.; Otterphol, R. Factors affecting the performance and cost efficiency of sand storage dams in South-Eastern Kenya. In Proceedings of the 36th IAHR World Congress, Hague, The Netherlands, 28 June–3 July 2015.
62. Borst, L.; De Haas, S. *Hydrology of Sand Storage Dams: A Case Study in the Kiindu Catchment, Kitui District, Kenya*; VU University Amsterdam: Amsterdam, The Netherlands, 2006.
63. El-Fakharany, Z. Environmental Impact Assessment of Artificial Recharge of Treated Wastewater on Groundwater Aquifer System Case study Abu-Rawash, Egypt. *J. Am. Sci.* **2013**, *9*, 309–315.
64. Botha, F.; Maleka, L. Results show that man-made aquifers within the platinum mining industry in South Africa can provide a solution for future water demands. In Proceedings of the International Mine Water Conference, Aachen, Germany, 4–11 September 2011.

65. Page, D.; Dillon, P.; Vanderzalm, J.; Bekele, E.; Barry, K.; Miotlinski, K.; Levett, K. *Managed Aquifer Recharge Case Study Risk Assessments*; Water for a Healthy Country National Research Flagship Report Series ISSN: 1835-095X; CSIRO: Canberra, Australia, 2010.
66. NRMMC-EPHC-NHMRC. National Water Quality Management Strategy Document No 24. In *Australian Guideline for Water Recycling Managing Health and Environmental Risks (Phase 2), Managed Aquifer Recharge Natural Resource*; Biotext: Canberra, Australia, 2009.
67. RAIN. *A Practical Guide to Sand Dam Implementation Water Supply through Local Structures as Adaptation to Climate Change*; RAIN Foundation: Amsterdam, The Netherlands, 2008.
68. Demenge, J.; Alba, R.; Welle, K.; Manjur, K.; Addisu, A.; Mehta, L.; Woldearegay, K. Multifunctional roads: The potential effects of combined roads and water harvesting infrastructure on livelihoods and poverty in Ethiopia. *J. Infrastruct. Dev.* **2015**, *7*, 165–180. [CrossRef]
69. Puertas, D.G.-L.; Woldearegay, K.; Mehta, L.; Beusekom, M.; Agujetas, M.; Van Steenbergen, F. Roads for water: The unused potential. *Waterlines* **2014**, *33*, 120–138. [CrossRef]
70. Van Steenbergen, F.; Haile, A.M.; Alemehayu, T.; Alamirew, T.; Geleta, Y. Status and potential of spate irrigation in Ethiopia. *Water Resour. Manag.* **2011**, *25*, 1899–1913. [CrossRef]
71. Boers, T.M.; Ben-Asher, J. A review of rainwater harvesting. *Agric. Water Manag.* **1982**, *5*, 145–158. [CrossRef]
72. Helmreich, B.; Horn, H. Opportunities in rainwater harvesting. *Desalination* **2009**, *248*, 118–124. [CrossRef]
73. Pandey, D.N.; Gupta, A.K.; Anderson, D.M. Rainwater harvesting as an adaptation to climate change. *Curr. Sci.* **2003**, *85*, 46–59.
74. Woldearegay, K.; van Steenbergen, F.; Alemayehu, T.; Manjur, K.; Perez, M.A. The Beneficial Use of Road Water for Climate Resilience and Asset Management. In Proceedings of the 2nd IRF Africa Regional Congress, Windhoek, Namibia, 11–13 July 2017.
75. DWA. *A Check-List for Implementing Successful Artificial Recharge Projects*; PRSA 000/00/11609/2—Activity 12 (AR02); Department of Water Affairs: Pretoria, South Africa, 2009.
76. WorldPop (School of Geography and Environmental Science, University of Southampton; Department of Geography and Geosciences, University of Louisville; Departement de Geographie, Universite de Namur) and Center for International Earth Science Information Network (CIESIN), Columbia University (2018). Global High Resolution Population Denominators Project—Funded by The Bill and Melinda Gates Foundation (OPP1134076). 2020. Available online: https://dx.doi.org/10.5258/SOTON/WP00647 (accessed on 8 June 2020).
77. Martin, R. *Clogging Issues Associated with Managed Aquifer Recharge Methods*; IAH Commission on Managing Aquifer Recharge: Australia, 2013. Available online: https://recharge.iah.org/files/2015/03/Clogging_Monograph.pdf (accessed on 8 June 2020).
78. Hanson, G.; Nilsson, Å. Ground-Water Dams for Rural-Water Supplies in Developing Countries. *Groundwater* **1986**, *24*, 497–506. [CrossRef]
79. Gale, I. *Strategies for Managed Aquifer Recharge (MAR) in Semi-Arid Areas*; UNESCO IHP: Paris, France, 2005.
80. Foster, S.; Tuinhof, A. Subsurface Dams to Augment Groundwater Storage in Basement Terrain for Human Subsistence-Brazilian and Kenyan Experience. *World Bank Groundw. Manag. Advis. Team* **2004**, *5*, 78–92.
81. Sobowale, A.; Ramalan, A.; Mudiare, O.; Oyebode, M. Groundwater recharge studies in irrigated lands in Nigeria: Implications for basin sustainability. *Sustain. Water Qual. Ecol.* **2014**, *3*, 124–132. [CrossRef]
82. Department of Water Affairs. *Artificial Recharge Strategy Version 2*; 2010 November. Available online: http://www.artificialrecharge.co.za/strategydocument/ARTIFICIAL_RECHARGE_STRATEGY_Ver_2_071210_update_010211_1.pdf (accessed on 8 June 2020).
83. Bachtouli, S.; Comte, J.-C. Regional-Scale Analysis of the Effect of Managed Aquifer Recharge on Saltwater Intrusion in Irrigated Coastal Aquifers: Long-Term Groundwater Observations and Model Simulations in NE Tunisia. *J. Coast. Res.* **2019**, *35*, 91–109.
84. Chaieb, H.; Moncef, R.; Ouerfelli, N.; Laghi, M.; Magagnini, L.; Tosatto, O.; De Angelis, A.; Sollazzo, F.; Teatini, P. On the effectiveness of reusing treated wastewater by infiltration ponds in coastal farmlands. Preliminary investigation on insights from the Korba site, Tunisia. In Proceedings of the 1st CIGR Inter-Regional Conference on Land and Water Challenges, Bari, Italy, 10–14 Septemebr 2013.

85. Ouelhazi, H.; Lachaal, F.; Charef, A.; Challouf, B.; Chaieb, H.; Horriche, F.J. Hydrogeological investigation of groundwater artificial recharge by treated wastewater in semi-arid regions: Korba aquifer (Cap-Bon Tunisia). *Arab. J. Geosci.* **2014**, *7*, 4407–4421. [CrossRef]
86. Rekaya, M.; Bedmar, A.P. The Tunisian Experience in Ground Water Artificial Recharge by Treated Wastewater. In Proceedings of the Artificial Recharge of Ground Water, Anaheim, CA, USA, 23–27 August 1988; pp. 612–627.

 © 2020 by the authors. Licensee MDPI, Basel, Switzerland. This article is an open access article distributed under the terms and conditions of the Creative Commons Attribution (CC BY) license (http://creativecommons.org/licenses/by/4.0/).

Perspective

An Overview of Managed Aquifer Recharge in Brazil

Tatsuo Shubo [1,*], Lucila Fernandes [2] and Suzana Gico Montenegro [2]

1. Departamento de Saneamento e Saúde Ambiental, Escola Nacional de Saúde Pública, Fundação Oswaldo Cruz, 21040-361 Rio de Janeiro, Brazil
2. Centro de Tecnologia e Geociências, Universidade Federal de Pernambuco, 50670-901 Recife, Brazil; lucila.araujo@gmail.com (L.F.); suzanam.ufpe@gmail.com (S.G.M.)
* Correspondence: tatsuo.shubo@ensp.fiocruz.br; Tel.: +55-21-2598-2469

Received: 1 December 2019; Accepted: 7 April 2020; Published: 9 April 2020

Abstract: In order to face the severe climate conditions in semiarid regions, many managed aquifer recharge (MAR) and rainwater storage systems have been implemented by local communities. Governmental programs have helped to propagate the concept of MAR. Based on a systematic review, popular initiatives, current legislation, and research lines and programs were compiled and analyzed. Although the MAR global inventory points to the prevalence of in-channel modifications among ninety MAR sites, the Barraginhas Project alone has been responsible for the construction of more than 500,000 infiltration ponds up to 2013. In urban areas, aquifer recharge initiatives mostly aim to reduce runoff peak flows. In some cases these initiatives have been stimulated by urban drainage public policies. Compared to countries such as the USA and Australia, Brazil is still at an early stage in MAR initiatives and needs to overcome technical, legal, and socio-cultural challenges to adopt MAR approaches, in order to help in facing water security challenges in a future climate change scenario. This article aims to provide an overview of the state of the art concerning technological, scientific, and legal issues around MAR in Brazil and the respective challenges for the adoption of this approach at a national level.

Keywords: water security; urban water management; semiarid; Social Technology; Managed Aquifer Recharge; developing countries

1. Introduction

1.1. Historical Background

Although Brazil has a huge water availability, about 30,342 m^3/inhab./year in 2015 [1], it is not evenly distributed across the country, with 80% of the surface water concentrated in the Amazon region [2]. Besides this, Brazil has been struggling with many water crises since the beginning of its settlement by Europeans. A priest called Fernando Cardin recorded the first drought in 1583. Since then, more than 120 droughts have been recorded in the northeastern semiarid region alone. A seven-year drought recorded in the 18th century (1720–1727) that struck the region currently known as the states of Ceará, Rio Grande do Norte, Paraíba, and Pernambuco has been considered the worst one on record. During that particular event, most livestock perished, rivers and springs dried up, and widespread starvation devastated the region [3]. The time period between the years of 1877 and 1879, recorded as the hottest and driest of the 19th century, imposed severe hardships and suffering on the local populations. During that period, approximately five hundred thousand people starved to death, and crops and cattle suffered devastating losses. This scenario triggered massive waves of migration of people moving towards coastal cities, bringing a demographic explosion to areas that did not yet have the appropriate infrastructure in place to support these migrations. Poor living conditions, such as the lack of proper sanitation systems, have been associated with a smallpox epidemic which contributed to the hardships [4].

From the 1980s up to the present time, Brazil has been experiencing many of its worst droughts on record and struggling with their consequences. As a consequence of yet another seven-year drought period (1979–1985), more than 3.5 million people ended up starving to death, with most of the victims being children who perished from undernourishment. Crops and cattle were lost, forcing desperate farm people to loot local markets seeking food [5]. In 2002, Brazil also faced an energy crisis, known as "the apagão" (The Big Blackout), mostly caused by a series of dry periods. In 2007, the northern part of the state of Minas Gerais suffered a fifteen-month dry spell with virtually no rainfall.

In the Brazilian countryside, especially in the semiarid region, there was a lack of rainfall during the time period of 1981–2019. On the other hand, in the northern Region, high rates of rainfall still occur. Climate models show a trend towards increased frequency and intensity of droughts and length of dry periods in the northeast, as already has occurred in some Brazilian regions [6]. From 2012 to 2017 another major drought affected the semiarid region and 2015 was considered the most critical year of that period. Figure 1a shows the average precipitation for the early dry season in Brazil (April–May), from 1981 to 2010, and the Figure 1b shows the total precipitation in 2015. In the Southern Hemisphere, autumn is the transitional period from the wet to the dry season. As can be seen, in 2015, the total precipitation was far lower than the historical average [7]. Although 2017 has not been the driest year in the northeastern region of Brazil, the rainfall amounts there were far below the historic average, and can be counted as an extension of the 2012 drought. During this period (2012–2017), some of the São Francisco River Hydrographic Region gauging stations recorded zero flow, and the flow released by the reservoirs had to be reduced in order to avoid water supply failures. In 2017, 38 million people were affected by drought and 51% (2.839) of all Brazilian municipalities declared a state of emergency [8]. Brasília, the capital of the nation, and São Paulo, the largest and wealthiest city, endured water rationing during this period.

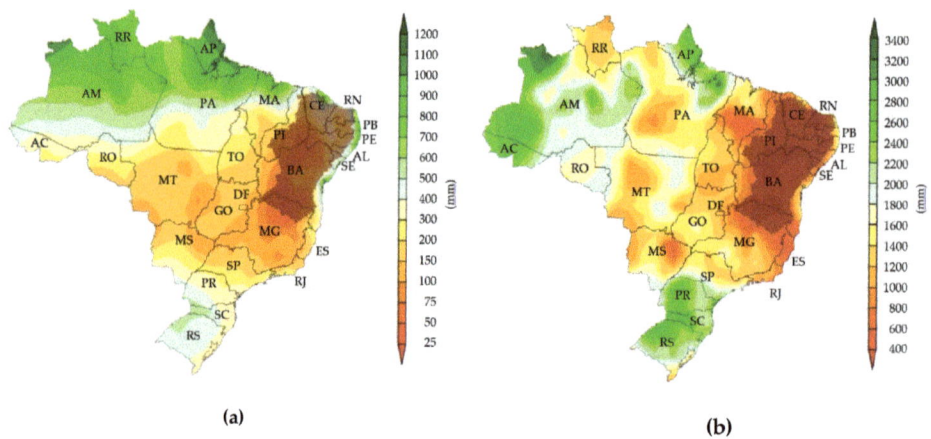

(a) (b)

AC: Acre   AL: Alagoas   AP: Amapá   AM: Amazonas   BA: Bahia   CE: Ceará   DF: Distrito Federal
ES: Espírito Santo   GO: Goiás   MA: Maranhão   MT: Mato Grosso   MS: Mato Grosso do Sul   MG: Minas Gerais
PA: Pará   PB: Paraíba   PR: Paraná   PE: Pernambuco   PI: Piauí   RJ: Rio de Janeiro   RN: Rio Grande do Norte
RS: Rio Grande do Sul   RO: Rondônia   RR: Roraima   SC: Santa Catarina   SP: São Paulo   SE: Sergipe   TO: Tocantins

Figure 1. (a) Average rainfall in the early dry season (1981–2010); (b) total precipitation in 2015. Source: Adapted from Instituto Nacional de Meteorologia (INMET) [7]. The semiarid region (shaded in the Figure 1a,b) encompasses parts of the States of Alagoas, Bahia, Ceará, Minas Gerais, Paraíba, Pernambuco, Piauí, Sergipe, and the whole State of Rio Grande do Norte.

1.2. Water Availability in Brazil

According to the ANA [8], close to 90% of all Brazilian rivers rely on a base flow from aquifers that feed these rivers during dry periods, keeping them perennial. The exception occurs in the northeastern region where the ground is composed of a thin layer of soil and fractured rocks known to be crystalline, unable to feed water back to the rivers.

Although surface water dams are the main plan of action against droughts in the northeast, corresponding to 67% of the solutions adopted by the government [8], in actuality, at the national level, 47% of all municipalities have adopted surface water sources, while 39% consume groundwater, and 14% supply their systems with a mix of both [9]. In 2013 there were 225,868 registered tubular wells across the country. It is estimated, however, that this number could be much larger as a consequence of the proliferation of non-registered wells, possibly close to 477,000 wells in 2013 [10].

In 2014, there were around 21 million people living in the Alto Tietê Hydrographic Basin; 97% of the São Paulo Metropolitan Region population. In this basin, the water demand far exceeds the natural water availability, making it a necessity to import from other basins almost half of all water consumed. As a result of a lack of adequate water resources management, during the 2014–2015 water shortage, the government of the São Paulo State was forced to impose water rationing [11].

Traditionally, investments to fulfill water demands in large Brazilian cities are exclusively allocated to the discovery of new water sources, generally without considering important alternatives, such as water reuse [12]. During the 2014–2015 water crisis, the number of private wells for groundwater extraction increased exponentially as a popular response to water access restrictions, ultimately depleting regional underground water sources.

Groundwater sourcing represents, thus, a very important issue in the national and international context of water management both for rural and as well as urban areas. Given this, the introduction of the managed aquifer recharge (MAR) concept is an innovation in integrated water resources management in Brazil. In a climate change scenario and ever-increasing demand, it is considered both as an adaptation measure regarding extreme events, such as droughts, as well as a mitigation strategy for future water crises.

1.3. MAR as Solution

MAR is defined as "the purposeful recharge of water to aquifers for subsequent recovery or environmental benefit" [13]. MAR can take many forms, including recharge weirs, infiltration basins, riverbank filtration, recharge releases from dams, and recharge wells. The applications of MAR have been implemented since the 1950s for various purposes, such as to increase groundwater storage, improve quality, restore groundwater levels, prevent saline intrusion, and increase ecological benefits [14]. In many arid or semiarid areas, where groundwater is usually already overexploited or saline, recharge has the potential of storing excessive runoff, including in fractured rocks aquifers [15]. The most common MAR applications are for maximizing natural storage, representing 45% of the case studies in Australia, 62% in South America, and 84% in Africa. The main application of MAR in Europe is for water quality management, where approximately 200 riverbank filtration schemes are used for the production of drinking water [14].

In China there are reports of canals dug in the middle of the 5th century BC close to rivers in regions periodically flooded by storm water. These channels were intended to facilitate the infiltration of surface water into groundwater, changing the quality of groundwater and transforming saline land into fertile soil [16]. The USA today stands out in MAR capacity, second only to India [17]. Arizona has implemented MAR facilities that are able to recharge up to 173 M m^3 of the Colorado River, the water of which is considered a renewable resource. In the same state, another project consists of spreading basins through a flood plain producing an annual recharge of up to 37 M m^3 [18]. Arizona has a policy for the specific types of MAR that may be implemented in the state. The Underground Storage and Recovery Act, 1986, and the Underground Water Storage, Savings and Replenishment Program, 1994, provide guidelines for allowing state-supported aquifer recharge. This legislation involves three permissions [17].

This paper aims to provide an overview of the current use of MAR, and the potential and challenges of the adoption of MAR in Brazil, by surveying practices adopted to mitigate the effects of droughts, relevant research, and current policy and legal frameworks.

2. Methodology

A systematic literature review was carried out with the goal of understanding the challenges and opportunities of MAR adoption to strengthen integrated water resources management (IWRM) to confront climate changes in Brazil. Initially, the main social technologies (ST) to deal with droughts were identified, which were simple, low-cost, and easily applicable technologies. Information was sought about their main uses and relevant constructive features. A free term search was applied aiming to eliminate any misunderstandings about these technologies, as a function of the idiomatic diversity among those who share the knowledge construction base.

Then, further research was carried out, in order to provide a brief overview of stormwater/rainwater infiltration and retention techniques increasingly put to use in Brazilian urban areas. Considering that the sustainable urban drainage systems (SUDS) are solutions used worldwide, we decided that it is not necessary to delve into these technologies with as much detail as in the national rural ST.

Finally, environmental and groundwater legislation at national and state levels were gathered. In possession of these documents, the keywords managed aquifer recharge, artificial recharge, and groundwater were used to find the laws that provide specific guidelines on this subject.

The data were obtained from papers, conference proceedings, academic theses and, mainly, from official reports of state and federal institutions that handle this topic directly, e.g., Embrapa and Articulation for the Semi-arid (ASA).

3. Results

3.1. Strategies to Deal with Drought in Brazil

When starting a study in any field of knowledge shared by people whose languages are different, it is essential to define clearly basic concepts and terminology. Thus, in this section, the concepts of MAR are discussed, along with the complementary measures of rainwater and stormwater harvesting techniques found in Brazilian rural areas.

As already defined, MAR is a set of measures that aim to artificially increase the recharge of water in an aquifer. In the Brazilian semiarid region, some of the most common techniques for increasing the natural water reserve are underground dams and infiltration ponds. A detailing of these and other techniques and specific examples applied in Brazil are reported below.

The MAR Global Inventory, available on the International Groundwater Resources Assessment Centre (IGRAC) portal, has gathered ninety MAR applications located in Brazil divided in eight specific MAR types: ditch and furrow, dug well–shaft–pit injection, excess irrigation, induced bank filtration, infiltration ponds and basins, rooftop rainwater harvesting, subsurface dam, and trenches [19]. According to this inventory, MAR solutions are mostly concentrated in the northeast region, as can be seen in Figure 2 below. With respect to the specific MAR type, this inventory points to a predominance of subsurface dam technology (64%), mainly located in northeastern Brazil, 100% of which aim to maximize natural storage to be used for agricultural purposes. In reality, what has been called "subsurface dam" technology in Brazil should be named "underground dam", which is divided into two main types: submerged dams (Costa & Mello type) and submersible dams [20,21]. These methods will be described later. The main influent sources used in Brazil are river water, representing 54% of the cases and stormwater, representing 40%. Regarding the final uses of MAR, 60 applications are used for agriculture, 20 cases use water for domestic use, 8 have ecological uses, and only 2 applications are for research purposes [19].

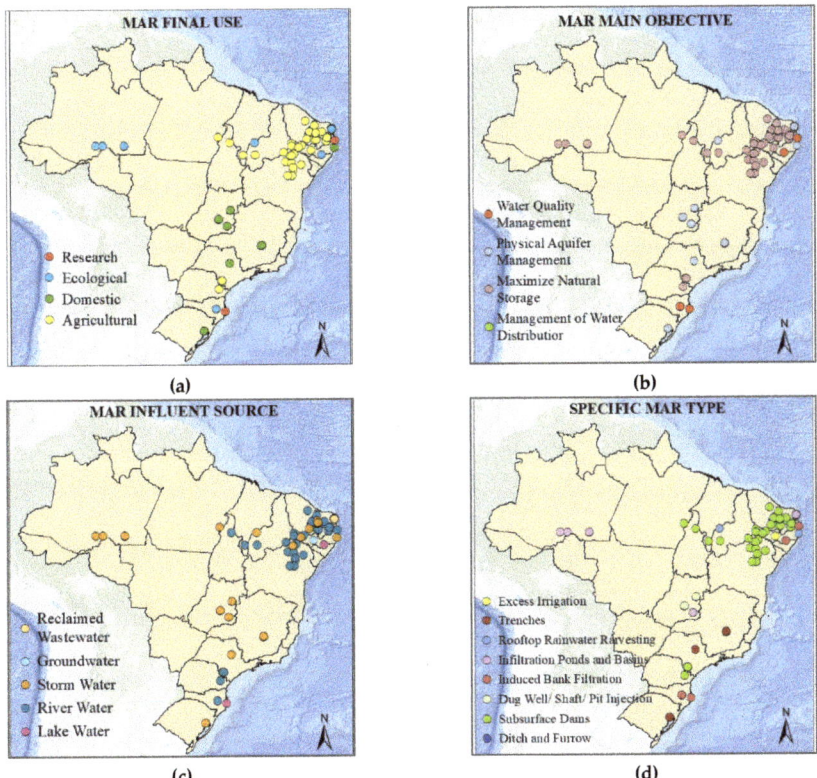

Figure 2. Localization of managed aquifer recharge (MAR) solutions in Brazil, sorted by (**a**) MAR final use; (**b**) MAR main objective; (**c**) MAR influent source; (**d**) Specific MAR type. Source: Adapted from International Groundwater Resources Assessment Centre (IGRAC) website [19].

The applications of underground dams in the Brazilian semiarid region have been very important for transforming the reality of the region's farming population, as they are sustainable and easy to apply technologies. In Alexandrina-RN, the construction of an underground dam of approximately 2.0 ha provided a significant increase in the production of maize, beans and rice, allowing an increase in family income through the sale of the surplus [22]. In Paraíba, the Local Development Training Project constructed two underground dams in two communities in the municipality of Texeira. In one of the communities it was possible to harvest fruit even in the dry season. In addition, the owners reported a better quality of the harvested fruits, as bigger and better looking. The participation of the community itself in the construction of these dams is highlighted [23]. Silva et al. [24] analyzed the use of four underground dams, one in Pernambuco, one in Paraíba, and two in Bahia. In all of them, the importance of underground dams was verified for the food security of the families, as well as the food security of their animals.

The effort in water quality monitoring varied from site to site. Samples of water taken from 8 underground dams located in the states of Pernambuco and Bahia were analyzed [25]. Among them, six dams presented low salinity and low sodicity water, which made them suitable for crop irrigation. The other two dams presented water with some risk for use in irrigation, requiring careful monitoring and soil and water management actions.

Infiltration ponds are widespread in Brazil and can be found in thirteen states, and the Federal District as well [26]. Up to 2013, the Barraginhas Project alone has been responsible for implementing around 500,000 small ponds, all around the country [27]. According to Embrapa Generated Technologies Impacts Evaluation Report, the project has also been responsible for ensuring water and food security and income generation for thousands of families in the semiarid region. Social Technology has also promoted environmental benefits such as soil conservation/restoration, headwaters recovery, and groundwater replenishment [28].

The program P1MC—A Million Cisterns ("Um Milhão de Cisternas") is a program promoted by ASA, whose objective is to promote and ensure access to drinking water for communities in the semiarid region. Based on the principle of stocking up in times of plenty to have enough in times of shortage, this project was awarded in the Future Policy Award in 2017. The project started in 2003; and the goal of building 1 million cisterns was achieved in 2014. Another program of great relevance for living with the semiarid region was the P1+2—One Land and Two Waters Program ("Uma Terra Duas Águas"), whose objective was to increase water security, as well as to promote land management, food security, and income generation [29].

Most of the rural makeshift schemes implemented to deal with drought in Brazil are low-cost measures focused on storing rainwater and stormwater in buried or semi-buried tanks, aiming to ensure small farmers' food and water security. A large portion of these methods are recorded in Government Documents, in free magazines distributed by NGOs and their websites, and referenced in scientific papers.

In the next section, there is a description of some examples of MAR technology that are based on water infiltration into the soil. Following this, some technologies are presented that are considered as MAR, but do not involve aquifer recharge, being technologies for storing water in buried or semi-buried tanks.

3.1.1. MAR based Solutions in Brazil

- Infiltration Pond (Small Dam)

Infiltration ponds are well-known in Brazil as small dams (Barraginhas). The main objective in utilizing this ancient technique is soil restoration and conservation. The first experiment with a small dam was performed at Embrapa Milho and Sorgo, in Sete Lagoas—Minas Gerais, 1991. The success of this experiment triggered the widespread use of this technique, extending it to the Brazilian semiarid region with the objective of helping small farming communities to deal with degradation and water shortages [30]. Aragão [31] states that, although widely used in Brazil, there is a lack of studies on the most suitable areas that allow the best performance of this technique.

Small dams are small half-moon-shaped dams. Their dimensions range from 1.5 m to 2.0 m in depth, and 15 m to 20 m in diameter. They are scattered and successively constructed in the main thalwegs of pastureland and degraded fields (Figure 3a), as well as along roadsides, aiming to prevent soil erosion by surface runoff [32]. Figure 3b shows a cross section of a small dam scheme. In steeper and dry thalwegs the dam cross-sectional shapes need to be trapezoidal. On smooth level thalwegs and roadsides, a triangular shape is the format usually utilized [30,33]. The dams are equipped with spillways in both sides to squirt around the excess of runoff, protecting the structure [34].

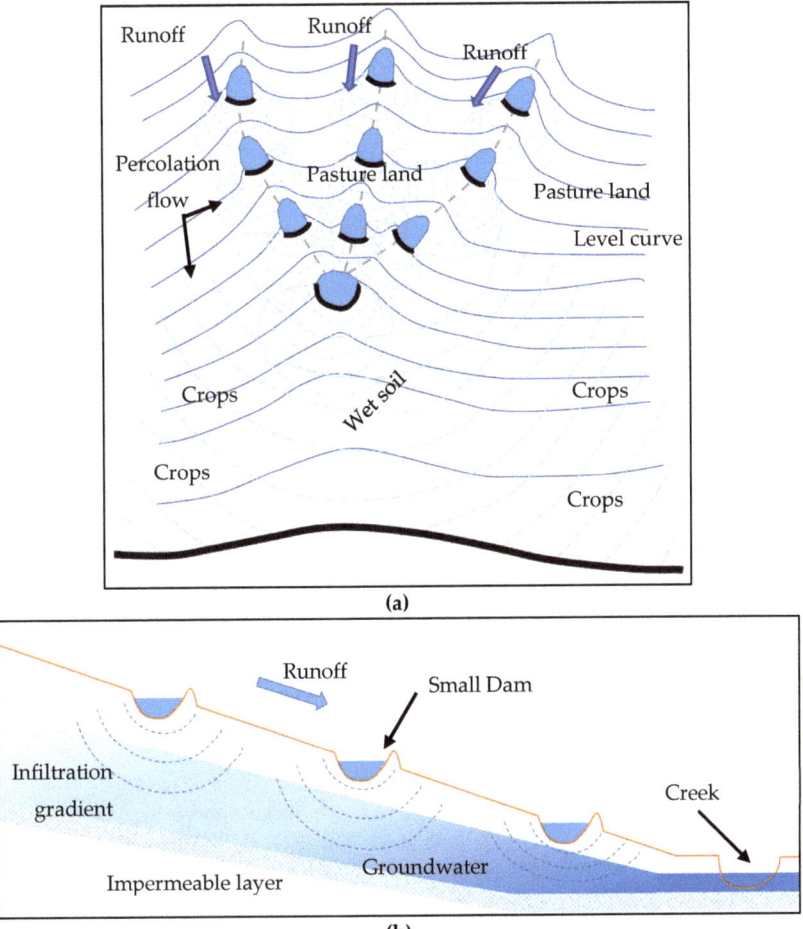

Figure 3. Small Dam scheme: (**a**) example of a scattered located small dam scheme; (**b**) cross section of a small dam scheme.

Environmental features, such as topography, soil type, and land use as well, are fundamental elements that affect its location and functioning. A multicriteria analysis carried out on pre-existing small dams, to check if the locations are the most suitable, concluded that slopes less than 3% are not suitable for the implementation of small dams, slopes between 3% and 8% are moderately adequate, and slopes between 8% and 20% are the most suitable [31]. However, field practice points to the fact that slopes greater than 12% should be avoided [35]. Regarding land use and vegetation cover, anthropized areas with sparse vegetation are the most appropriate [31].

Although the soil features such as infiltration and percolation rates are critical for the analysis of rainwater infiltration systems [36], there is a lack of investigation correlating these points to the efficiency of the small dams. A study carried out on eight small dams built in the north of the state of Minas Gerais has highlighted that the recharge capacity increases with soil porosity. Regarding maintenance, silting processes are the main cause of efficiency loss. Around over eleven years of use, the radiuses of the small dams have decreased 1.26 m on average, and the depth loss was up to 1.32 m. Based on these data, the study recommends that every five years, the owners remove the silting from the small dams until recovering the original dimensions (diameter 6 m, depth 1.5–2.0 m) [37].

In order for this technique to be effective, annual precipitation can be up to 1800 mm. Depending on soil type, each small dam can percolate from 800 m^3 up to 1200 m^3 in a wet season [31]. The notion of successive dams actually helping with retention of pollutants, and the soil acting as a filtering medium, and thus improving water quality, is approached as a consequence of the application of this method, not as its intended goal.

- Underground Dam

An underground dam is an ST that allows rainwater to be stored under riverbeds during the rainy season, making it available for the dry season. Although simple, its construction has to comply with some technical requirements regarding the local where it is built: the alluvium must be predominantly sandy; the slope has to be as level as possible; the depth of the impermeable layer must be greater than 1.5 m; the construction site must be at the narrowest part of the riverbed; and the river head should be avoided, where there is less water. Regarding water quality, low salinity rates are essential to make its implementation feasible [21].

The core idea of its operation is to restrict the flow of the alluvial aquifer by building an impermeable transverse septum, thus raising the level of the upstream water table. In Brazil, there are two types of underground dams suitable to local features: the submerged type, also known as the Costa & Mello type and submersible type. Both of them make use of a buried impermeable septum to restrict the underground flow and are equipped with an Amazon type well to allow the use of the accumulated water in the saturated zone. To build the septum, a trench is dug down to the impervious layer. Then, a plastic blanket is placed over the septum and covered with the excavated material, to block the groundwater flow [21,38,39].

The first type, known as a Costa & Mello-type underground dam (submerged dam), is suitable for the bed river of temporary creeks where the thickness of the sedimentary layer is greater than 1.5 m. This style of construction uses an impermeable septum that is totally buried, retaining only the groundwater flow, making the water table in the alluvium rise upstream of the barrier. There is no physical constraint to the runoff [21,38,39].

There are records of this type of technology in India, Turkey, and Japan, for both irrigation and saline intrusion containment [40–42]. In the Brazilian semiarid region, mainly in the states of Pernambuco, Ceará, and Rio Grande do Norte, this is one of the most applied techniques to deal with water shortages [39]. Figure 4 shows a schematic ground plan (Figure 4a) and cross-section (Figure 4b) of a Costa & Mello type underground dam.

In the second type of underground dam, the submersible underground dam, apart from the buried septum, there is another one made of rocks, bricks, or clay over the riverbed. This barrier makes the superficial flow spread over the land, creating a water pond that lasts up to two to three months after the end of the wet season [21,38,39]. The process of lake formation generates a gradual accumulation of sediments, increasing the thickness of the soil upstream of the dam, thus providing an increase of the storage capacity over time, as happens in sand dams [20,43]. However, some authors warn that it only happens in some cases [44]. This technique is suitable for small rivers and water pathways. This dam over the riverbed is equipped with a spillway made of concrete to spill over excess water and preserve the barrage above the ground, limiting the water level. Upstream, close to the dam, an Amazon-type well is built to recover water for irrigation and for other uses, such as livestock water supply when the water level falls below to the ground level [21,38,39]. Figure 5a shows a schematic ground plan of the submersible underground dam. Figure 5b shows a schematic cross-section of the submersible underground dam.

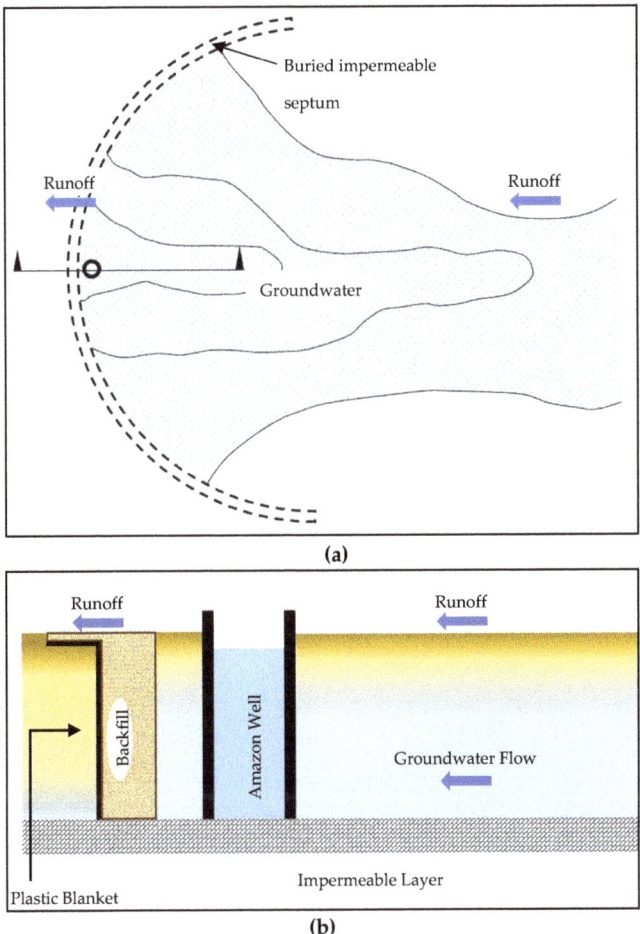

Figure 4. Costa & Mello type underground dam—submerged dam (**a**) Ground plan; (**b**) Cross section.

To implement an underground dam, a set of several factors affects the costs. Among them, the length of the impermeable septum, the raw material, the depth of the impermeable layer, and the workforce available. Based on an underground dam with a septum length of 100 m, using a plastic blanket, and a maximum depth of up to 3.5 m, a study performed by Semiárid Embrapa (CPATSA)/Farming Brazilian Research Company (EMBRAPA) estimated the costs to be around 1300 USD if using heavy machinery [45].

- Dry Well (Caixa Seca)

This is a rainwater harvesting method used to reduce soil erosion, preventing unpaved road deterioration and the silting of rivers and streams. These structures are normally built in a series connected by trenches dug along the roadsides. Based on the builder's empirical knowledge of runoff speed, they are associated with ditches dug diagonally to the axis of the roads, aiming at reducing the surface runoff speed and to covey the rainwater to the dry wells [46]. For safety reasons in the reduction of erosion risks, Bertoni and Neto [47] have recommended that the spacing between the dry wells should be within specific limits, as shown in the Table 1 below.

Table 1. Maximum spacing between the Dry Wells as a function of road slope.

Road Slope (%)	Distance between Dry Wells (m)
5–10	50
10–15	30
>15	20

Source: adapted from Bertoni and Lombardi Neto [47].

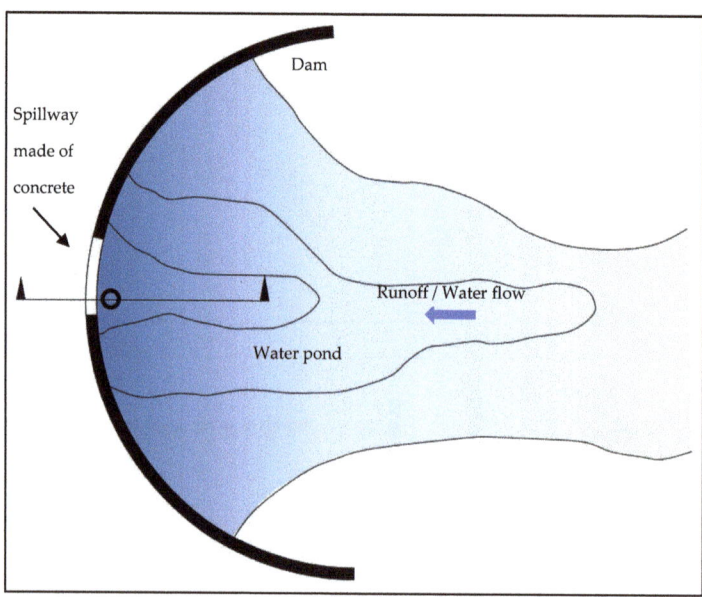

(a)

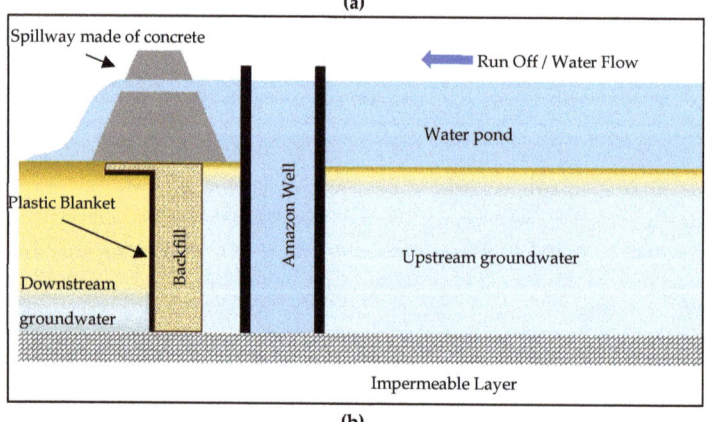

(b)

Figure 5. Submersible underground dam (**a**) ground plan; (**b**) cross section.

The dry well dimensions are defined in the field, based on the builders' experience, never being less than 1 m deep, 1 m wide, and 1 m long (1 m^3). Caixa Seca translates literally to English as dry box. The trenches are 0.30 m in depth, and their width is defined by the width of a hoe. These structures also function as infiltration devices, helping to distribute the water into the ground [48]. Whenever the silt in a dry well reaches 50% of its volume, the silt must be removed [49]. Figure 6 shows a dry well scheme.

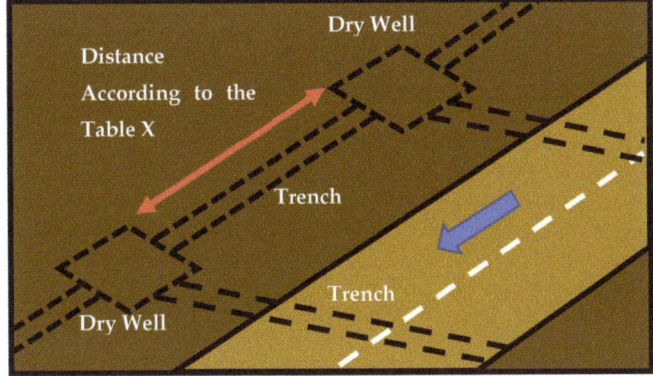

Figure 6. Dry well (caixa seca) scheme.

- Bank Filtration

Bank filtration is a simplified water treatment technique developed in Europe more than a hundred years ago [50]. It consists of groundwater abstraction by a constructed well located close to a river or a lake, aiming to induce a groundwater gradient, forcing the infiltration of the surface water towards the well, thereby improving water quality [51]. Depending on the soil, underground, and water source features, bank filtration could be the only water treatment before a final chlorination step, or at least used as a pre-treatment [52]. The use of this technique is not limited to the Brazilian semiarid region. It can also be found in the Southeastern Region, especially in the State of Santa Catarina, in the south of Brazil. The bank filtration scheme is shown in Figure 7.

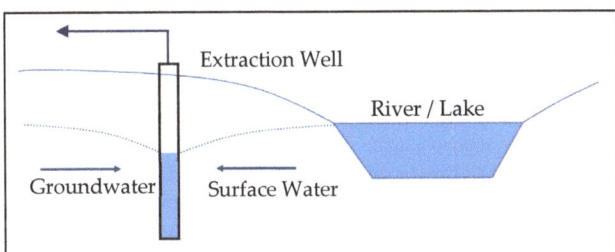

Figure 7. Bank filtration scheme.

3.1.2. Other Solutions

- Rainwater Tank

A rainwater tank is a covered, semi-buried cistern connected to the house gutters and equipped with a first flush device. Its capacity ranges between 16,000 and 21,000 L. The most common storage capacity is the 16,000 L with a 3.45 m diameter and 2.4 m depth. Since it is covered and made from concrete, it is impervious, preventing evaporation and debris contamination [53,54].

- Boardwalk Cistern (Cisterna calçadão)

This method was conceived of as a way to help scattered small farming communities not being served by a large-scale hydraulic infrastructure as a way to help them sustain their food production in their backyards.

It is a covered, semi-buried reservoir 6.4 m in diameter and 1.8 m in depth. Connected to a 200 m^2 concrete floor by a 100 mm pipe, this system is able to collect and store around 52 m^3 of rainwater even under yearly rainfall rates around 350 mm. Since it is covered and made of concrete, it is impervious, preventing evaporation, and debris contamination.

The harvesting area known as boardwalk (calçadão) is surrounded by a small concrete fence, and has a slight slope towards the cistern. The linking device between the boardwalk and the cistern is provided with a small sedimentation tank to avoid particles entering the cistern [55].

- Stormwater Tank

This tank has a 6.2 m diameter, a depth of 1.8 m, and is a covered, totally buried, concrete-plate made cistern, capable of storing up to 52 m^3 of stormwater. The main difference between this technique and the boardwalk cistern is the catchment area. In this model, there is no specific design for the harvesting area. It must be implemented in a slightly sloped ground (<5%) where a water stream naturally flows. A vegetated catchment should be used to avoid the erosion of coarse sediments. This ST uses two serially-placed sedimentation tanks before the water reaches the cistern, in order to remove sand and small stones from the harvested water [56].

- Retention Trench

The standard retention trench (barreiro trench) is a narrow, deep, inverse-trapezoidal shaped, dug reservoir, capable of storing around 500 m^3 of stormwater. The trapezium's larger base measures approximately 24 m with the smaller base at around 16 m, the width at 5 m, and the depth varying from 3 m to 5 m. It should be implemented in the runoff natural pathway, where the terrain slope is as low as possible, avoiding silting and consequent storage volume losses. Its feasibility must be checked by at least three survey boreholes along to the reservoir axis with the objective of identifying the impermeable layer depth [57].

3.2. Early MAR Initiatives in Brazil: Projects and Researches

In larger Brazilian cities, the main rainwater and stormwater management concern is the runoff peak flow attenuation and flooding avoidance. Towards this, various sustainable drainage techniques have been encouraged, but only a few of these focus on stormwater infiltration. Among them, the most widely used techniques are rain gardens, infiltration trenches, and permeable pavements [58]. None of these have water quality control concerns before infiltration and nor water recovery methods.

Nevertheless, a few MAR initiatives have been evaluated, mainly in the academic environment, as a way of improving integrated water resource management processes. They contribute to the assessment of the technical feasibility and were not implemented on a large scale. Okpala [59] carried out the first specific MAR academic assessment in Brazil in 2010. The study assessed the soil aquifer treatment (SAT) feasibility at the Governador André Franco Montoro International Airport (State of São Paulo), in order to help the Brazilian National Airport Infrastructure Company—Infraero to discover an alternative water source for the growing water demand. In this pilot experiment, the most suitable area was chosen from a group of previously assessed areas. The undisturbed samples taken at the unsaturated layer of the selected area were characterized and subsequently assembled into special testing columns, through which secondary treated effluent was infiltrated. Another set of undisturbed samples were taken from a second area similar to the first one. Based on the results, the authors concluded that for SAT be feasible, the first layer of the soil should be removed or replaced by coarse sand for adequate treatment.

Rayis [60] has assessed water quality requirements and costs for MAR implementation in the São Paulo Metropolitan Region as a response to the 2015 water crises. The most suitable treatment process, recharge methods, and water-treated sewage quality features were based on international experiences, such as in the cities of Shafdan (Israel), Atlantis (South Africa), Sabadell (Spain), and Adelaide (Australia). The study adopted influent water quality standards based on the legislation from the US and Spain that specify the required sewage treatment plant effluent treatment level, which is the tertiary treatment level for nitrogen removal. The MAR unitary operation cost was estimated at 1.41 USD/m^3, almost twice the cost of a water mains unitary operation.

The BRAMAR Project, cooperation between Brazil and Germany, intended to help face water shortage by planning and preparing MAR operation schemes. Thus, the Gramame river coastal basin, in the state of Paraíba, was chosen as a study area, aiming to assess the soil infiltration capability and the necessary treatment efficiency of the effluent from the stabilization pond. For this purpose, a 10 mm h^{-1} average input flow was infiltrated into two undisturbed soil columns collected from the study area. Over 72 days, the researchers assessed a set of six physicochemical parameters (BOD$_5$, COD, DOC, TSS, NH$_3$, and NO$_3$). The results showed a reduction in organic matter, suspended soil, and ammoniacal nitrogen greater than 60%. Clogging problems were observed, and the feed procedures were changed. From the 42nd day, wet and dry cycles were implemented aiming to restore soil infiltration capability [61].

A master thesis developed at Campina Grande Federal University aimed to propose a transition from an existing non-managed aquifer recharge reality to an intended MAR scenario in the semiarid region [62]. The study area was the Surucucu alluvial aquifer, located at Paraíba River Basin, in the Sumé Municipality, Paraíba State. The lithologic characterization of the study area was carried out based on data of 117 intrusive investigation boreholes made by the BRAMAR Project. The water table level was monitored through a set of 40 wells distributed along 12 km of the Surucucu riverside, from April 2016 to October 2017. From May 2016 to October 2017, the study has used the chloride ion as a tracer to indicate aquifer contamination by sewage. The results showed that the chloride concentrations kept in high levels, especially in the urban area due to the lack of sanitation. Although decreasing along with the underground flow, the results showed the negative impacts of the non-managed aquifer recharge, pointing out to the risk of salinization. The study has proposed a set of actions to work from non-managed aquifer recharge towards MAR, such as to infiltrate treated sewage by using infiltration ponds, where there is no restriction of space, and to recharge surface water resources when it is possible by using an aquifer storage transfer and recovery systems (ASTR).

3.3. Groundwater Legal Framework

This section aims to present the main groundwater legal documents from both federal and state levels. Although not the same, artificial recharge is the closest in meaning to the MAR expression. Therefore, it was used as a keyword in the place of MAR when searching on the internet for documents related to the Brazilian legal framework for MAR. From now on, the acronym "MAR" will be used in place of "artificial recharge".

3.3.1. Federal Level

From a set of seventeen groundwater legal documents at the federal level, four (23.5 %) mention MAR in their content. Up to the early 2000s, there was no legislation clearly addressing MAR. In 2001, the Water Resources National Council Resolution n° 15 encouraged municipalities to adopt MAR. In the following year, Water Resources National Council Resolution n° 22 established that withdraw and recharge estimates should be included in the water resources plans. In 2008, the Environment National Council Resolution n° 396 permitted establishment of MAR to avoid saline intrusion, providing that there were no changes in water quality, and established water quality mandatory monitoring. At the end of the same year, the Water Resources National Council Resolution n° 92 made prior authorization and mandatory monitoring a condition of aquifer recharge (Table 2).

Table 2. Legal framework at the federal level.

Legal Documents Addressing Underground Water	17
Underground Water Legislation Addressing MAR	4
Legal Document	Article
CONAMA N° 396/2008	Art. 23 Allows MAR to avoid saline intrusion Art. 25 Made water quality monitoring mandatory
Resolution n° 15, 2001/01/11	Art. 6 Encourages MAR
Resolution n° 22, 2002/05/24	Art. 3 Establishes discharge and recharge estimate mandatory in Water Resources Plans
Resolution n° 92, 2008/11/05	Art. 8 Conditions aquifer recharge on prior authorizationArt. 10 VIII Makes water quality monitoring mandatory

3.3.2. State Level

Of all state water resources laws that address underground water, around a fifth (20.3%) mention MAR. Most of them have been promulgated by the states located in the semiarid region, and follows the federal level regulations. Concerning treated wastewater, the Tocantins State legal framework prohibits its discharge into groundwater, although it allows MAR under technical, economic and sanitary assessment, and prior authorization by the Tocantins' Nature Institute. The States of Pernambuco, Ceará, and Maranhão define MAR clearly as water injection through underground dams or injection wells. Santa Catarina State's definition of MAR is generic, defining it as any intentional infiltration technique. It should be highlighted that, among several states, only Pernambuco and Ceará encourage MAR adoption by citizens and companies through rebate schemes on sanitation taxes. Both Pernambuco and Ceará States condition water withdraws to natural recharge features maintenance or MAR. Table 3 shows the basic principles of the state laws addressing MAR.

Table 3. Legal Framework at the State Level.

	Legal Documents Addressing Underground Water	74
	Underground Water Legislation Addressing MAR	15
State	Legal Documents	Basic Contents Related to MAR
Pará	State Law n° 6.381, 2001/07/25	The State Water Resources Council allows MAR under technical, economic, and sanitary assessment, preserving groundwater quality
Tocantins	State Law n° 1.307, 2002/03/22	The State Water Resources Council allows MAR under technical, economic, and sanitary assessment, preserving groundwater quality
	Decree n° 2.432, 2005/06/06	Prohibits treated wastewater discharge into groundwater, although The State Water Resources Council allows MAR under technical, economic and sanitary assessment, and prior authorization by Tocantins' Nature Institute
Roraima	State Law n° 547, 2006/06/23	The State Water Resources Entity along the with Watershed Council allow MAR under technical, economic, and sanitary assessment
Pernambuco	Decree n° 20.423, 1998/03/26	Defines MAR clearly as water injection through underground dams or injection wells Encourages MAR adoption by citizens and companies through rebate schemes on sanitation taxes Conditions water withdraws to natural recharge features maintenance or MAR, and prior authorization
Ceará	State Law n° 14.844, 2010/12/28	Supports or carries out MAR projects to ensure groundwater quality and quantity Defines MAR clearly as water injection through underground dams or injection wells Encourages MAR adoption by citizens and companies through rebate schemes on sanitation taxes
Maranhão	Decree n° 31.077, 2012/12/12	Conditions water withdraws to natural recharge features maintenance or MAR, and prior authorization
	Decree n° 28.008, 2012/01/30	Defines MAR clearly as water injection through underground dams or injection wells Conditions water withdraws to natural recharge features maintenance or MAR, and prior authorization
Bahia	State Law n° 11.612, 2009/10/08	Supports or carries out MAR projects to ensure groundwater quality and quantity
Alagoas	State Law n° 7.094, 2009/09/02	Encourages MAR projects to ensure groundwater quality and quantity Conditions water withdraws to natural recharge features maintenance or MAR, and prior authorization
Piauí	State Law n° 5.165, 2000/08/17	The State Water Resources Council allows MAR under technical, economic, and sanitary assessment, preserving groundwater quality
Minas Gerais	State Law n° 13.771, 2000/12/11	The State Water Resources Council allows MAR under technical, economic, and sanitary assessment
Espírito Santo	State Law n° 6.295, 2000/07/27	The State Water Resources Council allows MAR under technical, economic, and sanitary assessment
Santa Catarina	Resolution CERH n° 02, 2014/08/14	Defines MAR as any intentional infiltration technique
		The State Water Resources Council allows MAR under technical, economic, and sanitary assessment, preserving groundwater quality
Rio Grande do Sul	Decree n° 42.047, 2002/12/26	Conditions MAR to prior authorization by State Agencies

4. Discussion

Brazilian folk wisdom has been the generator of several initiatives to the fight against droughts. At least two of them have been developed and applied only in Brazil: small dams (Barraginhas) and dry wells (Caixa Seca). However, regarding water quality and volume monitoring, there are no systematic records that allow a scientific MAR approach. Although the MAR global inventory points to the prevalence of in-channel modifications (underground dams) over the infiltration ponds and basins, there is no registration of small dams (Barraginhas) on the platform, resulting in underreporting of this type of technology. However, there are governmental reports that state there have been more than 500,000 small dams constructed. It is possible to estimate the number of schemes implemented and costs by mining grey literature, but the reliability of these information should be checked in field research. Some of these technologies have been implemented into government programs, such as P1MC and P1+2. It must be highlighted that some STs, such as rainwater tanks, stormwater cisterns, boardwalk cisterns, and retention trenches may be misinterpreted as an aquifer recharge practice. Although using buried or semi-buried reservoirs, these STs use the impervious tanks to store, not to infiltrate the harvested rainwater. In a large portion of the northeastern region, the soil features, such as salinity and low infiltration rates, have led to these technologies instead of aquifer recharge [62]. Despite being widely encouraged by legal frameworks at both federal and state levels, and being used all around the country, MAR technologies have no water quality monitoring programs in Brazil due to both the lack of a custom of monitoring and government underfunding.

Although federal and state legislation cites water reuse and MAR, there is no clear link between them [60]. In addition, the use of alternative sources of water is generally viewed with some suspicion, as a result of an intricate set of social and institutional barriers resulting from the perception of risks related to the various possible uses of this water [60]. This lack of information and long-term studies is one of the factors that hinder the development of appropriate legal framework supporting on MAR in Brazil. In view of this, projects such as the international cooperation Brazil-Germany (BRAMAR), which helped to face water shortage by planning and preparing MAR operation schemes, should be encouraged, and their results should be further disseminated to promote the acceptance and encouragement of this type of technology.

The main concern related to rainwater/stormwater urban management is the runoff peak flow attenuation and flood avoidance by sustainable drainage techniques. Most of these techniques focus on rainwater retention close to where it precipitates. None of them have aquifer recharge for further water recovery purposes. Consequently, they cannot be considered MAR applications. This delay in adopting MAR initiatives, as happens also in India, induces the risk of loss of economic and social benefits [63]. However, since 2010, there have been some academic MAR assessment initiatives for both urban and rural areas as well. Regarding economic assessment of MAR in urban areas, preliminary assessment points to the MAR unitary operation cost to be almost twice the unitary water mains operation cost (US$ $1.41/m^3$ versus US$ $0.75/m^3$) [60]. It should be highlighted that "[t]he averaged costs of water supplies from four desalination plants in eastern and southern Australia built since the 'millennium drought' [64] to secure capital city supplies is more than 10 times the long-run marginal costs of normal supplies in those cities" [65].

5. Conclusions

The semiarid region of Brazil and big cities have been identified as the main areas that need strategies to combat water scarcity and measures that ensure water security. Identifying aquifers that are suitable for MAR application sand the availability of water sources for recharge are strategic actions to enable the selection of projects that present best cost benefits and that are alternatives for the use of traditional sources.

Although there are some studies being developed in the academic environment, Brazil is still at an early stage in MAR initiatives and needs to overcome technical, legal, and socio-cultural challenges to adopt MAR. Adoption of MAR should be considered a strategy for facing future droughts under a climate change scenario [65]. However, the lack of awareness concerning MAR solutions and non-specific local policies linking MAR schemes to water demand have delayed its development in Brazil. In this sense, it is necessary to identify areas, such as the semiarid region in the northeast, urban settlements, and water-intensive agricultural lands, where the demand for water overcomes the availability, threatening water security. In these areas, the government should encourage the search for aquifers suitable for MAR, on which its adoption could be a more cost-effective alternative than traditional sources, with the perspective of improving the integrated management of water resources.

Most Brazilian MAR scheme records are dispersed in grey literature, such as non-governmental organization (NGO) websites and governmental reports. The MAR Global Inventory is, therefore, a helpful initiative to gather them in a reliable source. However, the majority of the information must be checked in the field by research projects.

As has happened in a number of countries where MAR schemes have been widely adopted, there are no consistent regulations for the implementation of MAR in Brazil. Seeking to establish a MAR regulation framework, to be based on the Australian guidelines, is a sound and less difficult attempt to provide principles for safe implementation of MAR schemes [66]. Taking advantage of state autonomy in groundwater management [60], guidelines on MAR suitable for local features must be developed to help improving integrated water resources management in Brazil based on scientific approaches.

Author Contributions: T.S.: Writing—original draft; L.F.: Writing—review & editing; S.G.M.: Writing—review & editing. All authors have read and agreed to the published version of the manuscript.

Funding: This research was partially funded by CNPq (Conselho Nacional de Desenvolvimento Científico e Tecnológico) through the DIGIRES project, grant no. 400128/2019-5. Also, the authors would like to acknowledge the financial support from CNPq for the PQ research grant, and from FACEPE (Fundação de Amparo à Ciência e Tecnologia do Estado de Pernambuco) for granting PhD scholarship.

Acknowledgments: The authors would like to thank the Vice-Presidency of the Environment, Attention and Health Promotion - Fiocruz Brasil, for supporting the research and Rogério Silva, for his contribution to the English review. Our special thanks to Warish Ahmed for kindly transfer his waiver quote to publish this manuscript. Finally, we thank the reviewers and editors for their valuable comments and suggestions.

Conflicts of Interest: The authors declare no conflict of interest.

Abbreviations

The following abbreviations are used in this manuscript:

ANA	National Water Agency
ASA	Articulation for the Semi-arid
ASTR	Aquifer Storage Transfer and Recovery
BRAMAR	Brazil Managed Aquifer Recharge
CPATSA	Semiárid Embrapa
Embrapa	Farming Brazilian Research Company
Infraero	Brazilian Airport Infrastructure Company
IGRAC	International Groundwater Resources Assessment Centre
IWRM	Integrated Water Resources Management
MAR	Managed Aquifer Recharge
NGO	Non-governmental Organization
P1+2	One Land, Two Waters Program
P1MC	One Million Cisterns Program
SAT	Soil Aquifer Treatment
SPMR	São Paulo Metropolitan Region
ST	Social Technologies
SUDS	Sustainable Urban Drainage Systems

References

1. ANA—Agência Nacional de Águas. *Contas econômicas ambientais da água no Brasil 2013–2015/Agência Nacional de Águas, Instituto Brasileiro de Geografia e Estatística, Secretaria de Recursos Hídricos e Qualidade Ambiental*; ANA: Brasília, Brazil, 2018; p. 79. Available online: https://www.ana.gov.br/todos-os-documentos-do-portal/documentos-spr/contas_economicas.pdf (accessed on 15 February 2020).
2. ANA—Agência Nacional de Águas. *Conjuntura dos Recursos Hídricos no Brasil 2018: Informe Anual*; Agência Nacional de Águas: Brasília, Brasil, 2018. Available online: http://arquivos.ana.gov.br/portal/publicacao/Conjuntura2018.pdf (accessed on 15 February 2020).
3. Centro de Estudos e Pesquisa em Engenharia e Defesa Civil. Available online: http://www.ceped.ufsc.br/historico-de-secas-no-nordeste-do-brasil/ (accessed on 10 September 2019).
4. Costa, M.C.L. Medical Theories and Urban Management: Fortaleza 1877-79 Drought. *Hist. Cienc. Saude Manguinhos* **2004**, *11*, 57–74. [CrossRef] [PubMed]
5. Banco de Dados Folha—A Seca já Preocupa Brasília. Available online: http://almanaque.folha.uol.com.br/brasil_08mar1981.htm (accessed on 10 September 2019).
6. Denys, E.; Engle, N.L.; Magalhães, A.R. *Secas no Brasil: Política e Gestão Proativas*; Centro de Gestão e Estudos Estratégicos—CGEE: Brasília, Brazil, 2016; ISBN 978-85-5569-116-4.
7. INMET—Instituto Nacional de Meteorologia. Available online: http://www.inmet.gov.br/portal/index.php?r=bdmep/bdmep (accessed on 5 October 2019).
8. Brasil. Agência Nacional de Águas. Conjuntura dos recursos hídricos no Brasil 2017: Relatório pleno (Brazilian Water Resources Report 2017: Full Report). Available online: http://www.snirh.gov.br/portal/snirh/centrais-de-conteudos/conjuntura-dos-recursos-hidricos/conj2017_rel-1.pdf (accessed on 2 November 2019).
9. Atlas Brasil: Abastecimento Urbano de Águas. Available online: http://atlas.ana.gov.br/Atlas/forms/Home.aspx/ (accessed on 13 October 2019).
10. ANA—Agência Nacional de Águas. Conjuntura dos recursos hídricos no Brasil 2013 (Brazilian Water Resources Report 2013). Available online: http://arquivos.ana.gov.br/institucional/spr/conjuntura/webSite_relatorioConjuntura/projeto/index.html (accessed on 3 October 2019).
11. Hirata, R.; Conicelli, B.; Pinhatti, A.; Luiz, M.; Porto, R.; Ferrari, L. O Sistema Aquífero Guarani e a Crise Hídrica nas Regiões de Campinas e São Paulo. *Revista USP* **2015**, *106*, 59–70. [CrossRef]
12. Cirilo, J.A. Crise Hídrica: desafios e superação. *Revista USP.* **2015**, *106*, 45–58. [CrossRef]
13. NRMMC; EPHC; NHMRC. *Australian Guidelines for Water Recycling, Managing Health and Environmental Risks, Vol 2C: Managed Aquifer Recharge*; Natural Resource Management Ministerial Council, Environment Protection and Heritage Council National Health and Medical Research Council: Camberra, Autralia, 2009; p. 237.
14. Stefan, C.; Ansems, N. Web-based global inventory of managed aquifer recharge applications. *Sustain. Water Resour. Manag.* **2018**, *4*, 153–162. [CrossRef]
15. Dillon, P. Future management of aquifer recharge. *Hydrogeol. J.* **2005**, *13*, 313–316. [CrossRef]
16. Wang, W.; Sun, X.; Xu, Y. Recent advances in managed aquifer recharge in China. In Proceedings of the International Conference on Challenges in Environmental Science and Computer Engineering (CESCE), Wuhan, China, 6–7 March 2010; Volume 2, pp. 516–519.
17. Dillon, P.; Stuyfzand, P.; Grischek, T.; Lluria, M.; Pyne, R.D.; Jain, R.C.; Bear, J.; Schwarz, J.; Wang, W.; Fernandez, E.; et al. Sixty years of global progress in managed aquifer recharge. *Hydrogeol. J.* **2018**, *27*. [CrossRef]
18. Portal Tucson City. Available online: https://www.tucsonaz.gov/water/recharged-water (accessed on 18 February 2020).
19. IGRAC Portal. Available online: https://apps.geodan.nl/igrac/ggis-viewer/viewer/globalmar/public/default (accessed on 9 February 2020).
20. Da Silva, M.S.L.; Lima, A.O.; Moreira, M.M.; Ferreira, G.B.; Barbosa, A.G.; Melo, F.R.; Neto, M.B.O. Barragem subterrânea. In *Ximenes*; da Silva, L.F., Brito, M.S.L., Org, L.T.L., Eds.; Tecnologias de Convivência com o Semiárido Brasileiro; Banco do Nordeste do Brasil: Fortaleza, Brasil, 2019; pp. 223–281.
21. Costa, W.D.; Cirilo, J.A.; Pontes, M.; Maia, A.Z.; Sobrinho, P.O. Barragem Subterrânea: Uma Forma Eficiente de Conviver com a Seca. In Proceedings of the X Congresso Brasileiro de Águas Subterrâneas, São Paulo, Brazil, 9–11 September 1998.

22. Brito, L.D.; da Silva, D.A.; dos Anjos, J.B. Barragem subterrânea: Um estudo de caso. In Proceedings of the Congresso Brasileiro de Engenharia Agricola, Campina Grande, Brazil, 21–25 July 1997. SBEA/UFPB.
23. Campos, J.D.; Neto, J.R.; Sampaio, O.B.; Sonda, C. Barragem subterrânea: Uma alternativa de captação e barramento de água da chuva no semi-árido. In Proceedings of the 3o Simpósio Brasileiro de Captação de Água de Chuva no Semi-árido, Campina Grande, Brazil, 21–23 November 2001; ABCMAC - Brazilian Rainwater Catchment and Management Association. Available online: http://www.abcmac.org.br/files/simposio/3simp_josedias_barragemsubterranea.pdf (accessed on 12 February 2020).
24. da Silva, M.S.; de Araújo, A.H.; Ferreira, G.; Cunha, T.; Oliveira Neto, M.B. Barragem subterrânea: Contribuindo para a segurança alimentar e nutricional das famílias do Semiárido brasileiro. *Cad. Agroecol.* **2013**, *8*. Available online: http://www.aba-agroecologia.org.br/revistas/index.php/cad/article/view/15261 (accessed on 24 February 2020).
25. de Oliveira, A.K.; da Silva, M.S.; Mendonça, C.C.; Ferreira, G.B.; Chaves, V.C.; Silva, D.J. Avaliação qualitativa da água de barragens subterrâneas no semi-árido nordestino brasileiro. *Rev. Bras. Agroecol.* **2007**, *2*. Available online: http://www.aba-agroecologia.org.br/revistas/index.php/rbagroecologia/article/view/7092 (accessed on 24 February 2020).
26. Embrapa Portal. Available online: https://www.embrapa.br/busca-de-noticias/-/noticia/38576935/barraginhas-e-seus-beneficios-sao-tema-de-exposicao-no-shopping-sete-lagoas (accessed on 30 March 2020).
27. Embrapa Portal. Available online: https://www.embrapa.br/busca-de-noticias/-/noticia/1489846/convenio-ira-assegurar-construcao-de-mais-de-17-mil-barraginhas-no-norte-de-minas (accessed on 30 March 2020).
28. Miranda, R.A. *Relatório de Avaliação dos Impactos de Tecnologias Geradas pela Embrapa—Mini-Barragens de Contenção de Águas Superficiais de Chuva—Barraginhas*; EMBRAPA. Empresa Brasileira de Pesquisa Agropecuária. Embrapa Milho e Sorgo: Sete Lagoas, Brasil, 2019; p. 32. Available online: https://bs.sede.embrapa.br/2018/relatorios/milhoesorgo_2018_minibarragens.pdf (accessed on 30 March 2020).
29. ASA—Articulação Semiárido Brasileiro. Available online: https://www.senado.gov.br/comissoes/CMMC/AP/AP20090924_ASA_Vida%20Semiarido.pdf (accessed on 28 January 2020).
30. de, Barros, L.C. *Captação de Águas Superficiais de Chuvas em Barraginhas*; EMBRAPA. Empresa Brasileira de Pesquisa Agropecuária. Circular Técnica, 2. Embrapa Milho e Sorgo: Sete Lagoas, Brasil, 2000; p. 16. Available online: https://www.embrapa.br/milho-e-sorgo/busca-de-publicacoes/-/publicacao/484688/captacao-de-aguas-superficiais-de-chuvas-em-barraginhas (accessed on 3 October 2019).
31. Aragão, V.R.; de Souza, A.C.S.; Paoliello, T.; Silva, T.G.; Lima, F.A. Identificação de áreas aptas a barraginhas na bacia do rio São Lamberto. *Holos Environ.* **2019**, *19*, 304–319. [CrossRef]
32. De Barros, L.C.; de Souza Tavares, W.; de Resende Barros, I.; de Aquino Ribeiro, P.E. Integração das tecnologias sociais barraginhas e lago de múltiplo uso. *Revista Brasileira de Agropecuária Sustentável (RBAS)* **2011**, *1*, 1–5.
33. de Barros, L.C. Amenização de Veranicos Através da Captação de Água de Chuvas por barraginhas, garantindo Safras na Agricultura Familiar, em Minas Novas, MG. In Proceedings of the Congresso Nacional de Milho e Sorgo, 26; Simpósio Brasileiro Sobre a Lagarta-do-Cartucho, Spodoptera Frugiperda, 2.; Simpósio Sobre Colletotrichum Graminicola, 1., Belo Horizonte, Brasil, 27–31 August 2006; ABMS. Available online: https://www.embrapa.br/busca-de-publicacoes/-/publicacao/490167/amenizacao-de-veranicos-atraves-da-captacao-de-agua-de-chuvas-por-barraginhas-garantindo-safras-na-agricultura-familiar-em-minas-novas-mg (accessed on 3 January 2019).
34. Hipólito, M.T.; Costa, T.C.C.; Barros, L.C.; Viana, A.A.O.N. *Alocação de Barraginhas com uso de Modelagem Hidrológica e Geoprocessamento*; Embrapa Milho e Sorgo: Sete Lagoas, Brazil, 2019; 28p, ISSN 1679-0154.
35. Cordoval, L.; Ribeiro, P.E.A. Barraginhas, Alimentação de aquíferos. Embrapa. Available online: https://www.infoteca.cnptia.embrapa.br/infoteca/bitstream/doc/1099353/1/Barraginhasrealimentacao.pdf (accessed on 24 March 2020).
36. Leal, M.S.; Tonello, K.C. Análise da morfometria e do uso e cobertura da terra da microbacia do córrego Ipaneminha de Baixo. *Floresta* **2016**, *46*, 439–446. [CrossRef]
37. Aragão, V.R.; Brito, A.F.; Souza, A.C.S.; Versiani Junior, E.R. Avaliação do funcionamento de barraginhas em solos de textura média e arenosa. *Rev. Int. Ciências* **2019**, *9*, 115–126. [CrossRef]
38. Lima, A.D. Nova abordagem metodológica para locação, modelagem 3D e monitoramento de barragens subterrâneas no semiárido Brasileiro. Ph.D. Thesis, Federal University of Rio Grande do Norte, Natal, Brazil, 2013.

39. Ministério da Cidadania—Secretaria Especial do Desenvolvimento Social—Programa Cisternas—Modelo Da Tecnologia Social De Acesso À Água N° 05—Barragem Subterrânea. 2015. Available online: http://www.mds.gov.br/webarquivos/arquivo/seguranca_alimentar/cisternas_marcolegal/tecnologias_sociais/Barragem%20Subterranea05/IO_SESAN_n4_09072015_ANEXO.pdf (accessed on 3 January 2019).
40. Senthilkumar, M.; Elango, L. Modelling the impact of a subsurface barrier on groundwater flowing in the lower Palar River basin, southern India. *Hydrogeol. J.* **2011**, *9*, 917–928. [CrossRef]
41. Apaydin, A.M. Groundwater dam: An alternative model for semi-arid regions of Turkey to store and save groundwater. *Environ. Earth Sci.* **2009**, *59*, 39–345. [CrossRef]
42. Ishida, S.; Koturu, M.; Abe, E.; Fazal, M.A.; Tsuchihara, T.; Imaizumi, M. Construction of subsurface dams and tehir impacts on the environment. *Mater. Geoenviron.* **2003**, *50*, 149–152.
43. Hansson, G.; Nilsson, A. Groundwater dams for rural water supplies in developing countries. *Groundwater* **1986**, *24*, 497–506. [CrossRef]
44. Van Haveren, B.P. Dependable water supplies from valley alluvium in arid regions. *Environ. Monit. Assess.* **2004**, *99*, 259–266. [CrossRef]
45. da Silva, M.S.L.; Mendonça, C.E.S.; dos Anjos, J.B.; Honório, A.P.M.; de Silva, A.S.; de Brito, L.T.L. Barragem subterrânea: Água para produção de alimentos. In *Potencialidades da Água de chuva no SemiÁrido Brasileiro*; de Brito, L.T.L., de Moura, M.S.B., Gama, G.F.B., Eds.; Embrapa SemiÁrido: Petrolina, Brazil, 2007; pp. 121–137.
46. Canazart, A.C. Plantio de Água: Um Processo de Formação, Desenvolvido em Conjunto à Comunidade de Camões- Sem Peixe. Bachelor's Thesis, Viçosa Federal Univerty, Viçosa, Brazil, 2017.
47. Bertoni, J.; Lombardi Neto, F. *Conservação de Solo*; Embrapa Milho e Sorgo; Livroceres: Piracicaba, Brazil, 1985; 368p.
48. Porto, E.R.; Vivallo Pinare, A.G.; Williams Fuentes, C.O.; Silva, A.D.; Lopes, L.D. *Pequenos Agricultores V - Métodos De Execução De Sistemas Integrados de Produção Agropecuária (SIP)*; EMBRAPA-CPATSA: Petrolina, Brazil, 1990; 70p, Available online: http://www.infoteca.cnptia.embrapa.br/infoteca/handle/doc/132911 (accessed on 3 January 2019).
49. Dalto, G.G.; Paye, H.S.; Comério, A.; Batista, R. Tecnologias de conservação e armazenamento de água em propriedades rurais. *Incaper Rev.* **2016**, *6*, 42–50.
50. Schubert, J. German Experience with Riverbank Filtration Systems. In *Riverbank Filtration—Improving Source-Walter Quality*; Ray, C., Linsky, R.B., Melin, G., Eds.; Kluwer Academic: Dordrecht, Denmark, 2002; pp. 35–48.
51. Dillon, P.; Pavelic, P.; Page, D.; Beringen, H.; Ward, J. *Managed aquifer recharge: An Introduction*; Waterlines Report Series No. 13; National Water Commission: Canberra, Australia, 2009; ISBN 978-0-9807727-6-0.
52. Ferreira, A.L.N. Diretrizes de Gestão Ambiental para o Sistema de Captação de Água por Filtração em Margem. Master's Thesis, Pernambuco Federal University, Recife, Brazil, 2014.
53. Brasil. Ministério do Desenvolvimento social. Modelo de Tecnologia Social de Acesso a Água n° 1: Cisternas de Placas de Concreto de 16 Mil Litros (Social Technology for Water Access n° 1: 16 KL Concrete Plate Cistern). Available online: http://www.mds.gov.br/webarquivos/arquivo/segurança_alimentar/cisternas_marcolegal/tecnologias_sociais/Cisterna%20de%20Placa%20de%2016%20mil%20litros01/ (accessed on 13 October 2019).
54. Governo do Estado do Ceará. Secretaria de Recursos Hídricos. Cisternas de Placas: Construção, Uso e Conservação. Available online: http://prodham.srh.ce.gov.br/index.php?option=com_phocagallery&view=category&id=3&Itemid=188 (accessed on 15 October 2019).
55. Brasil. Ministério do Desenvolvimento social. Modelo de Tecnologia Social de Acesso a Água n° 2: Cisterna Calçadão de 52 mil Litros (Social Technology for Water Access n° 2: 52 KL Calçadão Cistern). Available online: http://www.mds.gov.br/webarquivos/arquivo/seguranca_alimentar/cisternas_marcolegal/tecnologias_sociais/ (accessed on 13 October 2019).
56. Brasil. Ministério do Desenvolvimento social. Modelo de Tecnologia Social de Acesso a Água n° 3: Cisterna de Enxurradas de 52 mil Litros (Social Technology for Water Access n° 3: 52 KL Enxurrada Cistern). Available online: http://www.mds.gov.br/webarquivos/arquivo/seguranca_alimentar/cisternas_marcolegal/tecnologias_sociais/ (accessed on 13 October 2019).

57. Brasil. Ministério do Desenvolvimento social. Modelo de Tecnologia Social de Acesso a Água n° 4: Barrreiro Trincheira Familiar (Social Technology for Water Access n° 4: Familiar Barreiro Trench). Available online: http://www.mds.gov.br/webarquivos/arquivo/seguranca_alimentar/cisternas_marcolegal/tecnologias_sociais/ (accessed on 13 October 2019).
58. Brasil. Ministério das Cidades. Manual para Apresentação de Propostas para Sistemas de Drenagem Urbana Sustentável e de Manejo de Águas Pluviais-Programa—2040: Gestão de Riscos e Resposta a Desastres. Available online: https://contas.tcu.gov.br/etcu/ObterDocumentoSisdoc?seAbrirDocNoBrowser=true&codArqCatalogado=5986406&codPapelTramitavel=49928716 (accessed on 2 November 2019).
59. Okpala, W.O. Recarga Gerenciada do Aquífero do Aeroporto Internacional Governador André Franco Montoro, Guarulhos/SP: Estudo Piloto do Sistema Solo-aquífero. Ph.D. Thesis, São Paulo University, São Paulo, Brazil, 2010.
60. Rayis, M.W.A. Avaliação da Viabilidade do Reúso de Água para Recarga de Aquíferos na Região Metropolitana de São Paulo. Master's Thesis, São Paulo University, São Paulo, Brazil, 2018.
61. Coutinho, J.; Almeida, C.; Bernardo Da Silva, E.; Stefan, C.; Athayde Júnior, G.; Gadelha, C.L.M.; Walter, F. Managed aquifer recharge: Study of undisturbed soil column tests on the infiltration and treatment capacity using effluent of wastewater stabilization pond. *Braz. J. Water Resour.* **2018**, *23*, 1–9. [CrossRef]
62. Pontes Filho, J.D.A. Da Recarga não Gerenciada à Recarga Gerenciada: Estratégia para Aquífero Aluvial no Semiárido Brasileiro. Master's Thesis, Campina Grande Federal University, Campina Grande, Brazil, 2018.
63. Sakthivel, P.; Elango, L.; Amirthalingam, S.; Pratap, C.E.; Brunner, N.; Starkl, M.; Thirunavukkarasu, M. Managed aquifer recharge: The widening gap between law and policy in India. *Water Supply* **2015**, *15*, 1159–1165. [CrossRef]
64. Radcliffe, J.C. Water Recycling in Australia—During and After the Drought. *Water Res. Technol.* **2015**, *1*, 554–562. [CrossRef]
65. Dillon, P. Australian Progress in Managed Aquifer Recharge and the Water Banking Frontier. *Water AWA J.* **2015**, *42*, 53–57.
66. Capone, F.; Bonfanti, M.E. Legislative Framework Review and Analysis. Deliverable D17.1 MARSOL Project. 2015. Available online: http://www.marsol.eu/files/marsol_d17-1_legislative_20150321.pdf (accessed on 2 November 2019).

© 2020 by the authors. Licensee MDPI, Basel, Switzerland. This article is an open access article distributed under the terms and conditions of the Creative Commons Attribution (CC BY) license (http://creativecommons.org/licenses/by/4.0/).

Article

An Overview of Managed Aquifer Recharge in Mexico and Its Legal Framework

Mary Belle Cruz-Ayala [1,*] and Sharon B. Megdal [2]

1   The University of Arizona, 1064 E. Lowell St., Tucson, AZ 85721, USA
2   Water Resources Research Center, The University of Arizona, Tucson, AZ 85721, USA; smegdal@email.arizona.edu
*   Correspondence: marybelca@email.arizona.edu

Received: 18 November 2019; Accepted: 4 February 2020; Published: 10 February 2020

Abstract: In Mexico, one hundred of the 188 most important aquifers dedicated to agriculture and human consumption are over-exploited and 32 are affected by seawater intrusion in coastal areas. Considering that Mexico relies on groundwater, it is vital to develop a portfolio of alternatives to recover aquifers and examine policies and programs regarding reclaimed water and stormwater. Managed Aquifer Recharge (MAR) may be useful for increasing water availability and adapting to climate change in semi-arid regions of Mexico. In this paper, we present an overview of water recharge projects that have been conducted in Mexico in the last 50 years, their methods for recharge, water sources, geographical distribution, and the main results obtained in each project. We found three types of MAR efforts: (1) exploratory and suitability studies for MAR, (2) pilot projects, and (3) MAR facilities that currently operate. This study includes the examination of the legal framework for MAR to identify some challenges and opportunities that Mexican regulation contains in this regard. We find that beyond the technical issues that MAR projects normally address, the regulatory framework is a barrier to increasing MAR facilities in Mexico.

Keywords: MAR; Mexico; legal; regulatory; framework; LAN (Law of the Nation's Waters); reclaimed water; arid; semi-arid

1. Introduction

Groundwater overdraft in Mexico is a serious problem. One hundred of the 188 most important aquifer dedicated to agriculture and human consumption are overexploited, and 32 are affected by seawater intrusion in coastal areas [1,2]. Overexploitation of groundwater refers to the excessive withdrawal beyond the annual average recharge [1]. Expansion of irrigated agriculture, the growth in population, changes in consumption habits, and urbanization are the primary drivers of the increase in water use, resulting in water scarcity, a condition that is becoming a threat to sustainable development in Mexico [3]. Water scarcity refers to the geographic and temporal mismatch between freshwater demand and availability [3,4]. In Mexico, 58% of its national territory has dry lands—semi-arid, arid, and hyper-arid ecosystems [5]. Hence, natural conditions and overexploitation of groundwater [5] have led to a water-stress state in the country. Water stress refers to the ratio of total annual water withdrawals to the total available annual renewable supply [6].

Climate change is projected to affect rainfall patterns in Mexico; in particular, the Northern and Northwest regions may be facing severe water scarcity all year round by the year 2040. Mexican researchers analyzed weather databases to project rainfall patterns and temperature for the next four decades [7]. The information obtained was merged with data of water availability for overexploited aquifers; the outcomes showed that in the Northwest, Northern, and Northeast regions, water availability would decline [7,8]. For the US–Mexico border region, the results obtained by US and Mexican scientists using several climate models, have shown "a continued high degree of annual precipitation variability" [9,10]. Regarding the temperature, average annual and seasonal temperatures, also nighttime temperature are projected to increase for the border US–Mexico region, the western Sonoran Desert, and the northern region of the Chihuahuan Desert [7]. The Working Group II Contribution to the IPCC Fifth Assessment Report pointed out that over North America, which includes Northwest Mexico, exhibit very likely increases in mean temperature [11].

Future climatic scenarios might also affect groundwater recharge because higher temperatures will increase potential evapotranspiration [9,12]. Groundwater recharge includes the process of recharge as a natural part of the hydrological cycle, human-induced recharge that is also named artificial recharge, and recharge due to human activities such as irrigation and waste disposal [13]. Considering that groundwater supplies 70% of the water used for industry, agriculture, and human consumption in Mexico and several aquifers are overexploited [1,14], alternatives must be evaluated to recover aquifers.

In Northwest Mexico, the states of Sonora and Baja California Sur are annually impacted by tropical storms that provide them with a large volume of water in a short period. The rainfall generates floods because there is no infrastructure to manage stormwater or store it. However, several cities in the same region lack water during periods of the year. Another example is Mexico City, which imports water from distant basins [15–18]. Nevertheless, this city's precipitation ranges between 500 and 1200 mm per year, and almost every summer it suffers floods [1,19]. Mexico City and its metropolitan area combine reclaimed water, raw water, and stormwater to discharge into streams without evaluating whether this water can be recycled for other activities [20]. Reusing water is particularly relevant in water-stressed countries like Mexico. Therefore, it is crucial to examine policies and programs regarding reclaimed water and stormwater, which might be useful resources for increasing water availability in this country.

Artificial recharge, as it is termed in Mexican law, now called Managed Aquifer Recharge (MAR), where water quality management is also explicitly addressed, is a concept applied to describe diverse methods with the aim of both augmenting groundwater resources during times when water is available and recovering the water from the same aquifer in the future when it is needed [21,22]. There are a number of systems that could be categorized as MAR [21,23,24] and to implement one of them depends on local conditions such as financial budget, water quality, land availability, hydrogeological characteristics of the aquifer, water availability, and type of soil, among other considerations [24–26]. Although MAR is not a cure for overexploited aquifers, it can be useful for restoring groundwater balance [21]. In Mexico, reclaimed water and stormwater are the primary sources for MAR. In several countries, Managed Aquifer Recharge using reclaimed water has been growing as an option to recycle this resource, to replenish aquifers, and to stop seawater intrusion [20,24,27–29]. Globally, metropolises that have successfully used MAR to increase water supply include Adelaide, Australia [30], Los Angeles, United States [21], Tucson, United States [31,32], and Orange County, United States [33].

MAR in Mexico and Legal Framework

Since the 1950s, researchers, consultants, and local governments have designed and implemented water recharge projects [25]. Even though some cases were not designed specifically for water recharge, it has been pointed out that these experiences revealed the potential of harvesting and storing water to be used in the future [34,35]. In the early 2000s, in the states of Chihuahua, Sonora, and Mexico City, facilities for MAR using reclaimed water were proposed.

The need to increase water supply in semi-arid lands has triggered efforts for MAR. For instance, in 2007, in San Luis Río Colorado, Sonora facilities for MAR began operating [36], although there were no regulations in place. Therefore, this case was used as a model to define national standards for MAR. Although academics have stressed that public policies for MAR are needed [37–39], Mexican federal agencies have made slow progress in this regard. Moreover, the role that national policies have played in promoting or impeding MAR applications in Mexico has not been analyzed.

In Mexico, the water administration is centralized in the federal government and its regulation is based on Article 27 of the Political Constitution of the United States of Mexico [40]. The municipal responsibilities for water are established in Article 115. Regarding MAR, the Law of the Nation's Waters establishes that a permit is required to build a MAR pilot project [41]. The Federal Duties Law includes the charges that users must pay to extract groundwater; this legislation contains financial incentives for artificial recharge, as well. Furthermore, there are two official Mexican standards for MAR [42,43]. These regulations establish guidelines and water quality parameters that MAR projects, using reclaimed water and stormwater, must meet.

Throughout the world, water policies have been grounded in hard-path solutions, creating large scale infrastructure to deliver water, capital-intensive supply sources and centralized management [44]. Mexico is no exception. Projects to manage water demand, recycle water, and build small-scale infrastructure for water recharge with decentralized management (soft-path solutions) have not often been incorporated in governmental water project portfolios. Particularly in water-stressed regions, hard and soft-path solutions can be simultaneously implemented to improve water management [44,45]. In Mexico, there are examples of small-scale facilities for MAR in rural communities and medium-sized cities that are helping to increase water supply. Hence, it is beneficial to evaluate the role that small infrastructure for MAR in Mexico could play to increase water availability in rural and peri-urban areas. In this paper, we are using the definition of water availability provided by Tydwell et al. [46]: water availability is the supply of water in excess of that currently allocated for consumptive use in a particular basin, that is, the amount of water available for new development.

In Mexico, researchers have conducted an inventory of projects and estimated a significant potential for MAR in Mexico [35,47]. This study provides additional and updated information, including operational status. We present an overview of water recharge projects that have been conducted in Mexico in the last 50 years, their methods for recharge, water sources, geographical distribution, and the main results obtained in each project. Previous research has been focused only on assessing the performance of MAR projects or suitability studies. Our study also includes the examination of the legal framework for MAR to identify some challenges and opportunities that the Mexican legislation contains in this regard.

2. Materials and Methods

We carried out a systematic literature review of peer-reviewed publications in English and Spanish language, using such keywords as MAR, Mexico, artificial recharge, and water recharge. In addition, considering that in Mexico several studies for MAR have been conducted by universities and private consultants, the search included grey literature and official reports. These reports are mainly the products that consultants have delivered to state and municipal authorities. Additionally, we obtained information from publicly available conference proceedings and academic theses and reviewed the Mexican regulatory framework for groundwater and wastewater.

Recharge projects were grouped based on whether reclaimed water or stormwater was the source of water. In this paper, we will refer to reclaimed water to describe water that has been treated to meet quality standards; the term effluent will be used for water that has not been treated. We also consider each project's objectives. Finally, we analyzed the Mexican framework regulating MAR to identify gaps or aspects that could be clarified. The Legislation analyzed included the Political Constitution of the United States of Mexico, the Law of the Nation's Waters, Federal Duties Law, and Official Mexican Standards.

3. Results

3.1. MAR Projects, Types, Objectives, and Geographical Distribution

On the whole, we found that there are three types of MAR efforts (Table 1).

a. The first group includes exploratory and suitability studies for MAR, mainly proposed by researchers.

b. The second group is composed of pilot projects. In some cases, pilot projects have operated for a short period and later were stopped or canceled, and others are still operating. We use the definition of a pilot project for artificial recharge included in the Mexican regulation [40], where a pilot project is a temporary project that has been built to evaluate a recharge system, to assess its technical feasibility, to monitor hydraulic variables and water quality, and to identify possible impacts generated by artificial recharge in the aquifer and the environment.

c. The third group includes MAR facilities that currently operate.

Overall, MAR pilot projects and facilities have been planned with four objectives: to restore contaminated or depleted aquifers, to reduce land subsidence, to replenish aquifers and increase water availability, and to manage floods. Moreover, the infrastructure is mainly composed of small-scale facilities that are using stormwater or reclaimed water. In addition to the identified MAR projects and facilities, we include information on the most significant case of incidental recharge registered in Mexico, in the State of Hidalgo. Currently, only a portion of the water discharged meets quality standards; however, if this water is adequately treated, it could be used to recharge local aquifers. Finally, we present information regarding a MAR project that was created to recover the aquifer and to maintain environmental services in a riparian ecosystem. This information is presented and discussed in the sections below.

3.1.1. MAR Projects to Restore Contaminated or Depleted Aquifers

In a region between the states of Durango and Coahuila (Figure 1), a MAR pilot project was designed to find alternatives to halt arsenic contamination in aquifers used for human consumption and agriculture [48,49]. Using an infiltration reservoir that had previously been constructed in the basin, several experiments were conducted in 1991 and 2000. In those pilot projects, adequate rates of water infiltration and acceptable quality standards were obtained. However, this pilot project was not transformed into a large-scale installation because local farmers had provided most of the volume of water used for the project, and they were not interested in maintaining this experiment. In addition, the project lacked financial support for monitoring activities.

In Baja California Sur, Cardona et al. [50] analyzed the geological conditions in the municipality of Comondú to reduce seawater intrusion in its aquifers. Based on the information from Cardona [50], Wurl [51] evaluated three alternatives for recovering aquifers in this municipality. The first option is reducing 40 percent of withdrawals for agriculture; under this scenario, the aquifer would be recovered in more than 100 years [51]. The second option is to build small reservoirs to catch stormwater and slowly release it to increase natural recharge. The third option is the construction of storage dams for artificial recharge in areas with an adequate infiltration rate. Based on their hydrological models and maintaining the pumping rates registered in 2007, the last option would restore the aquifer at a faster pace. It would mean that in 40 years, 18% of the depleted volume of groundwater would have been restored [51,52].

3.1.2. MAR to Reduce Land Subsidence

Soil subsidence caused by depletion of groundwater has been widely documented in the United States, Mexico, and other countries [53–56]. In Mexico City, land subsidence has been recorded since 1925 and is one of the most notorious cases in the world [57,58]. Mexico City was built upon a lake over the ruins of the former Aztec city Tenochtitlan [59]. During the period 1991–2006 in some areas of this city, the subsidence rate has ranged 0.10–0.40 m per year [60,61]. Several studies argue that one of the leading causes of subsidence is that groundwater extraction exceeds the natural recharge and due to depressurization of the lacustrine aquitard [34,60–62]. It has been estimated that at least 23% of Mexico's City total surface area is placed on the former lakebed [57,59]. Therefore, to address the subsidence problem in Mexico City, some pilot projects using MAR methods have been proposed [34,63,64]. Since the 1960s, Figueroa Vega [62] has conducted studies and developed pilot projects for water recharge in Mexico City. Recently, Figueroa Vega [34] proposed a MAR project to collect water from local streams and runoff during the rainy season. This proposal may be developed where an ancient lake was located, and it would be linked to the hydraulic facilities for water management that has been built in the metropolitan area of Mexico City [65]. Water collected could be used to recharge the aquifer and would help to reduce the subsidence rate.

Figure 1. States in Mexico with the presence of Managed Aquifer Recharge (MAR) projects.

In Mexico City, another effect associated with over-pumping of groundwater and subsidence is the seismic damage in buildings and urban infrastructure. This city is a hot spot for earthquakes because it is situated between the North American Plate and the Cocos Plate [66]. Carreón-Freyre et al. [57] suggest that there is a relationship between seismic damage and land subsidence. Similarly, other scholars have developed a simulation model to identify urban areas located on top of the ancient lake and found that the most energetic shock waves during an earthquake are recorded in these zones [67]. Although more research is required to characterize specific zones where artificial recharge might be applied to diminish the negative impacts caused by earthquakes, it is a promising line for future research.

3.1.3. To Replenish Aquifers and Increase Water Availability

MAR projects in Mexico have been primarily designed to replenish aquifers and increase water availability using two types of water: stormwater and reclaimed water. Stormwater is used in small-scale facilities, mostly in rural environments, whereas in cities reclaimed water is used. In states located in the northern region, such as Baja California Sur, Chihuahua, and Sonora characterized by having a short but intense rainy season and a long dry season, there are proposals to build small-scale infrastructure to collect stormwater for replenishing aquifers [34,51,68,69]. In Chihuahua, a state located in northern Mexico, researchers from the Autonomous University of Chihuahua developed a pilot project in collaboration with the local government. This project has been ongoing for ten years [69]. In Caborca, Sonora, Minjarez et al. [68] conducted studies to develop a numerical groundwater model using stormwater as a source for MAR. Based on the model, they proposed sites for future MAR implementation. This proposal has not been transformed into a pilot project. In Southern Mexico, in a semi-arid region of the State of Oaxaca, there is a pilot project that uses stormwater for recharge [70].

In Mexico, in cities in which the population ranges from less than a million inhabitants to several million, reclaimed water is the primary source for MAR [15,37,63,71–75]. Based on the results registered by water managers, MAR using reclaimed water is more reliable because the water flow is constant, and its performance could be monitored as a component of a wastewater treatment plant.

In Mexico City, the Water System Agency has pursued a long process to create facilities for MAR using reclaimed water [63]. At the beginning of the 1960s, injection wells were used to infiltrate water from a reservoir. Even though their performance was adequate for more than nine years, these wells were closed because the water in the reservoir became contaminated [34]. In the late 1980s and early 1990s, studies were conducted to identify recharge areas in several areas of the city and the potential impacts of recharging reclaimed water [63,76]. In the eastern side of Mexico City, a suitable site for water recharge was identified. Therefore, a pilot project for MAR was built; it was running from 1992 to 2000 [63]. This installation did not operate for five years, and it was re-opened in 2005. Since 2010, these facilities located in the most populated area of Mexico City recharge reclaimed water using injection wells. This facility includes a program to monitor that the recharged water complies with the national water quality standards [63].

Table 1. MAR (Managed Aquifer Recharge) efforts in Mexico.

Author/Year/State in Mexico	Objective	Specific MAR Method *	Water Source
Research projects			
Figueroa-Vega [34] Estado de Mexico	Reduce land subsidence and replenish the aquifer	Dug well/shaft/pit injection	Stormwater
González & Juárez [77] Estado de Mexico	Replenish aquifer and increase water availability	Dug well/shaft/pit injection	Stormwater
Minjarez et al. [68] Sonora	Replenish aquifer	Dug well/shaft/pit injection	Stormwater
Palma et al. [74] Hermosillo, Sonora	Replenish aquifer and increase water availability	Infiltration ponds and basins	Reclaimed water
Wurl [51] Wurl and Imaz-La Madrid [52] Baja California Sur	Replenish aquifer	Recharge dams	Stormwater
Pilot projects			
Gutiérrez-Ojeda [48], Gutiérrez-Ojeda and Ortiz [49] Coahuila	Restore contaminated aquifers	Infiltration ponds and dug well/shaft/pit injection	Water from a dam
Morales-Escalante [73] Baja California	Replenish aquifer and increase water availability	Dug well/shaft/pit injection	Reclaimed water
Ojeda-Olivares et al. [70] Oaxaca, Oaxaca	Replenish aquifer and increase water availability	Subsurface dams	Stormwater
Palma et al. [37] Chihuahua, Chihuahua **	Replenish aquifer and increase water availability	Infiltration ponds and basins	Reclaimed water
Silva-Hidalgo et al. [69] Chihuahua, Chihuahua **	Replenish aquifer and increase water availability	Subsurface dams and dug well/shaft/pit injection	Stormwater
Facilities			
Álvarez, M. [78] Querétaro	Replenish aquifer	Subsurface dams	Stormwater
Ávila et al. [63] Mexico City	Replenish aquifer, reduce land subsidence and increase water availability	Dug well/shaft/pit injection	Effluent
Briseño-Ruíz et al. [79] San Luis Potosí	Manage floods	Infiltration ponds and basins	Stormwater
Escolero et al. [80] San Luis Potosí	Manage floods and replenish aquifer	Infiltration ponds and dug well/shaft/pit injection	Stormwater
Hernández-Aguilar [36] Sonora **	Replenish aquifer and increase water availability	Infiltration ponds and basins	Reclaimed water
Korenfeld & Hernández [15]	Replenish aquifer	Dug well/shaft/pit injection	Reclaimed water
Mendoza-Cázares et al. [64] Mexico City	Reduce land subsidence and increase water availability for environmental services	Recharge dams	Stormwater

* International Groundwater Resources Assessment Centre-MAR classification ** Facilities/projects functioning.

The Valle de Toluca aquifer in Estado de Mexico (which is the state bordering Mexico City) is exporting water to Mexico City, even though this aquifer is experiencing overdraft. In order to increase recharge in the aquifer and reduce water stress in this basin, the government of the Estado de

Mexico built facilities for MAR using reclaimed water [15]. The effluent is treated, including ultraviolet radiation to reach potable water quality standards and then injected into the ground using injection wells [15,71]. The capacity of this facility is 0.63 million cubic meters per year.

In 2007, in San Luis Río Colorado (SLRC), Sonora MAR facilities were built (Figures 2 and 3). The method used for recharge is infiltration ponds and the source of water is reclaimed water [36,72]. Annually 8.2 million cubic meters are being recharged [36]. The SLRC experience is the most successful example of MAR facilities in Mexico, using reclaimed water with the municipal government managing it.

Figure 2. Infiltration basins in San Luis Rio Colorado, Sonora.

Figure 3. Infiltration basins in San Luis Rio Colorado, Sonora, when empty.

In the city of Chihuahua, Palma et al. [37] propose a model to recharge the unallocated effluent (about 30 million cubic meters) that is currently being discharged into the river. The water recharged would alleviate water stress in the aquifer and increase water availability for the city. In Hermosillo, Sonora, Palma et al. [74] evaluate different scenarios to replenish the aquifer and balance the growing demand for the city's water supply. The proposal includes modifying the recharge in the irrigation districts and building MAR facilities to reduce both water loss and pollution in the system, providing additional water supply to urban areas. Based on their models, Palma et al. [74] identify the best option would be to reduce the effluent for irrigation and transfer this water to MAR facilities. This scenario means a reduction in the irrigated area, cultivating with less water, and growing a single crop each year. However, this plan would require a robust communication program with farmers and, perhaps, financial incentives to attract their participation. Another scenario would be to share the effluent between MAR and agriculture. The recharging rate will be lower compared to the other scenarios evaluated, but the recharge will be continuous [74]. The last option is the most viable because it will maintain agriculture activities and increase the groundwater recharge.

A different water source for artificial recharge was evaluated in Los Cabos, a municipality of the State of Baja California Sur. Saval [81] conducted a study using water from the San Lázaro dam. This researcher found that dam facilities were not adequate for monitoring the water recharge rate, and water quality was not satisfactory because there were anthropogenic sources of pollution. However, Saval [81] did not propose any options to eliminate these contaminants. Although, based on the information described, pollution was originated from domestic discharges which could be treated to remove contaminants.

Although there are several benefits of MAR using reclaimed water, this type of water has also generated some concerns because this water could include pollutants that cannot be removed by the soil [28,82]. To guarantee the recharged water will not affect native groundwater, the first step is to characterize the reclaimed water and include a monitoring program in the MAR project. Moreover, water composition and residence time before water is extracted for final use warrant evaluation [27,83,84]. Furthermore, new policies might be created to match water quality with native groundwater quality and the final intended use of the recharged water. For instance, water for the mining industry or farming forage crops does not require high-quality standards.

3.1.4. To manage Floods

Located in a semi-arid region, San Luis Potosí, capital of the state of San Luis Potosí, has faced flooding problems and water scarcity; therefore, a master plan to reduce flooding and increase water recharge was designed. For artificial recharge, the alternatives evaluated were building dams, recharge with reclaimed water, and injection wells. Based on studies conducted previously [80], a system was designed that includes a storm drain to collect water and dispose of it in ponds to infiltrate slowly [79]. This system is using facilities that previously were dedicated to sand mining, which meant a lower cost for the project. This artificial recharge project works very well; however, it does not have a program to monitor water quality.

3.1.5. Unintentional Recharge

Unintentional recharge refers to water discharged into a surface body or infiltrated into an aquifer, not as part of a planned project [85,86]. The Official Mexican Standard (NOM), NOM-014-CONAGUA-2003 [42] suggests that water infiltrated can be reclaimed water or non-treated water. In Mexico, one of the largest examples of unintentional recharge is in the Mezquital Valley, in the State of Hidalgo. This valley is located 80 km north from Mexico City. Since 1896, to avoid floods and to drain the sewage, a system was built for delivering non-treated water from Mexico City into the Mezquital Valley [58,87]. According to Jiménez and Chavez [87], 60 cubic meters per second are discharged daily, while during the rainy season, the peak flows may reach 300 cubic meters per second, for a few hours. Although this water contains a high percentage of contaminants that can impact the aquifer, scholars have reported that the soils have been effective in cleaning wastewater, maintaining an aquifer of acceptable quality [87]. If the water is satisfactorily treated, a portion of this resource can be dedicated to replenishing the aquifer [88]. Hence, this environmental problem can be transformed into a solution to increase the water supply in the region.

3.2. Water Governance and MAR

In Mexico, the water administration is centralized in the federal government, and its regulation is grounded in Article 27 of the Political Constitution of the United States of Mexico [89]. The governance of groundwater could be summarized as having the following components:

- Legal framework: Political Constitution of the United States of Mexico, Law of the Nation´s Waters, Official Mexican standards, Federal Duties Law.
- Managers: National Water Commission and Municipalities.
- Users: Basin Councils, Groundwater Technical Councils.

3.2.1. National Regulations: Political Constitution of the United States of Mexico, Law of the Nation's Waters, and Federal Duties Law

Article 27 of the Political Constitution of the United States of Mexico (first paragraph) establishes: "The property of all land and water within national territory is originally owned by the Nation, who has the right to transfer this ownership to particulars" [89]. The sixth paragraph defines "the dominion by the State shall be inalienable and imprescriptible, and the exploitation, use or development of those resources, be that by individuals or by corporations incorporated in accordance with Mexican laws, shall not be carried out but through concessions granted by the Federal Executive in accordance with the rules and requirements so established by the laws" [89].

In 1992, Mexico adopted the Law of the Nation's Water (LAN), which together with constitutional regulations, contains specific provisions for groundwater management and the role of the National Water Commission (CONAGUA) as the federal agency responsible for water administration [41]. Article 3 of the LAN defines the types of water entitlements: appropriation, allocation, and concession.

Appropriation: can be consumptive or non-consumptive (for example for hydropower generation).

Allocation: water rights assigned to municipal and state governments for exploitation, use, and extraction of national water to be used for public services or domestic purposes.

Concession: water title assigned to public or private enterprises and citizens for exploitation, use, and extraction of national waters.

Municipal governments, state governments, and Mexico City have allocations for exploitation, use, and extraction of national water for public services and domestic utilities. Furthermore, the LAN describes that a permit issued by CONAGUA will be required to recharge reclaimed water into aquifers (article 91). Additionally, water recharged must fulfill quality parameters established in national standards.

The Federal Duties Law (LFD) contains most federal taxes and financial incentives. It is analyzed and voted on every year as a part of the national budget. The LFD includes the charges that municipal and state governments must pay to extract water (articles 223 and 223b). These charges vary according to the region where water is extracted. There are four zones, based on water availability, charges are higher where water availability is lower and vice versa. Furthermore, the LFD includes financial incentives for water recharge. Article 224, paragraph V mentions: "(Users) do not have to pay the extraction fees if the water is recharged to its original source. The water should comply with a certificate of quality recognized by CONAGUA and all physical, chemical, and biological parameters listed in Article 225". Consequently, a public or private agency that recharges water in the same aquifer, where the water was obtained, can request this payment exception or its reimbursement. In general, municipal and state governments pay water extraction fees and later demand a refund from CONAGUA equal to the fees paid.

3.2.2. Official Mexican Standards (NOMs)

NOMs are specialized regulations which contain technical details for products, processes, and services. However, they are not legislative ordinances because they were not created by both parliamentary bodies, the chamber of deputies (like the US House of Representatives) and the chamber of senators [90]. There are two NOMs concerning water quality for MAR: NOM-014-CONAGUA-2003 [42] and NOM-015-CONAGUA-2007 [43]. NOM-014 includes regulations for artificial recharge using treated wastewater. However, the NOM-014 does not include the legal definition of reclaimed water. This NOM establishes that to obtain a permit for a large-scale project, the following information is required: location, hydrogeology (piezometric and stratigraphic profile), physical and chemical characteristics of water (pH, biological oxygen demand, chemical oxygen demand, and total organic carbon) and microbiological analysis. On the other hand, NOM-015-CONAGUA-2007 describes the requirements to carry out infiltration activities using runoff to recharge groundwater. For direct aquifer recharge, the water used should meet drinking water standards.

3.2.3. Managers, CONAGUA and Municipalities

Article 4 of the LAN defines that CONAGUA as the federal agency responsible for water administration [41]. CONAGUA assumes the primary obligation to enforce laws, create a specific regulatory framework for water allocation and water quality standards that MAR projects must fulfill. Constitutional article 115 defines the responsibilities for municipal governments. Municipalities are responsible for the following functions and public services: drinking water, drainage, sewerage system, treatment, and disposal of sewage [89]. The federal government is responsible for enforcing the quality standards for drinking water. There are some participatory bodies included in the LAN, such as committees and councils that collaborate with national and state governments on water-related policies.

After examining the applicable regulations for groundwater and MAR (Table 2), we found that the LAN does not include a definition of reclaimed water and the procedures that must be followed to use it for other activities neither. There is no reference to the legal rights that public agencies have on water recharged when conducting a MAR project using wastewater and how this "new water" would be allocated. Today, municipal water agencies can use, sell, and exchange the wastewater produced in treatment plants managed by them. However, neither the CPEUM nor the LAN defines the rights and responsibilities that municipal agencies have on water recharged using a MAR method. Another omission is concerning stormwater. The LAN does not include a definition of stormwater and lacks information regarding the requirements that should be fulfilled to obtain an authorization for MAR using this type of water. Finally, we want to remark that the regulations do not establish how the stormwater recharged would be allocated or managed.

In summary, LAN lacks specific regulations for MAR. NOMs can establish directions regarding the steps to be followed for building MAR projects and define water quality standards for water recharged. However, NOMs cannot determine the requirements to obtain a permit for a MAR project or how to allocate water recharged. NOMs are complementary regulations of the laws [90], but they cannot regulate water rights.

Table 2. Regulations for Groundwater and MAR in Mexico, and proposals.

	Native Groundwater	Artificial Recharge	Proposals
Definition of water rights and institutional arrangements for allocations	Concessions Assignations The Law of the Nation's Water (LAN) establishes procedures for allocations.	There is no procedure to allocate water recharged or a priority list to issue new permits.	LAN must define the procedures that must be followed to allocate water from MAR.
Identification of in situ requirements of available water	The National Water Commission (CONAGUA) publishes periodically water availability studies.	There is no a comprehensive study about potential water recharge projects.	CONAGUA and researchers can integrate a portfolio of viable MAR projects.
Abstraction limit	Entitlements are linked to a volumetric measure of water that can be extracted	The Official Mexican Standard, NOM-014-CONAGUA-2003 establishes that water recharged can be extracted after six months (surficial recharge) and 12 months of residency (direct recharge). However, there is not reference regarding the volume that can be extracted.	Based on the water rights in the basin, it could define a volume that MAR managers might extract.
Definition of priority uses	LAN includes a list of priority uses: 1. Domestic 2. Urban 3. Livestock 4. Agriculture 5. Aquaculture 6. Environment 7. Energy production for public service 8. Industry 9. Energy production for private service	None	The priority list set in the LAN might be used to define water rights to recover water recharged when using a MAR method.
Mechanisms for monitoring and enforcement	LAN and other federal regulations define what are the conditions under concessions can be cancelled.	LAN doesn't define penalties or mechanisms for enforcement for MAR facilities.	LAN must include mechanisms for monitoring MAR facilities.

4. Discussion

In Mexico, there are several MAR pilot projects with favorable results [35]. However, only a few have been transformed into large scale facilities. In general, MAR projects have been created in arid and semiarid regions of Mexico, which have a short but intense rainy season and a long dry season. On the other hand, there are facilities in Mexico City, which is located in a region with an oceanic climate (Cwb). This city registers precipitation throughout the entire year [91] and on average is 600 mm per year.

We found examples of MAR projects in 10 states; these projects have primarily been designed to recover aquifers and to prevent floods [70], and local water agencies are responsible for them [25,34,68,69]. The experience and knowledge that local authorities have acquired are valuable resources and might be integrated by the federal government when designing MAR policies. Local agencies know how to create these types of facilities and are aware of the challenges they face in maintaining them.

Three MAR facilities using reclaimed water were operating. These facilities are located in Toluca, Estado de Mexico, Iztapalapa, Mexico City, and San Luis Río Colorado (SLRC), Sonora. To date, only San Luis Rio Colorado MAR facility is functioning. Another example is a pilot project located in the city of Chihuahua that uses reclaimed water [37,92]. Several cities in the United States of America use effluent as an additional source for water supply [31–33]. For instance, in Tucson, Arizona, Tucson Water (the principal water utility in the city) has integrated reclaimed water as a source for future water supply [32,93,94].

One of the largest facilities for MAR in the world, that uses reclaimed water, is located in Orange County, California. The effluent is treated to reach potable water quality standards and then injected into the aquifer [33]. MAR is helping to recover the aquifer and diminish saline intrusion [95]. In the northwest region of Mexico and other states located in the Yucatan Peninsula, several coastal aquifers are affected by seawater intrusion [1,8]. Cardona et al. [50] and Wurl [51] have studied seawater intrusion in Baja California Sur (located in the northwest region) and proposed that MAR can be a tool to address this problem. It is a promising research field for MAR in Mexico that might be explored in the future.

4.1. MAR in the Metro Area of Mexico City

Mexico City and its metropolitan area, the so-called Metropolitan Zone of the Valley of Mexico (MZVM), faces floods almost every summer because there is no infrastructure to manage and store stormwater. Since the early 1940s, the MZVM began importing small volumes of water from the Estado de Mexico [1]. This decision was made because of the rapid increase in the population that demanded more water and the subsidence problems caused by overexploitation of groundwater from the local aquifer [1,66,76]. To the present, more than 25 percent of the total volume used for human consumption and productive activities is imported. About 65 percent of the water used is groundwater pumped from the local aquifer [1]. At the MZVM, several researchers and consultants have carried out experiments and designed MAR pilot projects to reduce land subsidence [34,63,76]. Given that the rate of subsidence is high, small-scale infrastructure could be insufficient to solve this problem in the near future. However, even little steps can help to reduce the subsidence problem.

For more than 100 years, in Mexico City and the MZVM, the stormwater reclaimed water and untreated water have been blended to be discharged into streams outside of the area, producing positive and negative impacts [1,59,76,88]. On the one hand, the aquifer has been recharged, and the water table has increased. On the other hand, the soil has been contaminated because a portion of the water is not adequately treated. Jiménez [88] found that the water in the aquifer has better quality compared to the inflow; it means that the soil has functioned as a cleaning system. To date, farmers are entitled to a large percentage of this water. Jiménez [88] mentioned that significant volumes of water are wasted because there is no infrastructure for irrigation. If the water is satisfactorily treated, a portion of this resource can be dedicated for MAR [88]. In the future, this scenario would be possible because a new facility was built and, now, around 60 percent of the water from the MZVM is being treated [96,97]. The new wastewater treatment plant is designed to treat 35,000 L per second. Considering that the MZVM lacks water, part of the reclaimed water can be delivered to agricultural activities, and another percentage might be utilized for MAR. It would mean an additional source of water for this thirsty metropolis.

4.2. MAR Using Stormwater

Generally, it has been suggested that, before starting a MAR project using stormwater, the following actions must be conducted: characterize the water, create an inflow map, and carry out pre-treatment procedures [98]. These steps are needed because stormwater, from urban areas, may contain contaminants such as oil, grease, metals, and pesticides [98,99]. Globally, there are successful examples of MAR using stormwater from urban environments. For instance, in Santa Cruz County, California, a pilot project composed of a network of basins for MAR has shown positive results. This program is managed by the Resource Conservation District of Santa Cruz County, and it is designed to recover groundwater and collect excess runoff for mitigating flooding as well [100,101]. For the city of Rome, Italy, La Vigna et al. [102] proposed building infrastructure for MAR using stormwater as source for recharge. In Adelaide, Australia MAR facilities have been constructed using stormwater from peri-urban regions [103]. In Mexico, we found that the infrastructure for MAR using stormwater is composed of small-scale installations constructed by local communities, non-profit organizations, and researchers [64,69,70]. Non-centralized systems like these examples offer several benefits relative to a centralized MAR system because these installations take advantage of natural precipitation and flow pathways, they can be developed and operated at relatively low cost [100].

In the cities of Chihuahua, Oaxaca, Mexico City (rural area), and San Luis Potosi, there are pilot projects and small-scale facilities that use stormwater to recharge aquifers [52,64,70,79]. MAR infrastructure has been constructed outside the urban grid or in rural areas, except in the case of San Luis Potosí, where the objective is to reduce floods in the city. Therefore, it is essential to acknowledge the role that these projects have to increase water availability, alleviate water scarcity, provide environmental services in riparian ecosystems, and improve water use in agriculture.

Regarding the role that reclaimed water can play to provide environmental services, there are limited examples. In Tucson, Arizona, there is an effluent-dependent riparian ecosystem that provides environmental services and generates social benefits such as recreation [104,105]. In Mexico City, in a rural community within this city, researchers from the Mexican Institute for Water Technology built infrastructure to recharge water [64]. Although this recharge is not directly increasing the water availability for human consumption, in the long-term, this water will help to recover the local aquifer. Besides, the recharging process by itself is providing water to maintain environmental services.

Soft-path solutions for water management include the creation of small-scale sources for supply, methods to increase efficiency to meet water demands, and the design of new policies [106,107] In Mexico, it must be acknowledged that the small-scale facilities can be part of the portfolio for better groundwater management. These facilities are part of the soft-path solutions that can be better integrated into national or regional policies to increase MAR projects.

4.3. Legal Framework and Public Policies for MAR

Worldwide, scientific knowledge in hydrology and geology has expanded, but the governing institutions responsible for making decisions about groundwater have been slowly improved [108,109]. Moreover, the Organisation for Economic Cooperation and Development (OECD) has expressed that "the current water crisis is not a crisis of scarcity but a crisis of mismanagement, with strong public governance features" [110–112]. It has been remarked that adequate policies and guidelines must be in place to ensure the benefits that MAR can provide [113]. In Mexico, there are more than 50 years of technical experience in MAR research [35,51]. Nevertheless, the information generated during this period has not been sufficiently incorporated into the national policy for MAR to launch long-term projects or propose public policies to increase its practice.

Mexican researchers have provided evidence to support MAR as a feasible option to increase water supply [114]. Recharging facilities are helping in increasing water availability in arid regions, such as the San Luis Rio Colorado, Sonora, and Chihuahua examples [36,69]. MAR facilities in Mexico City are aiding to diminishing the adverse effects caused by floods and reducing the rate of subsidence [65,80]. Two of the most successful MAR projects utilize reclaimed water, and there are still other MAR proposals that could be implemented [37,73]. However, the regulatory framework needs clarification and improvement to succeed in implementing MAR projects.

In Mexico, all the waters are owned by the nation; therefore, to be used or exploited the LAN includes three types of water entitlements: appropriation, allocation, and concessions [41]. In general, municipal operators have appropriations which cannot be transferred or changed to other uses than those for human provision (LAN). Since the Mexican legal framework lacks definitions for entitlements to recover stored groundwater using MAR methods, some options to protect investment for MAR projects must be evaluated. The security of recovery entitlements for MAR operators is an essential consideration for investment [115].

The Law of the Nation's Waters does not include a definition for reclaimed water and lacks procedures to define how reclaimed water can be managed and allocated. Gilabert-Alarcon et al. [116] highlight that policies and regulations for reclaimed water are limited. Considering that reclaimed water can be used for agricultural activities and MAR projects, the lack of a regulatory framework is hampering the opportunities for reusing this water. The responsible use of reclaimed water would generate economic and environmental benefits such as reducing water stress and preventing seawater intrusion in coastal aquifers [116]. In Mexico, the absence of a legal definition for reclaimed water and how it can be allocated generates uncertainty regarding water rights. This gap must be solved when creating national policies for MAR.

Another absence in the legal framework is that neither the Political Constitution of the United States of Mexico or the Law of the Nation's Water (LAN) defines the rights and responsibilities that municipal agencies have on water recharged, using a MAR method. Usually, municipal and state water agencies responsible for wastewater treatment plants exchange reclaimed water with farmers. Furthermore, these agencies sell reclaimed water to industries, golf courses, and construction companies. Nevertheless, if wastewater is recharged, municipal or state agencies lose their rights. This situation is the same if stormwater is used. A water right is a legal entitlement tied to a user and to a volume allocated by CONAGUA in each basin. Only the National Water Commission (CONAGUA) could define how the recharged water can be apportioned. However, Mexican legislation does not establish the guidelines that CONAGUA should follow in this regard. This lack of definition is limiting the participation of municipal water agencies in MAR projects.

Nowadays, it is recognized that scientific and technical information is not enough to address the complexity of groundwater management. It is vital to collaborate with society to solve real-world problems [117]. In Mexico, a single agency, CONAGUA, defines groundwater policies and guidelines for their implementation. There is a single policy for the whole country. The national government has a central vision for groundwater management and MAR that applies to the entire country, regardless of local conditions. An effective groundwater governance framework acknowledges and incorporates the local and regional socio-cultural values of water [118]. Groundwater resources in Mexico are facing tremendous pressure; 70 percent of the water used for agriculture and human consumption is groundwater [1]. It would be desirable that CONAGUA initiates a dialogue with local governments, users, and researchers to design MAR rules. The success or failure of natural resources management initiatives promoted by central authorities strongly depends on the support provided to local governments [119].

Regarding public policies for MAR and decentralization in Mexico, the challenge is how to incorporate in the decision-making process, the knowledge acquired by local governments, without amending the legal framework for groundwater. As Nagendra & Ostrom [119] described in forest management, the interactions between actors from several agencies such as federal, state, and municipal could be more important than the changes in the formal legal structure. In Mexico, it is not required to amend the legislation and decentralize responsibilities to increase the participation of both local government and citizens in strengthening public policies for MAR.

Mexican scholars investigating groundwater and MAR recognize that a multidisciplinary approach needs to be taken regarding groundwater management and MAR policies [22,120]. Furthermore, scholars suggest that financial incentives and new legislative rules for stimulating the creation of MAR facilities are required [34,69,120]. However, academics do not specify how the legislation might be improved or the type of incentives that are needed. Allegedly, the current legal framework in Mexico is limiting the creation of MAR projects. Therefore, beyond technical experience or knowledge, today, there is a need to evaluate and update the national legal framework to incorporate artificial recharge methods as part of the solutions to increase water availability.

4.4. Legal Framework That Might Be Revised and Proposals

Allocation of the recharged water. Inadequate institutional arrangements for aquifer storage and extraction can lead to a legal dispute and potentially negatively impact natural environments [115,121]. Allocation of water recharged is an issue that has been addressed using different tools. In the US, for instance, the states of Arizona, California, and Florida [113,122–124] have created a complex legal framework that establishes the permits required to recharge and extract water for MAR projects.

In Mexico, the Law of the Nation's Water (LAN) establishes that groundwater might be assigned based on the availability calculated by CONAGUA [41]. If there is unallocated water, it might be apportioned according to the priority list established in the LAN. The use of groundwater for domestic and urban use has primacy over other uses. However, the regulatory framework does not define how to allocate water that has been recharged using artificial methods.

The priority list set in the LAN would be used to determine water rights for recovering water recharged when using a MAR method. It would be possible to consider incorporating in the Mexican legal framework a general concept proposed by Pyne [125] that "if a water user has a right to the water before water recharge, then the user also has a right to recover that water". It would mean that the recharged water would be assigned using the same priority list that the LAN establishes when it is extracted from a natural source. This regulation would help to protect investors that are recharging water.

The institutional arrangement for MAR in Mexico is in its early stages, while federal regulations include some ordinances for recharge projects, there are no provisions for the recovery of stored water. In addition, considering that all waters are public property, there is limited space to create legal tools to protect water volumes for users who lack concessions or assignations. Under the current legal framework, an option that could be explored is allowing that those water users, entitled with allocations and that are recharging water, to extract a specific volume in addition to their allocation. In South Australia, for example, in some MAR projects carried out in overexploited aquifers, a recovery rate of 80% has been established [115]. In Mexico, the percentage of water that could be extracted would depend on the specific conditions of the aquifer and the type of water right (domestic, agriculture, industry).

Demonstration projects. One option to promote MAR methods could be to use successful experiences as demonstration projects. There is an example of this type of public policy in the United States (US). In 1984, the US Congress enacted the High Plains States Groundwater Demonstration Program Act. Federal agencies conducted studies in 17 states to identify appropriate sites for MAR. A report developed by Rogers [126] pointed out that "aquifer recharge programs are a good example of federal-local partnership that result in long-term investments."

Another example is Arizona because, as part of the legislation enacted in the 1990s, specific provisions for state demonstration projects were included [124,127]. Based on this legislation, MAR demonstration projects were built to show their effectiveness and technical aspects to operate them.

Arizona's demonstration project program incorporated both facilities permitting and operation provisions. Furthermore, in Arizona, the State Demonstration Recharge Program was funded through the Central Arizona Water Conservation District [122,124,127,128]. In the state of California [129,130], the legislation includes demonstration MAR projects. Furthermore, the Australian Guidelines for MAR include a commissioning phase for MAR projects as a tool to manage risk for projects in the development stage [86].

In Mexico, it could be feasible to incorporate demonstration projects into national policies. The infrastructure will be publicly owned; such has been proposed in Australia for stormwater harvesting [121]. Local demonstration projects that present technical and cost information are essential to draw the attention of a broad audience and increase the chance that MAR can be adopted [26]. Additionally, these facilities would help to create a link between MAR researchers and water managers. In summary, it would be desirable to include specific provisions in the legislation or create public programs to promote successful experiences, show their effectiveness, and technical aspects to operating them in Mexico.

5. Conclusions

Undoubtedly, one way to ensure water supply in the future is by reducing groundwater demand [23]. For instance, increasing the efficiency in agriculture and recycling water in industrial activities where it is feasible. Furthermore, programs to improve water replenishment, such as MAR projects and new public policies, are needed. In Mexico, previous research has been focused on assessing the performance of MAR projects or suitability studies. To the best of our knowledge, this paper represents the first academic effort to evaluate the role that governance plays in augmenting water supply sources in arid and semi-arid regions in Mexico, using MAR methods.

In Mexico, since the 1950s, researchers, consultants, and local governments have designed and implemented water recharge projects [25]. Even though some cases were not designed specifically for water recharge, it has been pointed out that these experiences revealed the potential of collecting and storing water to be used in the future [34]. We found that beyond the technical issues that MAR projects normally address, the regulatory framework is a barrier to increasing MAR facilities because there are no provisions for the recovery of stored water.

Two of the most successful MAR projects, in terms of the amount of water recharged, use reclaimed water, and local water agencies are responsible for these facilities. Recharging facilities are helping in recovering water-deficit in arid regions in Mexico, such as the San Luis Rio Colorado, Sonora, and Chihuahua cases. However, the Law of the Nation's Waters does not include a definition for reclaimed water and lacks procedures to define how reclaimed water can be managed and allocated. Gilabert-Alarcon et al. [116] highlight that policies and regulations for reclaimed water are limited. Considering that reclaimed water can be used for agricultural activities and MAR projects, the lack of a regulatory framework is hampering the opportunities for reusing this water.

MAR facilities in Mexico City using stormwater have been valuable tools for diminishing the adverse effects caused by floods and helping to reduce the subsidence rate [25,35]. Local governments are responsible for these facilities. Hence, the experience of municipal water agencies should be incorporated when designing policies for MAR.

We found several cases in rural communities and in medium-sized cities where small-scale projects for MAR have been built. These facilities have been created to recover aquifers, and local authorities managed them. It is essential to recognize the role of small-scale projects, as part of soft-path solutions, can play to increase water supply [131]; also, how these installations can be incorporated into regional water portfolios.

It is critical for water-stressed countries like Mexico to contemplate MAR as an option to increase water availability. It has been remarked that adequate policies and guidelines must be in place to ensure the benefits that MAR can provide [113]. An effective groundwater entitlement scheme is required to ensure investment in MAR projects [115]. In Mexico, beyond engineering knowledge regarding MAR, specific federal policies addressing gaps in the legislation are needed.

Author Contributions: This study was designed by M.B.C.-A., under the supervision of S.B.M. Writing—Original Draft Preparation, M.B.C.-A.; Review & Editing S.B.M. All authors have read and agreed to the published version of the manuscript.

Funding: Partial funding for this research was provided by the Inter-American Institute for Global Change Research CRN3056 Project (supported by NSF Grant No. GEO-1128040) via support from the Udall Center for Studies in Public Policy and the Tinker Field Research Grant (Summer 2019) via the Center for Latin American Studies at the University of Arizona.

Acknowledgments: M.B.C.A. thanks the Mexican Council for Science and Technology (CONACYT) for the fellowship to M.B.C.A. through a Ph.D. grant, the Graduate & Professional Student Council at the University of Arizona and The Herbert E. Carter Travel Award Program for providing partial funding to present this research at the ISMAR 10 conference. We want to thank M. Wilder for her helpful recommendations and the three anonymous reviewers who helped to improve this paper with their accurate and insightful suggestions.

Conflicts of Interest: The authors declare no conflict of interest. The founding sponsors had no role in the design of the study; in the collection, analysis, or interpretation of data; in the writing of the manuscript, and in the decisions to publish the results.

References

1. CONAGUA (2018). Estadísticas del agua en México, edición 2018. Available online: http://www.conagua.gob.mx/CONAGUA07/Publicaciones/Publicaciones/EAM2015.pdf (accessed on 21 June 2019).
2. Guevara-Sanginés, A. Subsidios para el bombeo de agua subterránea en México: Efectos perversos y opciones para su desacoplamiento. In *Memoria de las Jornadas del Agua, Proceedings of the UNAM Quinto Encuentro Universitario del Agua, Ciudad de México, México, 27–29 August 2013*; UNAM-CONACYT: México City, México, 2013; pp. 39–42.
3. Mekonnen, M.M.; Hoekstra, A.Y. Four billion people facing severe water scarcity. *Sci. Adv.* **2016**, *2*, e1500323. [CrossRef] [PubMed]
4. Gude, V.G. Desalination and Water Reuse to Address Global Water Scarcity. *Rev. Environ. Sci. Biotechnol.* **2017**, *16*, 591–609. [CrossRef]
5. Oswald, U.; Sánchez, I. Water Resources in Mexico: A Conceptual Introduction. In *Water Resources in Mexico. Scarcity, Degradation, Stress, Conflicts, Management, and Policy*, 1st ed.; Oswald Spring, U., Ed.; Hexagon Series on Human and Environmental Security and Peace, UNAM, CONACYT; Springer Berlin Heidelberg: Berlin, Germany, 2011; Volume 7, pp. 5–17. [CrossRef]
6. Gassert, F.; Landis, M.; Luck, M.; Reig, P.; Shiao, T. *"Aqueduct Global Maps 2.0." Working Paper*; World Resources Institute: Washington, DC, USA, 2013.
7. Arreguín, F.I.; Chávez, R.; Rosengaus, M. Impacto del cambio climático sobre los acuíferos mexicanos. In Proceedings of the VI Congreso Nacional de Aguas Subterráneas, Asociación Geohidrológica Mexicana, A. C., Mexico, 17–19 October 2007.
8. Arreguín, F.I.; López, P.M.; Marengo, M.H. Mexico´s water challenges for the 21st century. In *Water Resources in Mexico. Scarcity, Degradation, Stress, Conflicts, Management, and Policy*, 1st ed.; Oswald Spring, U., Ed.; Hexagon Series on Human and Environmental Security and Peace, UNAM, CONACYT. Springer Berlin Heidelberg: Berlin, Germany, 2011; Volume 7, pp. 21–38. [CrossRef]
9. Wilder, M.; Garfin, G.; Ganster, P.; Eakin, H.; Romero-Lankao, P.; Lara-Valencia, F.; Cortez-Lara, A.; Mumme, S.; Neri, C.; Muñoz-Arriola, F. Climate Change and U.S. Mexico Border Communities. In *Assessment of Climate Change in the Southwest United States: A Report Prepared for the National Climate Assessment*; Garfin, G., Jardine, A., Merideth, R., Black, M., LeRoy, S., Eds.; A report by the Southwest Climate Alliance; Island Press: Washington, DC, USA, 2013; pp. 340–384.
10. Shamir, E.; Megdal, S.B.; Carrillo, C.; Castro, C.L.; Chang, H.I.; Chief, K.; Corkhill, F.E.; Eden, S.; Georgakakos, K.P. Climate change and water resources management in the Upper Santa Cruz River, Arizona. *J. Hydrol.* **2015**. [CrossRef]
11. IPCC WGIIAR5 Report. 2014. Available online: https://www.ipcc.ch/report/ar5/wg2/ (accessed on 14 September 2019).

12. Meixner, T.; Manning, A.H.; Stonestrom, D.A.; Allen, D.M.; Ajami, H.; Blasch, K.W.; Brookfield, A.E.; Castro, C.L.; Clark, J.F.; Gochis, D.J.; et al. Implications of Projected Climate Change for Groundwater Recharge in the Western United States. *J. Hydrol.* **2016**, *534*, 124–138. [CrossRef]
13. Alley, W.M. Groundwater Resources: Sustainability, Management, and Restoration. *Groundwater* **2009**, *47*, 479. [CrossRef]
14. Scott, C.A.; Banister, J.M. The Dilemma of Water Management 'Regionalization' in Mexico under Centralized Resource Allocation. *Int. J. Water Resour. D* **2008**, *24*, 61–74. [CrossRef]
15. Korenfeld, F.D.; Hernández, L.O.J. Proyecto de Planta Piloto de tratamiento avanzado para la recarga artificial del acuífero. *Rev. Digit. Univ.* **2011**, *2*, 1067–6079.
16. Perló-Cohen, M.; González-Reynoso, A.E. *Guerra por el agua en el Valle de México? Estudio sobre las relaciones hidráulicas entre el Distrito Federal y el Estado de México*, 1st ed.; Friedrich Ebert Foundation: Mexico City, México, 2005; p. 139.
17. Scott, C.A.; Pablos, N.P. Innovating Resource Regimes: Water, Wastewater, and the Institutional Dynamics of Urban Hydraulic Reach in Northwest Mexico. *Geoforum* **2011**, *42*, 439–450. [CrossRef]
18. Wilder, M.; Lankao, P.R. Paradoxes of Decentralization: Water Reform and Social Implications in Mexico. *World Dev.* **2006**, *34*, 1977–1995. [CrossRef]
19. Ochoa, C.A.; Quintanar, A.I.; Raga, G.B.; Baumgardner, D. Changes in Intense Precipitation Events in Mexico City. *J. Hydrometeor.* **2015**, *16*, 1804–1820. [CrossRef]
20. Foster, S.; Garduño, T.A.; Kemper, K.; Nanni, M. Briefing Note Series Note 12 Urban Wastewater as. Available online: http://documents.worldbank.org/curated/en/239091468161650687/pdf/301040BRI0REVI101public10BOX353822B.pdf (accessed on 25 June 2019).
21. Dillon, P. Future Management of Aquifer Recharge. *Hydrogeol. J.* **2005**, *13*, 313–316. [CrossRef]
22. Khan, S.; Mushtaq, S.; Hanjra, M.A.; Schaeffer, J. Estimating Potential Costs and Gains from an Aquifer Storage and Recovery Program in Australia. *Agric. Water Manag.* **2008**, *95*, 477–488. [CrossRef]
23. Dillon, P.P.; Pavelic, D.; Page, H. Beringen. Available online: https://recharge.iah.org/files/2016/11/MAR_Intro-Waterlines-2009.pdf (accessed on 24 June 2019).
24. Yuan, J.; Dyke, M.I.V.; Huck, P.M. Water Reuse through Managed Aquifer Recharge (MAR): Assessment of Regulations/Guidelines and Case Studies. *Water Qual. Res. J.* **2016**, *51*, 357–376. [CrossRef]
25. Bonilla, J.P.; Stefan, C. Manejo de la Recarga de Acuíferos. In *Manejo de la recarga de acuíferos: Un enfoque hacia Latinoamérica*, 1st ed.; Escolero, O., Gutiérrez-Ojeda, C., Mendoza, E.Y., Eds.; Instituto Mexicano de Tecnología del Agua, Universidad Nacional Autónoma de México: Morelos, México, 2017; pp. 65–79.
26. Page, D.; Bekele, E.; Vanderzalm, J.; Sidhu, J. Managed Aquifer Recharge (MAR) in Sustainable Urban Water Management. *Water* **2018**, *10*, 239. [CrossRef]
27. Abiye, T.A.; Sulieman, H.; Ayalew, M. Use of Treated Wastewater for Managed Aquifer Recharge in Highly Populated Urban Centers: A Case Study in Addis Ababa, Ethiopia. *Environ. Geol.* **2009**, *58*, 55–59. [CrossRef]
28. Hartog, N.; Stuyfzand, P.J. Water Quality Considerations on the Rise as the Use of Managed Aquifer Recharge Systems Widens. *Water* **2017**, *9*, 808. [CrossRef]
29. Meehan, K.; Ormerod, K.J.; Moore, S.A. Remaking waste as water: The governance of recycled effluent for potable water supply. *Water Altern.* **2013**, *6*, 67–85.
30. Dillon, P.; Pavelic, P.; Toze, S.; Rinck-Pfeiffer, S.; Martin, R.; Knapton, A.; Pidsley, D. Role of Aquifer Storage in Water Reuse. *Desalination* **2006**, *188*, 123–134. [CrossRef]
31. Scanlon, B.R.; Reedy, R.C.; Faunt, C.C.; Pool, D.; Uhlman, K. Enhancing Drought Resilience with Conjunctive Use and Managed Aquifer Recharge in California and Arizona. *Environ. Res. Lett.* **2016**, *11*, 035013. [CrossRef]
32. Megdal, S.B.; Forrest, A. How a Drought-Resilient Water Delivery System Rose Out of the Desert: The Case of Tucson Water. *J. Am. Water Work. Assoc.* **2015**, *107*, 46–52. [CrossRef]
33. Hutchinson, A. Transforming Wastewater to Drinking water: How two Agencies Collaborated to Build the World´s Largest Indirect Potable Reuse Project. Available online: https://wrrc.arizona.edu/sites/wrrc.arizona.edu/files/BB-OCWD-OCSD-Collaboration-4-13-17.pdf (accessed on 23 August 2019).
34. Figueroa-Vega, G.E. El extinto lago de Texcoco y la infiltración artificial. In *Manejo de la recarga de acuíferos: Un enfoque hacia Latinoamérica*, 1st ed.; Escolero, O., Gutiérrez-Ojeda, C., Mendoza, E.Y., Eds.; Instituto Mexicano de Tecnología del Agua, Universidad Nacional Autónoma de México: Morelos, México, 2017; pp. 283–302.

35. González-Villarreal, F.; Cruickshank, C.; Palma Nava, A.; Mendoza Mata, A.; Recarga Artificial de Acuíferos en México. Rev. *H2O Del Sist. De Aguas De La Ciudad De México 2015. Año 2, enero-marzo*. Available online: https://issuu.com/helios_comunicacion/docs/h2o_-_5/30 (accessed on 4 January 2020).
36. Hernández-Aguilar, H.; Raúl, C.C.; Lorenzo, V.V.; Jorge, R.-H. Aquifer Recharge with Treated Municipal Wastewater: Long-Term Experience at San Luis Río Colorado, Sonora. *Sustain. Water Resour. Manag.* **2017**, *4*, 251–260.
37. Palma, A.; González, F.; Mendoza, A. The development of a managed aquifer recharge project with recycled water for Chihuahua, Mexico. *Sustain. Water Resour. Manag.* **2018**. [CrossRef]
38. Domínguez, J. Elementos para una nueva gobernabilidad del agua en México. In *Memoria de las Jornadas del Agua, Proceedings of the UNAM Quinto Encuentro Universitario del Agua, Ciudad de México, México, 27–29 August 2013*; UNAM-CONACYT: México City, México, 2013; pp. 19–20.
39. CONAGUA, Instituto de Ingeniería. ISMAR9 Call to Action, Sustainable Groundwater Management Policy Directives, June 2016, Mexico City. Available online: https://wrrc.arizona.edu/sites/wrrc.arizona.edu/files/SUSTAINABLE-GROUNDWATER-MANAGEMENT-POLICY-DIRECTIVES-102016.pdf (accessed on 14 September 2019).
40. Cossio-Díaz, J.R. Constitutional Framework for Water Regulation in Mexico. *Nat. Resour. J.* **1995**, *35*, 489–499.
41. Cámara de Diputados. Ley de Aguas Nacionales. Available online: http://www.diputados.gob.mx/LeyesBiblio/pdf/16_240316.pdf (accessed on 26 June 2019).
42. Diario Oficial de la Federación 2009. Available online: http://www.conagua.gob.mx/conagua07/contenido/documentos/NOM-014-CONAGUA-2003.pdf (accessed on 26 June 2019).
43. Diario Oficial de la Federación 2009. Available online: http://www.conagua.gob.mx/CONAGUA07/Contenido/Documentos/NOM-015-CONAGUA2007.pdf (accessed on 26 June 2019).
44. Gleick, P.H. Global Freshwater Resources: Soft-Path Solutions for the 21st Century. *Science* **2003**, *302*, 1524. [CrossRef] [PubMed]
45. Tortajada, C. (Ed.) Enhancing Water Governance for Climate Resilience: Arizona, USA—Sonora, Mexico Comparative Assessment of the Role of Reservoirs in Adaptive Management for Water Security. In *Increasing Resilience to Climate Variability and Change*; Springer: Singapore, 2016; pp. 15–40.
46. Tidwell, V.C.; Moreland, B.D.; Shaneyfelt, C.R.; Kobos, P. Mapping Water Availability, Cost and Projected Consumptive Use in the Eastern United States with Comparisons to the West. *Environ. Res. Lett.* **2018**, *13*, 014023. [CrossRef]
47. Bonilla-Valverde, J.P.; Stefan, C.; Palma, A.; Bernardo da Silva, E.; Pivaral, H.L. Vivar. Inventory of managed aquifer recharge schemes in Latin America and the Caribbean. *Sustain. Water Resour. Manag.* **2018**, *4*, 163–178. Available online: https://recharge.iah.org/swarm-vol-4-issue-2-june-2018 (accessed on 12 June 2019). [CrossRef]
48. Gutiérrez-Ojeda, C. Experiencias de recarga en el acuífero principal de la Comarca Lagunera. In *Memoria de las Jornadas del Agua, UNAM Quinto Encuentro Universitario del Agua, Ciudad de México, México, 27–29 August 2013*; UNAM-CONACYT: México City, México, 2013.
49. Gutiérrez-Ojeda, C.; Ortíz, F. Proyectos de Recarga MAR en el acuífero principal de la región lagunera. In *Manejo de la recarga de acuíferos: Un enfoque hacia Latinoamérica*, 1st ed.; Escolero, O., Gutiérrez Ojeda, C., Mendoza, E.Y., Eds.; Instituto Mexicano de Tecnología del Agua, Universidad Nacional Autónoma de México: Morelos, México, 2017; pp. 139–157.
50. Cardona, A.; Carrillo-Rivera, J.J.; Huizar-Alvarez, R.; Castro, H.G. Salinization in coastal aquifers of arid zones: An example from Santo Domingo, Baja California Sur, México. *Environ. Geol.* **2004**, *45*, 350–366. [CrossRef]
51. Wurl, J. Planificación de la recarga artificial del acuífero Valle de Santo Domingo. In *Memoria de las Jornadas del Agua, UNAM Quinto Encuentro Universitario del Agua, Ciudad de México, México, 27–29 August 2013*; UNAM-CONACYT: México City, México, 2013.
52. Wurl, J.; Imaz-Lamadrid, M.A. Coupled Surface Water and Groundwater Model to Design Managed Aquifer Recharge for the Valley of Santo Domingo, B.C.S., Mexico. *Sustain. Water Resour. Manag.* **2018**, *4*, 361–369. [CrossRef]
53. Donovan David, J.; Terry, K.; Kay, B.; Erin, C.; Michael, J. Cost-Benefit Analysis of Artificial Recharge in Las Vegas Valley, Nevada. *J. Water Resour. Plan. Manag.* **2002**, *128*, 356–365. [CrossRef]
54. Maliva, G.R. Economics of Managed Aquifer Recharge. *Water* **2014**, *6*, 1257–1279. [CrossRef]

55. Perrone, D.; Rohde, M.M. Benefits and Economic Costs of Managed Aquifer Recharge in California. Available online: https://escholarship.org/uc/item/7sb7440w (accessed on 27 June 2019).
56. Guide Book to Studies of Land Subsidence due to Ground-Water withdrawal Edited by Joseph, F. Poland. Available online: https://wwwrcamnl.wr.usgs.gov/rgws/Unesco/ (accessed on 27 June 2019).
57. Posible influencia de la subsidencia y fracturamiento en. Available online: http://www1.cenapred.unam.mx/SUBCUENTA/22aSESIÓNORDINARIA/VIII.ASUNTOSGENERALES/ReporteCDMXsismo2017.pdf (accessed on 26 June 2019).
58. Figueroa-Vega, G.E. Guidebook to studies of land subsidence due to ground-water withdrawal Edited by Joseph, F. Poland. Available online: https://wwwrcamnl.wr.usgs.gov/rgws/Unesco/PDF-Chapters/Chapter9-8.pdf (accessed on 27 June 2019).
59. National Research Council. Academia Nacional de la Investigacion Cientifica. In *Mexico City's Water Supply: Improving the Outlook for Sustainability*; National Academies Press: Washington, DC, USA, 1995. [CrossRef]
60. Ortiz-Zamora, D.; Ortega-Guerrero, A. Evolution of long-term land subsidence near Mexico City: Review, field investigations, and predictive simulations. *Water Resour. Res.* **2010**, *46*, W01513. [CrossRef]
61. Ruiz, G. Estimation of the groundwater recharge in the aquifer of the Mexico City. *Procedia Environ. Sci* **2015**, *25*, 220–226. [CrossRef]
62. Figueroa-Vega, G.E. *El Valle de Mexico y la infiltración artificial*; Comisión Hidrológica de la Cuenca del Valle de Mexico: México City, Mexico, 1970.
63. Ávila, F.A.; Correa, L.A.; Peralta, S.O.; Melchor, M. Recarga artificial del acuífero en el cerro de La Estrella, Iztapalapa, Ciudad de México. In *Manejo de la recarga de acuíferos: Un enfoque hacia Latinoamérica*, 1st ed.; Escolero, O., Gutiérrez-Ojeda, C., Mendoza, E.Y., Eds.; Instituto Mexicano de Tecnología del Agua, Universidad Nacional Autónoma de México: Morelos, México, 2017; pp. 383–429.
64. Mendoza-Cázares, E.Y.; Ramírez-León, J.M.; Puerto-Piedra, Z.Y. Recarga utilizando agua de lluvia en la cuenca del río Magdalena, Ciudad de México. In *Manejo de la recarga de acuíferos: Un enfoque hacia Latinoamérica*, 1st ed.; Escolero, O., Gutiérrez-Ojeda, C., Mendoza, E.Y., Eds.; Instituto Mexicano de Tecnología del Agua, Universidad Nacional Autónoma de México: Morelos, México, 2017; pp. 227–281.
65. PROYECTO HIDRÁULICO DEL -gob.mx. Available online: https://www.gob.mx/cms/uploads/attachment/file/101451/PRESENTACION_CONAGUA_NAICM_11sep14.pdf (accessed on 27 June 2019).
66. Cerca, M.; Carreón-Freyre, D.; López-Quiroz, P.; Ovando-Shelley, E.; Doin, M.P.; Gutierrez-Calderón, R.; González-Hernández, M.; Jimenez-Sánchez, A.; Blancas-Dominguez, D. Engineering Geology Approach to the Effects of Land Subsidence in Mexico City. *South. Cordill. Beyond* **2012**, *25*, 115.
67. Cruz-Atienza, V.M.; Tago, J.; Sanabria-Gómez, J.D.; Chaljub, E.; Etienne, V.; Virieux, J.; Quintanar, L. Long Duration of Ground Motion in the Paradigmatic Valley of Mexico. *Sci Rep.* **2016**, *6*, 38807. [CrossRef]
68. Minjarez, I.; Ochoa, A.; Tapia, E.; Montijo, A. *Estudio de recarga artificial del acuífero Caborca en el estado de Sonora*; No. OCNO-01-2012, Tomo 1; Comisión Nacional del Agua-Universidad de Sonora Convenio de Colaboración: Hermosillo, Sonora, México, 2012; Unpublished work.
69. Silva-Hidalgo, H.; González-Núñez, M.A.; Pinales, A.; Villalobos, A. Proyecto de manejo de recarga de acuíferos en los ojos de Chuvíscar, Chihuahua, México. In *Manejo de la recarga de acuíferos: Un enfoque hacia Latinoamérica*, 1st ed.; Escolero, O., Gutiérrez-Ojeda, C., Mendoza, E.Y., Eds.; Instituto Mexicano de Tecnología del Agua, Universidad Nacional Autónoma de México: Morelos, México, 2017; pp. 191–225.
70. Ojeda-Olivares, E.A.; Belmonte-Jiménez, S.I.; Ladrón de Guevara, M.A. Evaluación de obras de recarga hídrica construidas por comunidades autóctonas en la subcuenca del Valle de Ocotlán, Oaxaca, México. In *Manejo de la recarga de acuíferos: Un enfoque hacia Latinoamérica*, 1st ed.; Escolero, O., Gutiérrez-Ojeda, C., Mendoza, E.Y., Eds.; Instituto Mexicano de Tecnología del Agua, Universidad Nacional Autónoma de México: Morelos, México, 2017; pp. 103–137.
71. Asociación Nacional de Empresas de Agua y Saneamiento de México, A.C. aneas.com.mx. Available online: http://aneas.com.mx/wp-content/uploads/2016/04/SGAPDS-1-15-Libro38.pdf (accessed on 27 June 2019).
72. Hernández-Aguilar, M.H. Recarga artificial en el acuífero del valle de San Luis Río Colorado a través de lagunas de infiltración. In *Manejo de la recarga de acuíferos: Un enfoque hacia Latinoamérica*, 1st ed.; Escolero, O., Gutiérrez-Ojeda, C., Mendoza, E.Y., Eds.; Instituto Mexicano de Tecnología del Agua, Universidad Nacional Autónoma de México: Morelos, México, 2017; pp. 431–462.

73. Morales-Escalante, R. Estudio para evaluar la factibilidad de recargar el acuífero libre somero de Valle de Las Palmas, BC. In *Manejo de la recarga de acuíferos: Un enfoque hacia Latinoamérica*, 1st ed.; Escolero, O., Gutiérrez-Ojeda, C., Mendoza, E.Y., Eds.; Instituto Mexicano de Tecnología del Agua, Universidad Nacional Autónoma de México: Morelos, México, 2017; pp. 463–497.
74. Palma, A.; González, F.; Cruickshank, C. Managed Aquifer Recharge as a Key Element in Sonora River Basin Management, Mexico. *J. Hydrol. Eng.* **2015**, *20*, B4014004. [CrossRef]
75. UABC (Autonomous University of Baja California). *Reporte Técnico Final del estudio geohidrológico puntual para obtener las características hidráulicas del acuífero donde se pretende realizar el "Proyecto de Recarga Artificial de Acuífero mediante la infiltración con agua residual tratada"*; UABC (Autonomous University of Baja California): Mexicali, Mexico; p. 45, annex; Unpublished work.
76. Dirección General de Construcción y Operación Hidráulica-Secretaría General de Obras-DDF; Lesser & Asociados, S. A. de C.V. *Recarga artificial de agua residual tratada al acuífero del Valle de México, Ingeniería Hidráulica en México/mayo-agosto*; Instituto Mexicano de Tecnología del Agua: Morelos, México, 1991; pp. 65–71. Available online: http://www.lesser.com.mx/publicaciones.html (accessed on 10 January 2019).
77. González, S.; Juárez, M.A. Recarga artificial de acuíferos: un caso de estudio en la zona de El Caracol, ubicado en el municipio de Ecatepec de Morelos, en el Estado de México. In *Manejo de la recarga de acuíferos: Un enfoque hacia Latinoamérica*, 1st ed.; Escolero, O., Gutiérrez-Ojeda, C., Mendoza, E.Y., Eds.; Instituto Mexicano de Tecnología del Agua, Universidad Nacional Autónoma de México: Morelos, México, 2017; pp. 303–353.
78. Álvarez, M.J. Presa subterránea Aire No. 1, Charape de Los Pelones, Querétaro, México. In *Manejo de la recarga de acuíferos: Un enfoque hacia Latinoamérica*, 1st ed.; Escolero, O., Gutiérrez-Ojeda, C., Mendoza, E.Y., Eds.; Instituto Mexicano de Tecnología del Agua, Universidad Nacional Autónoma de México: Morelos, México, 2017; pp. 355–381.
79. Briseño-Ruiz, J.; Escolero-Fuentes, O.; Mendoza-Cázares, E.Y.; Gutiérrez-Ojeda, C. Infiltración de agua de tormenta al acuífero de San Luis Potosí, México: Colector Salk. In *Manejo de la recarga de acuíferos: Un enfoque hacia Latinoamérica*, 1st ed.; Escolero, O., Gutiérrez-Ojeda, C., Mendoza, E.Y., Eds.; Instituto Mexicano de Tecnología del Agua, Universidad Nacional Autónoma de México: Morelos, México, 2017; pp. 159–189.
80. Escolero, O.; Dillon, P.; Murillo, J.M.; Análisis de alternativas para la recarga artificial del sistema acuífero de San Luis Potosí. MIRH -OMM-MEX. Available online: https://sites.google.com/a/wmo.int/mx/infospremia/premiaesp/MIRH (accessed on 4 July 2019).
81. Saval, S. *Estudio de evaluación para la recarga artificial del acuífero de San José del Cabo*. In Memoria de las Jornadas del Agua, UNAM Quinto Encuentro Universitario del Agua, Ciudad de México, México, 27–29 August 2013; UNAM-CONACYT: México City, México, 2013.
82. Yuan, J.; Van Dyke, M.I.; Huck, P.M. Identification of Critical Contaminants in Wastewater Effluent for Managed Aquifer Recharge. *Chemosphere* **2017**, *172*, 294–301. [CrossRef]
83. Elkayam, R.; Michail, M.; Mienis, O.; Kraitzer, T.; Tal, N.; Lev, O. Soil Aquifer Treatment as Disinfection Unit. *J. Environ. Eng.* **2015**, *141*, 05015001. [CrossRef]
84. Vanderzalm, J.L.; Page, D.W.; Dillon, P.J. Application of a Risk Management Framework to a Drinking Water Supply Augmented by Stormwater Recharge. *Water Sci. Technol.* **2011**, *63*, 719–726. [CrossRef]
85. Jiménez, Blanca. Water Recycling and Reuse: An Overview. In *Water Reclamation and Sustainability*; Ahuja, S., Ed.; Elsevier: Boston, CA, USA, 2014; pp. 431–454. [CrossRef]
86. NRMMC, EPHC and NHMRC. *Australian Guidelines for water recycling, managing health and environmental risks, vol 2C: Managed Aquifer Recharge*; Natural Resource Management Ministerial Council, Environment Protection and Heritage Council National Health and Medical Research Council: Canberra, Australia, 2009; p. 237. Available online: https://clearwatervic.com.au/user-data/resource-files/WQ_AGWR_GL__Managed_Aquifer_Recharge_Final_200907%5B1%5D.pdf (accessed on 12 January 2020).
87. Jimenez, B.; Chávez, A. Quality Assessment of an Aquifer Recharged with Wastewater for Its Potential Use as Drinking Source: "El Mezquital Valley" Case. *Water Sci. Technol.* **2004**, *50*, 269–276. [CrossRef] [PubMed]
88. Jiménez, Blanca. Unplanned reuse of wastewater for human consumption. In *Water reuse: An International Survey of Current Practice, Issues and Needs*, 1st ed.; Jiménez, B., Asano, T., Eds.; IWA Publishing: London, UK, 2008; pp. 414–433.
89. Gómez, A. Available online: https://www.constituteproject.org/constitution/Mexico_2015.pdf?lang=en (accessed on 26 June 2019).

90. Huerta, C. Las Normas Oficiales Mexicanas en el ordenamiento jurídico mexicano. *Boletín Mex. De Derecho Comp.* **1998**, 2448–4873. Available online: https://revistas.juridicas.unam.mx/index.php/derecho-comparado/article/view/3543/4236 (accessed on 12 January 2020). [CrossRef]
91. Beck, H.E.; Zimmermann, N.E.; McVicar, T.R.; Vergopolan, N.; Berg, A.; Wood, E.F. Present and Future Köppen-Geiger Climate Classification Maps at 1-Km Resolution. *Sci. Data* **2018**, *5*, 180214. [CrossRef] [PubMed]
92. Espino, M.-S.; Navarro, C.-J.; Pérez, J.-M. Chihuahua: A Water Reuse Case in the Desert. *Water Sci. Technol.* **2004**, *50*, 323–328. [CrossRef] [PubMed]
93. Quanrud, D.M.; Hafer, J.; Karpiscak, M.M.; Zhang, J.; Lansey, K.E.; Arnold, R.G. Fate of Organics during Soil-Aquifer Treatment: Sustainability of Removals in the Field. *Water Res.* **2003**, *37*, 3401–3411. [CrossRef]
94. Zuñiga, A.; Staddon, C. Tucson Arizona–A Story of "Water Resilient" Through Diversifying Water Sources, Demand Management, and Ecosystem Restorarion. In *Resilient Water Services and Systems: The Foundation of Well-Being*; Juuti, P., Mattila, H., Rajala, R., Schwartz, K., Staddon, C., Eds.; IWA Publishing: London, UK, 2019. [CrossRef]
95. Herndon, R.; Markus, M. Large-Scale Aquifer Replenishment and Seawater Intrusion Control Using Recycled Water in Southern California. *Bol. Geol. Y Min.* **2014**, *125*, 143–155.
96. CONAGUA Home Page. Available online: http://www.conagua.gob.mx/CONAGUA07/Publicaciones/Publicaciones/SGAPDS-19-11.pdf (accessed on 28 June 2019).
97. Bello, J. Harvard University Graduate School of Design. Available online: https://research.gsd.harvard.edu/zofnass/files/2016/08/05_Atoltonico_SP_FinalDocument.pdf (accessed on 29 June 2019).
98. Alexander, K.; Moglia, M.; Gould, S.; Leviston, Z.; Tapsuwan, S.; Dillon, P. *Managed Aquifer Recharge and Stormwater Use Options: Public Perceptions of Stormwater Uses in Adelaide*; Water for a Healthy Country Flagship Report series: Australia, 2012; ISSN 1835-095X.
99. Managed Aquifer Recharge and Urban Stormwater Use Options. Available online: http://www.goyderinstitute.org/_r106/media/system/attrib/file/97/MARSUO-SummaryofResearchFindings-final_web.pdf (accessed on 29 June 2019).
100. Beganskas, S.; Fisher, A. Coupling Distributed Stormwater Collection and Managed Aquifer Recharge: Field Application and Implications. *J. Environ. Manag.* **2017**, *200*, 366–379. [CrossRef]
101. Fisher, A.T.; Lozano, S.; Beganskas, S.; Teo, E.; Young, K.S.; Weir, W.; Harmon, R. Regional Managed Aquifer Recharge and Runoff Analyses in Santa Cruz and northern Monterey Counties, California. Available online: https://escholarship.org/uc/item/5311s4wj (accessed on 29 June 2019).
102. La Vigna, F.; Martelli, S.; Bonfà, I. Stormwater Harvesting and Managed Aquifer Recharge (Mar) in the City of Rome: Possible Solution for a Better Management of Stormwater and Urban Floods. In Proceedings of the Resilience of Art Cities to Flooding: Success and Failure of the Italian Experience, Rome, Italy, 4–5 November 2014.
103. Northern Adelaide Plains Water Stocktake the Goyder. Available online: http://www.goyderinstitute.org/_r169/media/system/attrib/file/160/Final_NorthernAdelaidePlainsWaterStocktake-TechnicalReport2016.pdf (accessed on 29 June 2019).
104. Boyle, T.P.; Fraleigh, H.D. Natural and Anthropogenic Factors Affecting the Structure of the Benthic Macroinvertebrate Community in an Effluent-Dominated Reach of the Santa Cruz River, AZ. *Ecol. Indic.* **2003**, *3*, 93–117. [CrossRef]
105. Zugmeyer, C.; Steichen, S.; Martin, A. A Living River: Wetland Conditions on the Lower Santa Cruz River-2016. Available online: https://sonoraninstitute.org/resource/living-river-report-2016/ (accessed on 29 June 2019).
106. Brooks, D.B. Beyond Greater Efficiency: The Concept of Water Soft Path. *Can. Water Resour. J. Rev. Can. Des Ressour. Hydr.* **2005**, *30*, 83–92. [CrossRef]
107. Wolff, G.; Gleick, P.H. The Soft Path for Water. In *The World's Water 2002–2003*; Gleick, P.H., Ed.; Island: Washington, DC, USA, 2002; pp. 1–32.
108. Mukherji, A.; Shah, T. Groundwater Socio-Ecology and Governance: A Review of Institutions and Policies in Selected Countries. *Hydrogeol. J.* **2005**, *13*, 328–345. [CrossRef]
109. Villholth, K.G.; Elena, L.; Conti, K.; Garrido, A.; van der Gun, J.A.M. *Advances in Groundwater Governance*; CRC Press/Balkema: Leiden, The Netherlands, 2018.

110. Gerlak, A.K.; Wilder, M. Exploring the Textured Landscape of Water Insecurity and the Human Right to Water. *Environ. Sci. Policy Sustain. Dev.* **2012**, *54*, 4–17. [CrossRef]
111. Akhmouch, A. Water Governance in Latin America and the Caribbean. Available online: http://dx.doi.org/10.1787/5k9crzqk3ttj-en (accessed on 29 June 2019).
112. Woodhouse, P.; Muller, M. Water Governance—An Historical Perspective on Current Debates. *World Dev.* **2017**, *92*, 225–241. [CrossRef]
113. Dillon, P.; Stuyfzand, P.; Grischek, T.; Lluria, M.; Pyne, R.D.G.; Jain, R.C.; Bear, J.; Schwarz, J.; Wang, W.; Fernandez, E.; et al. Sixty Years of Global Progress in Managed Aquifer Recharge. *Hydrogeol. J.* **2018**, *27*, 1–30. [CrossRef]
114. Arreguín-Cortés, F.I.; López-Pérez, M.; Escolero, O.; Gutiérrez-Ojeda, C. Líneas de investigación y desarrollo tecnológico en materia de aguas subterráneas. In *Manejo de la recarga de acuíferos: Un enfoque hacia Latinoamérica*, 1st ed.; Escolero, O., Gutiérrez-Ojeda, C., Mendoza, E.Y., Eds.; Instituto Mexicano de Tecnología del Agua, Universidad Nacional Autónoma de México: Morelos, México, 2017; pp. 47–63.
115. Ward, J.; Dillon, P. Principles to Coordinate Managed Aquifer Recharge with Natural Resource Management Policies in Australia. *Hydrogeol. J.* **2012**, *20*, 943–956. [CrossRef]
116. Gilabert-Alarcón, C.; Salgado-Méndez, S.; Daesslé, L.; Mendoza-Espinosa, L.; Villada-Canela, M. Regulatory Challenges for the Use of Reclaimed Water in Mexico: A Case Study in Baja California. *Water* **2018**, *10*, 1432. [CrossRef]
117. Morehouse, B.J.; Ferguson, D.B.; Owen, G.; Browning-Aiken, A.; Wong-Gonzalez, P.; Pineda, N.; Varady, R. Science and Socio-Ecological Resilience: Examples from the Arizona-Sonora Border. *Environ. Sci. Policy* **2008**, *11*, 272–284. [CrossRef]
118. Varady, R.G.; van Weert, F.; Megdal, S.B.; Gerlak, A.C.; Iskandar, A.; House-Peters, L. *Thematic Paper No. 5: Groundwater Policy and Governance*; GEFFAO Groundwater Governance Project a Global Framework for Country Action: Rome, Italy, 2013; Available online: http://www.groundwatergovernance.org/resources/thematic-papers/en/ (accessed on 4 July 2019).
119. Nagendra, H.; Ostrom, E. Polycentric Governance of Multifunctional Forested Landscapes. *Int. J. Commons* **2012**, *6*, 104. [CrossRef]
120. Escolero Fuentes, O.; Gutiérrez Ojeda, C.; Mendoza, E.Y. (Eds.) *Manejo de la recarga de acuíferos: Un enfoque hacia Latinoamérica*; IMTA–Instituto Mexicano de Tecnología del Agua: Morelos, México, 2017; p. 978, 25 Chapters (in Spanish). Available online: https://www.gob.mx/imta/articulos/libro-manejo-de-la-recarga-de-acuiferos-un-enfoque-hacia-latinoamerica?idiom=es (accessed on 14 January 2019).
121. Ward, J.; Dillon, P. *Robust Policy Design for Managed Aquifer Recharge*; National Water Commission: Canberra, Australia, 2011.
122. Eden, S.; Joe, G.; Megdal, S.; Shipman, T.; Smart, A.; Escobedo, M. A Multi-Purpose Water Management Tool-wrrc.arizona.edu. Available online: https://wrrc.arizona.edu/sites/wrrc.arizona.edu/files/arroyo_winter_2007.pdf (accessed on 29 June 2019).
123. Megdal, S. Arizona´s Recharge and Recovery Programs. In *Arizona Water Policy: Management Innovations in an Urbanizing, Arid Region*, 1st ed.; Colby, B.G., Jacobs, K.L., Eds.; Resources for the Future: Washington, DC, USA, 2007; pp. 188–203.
124. Summary of Arizona Water Law. Available online: http://www.g-a-l.info/Water-Law.htm (accessed on 29 June 2019).
125. Pyne, R.D.G. *Aquifer Storage Recovery: A Guide to Groundwater Recharge through Wells*; ASR Systems: Gainesville, FL, USA, 2005.
126. The High Plains Groundwater Demonstration Program. Available online: http://www.usbr.gov/history/ProjectHistories/HighPlainsStatesGroundwaterDemonstrationProgram(1).pdf (accessed on 29 June 2019).
127. Megdal, B.S.; Dillon, P.; Seasholes, K. Water Banks: Using Managed Aquifer Recharge to Meet Water Policy Objectives. *Water* **2014**, *6*, 1500–1514. [CrossRef]
128. Silver-Coats, N.; Eden, E. Arroyo 2017-Arizona Water Banking, Recharge, and Recovery. Available online: https://wrrc.arizona.edu/publications/arroyo/arroyo-2017-arizona-water-banking-recharge-and-recovery (accessed on 29 June 2019).
129. Rodriguez, C.; Van Buynder, P.; Lugg, R.; Blair, P.; Devine, B.; Cook, A.; Weinstein, P. Indirect Potable Reuse: A Sustainable Water Supply Alternative. *Int. J. Environ. Res. Public Health* **2009**, *6*, 1174. [CrossRef]

130. California Water Code. Available online: http://leginfo.legislature.ca.gov/faces/codesTOCSelected.xhtml?tocCode=WAT&tocTitle=+Water+Code+-+WAT (accessed on 12 June 2019).
131. Brooks, D.B.; Holtz, S. Water Soft Path Analysis: From Principles to Practice. *Water Int.* **2009**, *34*, 158–169. [CrossRef]

© 2020 by the authors. Licensee MDPI, Basel, Switzerland. This article is an open access article distributed under the terms and conditions of the Creative Commons Attribution (CC BY) license (http://creativecommons.org/licenses/by/4.0/).

Article

Lessons from 10 Years of Experience with Australia's Risk-Based Guidelines for Managed Aquifer Recharge

Peter Dillon [1,2,3,*], Declan Page [1], Joanne Vanderzalm [1], Simon Toze [4], Craig Simmons [2], Grant Hose [5], Russell Martin [6], Karen Johnston [7], Simon Higginson [8] and Ryan Morris [9]

1. CSIRO Land and Water, Waite Laboratories, Waite Rd, Urrbrae, SA 5064, Australia; declan.page@csiro.au (D.P.); Joanne.Vanderzalm@csiro.au (J.V.)
2. National Centre for Groundwater Research and Training (NCGRT) & College of Science and Engineering, Flinders University, SA 5001, Australia; Craig.Simmons@groundwater.com.au
3. School of Civil, Environmental and Mining Engineering, University of Adelaide, SA 5005, Australia
4. CSIRO Land and Water, Ecosciences Precinct, Boggo Rd, Dutton Park, Qld 4102, Australia; Simon.Toze@csiro.au
5. Department of Biological Sciences, Macquarie University, Sydney, NSW 2109, Australia; grant.hose@mq.edu.au
6. Wallbridge Gilbert Aztec, Adelaide, SA 5000, Australia; RMartin@wga.com.au
7. Managed Recharge, Perth, WA 6000, Australia; karen.johnston@managedRecharge.com.au
8. Water Corporation, Perth, WA 6000, Australia; Simon.Higginson@watercorporation.com.au
9. RDM Hydro Pty Ltd., Tarragindi, Qld 4121, Australia; ryan@rdmhydro.com.au
* Correspondence: pdillon500@gmail.com; Tel.: +61-419-820-927

Received: 19 November 2019; Accepted: 11 January 2020; Published: 14 February 2020

Abstract: The Australian Managed Aquifer Recharge Guidelines, published in 2009, were the world's first Managed Aquifer Recharge (MAR) Guidelines based on risk-management principles that also underpin the World Health Organisation's Water Safety Plans. In 2015, a survey of Australian MAR project proponents, consultants and regulators revealed that in those states advancing MAR, the Guidelines were lauded for giving certainty on approval processes. They were also considered to be pragmatic to use, but there was feedback on onerous data requirements. The rate of uptake of MAR has varied widely among Australian state jurisdictions, for reasons that are not explained by the drivers for and feasibility of MAR. The states where MAR has progressed are those that have adopted the Guidelines into state regulations or policy. It was originally intended that these Guidelines would be revised after five to ten years, informed by experience of any hazards not considered in the guidelines, and by new scientific developments including advances in monitoring and control methods for risk management. As such revision has not yet occurred, this paper was prepared to give a precis of these Guidelines and review ten years of experience in their application and to identify issues and suggest improvements for consideration in their revision by Australian water regulators. This paper also discusses the factors affecting their potential international applicability, including the capabilities required for implementation, and we use India as an example for which an intermediate level water quality guideline for MAR was developed. This paper is intended to be useful information for regulators in other countries considering adopting or developing their own guidelines. Note that the purpose of these Guidelines is to protect human health and the environment. It is not a guide to how to site, design, build and operate a managed aquifer recharge project, for which there are many other sources of information.

Keywords: environment protection; health protection; safety; risk; ecosystems; contaminants; recycling; drinking water; regulation; governance

1. Introduction

The first Managed Aquifer Recharge Guidelines [1] based on risk-management principles that also underpin World Health Organisation's Water Safety Plans [2] were published in July 2009, within the framework of the Australian National Water Quality Management Strategy. These Australian Managed Aquifer Recharge (MAR) Guidelines are one of four documents in the Australian Water Recycling Guidelines [1,3–5] (Figure 1). The others address founding principles and non-potable applications of recycled water [3], recycling to augment drinking water supplies [4] and harvesting stormwater for non-potable use [5]. The MAR Guidelines cover all types of water intentionally recharged to aquifers for recovery and use or for environmental benefit. These Guidelines were developed consultatively over three years and approved by three Ministerial Councils of the Council of Australian Governments (COAG), that include all state and national ministers whose portfolios address natural resources management, environment and heritage protection and public health.

Figure 1. National Water Quality Management Strategy, showing the foundations for protecting human health and the environment, and innovation in Australian water management ([6] ARMCANZ-ANZECC (1994); [7] ANZECC–ARMCANZ (2000a), [8] WQA (2018); [9] ANZECC–ARMCANZ (1995); [10] WQA (2013); [11] ANZECC–ARMCANZ (2000b); [12] NHMRC–NRMMC (2004), [13] WQA (2011); Australian Water Recycling Guidelines - [3] NRMMC–EPHC–AHMC (2006); [4] NRMMC-EPHC–NHMRC (2008); [1] NRMMC-EPHC-NHMRC–(2009a); [5] NRMMC–EPHC–NHMRC (2009b)). Where Guidelines have been updated, both dates are given to show the evolution of guidelines but only the latter is applicable. All current guidelines are accessible from: https://www.waterquality.gov.au/guidelines.

The MAR Guidelines were immediately welcomed and implemented in the three states most active in MAR: South Australia, Western Australia and Victoria. In other states implementation of MAR was not progressing, despite obvious needs, opportunities and viability (as described later). Now that ten years have passed, a review of Australia's experiences with MAR Guidelines is warranted, with a view to informing revision which is normally expected on a five- to ten-year cycle to account for experience, such as any hazards that may have emerged that were not considered in the Guidelines and advances in science and technology, especially related to environmental monitoring and contaminant fate. This review could also assist other states where MAR Guidelines are still not in regular use and countries that currently do not have guidelines to consider the benefits of adopting or adapting risk-based guidelines.

2. Australian MAR Guidelines

The Australian Guidelines for MAR [1] define managed aquifer recharge as purposeful recharge of an aquifer using a source of water (including recycled water) under controlled conditions, in order to store for later use or for environmental benefit while protecting human health and the environment. It is not a method for waste disposal. The Guidelines allow for an attenuation zone beyond which at all times, all ambient environmental values (i.e., beneficial uses) of the aquifer are protected. This relies on information concerning inactivation rates of pathogens and degradation rates of degradable organic chemicals. The risk management framework common to drinking water and recycled water guidelines applies. In the MAR Guidelines, this is extended beyond water quality issues to also address aquifer pressures, discharges and leakages and impacts on groundwater-dependent ecosystems [14].

The Guidelines provide for staged development of projects. They are intended to provide a confident pathway forward for proponents, regulators and other stakeholders. The Guidelines also reinforce the need for public consultation processes where other people may potentially be impacted by managed aquifer recharge projects.

Hazards addressed in the Guidelines are:

1. Pathogens,
2. Inorganic chemicals,
3. Salinity and sodicity,
4. Nutrients,
5. Organic chemicals,
6. Turbidity/particulates,
7. Radionuclides,
8. Pressure, flow rates, volumes and levels,
9. Contaminant migration in fractured rock and karstic aquifers,
10. Aquifer dissolution and aquitard and well stability,
11. Impacts on groundwater-dependent ecosystems, and
12. Greenhouse gas emissions.

For each hazard, the Guidelines describe sources or causes, the effect on public health and environment, how it can be managed, including preventive measures, the proposed validation, verification and operational monitoring, and list the acceptance criteria for the various stages of risk assessment that parallel the stages of project development. The first seven hazards are common across all four recycled water guidelines, but the management of these is specific to MAR. The last five hazards are unique to the MAR Guidelines.

2.1. Reactions between Recharged Water and Aquifers

The MAR Guidelines were unique in that they were not simply a set of numerical standards for water quality parameters considered fit for recharge. The Guidelines reflect that aquifers are biogeochemical reactors and local information on aquifer mineralogy and structure and ambient groundwater quality are needed in order to determine the quality of recharge water that would result in acceptable quality of recovered water, protection of the aquifer and related ecosystems, and ensure sustainable operation. They also account for pressure, flow rate, volumes and levels in confined, semi-confined and unconfined aquifers, and address energy and greenhouse gas considerations. They regard clogging and recovery efficiency as matters for the proponent to address but provide advisory information on managing these operational issues that impact most on the proponent.

While proponents would have preferred that the MAR Guidelines specified maximum values of analytes in recharge water, such as by treating water to drinking standards before recharge, this was considered not to assure protection of the aquifer and recovered water. Experience had already shown that e.g., chlorination, which removes pathogens that would have been inevitably removed over time in a warm aquifer [15] can result in water recovered from some (oxic) aquifers containing persistent excessive chloroform [16]. In some locations, drinking water injected into potable aquifers has resulted in excessive arsenic concentrations on recovery due to reactions between injected water and pyrite containing arsenic [17]. Source water that has been desalinated to a high purity dissolves more minerals within the aquifer than water that has been less treated and can also react with dispersive clays to cause clogging [18–20]. Also, in some experiments biodegradable organic carbon in recharge water has been found to enhance microbial diversity and assist co-metabolism of some trace organics thereby enhancing their removal and supplementing removal processes under oligotrophic conditions deeper in the aquifer [21–23]. The American Water Works Association Research Foundation, along with Australian, European and American partners have supported much of the research in this area [15,17,24]. Consequently, the MAR Guidelines adopt a scientific approach that takes into account three ways that aquifers interact with recharged water [25]:

1. Sustainable hazard removal. The Guidelines allow for pathogen inactivation, and biodegradation of some organic contaminants during the residence time of recharged water in the soil and/or aquifer within an attenuation zone of finite size.
2. Ineffective hazard removal. These hazards need to be removed prior to recharge because they are either not removed (e.g., salinity) or removal is unsustainable (e.g., adsorption of any metals and organics that are not subsequently biodegraded, or excessive nutrients or suspended solids).
3. New hazards introduced by aquifer interaction (e.g., metal mobilization, hydrogen sulphide, salinity, sodicity, hardness, or radionuclides). There is a need to change the quality of recharge water to avoid these (e.g., change acidity/alkalinity, reduction/oxidation status or reduce nutrients).

In undertaking this 10-year review it was of interest to determine whether the Guidelines were sufficiently comprehensive to address all water quality deterioration processes encountered in MAR projects.

2.2. Zones of Influence of a MAR Operation

The response of an aquifer to any water quality hazard depends on specific conditions within the aquifer, including temperature, presence of oxygen, nitrate, organic carbon and other nutrients and minerals, and prior exposure to the hazard. The Guidelines indicate the state of knowledge in 2009 on attenuation rates of pathogens and organic compounds under a range of conditions. They also allow for new local knowledge to be taken into account in assessing risks and determining sizes of attenuation zones and siting of monitoring wells.

In most aquifers, with appropriate pretreatment of water to be recharged, the attenuation zone will generally be a small zone around the recharge area or well (see Figure 2). Water that travels further has had sufficient residence time in the aquifer for attenuation of pathogens and contaminants to below the relevant guideline values for native groundwater and intended uses of recovered water.

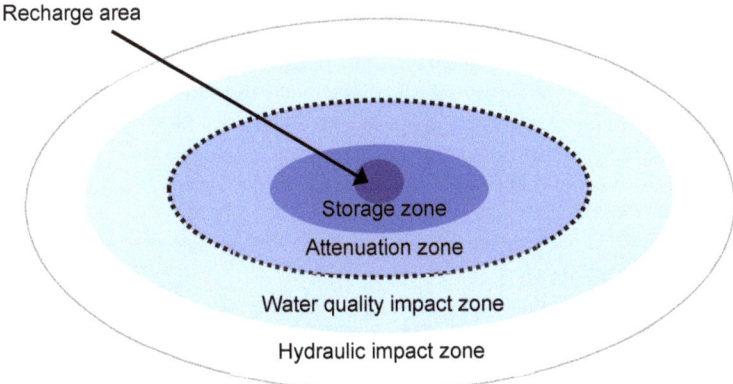

Figure 2. Schematic showing plan view of zones of influence of a managed aquifer recharge (MAR) operation.

The zone of aquifer in which water quality may be measurably affected by MAR may be larger, but in this outer domain the water quality should continuously satisfy the initial environmental values of the aquifer (Figure 2). The effects of managed aquifer recharge operations on hydraulic heads (pressures) may be measurable over a much larger area, especially in confined aquifers. If the aquifer is originally too saline for the uses of recovered water, a storage zone can be identified that contains water which, when recovered, is fit for its intended use (Figure 2).

The dotted line in Figure 2 marks the outer boundary of the attenuation zone. This represents the maximum separation distance between the recharge structure and well(s) for verification monitoring to ensure that the ambient groundwater quality is protected. As the attenuation zone is defined only for enduring attenuation processes, on cessation of the MAR operation, it will shrink and disappear as ultimately the whole aquifer will meet all its initial environmental values. Attenuation rates under various aquifer conditions as known in 2009, are summarised in the appendices of the Guidelines. These warrant updating with results of subsequent studies, in which aquifer environmental conditions and attenuation rates and mechanisms have been documented.

In the entry level assessment stage, the Guidelines refer proponents to water resources planning and management regulations, which require an ability to gain an entitlement to take water for recharge, to recharge an aquifer, to recover water from the aquifer and for appropriate uses of the recovered water. This may also require consideration of cumulative impacts on water level, pressure and quality in the aquifer of multiple recharge operations. For confined aquifer systems where the hydraulic impact zone can extend tens of kilometres from a recharge well, or where ambient groundwater is brackish and neighboring injection and recovery wells reduce the recovery efficiency of an aquifer storage and recovery (ASR) system, this may be a significant consideration for approvals [26].

3. Experience that Suggests Future Refinement of MAR Guidelines

Both experience with the application of the MAR Guidelines and recent research suggest ways in which the MAR Guidelines could be improved if they are to be revised in the near future. These two elements will largely be addressed separately here, starting with experience. The following list of experiences have provided sources of information that have resulted in changes in state regulations or suggest potential improvements for future inclusion in the MAR Guidelines. While there have been many MAR projects undertaken in Australia (see [27] ESM1), few have resulted in suggestions for changes, and so this list below largely focuses on these exceptions and related regulatory change:

- A survey on MAR in Australia by the National Centre for Groundwater Research and Training (NCGRT) in 2015 with 134 respondents from all states.

- An update of the Guidelines for Groundwater Protection in Australia [9,10] and the South Australian Environmental Protection Policy for Water Quality [28] to enable a pathway to pragmatically define the environmental values of an aquifer, where these have not already been defined or where default values were unsupported by the facts, driven by MAR.
- An update on the South Australian Environmental Protection Policy for Water Quality [28] in 2015 where a previous arbitrary requirement for zero concentrations of herbicides was revised to conform with Aust. and N.Z. Guidelines for Fresh and Marine Water Quality 2000 [7] as periodically updated [8].
- Victorian Civil and Administrative Tribunal ruling in 2017 on reinjection of spent geothermal water into a geothermal aquifer.
- Experience in reinjection of desalinated, deoxygenated associated saline water from coal seam gas wells into a fresh water aquifer capable for use as drinking water supplies in Queensland.
- Experience in Western Australia in reinjection of dewatering water from iron ore mines to protect a groundwater-dependent salina and replenish needed groundwater resources.
- Experience in injection of advanced-treated recycled water into deep aquifers beneath Perth that contribute to public drinking water supplies.
- Lack of confidence in managing water quality, quantity and reliability of a recharge project on the Darling River in New South Wales for a drinking water supply for Broken Hill, that resulted in an alternative project being selected at four times the cost and with higher vulnerability to drought.
- Review of responses made since 2015 to the detection of per- and poly-fluoroalkyl substances (PFAS) in stormwater and aquifers in aquifer storage and recovery projects.
- Experience with cumulative impacts of aquifer storage and recovery schemes resulting in uncapped third-party wells overflowing in South Australia.
- Potential for problems of rising water table due to expansion of water sensitive urban design with increasing reliance on stormwater infiltration systems as a means of stormwater management but currently not considering potential groundwater impacts.

The NCGRT survey on MAR in Australia in 2015 revealed the perceived main drivers for MAR (Figure 3) and perceived main deterrents (Figure 4) in each Australian jurisdiction. The three largest drivers were perceived as water security in drought, meeting demand for water and mitigating decline in groundwater levels. Main deterrents were seen to be lack of information on aquifer suitability, lack of confidence that MAR will work and lack of funding mechanisms (Figure 4). However, 'lack of definition of water quality requirements for health and environmental protection' (Figure 4, item 11) was among the least deterrents, along with 'onerous water quality requirements for health and environmental protection' (Figure 4, item 12). However, the states where these both received their highest rankings as deterrents were the ones where the MAR Guidelines had been quickly adopted. This seems to suggest that in other states where MAR had not progressed, potential proponents had yet to identify water quality concerns as an issue. Where MAR Guidelines were regularly applied, the monitoring requirements increased but, on the whole, this was seen as a very minor issue. This suggests that implementation of MAR Guidelines would be accelerated if there were water quality and risk management training programs for proponents and regulators in states where uptake is lagging, as drivers for MAR are not lacking in those states.

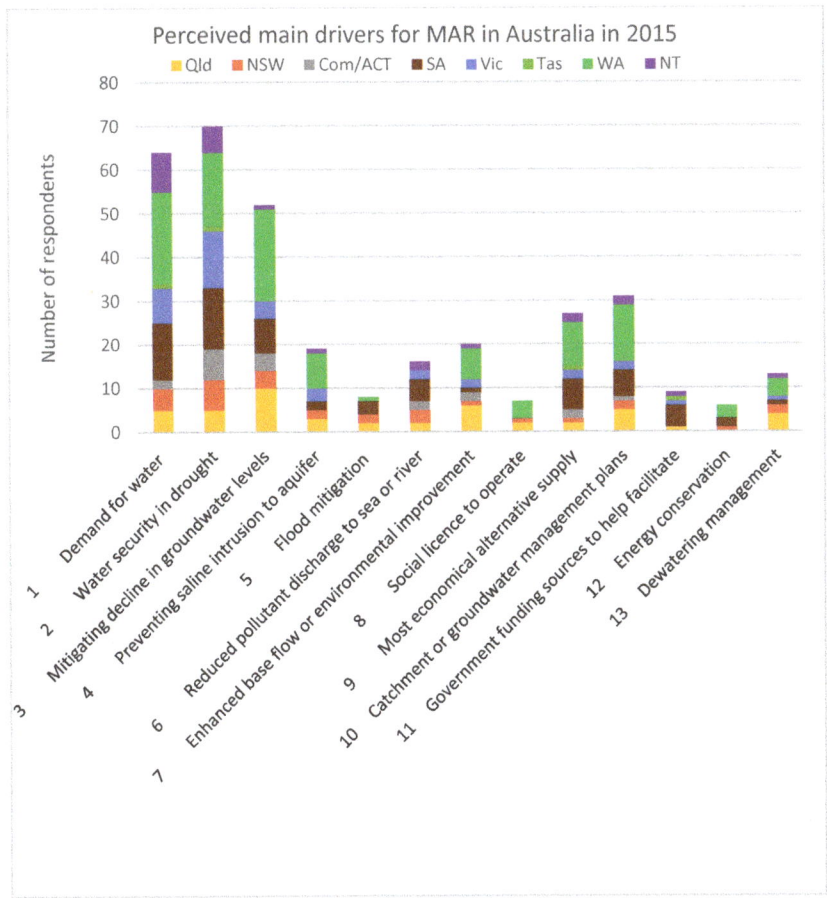

Figure 3. The main drivers for MAR perceived by 134 Australian respondents in a survey of the National Centre for Groundwater Research and Training (NCGRT) May–July 2015. This shows the number of respondents reporting perceived importance of each identified factor driving the development of MAR in their jurisdiction. (Total number of respondents = 134, total responses = 342, and average number of responses per category = 26.) (Also reported in supplementary info to [27].)

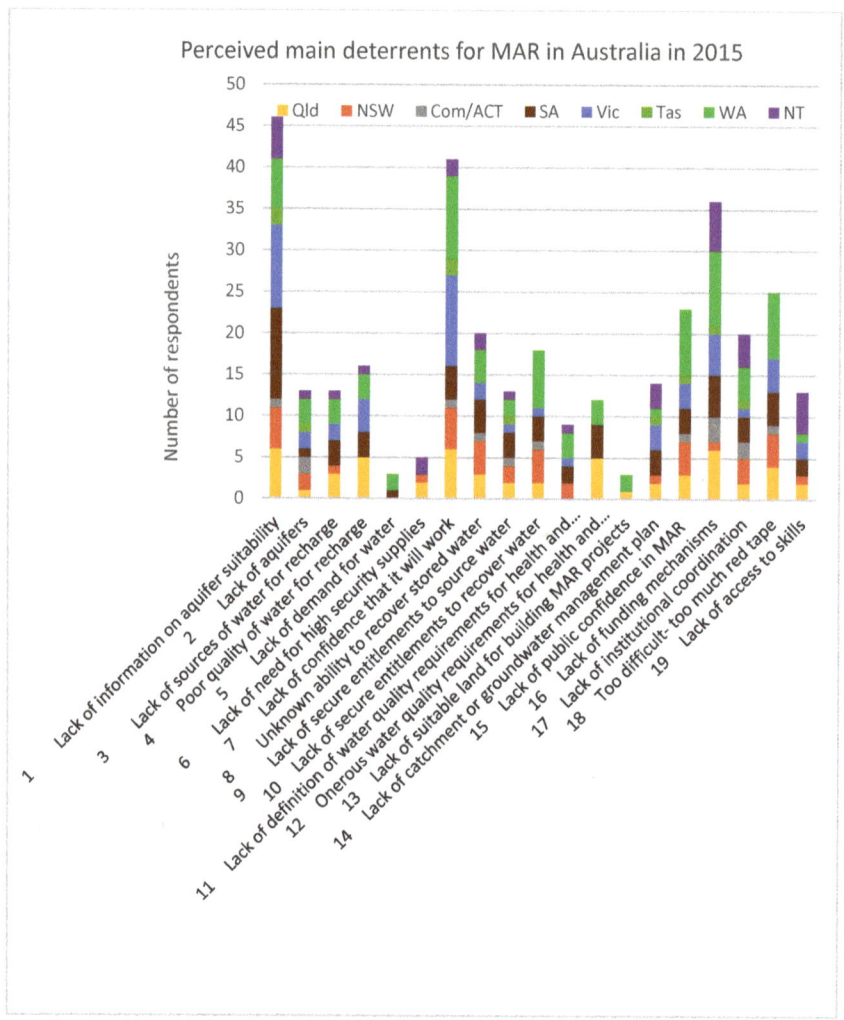

Figure 4. Number of respondents reporting perceived importance of each identified factor deterring MAR in their jurisdiction (NCGRT survey 2015). (Total number of respondents = 134, total responses = 343, and average number of responses per category = 18.).

The MAR Guidelines rely on other guidelines to specify the environmental values of the aquifers. These are for example; water for drinking, water for irrigation, water for livestock, and aquatic ecosystem protection, such as the aquifer itself and any connected rivers and wetlands that may be in pristine, good or degraded status. This differential protection policy therefore requires more effort to go into managing MAR operations where higher water quality requirements need to be sustained in the receiving aquifers and connected ecosystems. However, in Australia, as in other countries of the British Commonwealth, while there are common principles set nationally for managing water, each state enacts its own legislation and formulates its own policies. It was found in states aside from Victoria [29], that setting environmental values for aquifers could be a difficult task. Hence, in 2013, this process was laid out quite directly at national level [10]. In South Australia, this led to the SA Environment Protection Policy—Water Quality 2003 being revoked and replaced in 2015 [28] to enable the differential protection policy to be implemented within the intent of the original and revised Groundwater Protection Guidelines [9,10].

In this revised SA policy, the opportunity was taken to remove the requirement for zero concentrations of herbicides in groundwater. This was untenable scientifically due to ever lower detection levels being now well below the levels considered as acceptable for all environmental values, including drinking. This was replaced by reference to the guideline values for each beneficial use as recorded in the Aust. and N.Z. Guidelines for Fresh and Marine Water Quality 2000 [7] as periodically updated [8]. Until 2015 this had been a major impediment in ASR of stormwater in brackish aquifers where measured values for simazine and atrazine at lower than drinking water guideline values had resulted in shut-down of ASR under the former 2003 Water Quality Policy.

In 2017, the Victorian Civil and Administrative Tribunal made a ruling [30] in a case where a groundwater user was required to reinject spent geothermal water to its source aquifer if extraction exceeded a specified amount. The ruling supported increasing the licensed allocation of groundwater by the amount reinjected, within limits. (However in Western Australia geothermal waters reinjected into aquifers are currently not regarded as MAR and do not create allocation credits.) There were questions concerning impacts of reinjection reducing the temperature of extracted groundwater for both the licensed user and potential future users. Although not mentioned in the judgement, it is evident that temperature would be a useful inclusion in the MAR Guidelines as a hazard at least in relation to geothermal waters. Temperature could also be an important indicator of adverse interference effects between proximal aquifer thermal energy storage systems (ATES). Even in simple aquifer storage and recovery systems, where there is a large difference in temperature of source water and ambient groundwater, this could also have a bearing on ecosystem protection, inactivation rates of pathogens and biodegradation and sorption of contaminants [31]. Hence, consideration should be given to inclusion of water temperature as a hazard to be evaluated in MAR Guidelines and risk management strategies developed where needed.

In addition, the following points may be drawn from the ruling:

- A licenced allocation of groundwater for non-consumptive uses may be specified as an allowable net extraction (extraction minus reinjection);
- Reinjection of water is warranted to sustain the groundwater resource, even when the sustainable use limit is uncertain;
- Improving water use efficiency and reinjection are preferred to disposal of spent geothermal water to sea and much more so than disposal to leaky evaporation basins that increase the salinity of shallow groundwater;
- Reinjection gives benefits to groundwater users over a wide area by sustaining pressures whereas any residual risks of lowered temperature are primarily experienced by the holder of the licence;
- The benefits of carefully managed reinjection to the sustainability of the resource outweigh any residual risks to the resource attributes (such as reduced temperature);
- Good management of water quality and the reinjection system may make reinjection feasible, even in a complex, deep, aquifer used for geothermal operations;
- The costs of reinjection are considered commensurate with the benefits and not out of proportion with other costs of developing and utilising the groundwater resource;
- Reinjection is a part of the set of tools for adaptive management of groundwater for non-consumptive uses.

In the Surat Basin of Queensland, coal seam gas (CSG) production requires the extraction of groundwater. At two sites operated by Origin Energy, this water is desalinated, deoxygenated and reinjected into a fresh water aquifer in which pressure has been declining over many years due to extraction for irrigation, town water supplies and livestock water supplies (OGIA 2016) [32]. In Queensland, there is no clear State government guidance on MAR, and approval has been on a case by case basis. MAR associated with the CSG production is also encumbered with the need to obtain approval under the Commonwealth Environmental Protection Biodiversity Conservation Act. Despite significant investment by the CSG industry in MAR trials into multiple aquifers at several geographical sites with general adherence to the MAR Guidelines, the uncertainty in regulatory approval pathways has curtailed further CSG schemes. Key considerations for future updates to the MAR Guidelines include potential impacts to springs and groundwater-dependent ecosystems from rising heads/pressures due to multiple reinjection operations, and improved advice to operators on the effects of non-isothermal conditions on wellbore hydraulics.

Water from dewatering of several Western Australian iron ore mines has been reinjected to restore groundwater levels and thereby protect natural groundwater-dependent salinas, and to reduce net loss to the aquifer accounting also for extraction for mineral processing. These operations have been in use since 2012 and are now embedded in mine site management plans. To date no issues have arisen to suggest that a change in MAR Guidelines is warranted, although the considerations for decommissioning of MAR operations used for environmental benefits could be expanded.

Injection of advanced-treated recycled water into deep aquifers beneath Perth that contribute to public drinking water supplies was intensively monitored during a three-year pilot project. This initial extensive study expedited health and environmental approvals, increased knowledge of proponents and regulators, and communications of results created a well-informed public willing to trust and support the project [33]. The approval process was supported by the MAR Guidelines and at the time of approval revealed that no additional requirements for the MAR Guidelines were necessary. Approval processes also took account of the economics of MAR versus alternative water sources, effective and transparent monitoring and reporting by the proponent (Water Corporation) and ongoing community engagement and support for the project that enabled subsequent expansion to a full-scale scheme to proceed. Following the MAR Guidelines enabled all water quality deterioration processes to be determined. Some of these had not been anticipated such as minor release of fluoride and phosphate [34], but these require no specific revision of the Guidelines.

In contrast, the New South Wales government claimed that lack of confidence in managing water quality, quantity and reliability of a well-investigated MAR project in alluvial aquifers at Jimargil, on the Darling River for a drinking water supply for Broken Hill [35], led to selecting an alternative surface water pipeline from the Murray River at a cost of US $350 M. This was approximately four times the estimated cost of the MAR project and had higher vulnerability to drought than the MAR project. This was caused by a combination of NSW government's lack of experience in MAR and political posturing for conspicuous infrastructure. This demonstrates that in addition to comprehensive investigations and access to MAR Guidelines there is an element of capability building required at jurisdiction level to realise the potential for MAR.

In each Australian State, per- and poly-fluoroalkyl substances (PFAS) have been found in groundwater at airports, fire-stations, landfills and have also been detected in stormwater at some MAR sites. This group of man-made PFAS chemicals have been used since the 1950s, and although their use has ceased in Australia and exposure is declining, they are highly persistent and mobile in soils and aquifers. Their human health impacts have been studied but evidence is incomplete on whether this level of exposure is harmful to human health [36]. Page et al. (2019) [37] reviewed the risks of PFAS for sustainable water recycling via aquifers. PFAS substances can be removed from drinking water by sorption to Granular or Powdered Activated Carbon (GAC, PAC), or Ion Exchange Resins or by nanofiltration with reverse osmosis. At several MAR sites, where advanced monitoring was undertaken and PFAS detected, recharge was shut down as a safety precaution until a risk assessment could be performed. While the MAR Guidelines do not mention PFAS specifically, they do cover a process to address risks for all organic chemicals, and the procedure to estimate the threshold limit in drinking water, is contained in the Augmentation of Drinking Water Supplies Guideline [4] but not as yet in irrigation supplies.

In another South Australian stormwater recharge case it was found that multiple aquifer storage and recovery wells in close proximity, operated by different organisations, in some cases resulted in heads becoming artesian in third-party wells that were not equipped for this, and so water overflowed, causing a nuisance. The possibility for unacceptable cumulative impacts of multiple recharge operators had been forewarned and solutions proposed in a national document on robust policy design for MAR [26], but this had not as yet been taken up in South Australia. While such policies are not essential on the outset of MAR operations in an area, they warrant consideration to enable sustainable MAR operations. In Australia, the MAR Guidelines start with an entry level assessment that includes identifying (**a**) demand for the recharged water, (**b**) an allocation of source water, (**c**) a suitable aquifer for storage and recovery, (**d**) sufficient land for water detention and treatment, and (**e**) capability to design, construct and operate. Governance of MAR operations should be under the auspices of water resources departments, rather than environment protection authorities or health departments, because entry level issues need to be addressed first, and the primary drivers and constraints for MAR relate to ability to harvest, recharge and recover water. Some water quality matters that must be subsequently addressed would normally demand participation of environment protection and health authorities that contain the necessary skills for evaluation. This could for example include assessing MAR interactions with contaminated sites.

The expansion of water sensitive urban design in South Australia has a high reliance on stormwater infiltration systems. While this is ostensibly for increasing tree canopy cover, there is no requirement to balance increased infiltration with increased evapotranspiration. Such infiltration systems comply with the South Australia Planning Minister's Specification in 2003 for On-Site Retention of Stormwater [38], but are used as stormwater disposal, and fail to account for impacts on groundwater levels and quality. As unmanaged aquifer recharge, they do not need to comply with MAR Guidelines. Infiltration pits and trenches and permeable pavements clearly have benefits for stormwater systems and potentially for greening of cities. It is proposed that linking infiltration with tree water use in the Planning Minister's Specification is the most effective solution. However, if this is not done then the MAR Guidelines could potentially be brought to bear to prevent adverse impacts on groundwater.

4. Research that Suggests Future Refinement of MAR Guidelines

While there has been a plethora of research on MAR-related topics [27] in recent years, the vast majority of this adds new knowledge but does not require revision to the Guidelines. However, the following aspects were considered to potentially warrant revisions because they address gaps not previously covered or expose additional information that would otherwise not normally be taken into account in assessments of risk. It is recognized that a weight of evidence is needed before guidelines are changed. Hence, more than a single peer reviewed paper on a topic is needed, due to the range of circumstances to which guidelines apply. Guidelines usually follow a precautionary approach, and so guidelines may lag the latest scientific evidence.

- Research on deep well injection of brines from oil wells in USA suggests that fluid injection between 2 and 4 km in depth may be inducing seismicity with only marginal increases in pore pressure, suggesting that more explicit consideration of such risk for aquifer storage and recovery using deep wells in relevant geological settings.
- The update in 2018 of the Australian and New Zealand Environmental and Conservation Council (ANZECC) Water Quality Guidelines [8], based on research including that which has yielded improved genomics techniques to allow ecological impacts on aquifers and their connected ecosystems to be determined with higher reliability and reduced cost.
- Research has also resulted in improved methods to assess the sources and fate of pathogens recharged to aquifers to allow improved public health risk assessment.
- The advisory section of the Guidelines concerning the likelihood and extent of clogging and effectiveness of preventative and remedial strategies warrants updating.

Research on deep well fluid injection induced seismicity in Oklahoma associated with oil production and brine reinjection between 2 and 4 km in depth within 15 km of faults has reported induced seismic activity at a level significantly above background levels [39]. Increasing pore pressure reduces effective intergranular stress hence reducing resistance to shear stresses on faults and fractures to the point of instability and the possibility of earthquakes. Deep well fluid injection in Oklahoma was largely unregulated, however, this research raises an additional risk, not addressed in the current Australian MAR Guidelines, that warrants review for consideration of possible inclusions in the Australian MAR Guidelines. This could illuminate the factors affecting risk, and procedures to assess and mitigate risk for deep well injection and recovery in hard rock.

Advances in measuring and understanding ecosystem impacts of groundwater systems over the last decade have included genomics methods to allow changes in relative abundance of micro-organisms in aquifers and connected ecosystems to be assessed. The current Guidelines address microorganisms, stygofauna, phreatophytic vegetation and aquatic flora and fauna including discharge to marine ecosystems. They mention a range of tests for toxicity, genotoxicity and mutagenicity and biomarker methods, but reference to the increasing wealth of available -omics tools available and their applications is now warranted. During this evolution current knowledge has been incorporated into Guidelines for Groundwater Quality Protection in Australia (2013) [10] and Australian and New Zealand Guidelines for Fresh and Marine Water Quality (2018) [8,40,41]. New methods include those using fungi [42], and analyses of DNA in groundwater [43], that can be incorporated into new indices for groundwater health [44]. A recent review of available ecotoxicity data for subterranean fauna [45] may provide a starting point for establishing aquifer-specific water quality guideline values.

Research has also resulted in improved methods to assess the sources and fate of pathogens recharged to aquifers to allow improved methods for public health risk assessment. Research over several MAR sites within Australia demonstrated that pathogens are removed although there are local aquifer conditions that need to be considered, particularly low redox conditions for the removal of enteric viruses [46]. Sasidharan (2016) [20] identified factors affecting removal of virus surrogates. Viral indicators of sewage such as pepper mild mottle virus [47] and protozoan surrogates such as *Bacillus subtilis* spores [48] are becoming increasingly available as are methods to assess fate of antibiotic resistance genes [49]. It is anticipated that validation testing of pathogen removal in full scale systems using appropriate viral surrogates will become possible in the near future once a validation protocol has been established such as are used for engineered treatment systems.

Concerning the advisory section of the Guidelines that addresses clogging in MAR, there have been many papers since 2009 that have better-defined clogging processes (e.g., [50–53]) that would enable refinements to the current general guidance. Such refinements could include approaches to mitigate and remediate clogging, such as where statutory approvals may be required.

5. Discussion and International Relevance

Guidelines have reduced the uncertainty of approval processes, made clear what information was needed in order for project approval and shortened the time for decisions compared with the period before Guidelines, when in some cases years of discussion occurred before approval decisions were made. MAR has certainly advanced faster in States that adopted the Australian MAR Guidelines than those that had not [54] (Victoria [1,55], Western Australia [1,56] and South Australia [1,28]). While it cannot be claimed that Guidelines were the cause of increased uptake, there was a beneficial project that did not proceed due to lack of knowledge by regulatory authorities. There are to date no known examples of projects that have failed as a result of inadequacies of the MAR Guidelines having been followed throughout project investigations and commissioning. However, in parts of Adelaide the cumulative impacts of multiple aquifer storage and recovery sites caused artesian conditions in third party wells, which was in breach of the entry level requirements within the Guidelines. Cumulative impacts had not been addressed in awarding entitlements to recharge the aquifer. However, this was quickly identified as a potential issue and a national document to inform entitlement policy for MAR was published [26] that suggested pragmatic measures to avoid such problems. Due to the partitioning of water governance in Australia between water entitlements, and environmental and health protection, and institutional alignment with quantity and quality objectives, it is suggested that cumulative impacts of MAR are specifically accounted for in entitlement policies that form the basis for groundwater management plans.

There are also economic reasons for failures of MAR sites that relate to over-optimistic assessment of either demand for water or of water available for harvesting, or of inadequate understanding of operating costs, for example because trials were too short to reveal clogging and the processes necessary to manage it [57–59]. In short, there was no uncontestable business case. A tool to help assess the time series of costs and benefits of recycling schemes, and providing for the recovery efficiency of MAR systems has been developed and is freely available to help proponents to properly elucidate costs [60]. Public acceptance of stormwater and recycled water for MAR has been strong in Australia (e.g., [61]), and it is considered that the MAR Guidelines in conjunction with the other relevant guidelines do help to give the public confidence that there is an established auditable process to ensure human health and the environment are protected.

There are, however, a number of small improvements to the Guidelines that are warranted.

- adding temperature as a "hazard" in geothermal and open well ATES applications, and for explicit consideration of reactions within aquifers and contaminant removal processes (for organic and inorganic chemicals, microorganisms) and ecosystem impacts
- considering advances in scientific knowledge with respect to fluid-injection induced seismicity, fate of pathogens and organic chemicals, ecosystem monitoring methods, and clogging processes, which will make minor but warranted refinements to the Guidelines and extend their durability
- further elaborating project closure requirements, particularly where MAR is primarily for environmental benefit
- giving specific consideration of cumulative impacts of multiple MAR projects
- in the entry level section of the Guidelines, making more explicit the water entitlement arrangements for sourcing water, recharging aquifers, recovering from aquifers and end uses (e.g., [26]). In basins with groundwater levels in decline, groundwater management policies need to be strengthened to be effective in securing MAR entitlements.

Use of the Australian Guidelines depend heavily on having capabilities to monitor, sample and analyse water quality. In some countries with sparsity or absence of such capabilities, it is suggested that other forms of Guideline be considered. For example, Indian guidelines for water quality management in MAR [62] are an adaption of the sanitary survey within a WHO water safety planning approach, and are based on only visual observations, in order to improve the safety of MAR operations (Figure 5). Those Guidelines exclude MAR practices with urban stormwater, treated sewage or industrial wastewaters or waters likely to be contaminated by anthropogenic activity that are all expected to contain hazards that cannot be adequately managed without sufficient reliable quantitative analyses. This method is recommended only for MAR of natural waters where water is infiltrated through the unsaturated zone, mimicking natural recharge. This method therefore has limited utility and should not be used for recharge of confined aquifers or aquifers in the proximity of wells used for drinking water supplies.

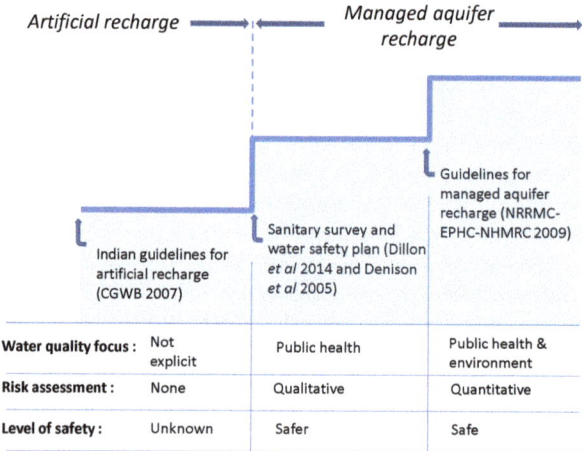

Figure 5. Implementing the Australian MAR Guidelines in India was not possible due to limitations in availability of analyses for viruses and protozoa in environmental water samples and difficulty in being able to get samples from field sites to the lab in time and at a correct temperature for reliable analytical results. Solid state extraction could enable viable delayed analyses of water samples. Hence, an Indian Guideline for MAR was prepared (Dillon et al. 2014) [62] as a step towards safer water supplies and improved groundwater protection than current practice, but without the rigor of data acquisition to support a risk assessment necessary to assure safety.

6. Conclusions

The Australian Guidelines address all types of source waters, all types of aquifers, all types of recharge methods, and all types of end uses of recovered water. While Australian experiences in implementing risk-based MAR Guidelines are currently unique, it is intended that awareness of those experiences in countries considering adopting or adapting risk based MAR Guidelines will become immediately relevant, and potential problems averted. Recent research output is also relevant internationally, as it advances understanding of biogeochemical, hydrogeological or geotechnical processes, improves measurement methods and affords greater awareness of risks that had previously not been as deeply explored. Hence, in aggregate the Australian MAR Guidelines and lessons learned are broadly applicable to a range of hydrogeological, climatic or legal conditions. Adaptation will most likely be required if existing groundwater protection policies do not acknowledge biogeochemical processes in aquifers or otherwise lack a scientific basis.

The Guidelines have served well to ensure protection of health and the environment in Australian MAR operations. Only minor revisions are suggested, and the largest of these is to account for temperature as a specific hazard to be managed in MAR in geothermal operations, for aquifer thermal energy systems and where reliance is placed on contaminant attenuation within the aquifer. Other small revisions recommended are to include more specifically cumulative hydraulic impacts of multiple MAR systems and expanding on decommissioning requirements where MAR is undertaken for environmental benefit. Research now offers a broader range of –omics tools and pathogen fate measures with a view towards future validation protocols for viral and trace organic removal and ecosystem protection. Guideline implementation appears more streamlined, and problems avoided, where they are primarily the responsibility of water resources management agencies, supported by environment protection and health agency expertise (rather than the reverse). Impacts of Guidelines would be accelerated by training at jurisdiction level on water quality and risk management, and on cumulative impact assessment.

Capabilities to measure and manage water quality and to fully understand and anticipate the risks are necessary to apply these Guidelines especially for aquifers used as drinking water sources. For this reason, an intermediate approach has been recommended for India, until such time as the full complement of capabilities is developed though demonstration projects initially with low inherent risk.

Author Contributions: Conceptualisation P.D. and D.P.; contributions to MAR experience, P.D., J.V., C.S., R.M. (Russell Martin), K.J., S.H. and R.M. (Ryan Morris); contributions to analysis of advances in research P.D., D.P., J.V., S.T. and G.H.; review and editing, all authors. All authors have read and agreed to the published version of the manuscript.

Funding: The NCGRT supported a distinguished lectureship in 2015 for the lead author that covered travel costs to facilitate the MAR survey. Otherwise there was no external funding for this research paper.

Acknowledgments: A precursor to this paper was presented at ISMAR10, Madrid, May 2019. Authors thank the three anonymous J. Water reviewers for their constructive comments on a draft of this paper.

Conflicts of Interest: The authors declare no conflict of interest. This paper does not contain any part of an organisational plan of any Australian jurisdiction. Recommendations in this paper should not be taken to represent the position of authors' organisations.

Abbreviations

The following abbreviations are used in this manuscript:

ASR	aquifer storage and recovery (injecting and recovering water from the same well)
ATES	aquifer thermal energy storage
MAR	managed aquifer recharge (the purposeful recharge of water to aquifers for subsequent recovery or environmental benefit. It is not a method for waste disposal.)

References

1. NRMMC; EPHC; NHMRC. *Australian Guidelines for Water Recycling: Managing Health and Environmental Risks (Phase 2) Managed Aquifer Recharge. National Water Quality Management Strategy Document 24*; National Water Resource Management Ministerial Council, Environment Protection and Heritage Council, National Health and Medical Research Council: Canberra, Australia, 2009; p. 237. Available online: http://www.waterquality.gov.au/guidelines/recycled-water (accessed on 19 January 2020).
2. WHO; IWA. *Water Safety Plan Manual*; World Health Organization: Geneva, Switzerland; International Water Association: London, UK, 2009; Available online: https://www.who.int/water_sanitation_health/publications/publication_9789241562638/en/ (accessed on 19 January 2020).
3. NRMMC; EPHC; AHMC. Australian Guidelines for Water Recycling: *Managing Health and Environmental Risks*. NWQMS Document 21; 2006. Available online: http://www.waterquality.gov.au/guidelines/recycled-water (accessed on 19 January 2020).
4. NRMMC; EPHC; NHMRC. Australian Guidelines for Water Recycling: Managing Health and Environmental Risks (Phase 2). In *Augmentation of Drinking Water Supplies*; NWQMS Document 22; 2008. Available online: http://www.waterquality.gov.au/guidelines/recycled-water (accessed on 19 January 2020).
5. NRMMC; EPHC; NHMRC. National Water Quality Management Strategy. Australian Guidelines for Water Recycling: Managing Health and Environmental Risks (Phase 2). In *Stormwater Harvesting and Reuse*; NWQMS Document 23; 2009. Available online: http://www.waterquality.gov.au/guidelines/recycled-water (accessed on 19 January 2020).
6. ARMCANZ–ANZECC. *National Water Quality Management Strategy: Policies and Principles—A Reference Document*; Paper No 2; Australian and New Zealand Environment and Conservation Council & Agriculture and Resource Management Council of Australia and New Zealand: Canberra, Australia, 1994.
7. ANZECC–ARMCANZ. *Australian and New Zealand guidelines for fresh and marine water quality. National Water Quality Management Strategy*; Paper No 4; Australian and New Zealand Environmental and Conservation Council, Agriculture Resource Management Council of Australia and New Zealand: Canberra, Australia, 2000.

8. Water Quality Australia. Australian and New Zealand Guidelines for Fresh and Marine Water Quality. 2018. Available online: http://www.waterquality.gov.au/guidelines/anz-fresh-marine (accessed on 19 November 2019).
9. ANZECC–ARMCANZ. *Guidelines for groundwater protection in Australia. National Water Quality Management Strategy*; Australian and New Zealand Environmental and Conservation Council, Agriculture and Resource Management Council of Australia and New Zealand: Canberra, Australia, 1995. Available online: https://www.water.wa.gov.au/__data/assets/pdf_file/0020/4925/8728.pdf (accessed on 19 January 2020).
10. Water Quality Australia. Guidelines for groundwater quality protection in Australia. In *National Water Quality Management Strategy*; Department of Agriculture and Water Resources, 2013. Available online: https://www.waterquality.gov.au/sites/default/files/documents/guidelines-groundwater-quality-protection.pdf (accessed on 19 January 2020).
11. ANZECC–ARMCANZ. Australian guidelines for water quality monitoring and reporting. In *National Water Quality Management Strategy*; Paper No 7; Australian and New Zealand Environmental and Conservation Council, Agriculture and Resource Management Council of Australia and New Zealand: Canberra, Australia, 2000.
12. NHMRC–NRMMC (National Health and Medical Research Council and Natural Resource Management Ministerial Council). *Australian Drinking Water Guidelines*; NHMRC, NRMMC: Canberra, Australia, 2004. Available online: http://www.nhmrc.gov.au/publications/_files/adwg_11_06.pdf (accessed on 31 March 2009).
13. Water Quality Australia. Australian Drinking Water Guidelines. 2011. Available online: https://www.waterquality.gov.au/guidelines (accessed on 19 January 2020).
14. Dillon, P.; Kumar, A.; Kookana, R.; Leijs, R.; Reed, D.; Parsons, S.; Ingleton, G. *Managed Aquifer Recharge—Risks to Groundwater Dependent Ecosystems—A Review. Water for a Healthy Country Report to Land and Water Australia*; CSIRO: Canberra, Australia, 2009. Available online: www.clw.csiro.au/publications/waterforahealthycountry/2009/wfhc-managed-aquifer-recharge-risks.pdf (accessed on 19 January 2020).
15. Dillon, P.; Toze, S. (Eds.) *Water Quality Improvements During Aquifer Storage and Recovery, Vol 1. Water Quality Improvement Processes. (347p). Vol 2. Compilation of Information from Ten Sites*; AWWARF Project 2618, Final Report; American Water Works Association Research Foundation: Denver, CO, USA, 2003; 347p.
16. Pavelic, P.; Dillon, P.J.; Nicholson, B.C. Comparative evaluation of the fate of disinfection by-products at eight aquifer storage and recovery sites. *Environ. Sci. Technol.* **2006**, *40*, 501–508. [CrossRef] [PubMed]
17. Vanderzalm, J.; Sidhu, J.; Bekele, G.-G.; Pavelic, P.; Toze, S.; Dillon, P.; Kookana, R.; Hanna, J.; Barry, K.; Yu, X.; et al. *Water Quality Changes During Aquifer Storage and Recovery*; Project #2974 Report; Water Research Foundation: Denver, CO, USA, 2009.
18. Dillon, P.; Pavelic, P.; Page, D.; Miotlinski, K.; Levett, K.; Barry, K.; Taylor, R.; Wakelin, S.; Vanderzalm, J.; Molloy, R.; et al. Developing Aquifer Storage and Recovery (ASR) Opportunities in Melbourne—Rossdale ASR Demonstration Project Final Report. *Water for a Healthy Country Report* to *Smart Water Fund*. June 2010. Available online: http://clearwater.asn.au/resource-library/smart-water-fund-projects/developing-aquifer-storage-and-recovery-opportunities-in-greater-melbourne.php (accessed on 19 January 2020).
19. Torkzaban, S.; Bradford, S.A.; Vanderzalm, J.L.; Patterson, B.M.; Harris, B.; Prommer, H. Colloid release and clogging in porous media: Effects of solution ionic strength and flow velocity. *J. Contam. Hydrol.* **2015**, *181*, 161–171. [CrossRef] [PubMed]
20. Sasidharan, S. Fate, Transport, and Retention of Viruses, Bacteria, and Nanoparticles in Saturated Porous Media. Ph.D. Thesis, Flinders University of South Australia, Adelaide, Australia, 2016; p. 258. Available online: https://flex.flinders.edu.au/file/951c9bd7-c036-4c5b-b36b-478fbae8c417/1/Thesis_Sasidharan_2016.pdf (accessed on 25 March 2019).
21. Rauch-Williams, T.; Hoppe-Jones, C.; Drewes, J.E. The role of organic matter in the removal of emerging trace organic chemicals during managed aquifer recharge. *Water Res.* **2010**, *44*, 449–460. [CrossRef] [PubMed]
22. Li, D.; Alidina, M.; Drewes, J.E. Role of primary substrate composition on microbial community structure and function and trace organic chemical attenuation in managed aquifer recharge systems. *Appl. Microbiol. Biotechnol.* **2014**, *98*, 5747–5756. [CrossRef] [PubMed]
23. Alidina, M.; Li, D.; Ouf, M.; Drewes, J.E. Role of primary substrate composition and concentration on attenuation of trace organic chemicals in managed aquifer recharge systems. *J. Environ. Manag.* **2014**, *144*, 58–66. [CrossRef]

24. Bouwer, H.; Pyne, R.D.G.; Brown, J.; St Germain, D.; Morris, T.M.; Brown, C.J.; Dillon, P.; Rycus, M.J. Design, operation and maintenance for sustainable underground storage facilities. In *American Water Works Association Research Foundation Report*; Denver, CO, USA, 2008; p. 235. Available online: https://websites.pmc.ucsc.edu/~{}afisher/post/MAR_Papers/Bouwer2009_DesignSustainMAR.pdf (accessed on 19 January 2020).
25. Dillon, P. Groundwater Replenishment with Recycled Water—An Australian Perspective. *Groundwater* **2009**, *47*, 492–495. [CrossRef]
26. Ward, J.; Dillon, P. Robust policy design for managed aquifer recharge. Waterlines Report Series No 38. 2011, p. 28. Available online: https://apo.org.au/sites/default/files/resource-files/2011/02/apo-nid23758-1179541.pdf (accessed on 19 January 2020).
27. Dillon, P.; Stuyfzand, P.; Grischek, T.; Lluria, M.; Pyne, R.D.G.; Jain, R.C.; Bear, J.; Schwarz, J.; Wang, W.; Fernandez, E.; et al. Sixty Years of Global Progress in Managed Aquifer Recharge. *Hydrogeology* **2019**, *27*, 1–30. [CrossRef]
28. South Australia. Environment Protection (Water Quality) Policy. 2015. Available online: https://www.legislation.sa.gov.au/LZ/C/POL/Environment%20Protection%20(Water%20Quality)%20Policy%202015.aspx (accessed on 19 January 2020).
29. Victoria EPA. State environment protection policy—Groundwaters of Victoria. In *Victoria Government Gazette*; No. S160; Victorian Govt. Printer: Melbourne, Australia, 17 December 1997. Available online: https://ref.epa.vic.gov.au/our-work/setting-standards/environmental-standards-reform/~{}/media/Publications/S160.pdf (accessed on 19 January 2020).
30. Victoria Civil and Administrative Tribunal. Ruling on Peninsula Hot Springs Pty Ltd v Southern Rural Water (2017) VCAT 2103 (19 December 2017) Planning & Environment List VCAT Reference Nos. P2730/2015 & P1843/2016. 2017. Available online: http://www7.austlii.edu.au/cgi-bin/viewdoc/au/cases/vic/VCAT/2017/2103.html#fnB41 (accessed on 25 March 2019).
31. Griebler, C.; Brielmann, H.; Haberer, C.M.; Kaschuba, S.; Kellermann, C.; Stumpp, C.; Hegler, F.; Kuntz, D.; Walker-Hertkorn, S.; Lueders, T. Potential impacts of geothermal energy use and storage of heat on groundwater quality, biodiversity, and ecosystem processes. *Environ. Earth Sci.* **2016**, *75*, 1391. [CrossRef]
32. OGIA. *Underground Water Impact Report for the Surat Cumulative Management Area. The Office of Groundwater Impact Assessment*; Department of Natural Resources and Mines: Brisbane, Australia, 2016. Available online: https://www.dnrme.qld.gov.au/__data/assets/pdf_file/0007/345616/uwir-surat-basin-2016.pdf (accessed on 19 January 2020).
33. Water Corporation. Groundwater Replenishment Trial Final Report. 2013. Available online: https://www.watercorporation.com.au/water-supply/our-water-sources/groundwater-replenishment (accessed on 31 March 2019).
34. Schafer, D.; Donn, M.; Atteia, O.; Sun, J.; MacRae, C.; Raven, M.; Pejcicg, B.; Prommer, H. Fluoride and phosphate release from carbonate-rich fluorapatite during managed aquifer recharge. *J. Hydrol.* **2018**, *562*, 809–820. [CrossRef]
35. Lawrie, K.C.; Brodie, R.S.; Dillon, P.; Tan, K.P.; Gibson, D.; Magee, J.; Clarke, J.D.A.; Somerville, P.; Gow, L.; Halas, L.; et al. Broken Hill Managed Aquifer Recharge (BHMAR) Project: Assessment of Conjunctive Water Supply Options to Enhance the Drought Security of Broken Hill, Regional Communities and Industries—Summary Report. In *Geoscience Australia: Record 2012/15, Canberra, Australia*; GeoCat #73823; 2012.
36. Water Research Australia. PFAS Resources and Information. 2019. Available online: https://www.waterra.com.au/research/knowledge-transfer/pfas-resources-and-information/ (accessed on 31 March 2019).
37. Page, D.; Vanderzalm, J.; Kumar, A.; Cheng, K.Y.; Kaksonen, A.H.; Simpson, S. Risks of Perfluoroalkyl and Polyfluoroalkyl Substances (PFAS) for Sustainable Water Recycling via Aquifers. *Water* **2019**, *11*, 1737 [CrossRef]
38. South Australia Planning. On-Site Retention of Stormwater. Minister's Specification SA 78AA. 2003. Available online: https://www.sa.gov.au/__data/assets/pdf_file/0017/7046/SA_78AA_Onsite_retention_of_stormwater.pdf (accessed on 25 March 2019).
39. Hincks, T.; Aspinall, W.; Cooke, R.; Gernon, T. Oklahoma's induced seismicity strongly linked to wastewater injection depth. *Science* **2018**, *359*, 1251–1255. [CrossRef] [PubMed]
40. Batley, G.E.; van Dam, R.A.; Warne, M.S.t.J.; Chapman, J.C.; Fox, D.R.; Hickey, C.W.; Stauber, J.L. *Technical Rationale for Changes to the Method for Deriving Australian and New Zealand Water Quality Guideline Values for*

Toxicants; Australian Government Department of Agriculture and Water Resources: Canberra, Australia, 2018.

41. Warne, M.St.J.; Batley, G.E.; van Dam, R.A.; Chapman, J.C.; Fox, D.R.; Hickey, C.W.; Stauber, J.L. *Revised Method for Deriving Australian and New Zealand Water Quality Guideline Values for Toxicants*; Australian Government Department of Agriculture and Water Resources: Canberra, Australia, 2018.

42. Lategan, M.J.; Klare, W.; Kidd, S.; Hose, G.C.; Nevalainen, H. The unicellular fungal tool RhoTox for risk assessments in groundwater systems. *Ecotoxicol. Environ. Saf.* **2016**, *132*, 18–25. [CrossRef] [PubMed]

43. Korbel, K.; Chariton, A.; Stephenson, S.; Greenfield, P.; Hose, G.C. Wells provide a distorted view of life in the aquifer: Implications for sampling, monitoring and assessment of groundwater ecosystems. *Sci. Rep.* **2017**, *7*, 40702. [CrossRef] [PubMed]

44. Korbel, K.L.; Hose, G.C. The weighted groundwater health index: Improving the monitoring and management of groundwater resources. *Ecol. Indic.* **2017**, *75*, 164–181. [CrossRef]

45. Castaño-Sánchez, A.; Hose, G.C.; Reboleira, A.S.P.S. The ecotoxicological effects of anthropogenic stressors in subterranean organisms. *Chemosphere* **2020**, *244*, 125422. [CrossRef]

46. Sidhu, J.; Toze, S.; Hodgers, L.; Barry, K.; Page, D.; Li, Y. Pathogen Decay during Managed Aquifer Recharge at Four Different Sites with Different Geochemical Characteristics and Recharge Water Sources. *J. Environ. Qual.* **2015**, *44*, 1402–1412. [CrossRef]

47. Kitajima, M.; Sassi, H.P.; Torrey, J.R. Pepper mild mottle virus as a water quality indicator. *NPJ Clean Water* **2018**, *1*, 19. [CrossRef]

48. Bradford, S.A.; Kim, H.; Head, B.; Torkzaban, S. Evaluating the Transport of Bacillus subtilis Spores as a Potential Surrogate for Cryptosporidium parvum Oocysts. *Environ. Sci. Technol.* **2016**, *50*, 1295–1303. [CrossRef]

49. Stange, C.; Sidhu, J.; Toze, S.; Tiehm, A. Comparative removal of antibiotic resistance genes during chlorination, ozonation, and UV treatment. *Int. J. Hygiene Environ. Health* **2019**, *222*, 541–548. [CrossRef]

50. Martin, R. (Ed.) *Clogging Issues Associated with Managed Aquifer Recharge Methods*; IAH Commission on Managing Aquifer Recharge, Monograph, 2013; Available online: https://recharge.iah.org/files/2015/03/Clogging_Monograph.pdf (accessed on 19 January 2020).

51. Wang, Z.; Du, X.; Yang, Y.; Ye, X. Surface clogging process modeling of suspended solids during urban stormwater aquifer recharge. *J. Environ. Sci.* **2012**, *24*, 1418–1424. [CrossRef]

52. Du, X.; Zhang, H.; Ye, X.; Lu, Y.Y. Flow velocity effects on Fe(III) clogging during managed aquifer recharge using urban storm water. *Water* **2018**, *10*, 358. [CrossRef]

53. Johnston, K.; Nelson, D. Hartfield Park water proofing project: Stormwater harvesting using aquifer storage and recovery. In Proceedings of the 10th International Conference on Water Sensitive Urban Design, Perth, Australia, 14 February 2018; p. 7, ISBN 978-1-925627-03-9.

54. Parsons, S.; Dillon, P.; Irvine, E.; Holland, G.; Kaufman, C. *Progress in Managed Aquifer Recharge in Australia*; National Water Commission Waterlines Report Series No 73; SKM & CSIRO: 2012; National Water Commission: Canberra, Australia, 2012; p. 107.

55. Victoria Environment Protection Agency. Guidelines for Managed Aquifer Recharge (MAR)—Health and Environmental Risk Management. Publication 1290: Melbourne, Australia. 2009. Available online: https://www.epa.vic.gov.au/about-epa/publications/1290 (accessed on 19 January 2020).

56. WA Department of Water. Operational Policy 1.01. Managed Aquifer Recharge in Western Australia. 2011. Available online: http://www.newwaterways.org.au/downloads/Resources%20-%20Policy%20and%20Guidelines/Water%20reuse%20and%20recycling/Operational%20Policy%201_01%20-%20Managed%20aquifer%20recharge%20in%20WA.pdf (accessed on 31 March 2019).

57. West, C.; Kenway, S.; Hassall, M.; Yuan, Z. Why do residential recycled water schemes fail? A comprehensive review of risk factors and impact on objectives. *Water Res.* **2016**, *102*, 271–281. [CrossRef] [PubMed]

58. West, C.; Kenway, S.; Hassall, M.; Yuan, Z. Expert opinion on risks to the long-term viability of residential recycled water schemes: An Australian study. *Water Res.* **2017**, *120*, 133–145. [CrossRef] [PubMed]

59. Kretschmer, P. Managed aquifer recharge schemes in the Adelaide metropolitan area. In *DEWNR Technical Report 2017/22*; Government of South Australia, Department of Environment, Water and Natural Resources: Adelaide, Australia, 2017; p. 48.

60. Marsden Jacob Associates. *Economic Viability of Recycled Water Schemes*; Australian Water Recycling Centre of Excellence: Brisbane, Australia, 2013; Available online: http://www.marsdenjacob.com.au/wp-content/uploads/2014/11/Economic-viability-of-recycled-water-schemes.pdf (accessed on 29 March 2019).
61. Mankad, A.; Walton, A.; Alexander, K. Key dimensions of public acceptance for managed aquifer recharge of urban stormwater. *J. Clean. Prod.* **2015**, *89*, 214–223. [CrossRef]
62. Dillon, P.; Vanderzalm, J.; Sidhu, J.; Page, D.; Chadha, D. A Water Quality Guide to Managed Aquifer Recharge in India. CSIRO Land and Water, UNESCO Report of AusAID PSLP Project ROU 14476. 34p+appendices. 2014. Available online: https://recharge.iah.org/files/2016/11/A-Water-Quality-Guide-to-MAR-in-India-2014.pdf (accessed on 25 March 2019).

© 2020 by the authors. Licensee MDPI, Basel, Switzerland. This article is an open access article distributed under the terms and conditions of the Creative Commons Attribution (CC BY) license (http://creativecommons.org/licenses/by/4.0/).

Article

The Effect of Soil Tillage Equipment on the Recharge Capacity of Infiltration Ponds

Ido Negev [1,*], Tamir Shechter [1], Lilach Shtrasler [1], Hadar Rozenbach [2] and Avri Livne [2]

[1] Mekorot, National Water Company, Tel Aviv PO 20128, Israel; tshechter@MEKOROT.CO.IL (T.S.); lshtrasler@MEKOROT.CO.IL (L.S.)
[2] Plagey Maim, Megido, Israel; Hadar@palgey-maim.co.il (H.R.); avri@palgey-maim.co.il (A.L.)
* Correspondence: nido@mekorot.co.il; Tel.: +972-50-5756042

Received: 19 November 2019; Accepted: 13 February 2020; Published: 15 February 2020

Abstract: The Dan Region Reclamation Project (Shafdan) reclaims ~125 millions of cubic meters per year (Mm^3/year) of treated wastewater from the Tel Aviv Metropolitan area. Following secondary treatment, the effluent is recharged into a sandy aquifer for soil aquifer treatment (SAT). Over the past three years, a decrease in recharge capacity was noticed. Several operational causes were considered including reservations regarding the tillage procedure of recharge ponds. Tillage of the recharge ponds facilitates aeration, breaking surface crusts and the removal of vegetation. The procedure includes deep (40–60 cm) plowing and shallow (10–20 cm) sweep-knives (SK) cultivator or discus. In this research, the existing tillage equipment was compared to a new equipment, which includes a deep subsoiler and a chisel-knives (CK) cultivator. The effects of each tool on the infiltration rate (IR), recharge capacity, and soil compaction were examined. The results suggest a significant improvement in the recharge capacity, up to 95% and 15% on average following subsoiler and CK cultivator treatments, respectively, with respect to the existing plowing treatment. In addition, the depth of the compacted soil layer increased from ~30 to ~55 cm after subsoiler treatment. It seems that this shallow layer, developed under an improper tillage regime, played a major role in the reduction of the recharge capacity. Essential understanding of other operational factors such as drying periods, preparation of the field, and soil micro-topography was also achieved.

Keywords: SAT; tillage; infiltration pond; infiltration rate; soil compaction

1. Introduction

The Dan Region Wastewater Reclamation Plant (Shafdan plant), established during the 1970s, provides a centralized, high-quality solution for sewage produced by the most populated area in Israel. The plant collects the sewage from the Tel-Aviv Metropolitan area (Dan Region) and neighboring municipalities, treats it and then recharges it into a defined coastal aquifer section for complementary soil aquifer treatment (SAT). In this manner, the Shafdan plant treats about 125 Mm^3/year of raw sewage from seven municipalities, industrial areas, and approximately 1.5 million inhabitants. The sewage is treated to a level of secondary effluents using a mechanical biological treatment plant (MBTP) prior to infiltration in the SAT ponds. The recharge and the SAT process take place in the coastal Quaternary sandstone aquifer, which also provides vast storage capacity that cannot be replaced by surface reservoirs. During the flow process of secondary effluent through the vadose zone and the aquifer, most of the biodegradable organic matter, suspended solids, bacteria, viruses, phosphorus, heavy metals, and other elements are removed from the effluents by a combination of geochemical, physical, and biological processes [1–13]. The recovered effluents, after SAT treatment, are characterized by excellent quality for most of the measured parameters, which comply with the levels allowed by the Israeli Health Ministry for unrestricted irrigation of any crop [1,10,11].

After more than 20 years of relatively stable operation, over the past three years, a trend of decreasing recharge capacity has been observed. Between the years 2015 to 2017, the recharge capacity gradually decreased from ~125 to ~115 Mm^3/year, with effluent volumes showing the opposite trend and increasing from ~130 to ~140 Mm^3/year [13]. The causes for the decrease in infiltration capacity are complicated. At least two different processes that occurred at approximately the same period including a decrease in effluent quality and a reduction in drying periods between recharge cycles could have led to a decrease in infiltration rates. However, it is difficult to exclude other related processes that could also have supported this trend. These include operational regimes such as tillage activity under increasing hydraulic loads, seasonal effects on infiltration rates, and a long-term decrease in the infiltration rates of recharge fields. These processes and their possible effects on infiltration capacity will be briefly discussed in the following.

Reduction of effluent quality: Several studies (e.g., [12,14–17]), along with the cumulative experience in the Shafdan plant, show the effect of effluent quality on infiltration rates. During 2016–2017, a major facility for sludge treatment operated in the Shafdan plant. This had some effect on the effluent quality, mainly by increasing the ammonium concentrations, and temporary periods of increased turbidity, phosphorous, and nitrogen concentrations. In some instances, these periods of low-quality effluents led to the development of a dark, thick crust on the infiltration ponds' soil surface, and to a decrease in infiltration rates. Even though the effect of effluent quality on infiltration rates is exceptionally important, it cannot explain the entire decrease in recharge capacity. This is primarily because the operation of the sludge treatment facility, and its effect on effluent quality, did not fully overlap with the process of decreasing recharge capacity.

Reduction of drying periods: The importance of an orderly and controlled recharge cycle that includes a minimal drying period is well known (e.g., [12,18,19]). The role of the drying stage between wetting periods is to enable drying and decomposing of the bio-crust developed on the basin floor to enable the aeration of the soil and maintain constant infiltration rates. The increase in effluent volumes, with no additional development of recharge fields, clearly leads to increasing loads on the existing recharge fields, to a shorter drying period between flooding cycles, and, in turn, to the decrease in infiltration rates. This mechanism can explain much of the decrease in recharge capacity. The reduction of the drying periods could also have a negative effect on soil aeration and on the redox conditions, which in turn could affect water quality due to manganese and ferrous reductive dissolution processes. The recharge cycle applied in this study included a drying period of 12 to 20 h for all of the recharge ponds (treatment and control) throughout the experiment. Thus, regardless of the importance of this factor, which will be discussed further, it is not expected to have any effect on the results of this experiment.

Seasonal effects: The effect of the climate on recharge capacity, and the decrease in infiltration rates during winter, is a well-known phenomenon in managed aquifer recharge (MAR) projects, and specifically in the Shafdan plant [15,20]. From 2015 to 2017, a significant decrease in the recharge capacity during the winter months (December to February) was observed, and in amounts, changes between 2 to 7 Mm^3/year could explain much of the annual volume losses. This decrease in recharge amounts is correlated with the general decrease in infiltration rates, which in turn may be directly related to the effects of the drying period and effluent quality, as discussed earlier. However, to eliminate this effect, the experiment was conducted during late spring and summer seasons.

Long-range effects: Many technical operational reports and design plans raised questions regarding the stability of the recharge ponds' infiltration rates, and the possibility of its gradual decrease. However, none of these were based on methodological research or surveys. On the other hand, monitoring of actual recharge capacity trends at specific fields of the Shafdan plant did not support this phenomenon, and could even be said to contradict it. For example, the Soreq field, which has operated since 1977, had the most stable and highest recharge capacity in comparison with the other recharge fields, whereas Yavne-4, which has only been in operation since 2003, had the lowest recharge capacity.

Long-range effects can also be explained by a slow accumulation of inorganic compounds arriving with the recharged effluents. Scraping of the upper layer is applied in some MAR projects around the world in order to remove this layer [18]. This was not the case, however, in the Shafdan plant. The Shafdan effluents (after secondary treatments) are characterized by relatively low suspended solids concentrations of about 3.9 mg/L, of which only 0.9 mg/L are inorganic compounds [13]. Under hydraulic loads of about 100 m/year, this equates to the addition of about 900 kg per hectare per year of inorganic compounds to the ponds, which (for bulk density 1500 kg/m^3) is only about 0.06 mm per year. For these loads, scraping operations with their compaction effect can cause more damage than benefit.

Tillage regime: Tillage of recharge ponds, which is a part of the routine operational activity of recharge fields, allows for the aeration of the upper layer, breaking surface crusts and removing vegetation. Even though its effect on the operation of recharge ponds and infiltration rates is well known, it appears that tillage regime in MAR projects is more of a local and operational practice rather than a global research issue. Thus, only a few and relatively generalized academic references are available, and most of them have focused on crust scraping instead of tillage regimes (e.g., [18,21,22]). The tillage procedure taking place in the Shafdan plant includes deep (40–60 cm) plowing every three months, and shallow (10–20 cm) tillage by a sweep-knives cultivator or discus every 2–3 weeks. However, over the past two years, plowing frequency has increased to every 3–4 weeks in an attempt to cope with the decline in the recharge capacity. The consequences of increased plowing frequency include not only an increase in operational costs, but also intensified soil compaction processes due to repeated passes with heavy (and maybe also improper) equipment over the field. In turn, it may lead to an additional decrease in infiltration rates.

The present research hypothesized that at least part of the decreased infiltration capacity was caused by the operation of tillage equipment, along with a tillage regime that is no longer suitable to the current conditions of the recharge fields. The objective of the research was to study the effect of different tillage tools on infiltration capacity and on soil compaction.

2. Materials and Methods

2.1. Research Outline

The research was conducted at Yavne-2, which is one of Shafdan plant's recharge fields. The Yavne-2 recharge field is located southwest of the city of Yavne, Israel, and is divided into nine operational recharge ponds of ~1.5 hectare each (Figure 1). In order to minimize the seasonal effects, the research was conducted during the late spring and summer periods only (i.e., between 29 May, 2019 to 3 September, 2019. During this period, four treatment stages were conducted, each of them lasting for a period of 3–4 weeks, as detailed in Table 1. The period between 1 May to 29 May was defined as a pre-experiment stage and was also used for comparative analyses. The first two stages of the experiment as well as the pre-experiment period included only the control treatment, that is, the operation of regular tillage equipment by the equipment's usual operator (contractor). The last two stages included a full comparison between the new tillage equipment that was employed in six recharge ponds (six replicates), and the regular tillage equipment that was employed in three recharge ponds (three replicates).

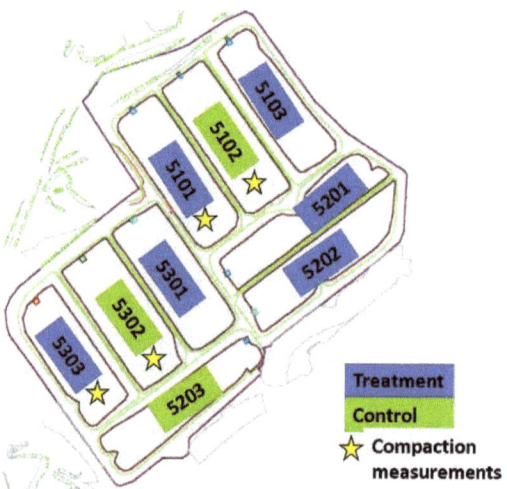

Figure 1. Schematic layout of the Yavne-2 recharge field with the operational pond's sub-divisions, tested treatments (new tillage tools), control treatments (existing tillage tools), and ponds with soil compaction measurements.

Table 1. The experiment's progress and the tillage treatments applied. Each stage began with a tillage operation, followed by 6–19 recharge cycles.

Stage	Dates	Tillage Treatment	Treatment Definition
Pre-treatment	1–28 May 2018	Plowing of all 9 ponds	(control)
Stage 1	28 May 2018–18 June 2018	Plowing of all 9 ponds	Control
Stage 2	18 June 2018–9 July 2018	SK [1] cultivator of all 9 ponds	Control
Stage 3	9 July 2018–6 August 2018	1. SK cultivator: 3 ponds 2. CK [2] cultivator: 6 ponds	1. Control 2. Treatment
Stage 4	6 August 2018–3 September 2018	1. Plowing: 3 ponds 2. Subsoiler: 6 ponds	1. Control 2. Treatment

[1] Sweep-knives cultivator. [2] Chisel-knives cultivator.

2.2. Tillage Equipment

Four different tillage tools were employed during the experiment (Table 1). Two of them, the sweep-knives cultivator and the plow, are tools routinely used by the Shafdan facility's regular tillage equipment operator (contractor). The other two tools were a chisel-knives cultivator and a subsoiler, which were tested as alternative tools and operated by another operator.

Sweep-Knives (SK) Cultivator: The SK cultivator is a shallow (~20 cm) tool routinely used in the Shafdan facility for breaking the surface crust and the removal of seasonal vegetation. The SK cultivator employed in the Shafdan facility and in this experiment was 6 m wide and equipped with 24 blades arranged on four iron beams at average intervals of 25 cm between each (Figure 2a). Each blade was 30 cm in length and 25 cm wide (at its widest).

Chisel-Knives (CK) Cultivator: The CK cultivator is a shallow to medium depth tool (30–40 cm) that was tested as an alternative to the SK cultivator and for plowing. In this experiment, a 7 m wide cultivator was employed that was equipped with 30 chisel blades arranged on four iron beams at average intervals of about 23 cm between blades (Figure 2b). Each blade was 55 cm long and 8 cm wide (at its widest).

Plow: The plow is a deep (40–50 cm) tool routinely used in the Shafdan facility for breaking, mixing, and aerating compacted layers up to a depth of 40–50 cm. The plow used in the Shafdan facility, and in this experiment, had four non-reversible curved blades (moldboards) arranged on a single 1.9 m wide diagonal beam (Figure 2c). Each moldboard blade was 115 cm long, and 55 cm high.

Subsoiler: The subsoiler is a deep (60–70 cm) tool that was tested as an alternative tool for the plow. In this experiment, a 4 m wide paraplow type subsoiler was employed, equipped with eight swept blades arranged on three iron beams at average intervals of about 50 cm between blades (Figure 2d). Each blade was 85 cm high and 25 cm wide (at its widest) and was equipped with special sand wings to increase the aeration channels.

The plow and the SK cultivator were mounted on a 250-hp Case International 7250 tractor equipped with balloon tires. The subsoiler and the CK cultivator were mounted on a 300 hp John Deere 8360R tractor, equipped with front and rear dual assembly tires and automated systems for tillage depth and GPS navigation control.

Figure 2. Tillage tools used during the experiment: (**a**) Sweep-knives (SK) cultivator; (**b**) Chisel-knives (CK) cultivator; (**c**) Plow; (**d**) Subsoiler (paraplow type).

2.3. *Measurements and Analyses*

Each tillage stage (cycle) lasted for 3–4 weeks and comprised of six to 19 recharge cycles, dependent on the individual infiltration capacity of each pond. Each recharge cycle was composed of three separate stages: flooding, infiltration, and drying stages. The drying stage was set for a minimum of 12 h. However, in some cases, it was prolonged for up to 20 h due to operational constraints. At the end of each tillage stage and before the next tillage operation, the ponds were allowed to dry for a minimum of 36 h to ensure complete drying of the soil.

An automated control system continuously measured and recorded water levels (L, cm), recharge volumes (V, m^3), flooding (T_f, h), infiltration (T_i, h), and drying (T_d, h) times as well as the infiltration rate (IR, cm/h) for each recharge cycle. Infiltration rate (IR) was calculated for each cycle by two independent methods: during infiltration (IR_i, Equation (1)) and during the wetting period (IR_{WB}, Equation (2)).

$$IR_i = -\frac{L(t_d) - L(t_i)}{t_d - t_i} \qquad (1)$$

$$IR_{WB} = \frac{V}{A \cdot (T_f + T_i)} \cdot 100 \qquad (2)$$

where L represents water levels (cm) and t_d and t_i represent the start times of the infiltration and the drying stages, respectively. In Equation (2), V represents the recharged volume (m^3); A represents the pond area (m^2); T_f and T_i represent the duration (hours) of flooding and infiltration stages, respectively; and 100 serves to transfer from m/h to cm/h. Using Equation (1) to calculate infiltration rates during the infiltration phase is based on the fact that the water level drop during this phase is completely linear for the Shafdan case, as shown in many measurements and studies (e.g., [12,20,21]).

Soil compaction measurements were conducted in representative ponds (Figure 1) with a penetrometer device (Spectrum Technologies) equipped with a 0.75-inch diameter tip adapted for sandy soils. Measurements were conducted before and after each tillage and between recharge cycles. Each measurement campaign included some 20 measurements made along a 10 m long line at intervals of 0.5 m. The depth of the compacted layer was defined by reaching a pressure of 1380 KPa (200 psi).

3. Results

3.1. Infiltration Rates and Recharge Capacity

Continuous infiltration rate (IR) measurements from two representative infiltration ponds—treatment and control—throughout the experiment period are presented in Figure 3. The relative change in IR for each stage of the experiment was calculated in comparison to the average IR measured after plowing during the first tillage stage (28 May 2018–18 June 2018). A summary of the relative changes in IR for every stage and treatment, in comparison to its parallel control treatment at the same stage, is presented in Figure 4.

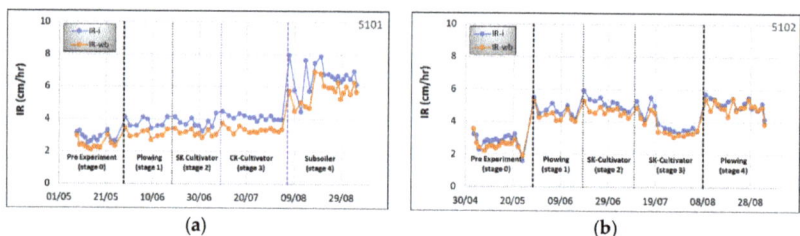

Figure 3. Infiltration rates measured in two representative ponds during the experiment: (**a**) Pond #5101 with new tillage tools at Stages 3 and 4 (treatment); (**b**) Pond #5102 with existing tillage tools only at all stages (control).

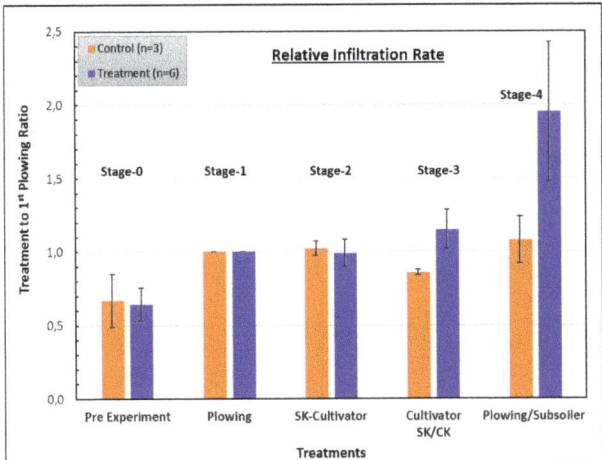

Figure 4. Relative infiltration rates (average ± standard deviation) calculated for each stage and treatment, relative to the average IR measured during Stage 1 (first plowing).

The results clearly show a significant increase in IRs in every recharge pond (replicate) where the new tillage equipment was applied (Figure 3) as well as in the integrated analyses for all replicates (Figure 4). During the experiment's third stage, relative IRs after the CK cultivator ranged between 1.01 to 1.39 with an average of 1.15 ± 0.13 in comparison with 0.84 to 0.88 and an average of 0.86 ± 0.02 after SK cultivation. The subsoiler showed an even greater effect. During the fourth stage of the experiment, relative IRs after use of the subsoiler ranged between 1.46 to 2.82 with an average of 1.95 ± 0.47, in comparison with 0.91 to 1.22 and an average of 1.08 ± 0.16 after plowing. Thus, we can show an average improvement of 15% and 95% for the CK cultivator and for the subsoiler, respectively, in comparison with the usual plowing regime. When compared with the routine SK cultivator, the improvement was even greater.

The relative change in recharged volumes for each stage and treatment during the experiment was calculated in comparison to the cumulative recharged volumes in each pond during the first tillage stage. Similarly, with regard to the improvement in IRs, the results clearly indicate a significant increase in recharge volumes of 58% (±16%) and 160% (±17%) for the CK cultivator and for the subsoiler, respectively, in comparison with the routine plowing regime. This significant improvement in recharge capacity was a result of the significant improvement in IRs, as indicated by the model calculations of the potential recharge capacity (not within the scope of this paper).

3.2. Soil Compaction

Measurements of soil compaction indicated the presence of a compacted layer at a depth of approximately 30 cm during the first to third stages of the experiment. The presence of this layer was also indicated in the soil profile measurements in open holes, which are not presented and discussed in this paper. The evaluation and the nature of the compacted layer could be related to different factors such as leaching and accumulation of organic and inorganic colloidal material in the soil pores, compaction processes under the recharged water weight, soil erosion processes, and the routine operations in the ponds [18]. In fact, the results from the soil profile measurements and from other research studies in the Shafdan (e.g., [4,7,21]) indicate that each of these factors can be relevant, or even dominant, in different areas of the recharge ponds. Regardless of these factors, it can be assumed that the development of this layer was enhanced with the utilization of improper tillage tools (e.g., narrow plow) and due to the improper tillage regime (e.g., working over too wet soil).

After subsoiler tilling, the depth of the compacted layer increased to approximately 55 cm (Figure 5a). The significant increase in the IR and in recharge capacity after subsoiler treatment, accompanied by increases in the compacted layer depth, indicates the importance of this tool and its ability to break subsoil barriers in the recharge fields. The compaction measurements were conducted along a grid with horizontal orientation relative to the tilling lines. The uneven, ridged, soil surface effects, mainly caused by the plow, but also by the SK cultivator, resulted in relatively high variations between measurements. This variation tended to decrease along the course of the experiment, particularly after the use of the CK cultivator and subsoiler in the treatment ponds (Figure 5b). The elimination of the ridges, which creates a micro-topography of the soil surface, leads to a more balanced and effective tillage operation, and contributes to an increase of IR in the new tool's treatment ponds.

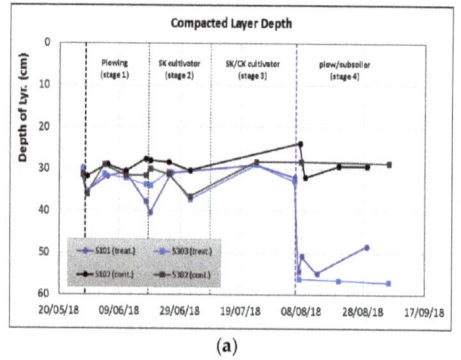 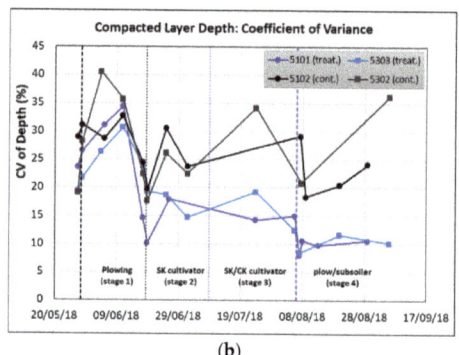

(a) (b)

Figure 5. Soil compaction measurements in representative ponds during the course of the experiment. (**a**) Depth of the compacted layer; (**b**) Coefficient of variance between measurements (replicates) along the grid line.

4. Discussion

The results of the experiment clearly demonstrate the positive effect of the new tillage tools in comparison with existing tools where there was an average increase in the recharge capacity of approximately 15% and 95% after the CK cultivator and subsoiler, respectively. In addition, the subsoiler managed to reduce soil compaction and break up the compacted layer that developed at a depth of 30 cm. The new tillage tools also reduced soil surface variations created by the existing tools, and thus increased the effectiveness of the tilling process.

However, during the experiment, changes were made not only in the tillage tools, but also in tractor power (300 hp as opposed to 250 hp) and in the control systems (the addition of a GPS navigation system and a depth controller). As a result, it could be difficult to separate the impact of these parameters, especially the tractor strength and the depth control system, from the tillage tool's effects. Nevertheless, as tillage activity was fully inspected during the experiment, with particular care regarding the working depths and overlap between tillage lines, we can assume that their effect was minor in comparison to the tool's effect.

A significant change was also made in the recharge regime, with respect to the recharge regime before the experiment. During previous years (2015–2018), the effluent load gradually increased due to an increase in effluent quantities, with no change in the recharge area. This situation led to a reduction in the recharge area's availability and to a gradual shortening of drying periods between flooding. In some cases, especially during the winter, this led to a total elimination of the drying stage. The reduced availability of the recharge area also led to a situation where ponds were incorrectly prepared before tillage, and to the execution of tillage operations on wet (sometimes even flooded) soil. These operational conditions were, in all probability, one of the main reasons for the resulting decrease in recharge capacity, which in turn led to an increased rate of tillage activity (especially plowing), an enhancement of the compacting processes (due to tillage on wet soil [23,24]), and a further decrease in the recharge capacity.

During the experiment, a steady recharge regime was enforced including minimal drying periods of 12 h between recharge cycles. Special attention was given to the complete drying of the soil for a minimum of 36 h before tillage operation. This was carried out in order to prevent soil destruction and compaction processes due to the tillage of wet soil [23,24]. Thus, the lack of a drying period cannot explain the difference in the results between the existing and new tillage tools during the experiment. However, with some degree of certainty, it can explain the significant increase in recharge capacity in comparison to the pre-experiment period (Figures 3 and 4) as well as in comparison to previous years (proven, but not included in this paper). The dramatic effect of the deep subsoiler (up to an average of 95% in IR, recharge capacity, and reducing soil compaction) led to the conclusion that the effect of this tool was due to the breaking-up of the sub-soil compacted layer at a depth of about 30 cm, which had developed during years of incorrect operation. The existing tools, particularly the narrow plow, which was operated too frequently, not only failed to break-up this layer, but also encouraged its development.

The effect of the deep subsoiler on the sub-soil compacted layer (at a 30 cm depth) and on the IR, in turn, can lead to the conclusion that it could be a short-term effect, and sooner or later, a deeper compacted layer will be developed, and the recharge capacity will be decreased. On the other hand, it is important to note that some of the Shafdan recharge ponds are active for more than 40 years. Thus, long-range effects such as the development of the shallow compacted layer, as described earlier in the Introduction, can only be related to decades of using the incorrect soil tillage regime with improper tillage tools (mainly plows or even discus plow at early stages). Hence, it could be concluded that proper tillage regime and tillage tools, which emphasize minimal compaction processes, will be able to maintain high recharge capacity for many years ahead.

5. Conclusions

This study takes a classic approach, which advocates a return to the fundamentals of recharge field maintenance. With no apparent innovations, and during a recharge crisis, it dramatically, immediately, and with almost no additional costs, increased the recharge capacity of the Shafdan plant. The reduction in recharge capacity, and its restoration on the other hand, are not solely related to the tillage equipment, but also to additional parameters that are required for correct maintenance procedures. These parameters include:

- Complete drying (not draining!) of the soil before tillage operations: the basic rule is to prevent or minimize puddles, wet spots, or moist soil spots during the tillage process;
- Employing tillage equipment with the maximum possible width to minimize the number of passes on the pond;
- Employing tillage equipment that keeps the soil surface even, with minimum ridges and micro-topography effects; and
- Strict adherence to the correct recharge regime, and especially to the drying stage before flooding.

This study also showed that long-range clogging processes that result from incorrect operation and maintenance are reversible and can be overcome by reverting to correct operation principles. Further study is planned to examine the effect of the new tillage equipment (and regime) on recharge capacity during the winter, and to develop alternative approaches for vegetation treatment. Further study is needed in order to gain a better understanding of the sub-soil compacted layers that develop in recharge ponds, and on the factors affecting it.

Author Contributions: I.N., T.S., and L.S. conceived the experiment; I.N., T.S., and A.L. designed the experiments; I.N., T.S., H.R., L.S and A.L. performed the experiments; and I.N. analyzed the data and wrote the paper. All authors have read and agreed to the published version of the manuscript.

Funding: This research was funded by Mekorot, The Israel National Water Co.

Acknowledgments: We wish to express our thanks to Dani Cohen and Roi Elkayam, who developed and enabled the automated measurement and analysis system (SATix system). To David Weisental, Efrat Cohen, and the Shafdan Unit for their help and support during this research. Special thanks go to Zvika Bitan for the great help, advice, and initiative in selecting the equipment and carrying out the soil tillage operations.

Conflicts of Interest: The authors declare no conflicts of interest.

Abbreviations

The following abbreviations are used in this manuscript:

SAT	Soil aquifer treatment
SK cultivator	Sweep knives cultivator
CK cultivator	Chisel knives cultivator
Mm^3/year	Millions of cubic meters per year
IR	Infiltration rate
MAR	Managed aquifer recharge

References

1. Icekson-Tal, N.; Blanc, R. Wastewater Treatment and Groundwater Recharge for Reuse in Agriculture: Dan Region Reclamation Project, Shafdan. In *Artificial Recharge of Groundwater*; Peters, J.H., Ed.; August Aime Balkama: Rotterdam, The Netherlands, 1998; pp. 99–103.
2. Lin, C.; Eshel, G.; Roehl, K.E.; Negev, I.; Greenwald, D.; Shachar, Y.; Banin, A. Studies of P Accumulation in Soil/Sediment Profiles Used for Large-Scale Wastewater Reclamation. *Soil Use Manag.* **2006**, *22*, 143–150. [CrossRef]
3. Amy, G.; Drewes, J. Soil Aquifer Treatment (SAT) as a Natural and Sustainable Wastewater Reclamation/Reuse Technology: Fate of Wastewater Effluent Organic Matter (EfOM) and Trace Organic Compounds. *Environ. Monit. Assess.* **2007**, *129*, 19–26. [CrossRef] [PubMed]
4. Lin, C.; Eshel, G.; Negev, I.; Banin, A. Long-Term Accumulation and Material Balance of Organic Matter in the Soil of an Effluent Infiltration Basin. *Geoderma* **2008**, *148*, 35–42. [CrossRef]
5. Goren, O.; Burg, A.; Gavrieli, I.; Negev, I.; Guttman, J.; Kraitzer, T.; Kloppmann, W.; Lazar, B. Biogeochemical Processes in Infiltration Basins and Their Impact on The Recharging Effluent, The Soil Aquifer Treatment (SAT) System of The Shafdan Plant, Israel. *Appl. Geochem.* **2014**, *48*, 58–69. [CrossRef]
6. Elkayam, R.; Michail, M.; Mienis, O.; Kraitzer, T.; Tal, N.; Lev, O. Soil Aquifer Treatment as Disinfection Unit. *J. Environ. Eng.* **2015**, *141*, 05015001. [CrossRef]
7. Sopliniak, A.; Elkayam, R.; Lev, O. Quantification of dissolved organic matter in pore water of the vadose zone using a new ex-situ positive displacement extraction. *Chem. Geol.* **2017**, *466*, 263–273. [CrossRef]
8. Elkayam, R.; Sopliniak, A.; Gasser, G.; Pankratov, I.; Lev, O. Oxidizer Demand in the Unsaturated Zone of a Surface-Spreading Soil Aquifer Treatment System. *Vadose Zone J.* **2015**, *14*. [CrossRef]
9. Elkayam, R.; Aharoni, A.; Vaizel-Ohayon, D.; Sued, O.; Katz, Y.; Negev, I.; Marano, R.B.M.; Cytryn, E.; Shtrasler, L.; Lev, O. Viral and Microbial Pathogens, Indicator Microorganisms, Microbial Source Tracking Indicators, and Antibiotic Resistance Genes in a Confined Managed Effluent Recharge System. *J. Environ. Eng. (United States)* **2018**, *144*, 05017011. [CrossRef]
10. Halperin, R. *Halperin Committee Report, Principles for Granting Permits to Irrigate Effluents*; Ministry of Health: Jerusalem, Israel, 2002. Available online: https://www.health.gov.il/hozer/bsv_Halperin.doc (accessed on 12 May 2019).
11. Inbar, Y. The Ministry of Environmental Protection, ISRAEL. Available online: http://www.sviva.gov.il/InfoServices/ReservoirInfo/DocLib2/Publications/P0301-P0400/P0321.pdf (accessed on 12 May 2019).
12. Elkayam, R. Shafdan Soil Aquifer Treatment System; Process Assessment & Improvement. Ph.D. Thesis, The Hebrew University of Jerusalem, Jerusalem, Israel, January 2019.
13. Aharoni, A.; Negev, I.; Cohen, E.; Bar, O.; Bar-Noy, N.; Khadya, N. Monitoring Shafdan Effluents Recharge and the Third Line Project. In *Analysis and Summary of Results: 2018 Yearly Report*; Mekorot, National Water Co.: Tel Aviv-Yafo, Israel, 2019. (In Hebrew, abstract written In English)

14. Aharoni, A.; Guttman, J.; Icekson-Tal, N.; Kraitzer, T.; Cikurel, H. SWITCH project Tel-Aviv Demo City, Mekorot's case: Hybrid Natural and Membranal Processes to Up-Grade Effluent Quality. *Rev. Environ. Sci. Biotechnol.* **2010**, *9*, 193–198. [CrossRef]
15. Pavelic, P.; Dillon, P.J.; Mucha, M.; Nakai, T.; Barry, K.E.; Bestland, E. Laboratory Assessment of Factors Affecting Soil Clogging Of Soil Aquifer Treatment Systems. *Water Res.* **2011**, *45*, 3153–3163. [CrossRef] [PubMed]
16. Sharma, S.K.; Hussen, M.; Amy, G. Soil Aquifer Treatment Using Advanced Primary Effluent. *Water Sci. Technol.* **2011**, *64*, 640–646. [CrossRef] [PubMed]
17. Rice, R.C.; Rice, C. Soil Clogging during Infiltration of Secondary Effluent. *J. Water Pollut. Control Fed.* **2013**, *46*, 708–716.
18. Bouwer, H. Artificial recharge of groundwater: Hydrogeology and engineering. *Hydrogeol. J.* **2002**, *10*, 121–142. [CrossRef]
19. Houston, S.L.; Duryea, P.D.; Hong, R. Infiltration Considerations for Ground-Water Recharge with Waste Effluent. *J. Irrig. Drain. Eng.* **1999**, *125*, 264–272. [CrossRef]
20. Lin, C.; Greenwald, D.; Banin, A. Temperature Dependence of Infiltration Rate during Large Scale Water Recharge into Soils. *Soil Sci. Soc. Am. J.* **2003**, *67*, 487–493. [CrossRef]
21. Nadav, I.; Tarchitzky, J.; Chen, Y. Soil cultivation for enhanced wastewater infiltration in soil aquifer treatment (SAT). *J. Hydrol.* **2012**, *470–471*, 75–81. [CrossRef]
22. Mousavi, S.F.; Rezai, V. Evaluation of scraping treatments to restore initial infiltration capacity of three artificial recharge projects in central Iran. *Hydrogeol. J.* **1999**, *7*, 490–500. [CrossRef]
23. Hamza, M.A.; Anderson, W.K. Soil compaction in cropping systems: A review of the nature, causes and possible solutions. *Soil Tillage Res.* **2005**, *82*, 121–145. [CrossRef]
24. Batey, T. Soil compaction and soil management—A review. *Soil Use Manag.* **2009**, *25*, 335–345. [CrossRef]

 © 2020 by the authors. Licensee MDPI, Basel, Switzerland. This article is an open access article distributed under the terms and conditions of the Creative Commons Attribution (CC BY) license (http://creativecommons.org/licenses/by/4.0/).

Article

Clogging Issues with Aquifer Storage and Recovery of Reclaimed Water in the Brackish Werribee Aquifer, Melbourne, Australia

Pieter J. Stuyfzand [1,2,*] **and Javier Osma** [3]

1. KWR Watercycle Research Institute, 3430 BB Nieuwegein, The Netherlands
2. Faculty of Civil Engineering and Geosciences, Delft University of Technology, 2628 CN Delft, The Netherlands
3. City West Water, Footscray 3011, Australia
* Correspondence: pieter.stuyfzand@kwrwater.nl; Tel.: +31-6-1094-5021

Received: 30 June 2019; Accepted: 24 August 2019; Published: 30 August 2019

Abstract: As part of an integrated water-cycle management strategy, City West Water (CWW) is conducting research to develop an aquifer storage recovery (ASR) scheme utilizing recycled water. In this contribution, we address the risk of well clogging based on two ASR bore pilots, each with intensive monitoring. Well clogging is a critical aspect of the strategy due to a projected high injection rate, a high clogging potential of recycled water, and a small diameter injection borehole. Microscopic and geochemical analysis of suspended solids in the injectant and backflushed water, demonstrate a significant contribution of diatoms, algae and colloidal or precipitating $Fe(OH)_3$, $Al(OH)_3$ and MnO_2. CWW is, therefore, testing additional prefiltration that includes a 20 μm spin Klin disc and 1–5 μm bag filter operating in series. In this paper, we present optimized methods to (i) detect the contribution of the injectant and aquifer particles to total suspended solids in backflushed water by hydrogeochemical analysis; and (ii) predict and reduce the risk of physical and biological clogging, by combination of the membrane filter index (MFI) method of Buik and Willemsen, a modification of the total suspended solids method of Bichara and an amendment of the exponential bacterial growth method of Huisman and Olsthoorn.

Keywords: ASR; recycled water; well clogging; geochemical analysis; filtration; biofouling; risk management

1. Introduction

The use of reclaimed or recycled water is on the rise worldwide, mainly because of (i) water scarcity due to exponential population growth and climate change, and (ii) the need to reduce the pollution load from waste water that ends in surface water systems [1,2].

As part of an integrated water-cycle management strategy for the fast growing city of Melbourne, City West Water (CWW) is conducting research to apply aquifer storage recovery (ASR) utilizing recycled water [3]. This non-potable water is to be injected and stored during winter in a brackish, anoxic sand aquifer at 220–250 m below ground level (BGL), and recovered during peak demand periods in summer. The purpose is to reduce drinking water consumption by supplying recycled water via a third pipe system. The infiltration water will be Class A recycled water, which is fit for high exposure uses (not for drinking water consumption), diluted with desalinated water (by reverse osmosis). Class A recycled water comes from Melbourne Water Corporation's Western Treatment Plant, which produces recycled water from a series of anaerobic and aerobic lagoons followed by UV disinfection and chlorine disinfection.

Several studies have confirmed the viability of ASR at the trial site (West Werribee, west of Melbourne; Figure 1), based on the acceptable overall quality of the infiltration water [4,5], the transmissivity and dispersivity of the aquifer and background flow [4], geochemistry of aquifer [6,7], the ambient water quality in the target aquifer [5], predicted water quality of recovered water [6,8], and economic, societal and water security implications [3].

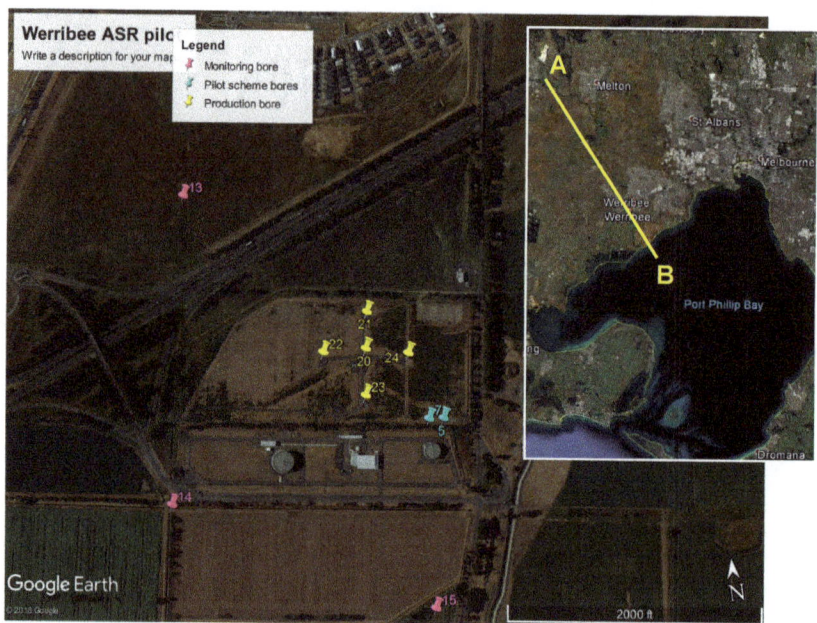

Figure 1. Site map showing the location of the aquifer storage recovery (ASR) (production) and monitoring wells. ASR well 5 was the production well during the trial in 2012, with its monitoring well 7 at ca. 40 m distance. Production wells 20–24 form together the future West Werribee ASR plant, but for now they are part of the second pilot, with well 20 as ASR well and 21–24 as monitoring wells. AB = geological section shown in Figure 2.

Nevertheless, two recognized main risks still warrant ongoing investigation of respectively the actual well clogging risk [7,9], and risk caused by undesired water quality changes in the target aquifer [5]. Injection well clogging can be extremely cumbersome leading to high costs due to monitoring, maintenance, premature well replacement, intensification of pretreatment or addition of more wells [9–12].

Common water quality issues with ASR application, especially when dealing with recycled water, consist of (i) mobilization of trace metals and arsenic by pyrite oxidation [13]; (ii) lack of full elimination of high concentrations of NO_3 and PO_4, (iii) the formation of trihalomethanes and haloacetic acids by chlorination [14]; (iv) the persistence of some pathogens and organic micropollutants in the aquifer [2,15]; and (v) the undesired admixing of ambient groundwater (total dissolved sSolids (TDS), H_2S, natural radionuclides) [5]. There are also concerns about potential changes in the microbiota and stygofauna (taxa that spend their whole life cycle in groundwater) by introducing among others oxygen, algae and xenobiotic compounds [16–18].

In this contribution, we evaluate the risks of ASR well clogging based on a short ASR pilot investigation in 2012, and an ongoing ASR pilot test which started in 2017. We present hydraulic, geochemical and microscopic data, diagnose the main clogging causes, and indicate how to mitigate or prevent the problem.

2. Materials and Methods

2.1. Aquifer Storage Recovery (ASR) Pilot Tests

Two pilot tests were conducted in 2012 and 2017–2019, respectively. The first used one injection bore (well 5) and well 7 served as a monitoring bore (Figure 1). The second test (still running) consists of one injection bore (well 20), wells 21–24 as nearby and wells 12–15 as remote monitoring bores (Figure 1).

Details of the ASR cycling scheme are summarized for both pilot tests in Table 1. The 5 clustered wells 20–24 are part of the future ASR system, which is expected to infiltrate 0.5–1.0 Mm^3/year during winter time and to recover ~80% of this volume during summer time, depending on demand. The ASR wells include a stainless steel (316) telescopic design, with internal diameter of 350 mm in the upper section, 200 mm in the lower section (incl. well screen), and screen aperture ranging from 0.4 to 0.8 mm. They are not supplied with a gravel pack nor a downhole flow control valve.

Table 1. Details of ASR cycling scheme during the first and second pilot. CSV = cumulative stored volume; NSV = net stored volume (injected minus recovered); ASV = actual remaining stored volume (bubble volume); RE = recovery efficiency. Red numbers = input data in spreadsheet.

Cycle No	Injection Period	Storage Period 1	Recovery Period	Storage Period 2	Total for Cycle	Time since Start	Injection Rate	Pumping Rate	RE per Cycle	RE cumul.	CSV	NSV	ASV
			day				m³/h	m³/h	%	%	m³ at End of Cycle		
1st Pilot 10 April–14 June 2012													
1	0.83	1.05	0.99	2.97	5.8	5.8	36.0	108.0	357.6	358	720	−1855	0
2	3.00	3.02	2.10	0.07	8.2	14.0	72.0	90.0	87.7	121	5904	−1216	639
3	14.88	20.00	16.17	0.00	51.0	65.1	72.0	108.0	163.0	155	31,608	−17,416	0
SUM (day)	18.71	24.07	19.26	3.04		SUM (m³)	31,608	49,024				R_END (m) = 0.0	
2nd Pilot 10 July 2017–likely September 2019													
1	1.00	4.00	1.00	5.00	11.0	11.0	54.0	104.4	193.3	193	1296	−1210	0
2	7.00	21.00	7.00	7.00	42.0	53.0	54.0	104.4	193.3	193	10,368	−9677	0
3A	13.00	386.00	0.00	0.00	399.0	452.0	54.0	0.0	0.0	74	27,216	7171	16,848
3B #	34.00	??	??	0.00		ONGOING	54.0	0.0	??	??	71,280	??	??
SUM (day)	21.00	411.00	8.00	12.00		SUM (m³)	27,216	20,045				R_END (m) = ??	

#: Including prefiltration over 20 μm and 1-5 μm; ?? = recovery cycle not yet started; R_END = radius injected water at end.

2.2. Characterization of Target Aquifer

The ASR target aquifer consists of the coarser grained sequences within the Lower Werribee Formation (LWF) of Early Tertiary age. It is formed by a heterogeneous sequence of sands, silts, clays and lignite, slightly dipping on average in SE direction. The LWF rests on nearly impervious Lower Paleozoic basement rock, and is overlain by 200 m of (un)consolidated sediments (Figure 2).

In the LWF, the sands are typically found between 220 and 240 m below sea level. They vary in grain size (0.1–0.8 mm) and thickness (3–22 m) over short distances. Most wells have two screen sections separated by a blind casing interval of 2–4 m in the middle, where the aquifer is too silty. Average aquifer characteristics are listed in Table 2, based on pumping tests [4], breakthrough curves [5], sedimentological and geochemical analysis of aquifer cores [7,19].

Table 2. Characteristics of principal wells in both ASR pilots, and of local aquifer. r_{MW} = radial distance ASR to monitoring well; D = thickness, T = transmissivity, S = storativity, ε = effective porosity, t_{50} = travel time, α_L = longitudinal dispersivity.

Well			Aquifer						Geochemistry			
Code	Screen m BGL	r_{MW} m	D m	T m²/day	S	ε	t_{50} day	α_L m	pyrite ppm	C_{ORG} % d.w.	$CaCO_3$ % d.w.	CEC meq/kg
ASR well 5	242–251	40	14	110	0.0003	0.3	0	0.5	-	-	-	-
MW 7	237–246	40	17	50	0.0002	0.3	13	0.4	22,214 #	1.7 #	<0.2 #	-
ASR well 20	220–236	100	15.5	71	0.00035	0.3	0	0.5	1020	<0.1	<0.4	20
MW 21	223–235	100	11	77	-	0.3	60	1.0	-	-	-	-

#: in WTP3, ca. 1720 m south of MW 7. Sample probably representative of silty layer, not sand.

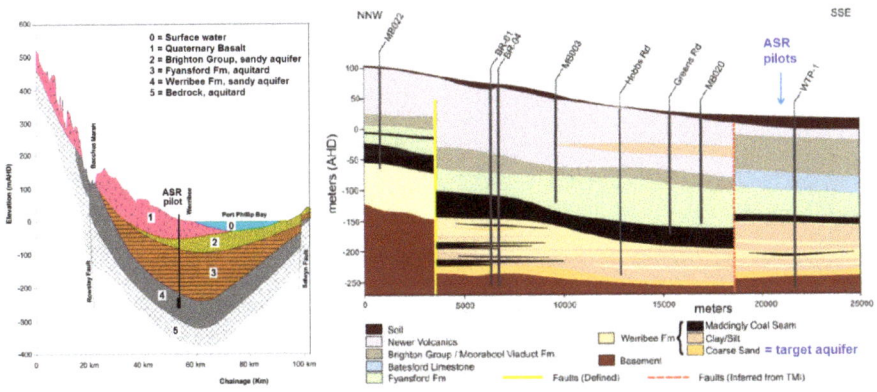

Figure 2. Generalised NNW-SSE geological section showing Werribee Formation and ASR pilot sites. (**Left**): large scale, modified after [20]; (**Right**): smaller scale, slightly modified after [21].

2.3. Water Quality Analysis

Samples of infiltration water and of water from wells were taken at the well head, with and without prior filtration over 0.45 µm in the field. The standing volume in well screen and riser was always evacuated 3 times prior to sampling. In the field, pH, electrical conductivity (EC), temperature, oxidation reduction potential (ORP) and O_2 were measured at selected intervals. Sensors of EC, temperature and water pressure, installed at screen depth in the wells, have been registered between 3 and 10 min.

Sample container type, preservation method, holding time and analytical method were as recommended by ALS Environmental [22], which performed the analyses on a wide spectrum of parameters: general chemistry (pH, EC, turbidity, total suspended solids (TSS), algae), major ions, nutrients, total and dissolved organic carbon (TOC and DOC, respectively), (trace) metals, gases (O_2, CO_2, H_2S, CH_4), disinfection by-products, industrial organic micropollutants, pesticides, pharmaceuticals, radionuclides, and microbiological parameters. Sampling frequency was on average on a weekly basis but fine-tuned to the duration of ASR cycles, the breakthrough in monitoring wells, and costs of analytical packages [8].

2.4. Analysis of Suspended Matter during Backflushing

Backflushing of the ASR well, with a 2–3 times higher pumping than injection rate, directly or after juttering, is a method to unclog the well screen or bore hole wall. Samples of the turbid backflush water were taken from ASR well 20 on 4 occasions (A–D): during injection cycle 2 (A), two well redevelopments after cycle 3A (B, C) and during injection cycle 3B (D), respectively. Samples of unfiltered water were taken after evacuation of 9–12 m³, corresponding with ~1 standing well volume. The samples (3.6–4.2 L) were transported to the lab in an upright position, in the dark and cooled at 4 °C, without shaking. In the lab, these samples were kept in the dark at 4 °C for 2 days in upright position without shaking/tremors, in order to let the suspended matter settle. Subsequently as much clear water as possible was slowly decanted or sucked out, containing on average 95% of all water with on average only 2.7% of all *TSS*: 1 mg *TSS*/L in 0.95 × 4 L and 690 mg *TSS*/L in 0.05*4 L yields indeed 100 × (0.95 × 4 × 1)/(0.95 × 4 × 1 + 0.05 × 4 × 690) = 2.7%, which is neglected in Equation (1). The decanted water was analyzed, after 0.45 μm filtration, for all main constituents (incl. pH and EC), DOC and trace elements. This yields the dissolved fraction.

The remaining fluid (0.2–0.5 L) with 97% of all *TSS* was first analyzed for *TSS*, and subsequently, after nearly complete destruction with strong suprapure HNO_3 at 95 °C, analyzed for TOC, total concentration of main constituents (excluding Cl, HCO_3, NO_3) and trace elements.

The analytical results of the dissolved and suspended fraction in mg/L were used to calculate the composition of the suspended material as follows:

$$X_{SS} = 10^6 \frac{X_T - X_{H2O}}{TSS},\tag{1}$$

where: X_{SS} = content of component X in suspended solids (ppm = mg/kg dry weight); X_T = total X concentration in water with suspended solids (mg/L); X_{H2O} = concentration of X dissolved in water (mg/L); *TSS* = total suspended solids in water (mg/L)

2.5. Clogging Predictor Based on Membrane Filter Index (MFI)

Buik and Willemsen [23] used the membrane filter index (*MFI*; [10,24,25]), well and hydrogeological parameters to predict the clogging rate of recharge wells, based on the following semi-empirical equation (without dimensional homogeneity):

$$v_{CLOG} = 2 \times 10^{-6} \times MFI \times t_{EQ} \times V_{DIR}^2/(K_h/150)^{1.2},\tag{2}$$

with:

$$K_h = 150(D_{50}/1000)^{1.65} \text{ or } K_h = T/D,\tag{3}$$

$$V_{DIR} = Q_{IN}/(2\pi r_B L),\tag{4}$$

where: v_{CLOG} = clogging rate (m/a); *MFI* = 0.45 μm membrane filter index (s/L²); t_{EQ} = total amount of equivalent full load hours per year (h); v_{DIR} = entrance velocity on borehole wall (m/h); D_{50} = median grain size diameter of aquifer (μm); Q_{IN} = mean infiltration rate when well is recharging (m³/h); r_B = radius of borehole (m); L = length of well screen (m); K_h = horizontal hydraulic conductivity (m/d).

2.6. Clogging Predictor Based on Total Suspended Solids (TSS)

Another way to predict the physical clogging rate, is to relate the decrease in flow rate of an injection well (Q_t/Q_0) to the ratio of the total input of suspended solids after a given injection time, to the open area of the external borehole wall (A_{OPEN}). On the basis of many experimental data Bichara [26] showed an interesting plot of that relation. Notwithstanding the scatter of his many data, an average linear trend can be used to estimate the decline in flow rate of an injection well by particle clogging:

$$Q_t/Q_0 = 100 - 17.5 TSS \times Q_{IN}t/A_{OPEN}, \quad \text{if} >0, \text{else } 0 \quad (5)$$

$$t_{10} = 10\, t/(100 - Q_t/Q_0), \quad \text{if} >0, \text{else } 0 \quad (6)$$

with:

$$A_{OPEN} = 2\pi r_B \varepsilon L, \quad (m^2) \quad (7)$$

where: Q_t, Q_0 = injection rate at time t since start and at start ($t = 0$), respectively (m^3/h); ε = aquifer porosity (-); t = time since injection start (day); t_{10} = injection period needed to reduce flow by 10% (day).

2.7. Predictor of Bioclogging

Bioclogging can be simulated by the following equation (modified after Huisman and Olsthoorn [27], by introducing a lag time), which calculates the buildup of bacteria in infiltration wells:

$$N_t = \frac{V_{DIR} N_0 T_D \left(e^{0.693 \frac{t-t_{LAG}}{T_D}} - 1 \right)}{0.693}, \quad (m^{-2}) \quad (8)$$

where: N_0 = average number of bacteria in input (n/m^3); N_t = average number of bacteria at and behind bore hole wall, at time t (n/m^2); t = time since injection start (day); t_{LAG} = lag time during which bacteria do not reproduce (day); T_D = doubling time for bacterial population (day); V_{DIR} = entrance velocity on borehole wall, see Equation (4) (m/day).

Thus, it is assumed that each input bacterium is filtered out at the borehole wall and starts multiplying after a specific lag time. The thickness of the biofouled layer ($D_{BAC,t}$) which fully occupies the aquifer pore space, can be calculated using Equation (9) by assuming that (i) the accumulation of cells with a specific cell volume is taking place on the borehole wall and behind that, both with open space A_{OPEN}, and (ii) there is no open space between bacteria:

$$D_{BAC,t} = \frac{1000 N_t V_{CELL}}{(1-\varepsilon)}, \quad (mm) \quad (9)$$

where: v_{CELL} = spherical volume of each cell (m^3); ε = aquifer or gravel pack porosity (-).

A more realistic model sets a limit to bacterial growth, based on e.g., the observed change in DOC or biodegradable DOC (BDOC; [28]). The available data show a very small decline of the injectant during short storage or recovery cycles (1–2 days) for DOC (<0.5 mg C/L) and BDOC (~0.2 mg C/L). This observed consumption of BDOC during a 1–2 days lasting storage or recovery cycle (Δ_{BDOC}; mg C/L), is the sum of bacterial respiration (BaR) and biomass production (BoP; [29]), so that:

$$BoP = 10^{-3} \Delta_{BDOC} Q_{IN} - BaR = f_{BoP} 10^{-3} \Delta_{BDOC} Q_{IN}, \quad (kg\,C/day) \quad (10)$$

where: Q_{IN} = injection rate (m^3/d); f_{BoP} = BoP/(BoP + BaR) – fraction of Δ_{BDOC} consumed by bacterial respiration (-); Δ_{BDOC} = consumption of biodegradable DOC (mg C/L).

We assume that Δ_{BDOC} represents steady state, and that it approaches consumption by the total number of active cells at maximum growth, so that:

+BDOC-limitation:

$$N_{tMAX} = \frac{10^{-3} f_{BoP} \Delta_{BDOC} Q_{IN}}{V_{CELL} C_{CELL}\, A_{OPEN}} = \frac{10^{-3} f_{BoP} \Delta_{BDOC} V_{DIR}}{V_{CELL} C_{CELL} \varepsilon}, \quad (m^{-2}) \quad (11)$$

where: C_{CELL} = Carbon bacterial mass, 220 kg dry weight/m³ according to [30].

N_{tMAX} can also be calculated with Equation (8) by replacing t by t_{MAX}, which is defined as the time since start of exponential bacterial growth to the maximum level of growth. Combination of this Equation (8) with Equations (11) and (4) yields after rewriting, for both limitation cases:

$$t_{MAX} = \frac{T_D}{0.693} \ln\left\{\frac{0.693 N_{tMAX}}{N_0 \, T_D \, V_{DIR}} + 1\right\} + t_{LAG}, \quad \text{(day)} \quad (12)$$

3. Results

3.1. Injection Well Clogging

A typical pattern of impressed head and drawdown, *TSS* (at well head) and turbidity (in central injection manifold) for ASR well 20 during ASR Cycle 3, is shown in Figure 3. The peaks of *TSS* and turbidity coincide with 9 short flow reversals due to backpumping (108 m³/h) and reinjection (54 m³/h). This triggered the mobilization of fines from the borehole wall and probably also from the well itself. *TSS* peaked during both switches (injection to backpumping, and vice versa), whereas turbidity only peaked during the switch from backpumping to injection, due to the position of the turbidity sensor right after the injection pump. The *TSS* sensor's output is maximum 20 mg/L in order to focus on lower level variability, but short peaks of up to 43.5 (=870/20) mg/L have been measured (Table 3).

Table 3. Summary of analytical results of suspended matter sampled during backflush events A–D. The column with heading 'A–C' contains the average of events A, B and C (without prefiltration of injectant). Also shown: composition of ASR input *TSS* (average 6 samples) and of an aquifer core (1 sample), and the calculated fraction of the ASR input contributing to the average composition of samples A–C (α). Numbers in red indicate a significant trend over time (see Figure 4).

Event # Vol. out	Unit	TSS Input ASR	SUSPENDED MATTER DURING BACKFLUSHING					Aquifer Core	α
			A 12.0	B 12.0	C 9.0	D 11.0	A–C		A–C
	m³								
TSS	mg/L	1	580[E]	870[E]	610[E]	690[E]	687[E]	-	-
C_{ORG}	ppm	0.23	-	-	7049	-	7049	-	-
Si	ppm	-	23,364	8433	15,091	35,500	15,629	-	-
S	ppm	1	13,811	9591	7113	2900	10,172	11,900	0.15
P	ppm	-	900	2180	2459	10,145	1846	140	-
Al	ppm	18,400	36,198	6632	8844	12,261	17,225	22,300	-
Fe	ppm	32,583	29,293	19,517	18,025	26,043	22,278	15,500	0.40
Mn	ppm	11,533	1462	3994	5007	31,842	3488	108	0.30
Co	ppm	100	35	29	34	121	33	2.2	0.31
Cr	ppm	87	144	42	58	111	81	193	1.05
Cu	ppm	200	55	62	38	157	52	11.4	0.21
Ni	ppm	-	53	54	41	155	49	6.4	-
Sr	ppm	-	157	52	56	142	88	25	-
Ti	ppm	217	1895	229	277	332	800	-	-
V	ppm	22	122	26	42	18	63	85	0.34
Sum $	%	6.3	11.3	5.7	6.9	17.8	8.4	5.3	-
Al/Fe	ppm	0.56	1.24	0.34	0.49	0.47	0.77	1.44	-
Fe/Mn	ppm	2.8	20.0	4.9	3.6	0.8	6.4	143.5	-
FeS_2	mmol/kg	0	215	150	111	45	159	186	0.15
Fe-FeS_2	ppm	1	12,028	8353	6194	2526	8858	10,363	0.15
Fe-rest	ppm	32,582	17,265	11,164	11,830	23,517	13,420	5137	0.30

#: A = 24 Aug. 2017; B = 3 Nov. 2017; C = 13 Dec. 2017; D = 19 Dec. 2018; $: Sum = (sum all elements)/10⁴·; [E]: *TSS* in ca. 20 times concentrated water sample; A–C injected water without prefiltration; D injected water with prefiltration over 20 μm + 5 μm.

The backflushing is successful in removing the bulk of clogging material, but a growing residual (chronic) clogging is remaining, as evidenced by the rising impressed head (Figure 3) during identical successive injection runs such as 3.7, 3.9 and 3.10 (not fully shown). The clogging process was linear during the 15 days lasting injection period 3 of the first ASR trial [4], which is typical for clogging mainly caused by particles [10].

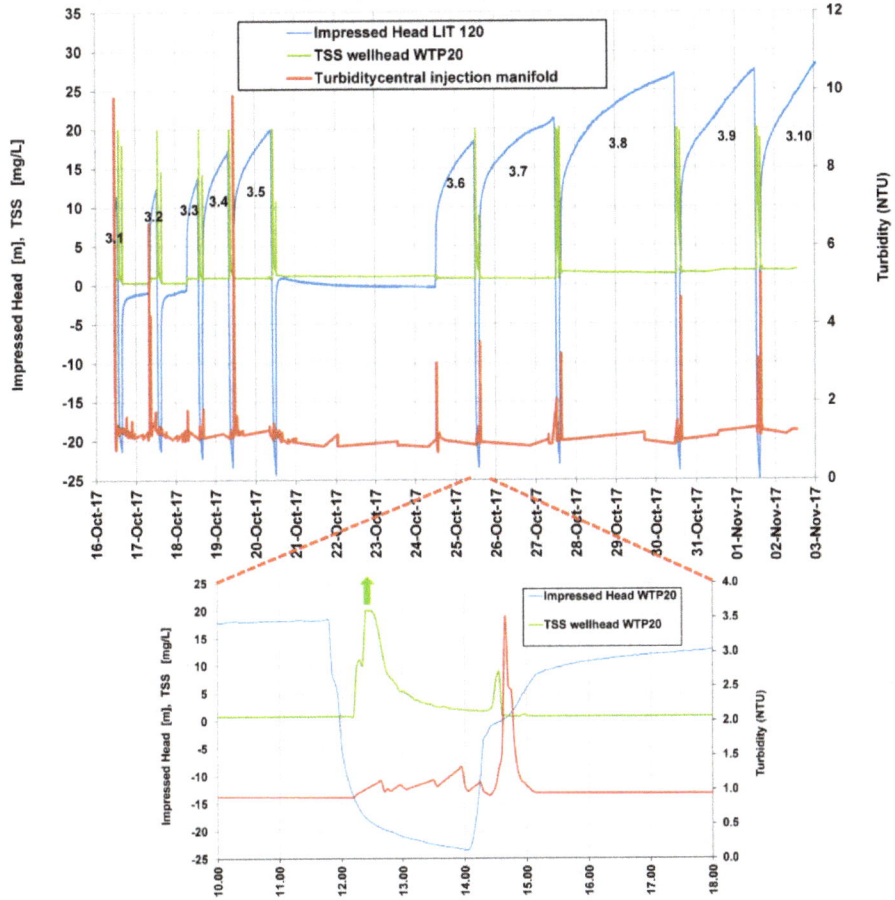

Figure 3. Impressed head/drawdown, total suspended solids (*TSS*, at well head) and turbidity (in central injection manifold) for ASR well 20, during ASR Cycle 3A (with 10 sub-cycles divided by backflush events), with zooming in on 25 October 2017 10–18 h. Note the *TSS* cut-off at 20 mg/L, and the logarithmic increase of impressed head.

3.2. Results of Hydrogeochemical Analysis of Suspended Material

The composition of the suspended solids was calculated for each sample taken during the 4 backflushing events A–D, by applying Equation (1). Each sample was taken after on average 1.5 days injection of 1950 m^3, and after ~5 min backflushing (corresponding with 11 m^3 and 1 standing volume in the well). The results are presented in Table 3, together with a summary of the geochemical analyses of aquifer cores and *TSS* of the injectant. Zero values mean that X_T equals X_{H2O}, so that practically all of X is in dissolved form. Empty cells in Table 3 (with '-') mean no data available.

3.2.1. Average Composition

In Table 3, the average composition is listed for samples of events A–C. The sample of event D was excluded because it refers to a different input (see Section 3.2.2). On average ~8.4% of the suspended material has thus been identified and quantified. This means that 92% is not covered by the analysis, due to (i) lack of dissolution in the acid and oxidizing leach (notably quartz), (ii) lack of the usual conversion of the elements in their oxide form, and (iii) no inclusion of H_2O which is present in the crystal lattice of some minerals such as amorphous $Fe(OH)_3$ and $Al(OH)_3$.

The highest contents are noticed for the following main constituents, in decreasing order: Si (underestimated due to incomplete extraction), Fe, Al, S, C, Mn and P. The true Si content is estimated at max 37% (not the 1.6% extracted; 37% = (100% − sum X_nO-bound H_2O)/MW_{SiO2}, where X_nO = all individual elements, SiO_2 excluded, each transformed in its oxide form, e.g., C in CH_2O and Fe in Fe_2O_3). Si is mainly linked to SiO_2 as quartz and diatoms. Concentrations of Na, K, Ca and Mg were <0.2%.

The high S content very likely indicates the presence of iron sulfide particles, because gypsum can be excluded and the alternative, organic matter, contains on average only 1.8% S (0.018 × C_{ORG} = 127 ppm S). Thus assuming the remaining 10,045 ppm (=10,172 − 127) to be FeS_2 (pyrite) requires the presence of enough Fe, namely 8858 ppm. This would mean that the *TSS* contains an excess of 13,420 ppm Fe, which could be $Fe(OH)_3$ or Fe_2O_3 deriving from the source water directly or indirectly (from biofilms on the transmission pipeline, a storage tank or ASR well casing), and from rust particles from corroding stainless steel parts of the transmission pipeline, storage tank or ASR well casing).

3.2.2. Trends

The composition of samples from backflushing events A–D varies over time, in a more or less systematic way for the following elements (red numbers in Table 3; Figure 4): (i) P and Mn steadily increasing, and (ii) S, FeS_2, Fe-FeS_2, Fe/Mn and V steadily decreasing. This indicates that over time, the infiltration water input is contributing more (P and Mn) and aquifer material less (pyrite and V) to *TSS* in the backflushed water. This agrees with expectations.

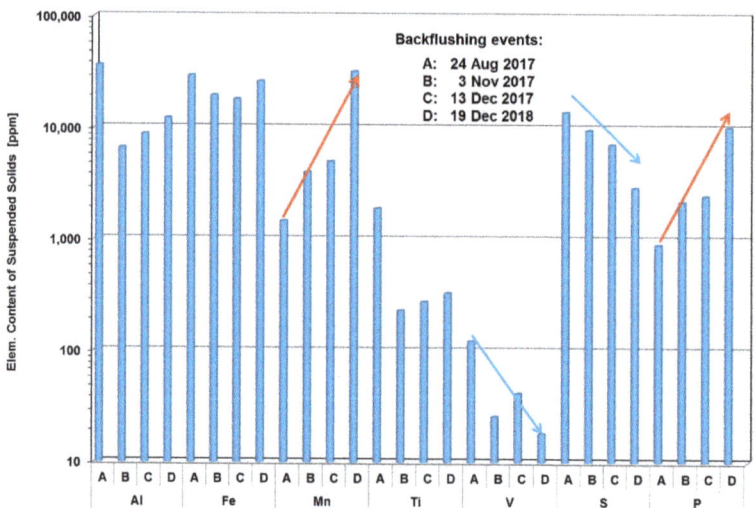

Figure 4. Changes in element content of suspended solids in a sample of backflushed water from ASR well 20 during backflushing events without (A–C) and with prefiltration (D). Based on Table 4. Arrows indicate significant trend over time.

Table 4. Results of microscopic analysis of suspended material in source water (SW) and backflushed water, in 2017. 1–7 = relative abundance low to very high. Samples 1, 2, 3 = after 5, 15 and 30 'equivalent' minutes of pumping. Samples coded 1 (taken after 5 min) correspond with events A, B and C in Table 3. Sample 1 on 3 November shown in Figure 5.

Parameter	SW	Backflush						
	13 Sept.	24 Aug.			3 Nov.			13 Dec.
Sample #		1	2	3	1	2	3	1
Time to filter 100 cc (s)	15	360	70	15	>600	98	20	140
Fine amorphous debris	3	7	5	5	4	5	5	4
Mainly non-filamentous bacteria	3	5	4	5	2	2	2	1
Diatom frustules	2	2	1	1	3	2	1	3
Viable diatoms					3	1	1	
Blue-green algae	1	1		1	1	1	1	1
Other algae	1	1	1	1	2	1	1	2
Zooplankton	1							

The sample from backflushing event D showed a remarkable overall concentration increase (S and V excluded). The reason for this increase is not clear, but could be related to the applied prefiltration step, which resulted in the accumulation of finer grained particles on the borehole wall and behind. These finer grained particles could contain less quartz particles and more soluble material.

3.2.3. Comparison with Aquifer Cores and Infiltration Water

In Table 3, the element content of *TSS* can be compared with the aquifer core data and with the average *TSS* composition of infiltration water. We selected the mean *TSS* composition of backflush events A–C, the injectant (without prefiltration) and the core data from monitoring well 3 (~1 km south of well 15, Figure 1). The core had about the same extraction and geochemical analysis as *TSS*, and showed a similar aquifer sedimentology as well 20. These data have been used in Table 3 to estimate the fraction of suspended particles in the infiltration water (α) and of aquifer particles ($1 - \alpha$) contributing to the mean *TSS* composition of the backflush sample. The following unmixing equation for 2 end-members was used:

$$\alpha = \frac{X_{SS} - X_{AQUIFER}}{X_{INPUT} - X_{AQUIFER}}, \tag{13}$$

Fraction α appears lowest (0.15) for S (and associated FeS_2 and $Fe\text{-}FeS_2$), so that practically all S (pyrite bound) in *TSS* is derived from the aquifer, as deduced earlier. Fraction α increases slightly for Cu, Mn, Co, V and Fe (from 0.21 to 0.40, respectively), which indicates that the aquifer is forming their mean supply during events A–C. Only P, Cr, Ni and Sr seem to be largely delivered by the injectant during events A–C, Cr perhaps even totally. The calculated α numbers are not very accurate due to temporal variations in the injectant and spatial variations in aquifer geochemistry.

An important assumption in this interpretation is, that the mentioned ions did not (co)precipitate in the water sample during transport from well to the lab and during 2 days of air-tight bottle detention prior to the separation of decantable fluid from the remaining high *TSS* liquid. We checked this by comparing the infiltration water with the decanted fluid, and the decanted fluid of the first with the last sample. Concentrations of dissolved Al, Fe, Mn NO_3, NH_4 and PO_4 did not show significant changes, indicating that the potential contribution of Fe or Mn flocs formed by oxidation of the backflushed water was rightly ignored.

3.3. Results of Microscopic and Particle Size Analysis of Suspended Material

Suspended material in source water and in backflushed water was examined by microscopic analysis. The results are summarized in Table 4, and a characteristic image is shown in Figure 5.

The diatoms and other algae in the backflushed water are clearly derived from the infiltration water input, because they are too undamaged and some are even still viable.

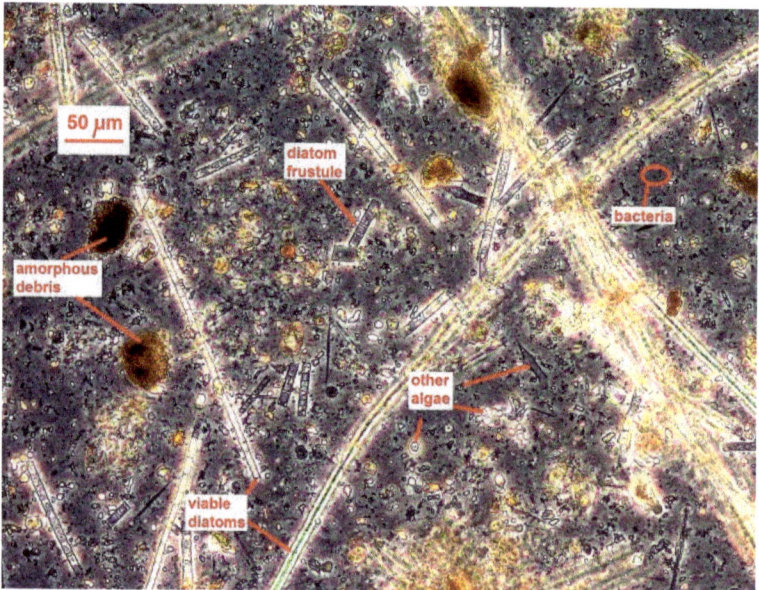

Figure 5. Microscopic image (×100 magnification) of backflushed material on 3 November 2017 (event B in Table 3), showing diatoms (incl. fragments), algae, mostly non-filamentous bacteria, and amorphous flocs.

Results of particle size analysis by Light-Scattering using a Mastersizer 3000 instrument, after sonication or addition of a dispersant, are shown in Figure 6. In the backflushed water, 3 peaks can be observed in the particle size distribution: around 0.5, 7 and 200 µm. In the infiltration water, there is a large peak around 13 µm with 2 small, ill-defined peaks around 0.8 and 800 µm. The difference between both particle size distributions could be not representative due to fluctuations in particle load and particle size distribution of either the infiltration water or backflush water. The peak around 200 µm in the backflush water could be due to either a rather high number of elongate diatoms >200 µm long (Figure 5), which the analyser underestimates due to their shape, or sand intake by the well.

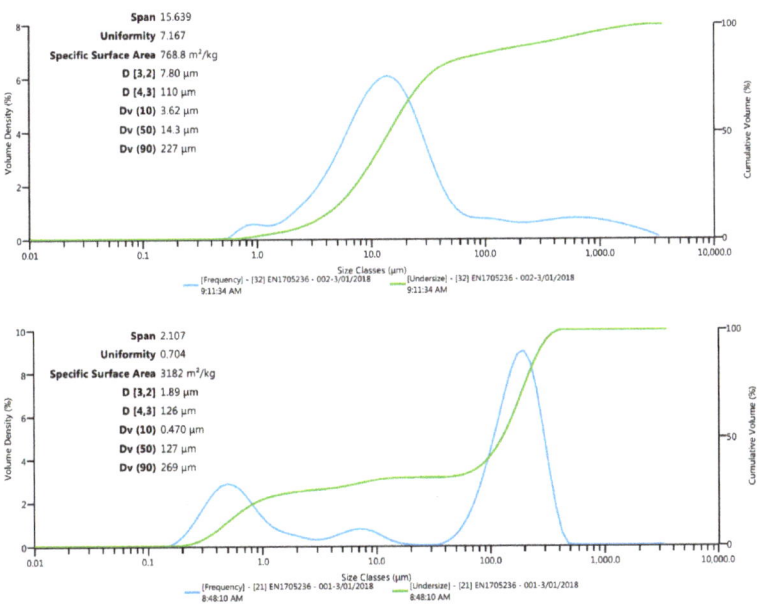

Figure 6. Particle size distribution of infiltration water (**top**) and backflushed water (**bottom**), both sampled on 13 December 2017 (event C in Table 3).

3.4. Evaluation of Clogging Risks

3.4.1. Global Evaluation

ASR well clogging can be caused by any of the following, potential clogging processes (or their concerted action): (i) suspended solids in the infiltration water such as clay, algae, diatoms and $Fe(OH)_3$ flocs, (ii) biofouling due to high concentrations of biodegradable organic matter (TOC, assimilable DOC (AOC), BDOC) and nutrients (PO_4, NO_3), (iii) air entrainment (gas bubbles formed during cascading in the well), (iv) chemical clogging by precipitates of e.g., $Fe(OH)_3$, MnO_2 or $CaCO_3$, (v) clay swelling and clay mobilization due to replacing brackish ambient groundwater with high SAR (sodium adsorption ratio) by fresher, lower SAR injectant, and (vi) permeability reduction by aquifer jamming or aquifer corrosion.

The last four processes are considered of minor importance. Well clogging by gas bubbles is unlikely because of a proper well design [4] and a non-corresponding clogging pattern [10]. The precipitation of minerals by mixing with ambient groundwater during injection phase can be ruled out, because the Fe concentrations in the ambient groundwater are low (0.16 mg/L) and the flow velocity near the well is sufficiently high and reactions are so slow, that reaction products such as iron hydroxide flocks will move relatively far away from the well before being deposited, while the mixing of (sub)oxic water with anoxic ambient groundwater will mainly take place far away from the well. Experiences with and calculations on intentional subterranean iron removal (SIR) [31] indicate that the accumulation of reaction products during both injection and recovery can be neglected on the time scale of one century. SIR is also an unintentional process during ASR [32], which prevents Fe(II) during recovery to reach the ASR well and mix there with any O_2 (if this survived the aquifer detention at some depth). The risk of $CaCO_3$ precipitation is also very low, because the injectant and ambient groundwater are undersaturated with respect to $CaCO_3$. Another potential cause of mineral precipitation is chemical instability of the injectant, which can lead to retarded, in-well flocculation of $Fe(OH)_3$ or MnO_2. This process is hard to distinguish from the advective transport of suspended particles, and therefore not considered further.

Barry et al. [7] investigated the clogging risk by dispersion of the clay minerals within columns filled with aquifer material, especially within the context of potential future changes in the mixing ratio of Class A water and RO desalinated water. They used a method based on the principles of the 'Emerson Crumb test' [33–35]. That approach characterizes dispersion using combinations of aquifer materials and source waters by measuring the turbidity increase of the supernatant, relative to that of the water alone, after a period of 48 h. Dispersion of clays in the batch tests was so low that no visible cloudiness was observed in any of the waters tested. Although kaolinite, the predominant clay mineral in the target aquifer, is a non-swelling clay and present within the aquifer storage zone at very low concentrations (<1%), its interaction with low salinity water can still contribute to a reduction in aquifer permeability by detachment from solid surfaces and migration through porous media where it can clog pore spaces [36]. In conclusion, the risk of clay mobilization is (very) low but cannot be ruled out completely, especially not if more RO water is contributing to its mixture with Class A recycled water.

The risk of aquifer permeability losses through aquifer jamming by repeatedly shifting from injection to backflushing, which creates shock waves, is considered low [10]. Aquifer dissolution (not only by the injectant but also by acids applied for well redevelopment) is potentially a risk factor in marley limestone or calcite cemented sandstone aquifers, because the carbonate dissolution will be accompanied by the mobilization of fines that may end up in the pore throats [37,38]. The target aquifer at Werribee is, however, not containing carbonates and thus not vulnerable to this dissolution.

In Table 5, an overview is given of potentially relevant clogging parameters and their levels in the infiltrated Class A recycled water with (second ASR pilot) and without RO water (first pilot), and with (first pilot and cycle 3B of second pilot) and without an onsite prefiltration step (cycles 1–3A of second pilot). The following conclusions are drawn, when comparing the injectant with the given guideline values:

TSS is relatively low (on average 1–2 mg/L), but still above the guideline value for a sandy aquifer (0.1 mg/L). This should lead to physical clogging. Particles containing Fe, Mn and Al probably play a significant role, as deduced from analyzing suspended solids during backflushing, even though the difference between total and dissolved concentrations (the particulate fraction), seems small. MFI is too high, suggesting a high risk of physical clogging. Concentrations of DOC, BDOC ([28]) and AOC ([39]) are high, far above their guideline values. This will lead to biofouling (biological clogging) if without regular backflushing, also because nutrients N + P are not limiting.

Table 5. Overview of parameters that have a potential impact on well clogging, and their average concentration or value in the infiltration water during the three cycles of each pilot. For several parameters a guideline value for deep well recharge is indicated.

ASR Pilot	TSS	Turb.	TOC	DOC	BDOC	AOC	MFI	pH	Temp	PO_4-T	NO_3	NH_4	O_2	Fe-T	Fe_{FILT}	Mn-T	Mn_{FILT}	Al-T	Al_{FILT}	$Cl_{2\,FREE}$	SAR
	mg/L	NTU	mg C/L	mg C/L		μg C/L	s/L²		°C						mg/L						meq$^{0.5}$
Guideline $	<0.1	<1	<2	<2	<0.2	<10	<3–5													>0.2	#
ASR 5	100% Class A recycled water																				
Cycle 1	<5	2.9	10	10	1.5	170		6.91		22.6	113.9	0.09	8.6	<0.05				<0.01			8.1
Cycle 2	<1	0.8	11	10	2.2	160		6.94		30.4	116.1	0.03	9.6	<0.05				<0.01			8.0
Cycle 3	1	<0.1	10	10	2.1	185		6.9		29.7	135.1	0.21	9.2	<0.05				0.020			7.9
ASR 20	33% Class A recycled water + 67% RO treated Class A recycled water																				
Cycle 1	<2	0.50	3.90	3.90	0.50		47	7.39	11.8	10.1	40.8	<0.13	8.0	0.027	0.012	0.015	0.012	0.071	0.002		5.1
Cycle 2		0.25	3.55	3.55	0.47			7.16	13.0	9.3	37.4		8.8	0.099	0.020	0.014	0.008	0.029	0.023		5.0
Cycle 3A		0.63	3.87	3.63	0.30			7.60	19.8	10.8	15.9		8.8	0.018	0.011	0.022	0.003	0.022	0.012		5.2
Cycle 3B	2.5 #	0.69	4.41	4.32	1.35			7.62	20.2	9.3	19.6	1.03	7.5	0.032	0.012	0.038	0.021	0.047	0.020		4.8

$ = values for deep well injection, partly according to Perez-Paricio & Carrera [40] and Barry et al. [10]; # = prior to prefiltration.

3.4.2. Risk of Clogging by Suspended Particles

We quantified this risk by calculating the clogging rate using the modified Bichara method and the Buik and Willemsen method (Section 2.6). The results of calculation are shown for several scenarios in Table 6 (Bichara's method) and Table 7 (Buik and Willemsen method).

Table 6. Predicted flow reduction (Q_t/Q_0 and Q_{10}) as function of injected particle mass ($\Sigma TSS = TSS$ Q_{IN} t) and initial open area of external borehole wall (A_{OPEN}). Based on average linearized relation between $\Sigma TSS / A_{OPEN}$ and Q_t/Q_0 trends in Bichara [26], see Section 2.6. t_{10} = injection period needed to reduce flow by 10%. φ_t, φ_{MEAS} = the impressed head in the injection well as predicted and measured, respectively. Red numbers: significant scenario changes.

Data Input						Model Output					Verif.
TSS	Q_{IN}	t_{INF}	L	r_B	n	ΣTSS	A_{OPEN}	Bichara	Q_t/Q_0	t_{10}	φ_t	φ_{MEAS}
mg/L	m³/h	d	m	m		kg	m²	g/cm²	%	day	m	m
1	54	0	13	0.1	0.3	0.0	2.5	0.000	100.0	-	0.0	0
1	54	1	13	0.1	0.3	1.3	2.5	0.053	90.7	1.1	22.0	21.3
1	54	2	13	0.1	0.3	2.6	2.5	0.106	81.5	1.1	24.5	24.9
1	54	3	13	0.1	0.3	3.9	2.5	0.159	72.2	1.1	27.7	27.1
1	54	5	13	0.1	0.3	6.5	2.5	0.264	53.7	1.1	37.2	-
1	54	10	13	0.1	0.3	13.0	2.5	0.529	7.4	1.1	268.6	-
1	54	10	13	0.2	0.3	13.0	4.9	0.264	53.7	2.2	37.2	-
1	20	10	13	0.1	0.3	4.8	2.5	0.196	65.7	2.9	30.4	-
0.1	54	10	13	0.1	0.3	1.3	2.5	0.053	90.7	10.8	22.0	-

f = 2; φ_0 (m) = 10.0.

Table 7. Predicted injection well clogging (v_{CLOG}) as function of the membrane filter index (MFI) and other parameters. Method based on Buik & Willemsen [23], see Section 2.6. Red numbers: significant scenario changes.

Data Input							Model Output			
MFi	Q_{IN}	t_{EQ}	L	r_B	D_{50}	D	K_h	KD	V_{DIR}	V_{CLOG}
s/L²	m³/h	day	m	m	mm	m	m/day	m²/day	m/h	m/a	m/day
47	54	2400	13	0.1	0.135	13	5.5	71.6	6.6	520	1.42
47	54	1200	13	0.1	0.135	13	5.5	71.6	6.6	260	0.71
47	54	27	2400	0.1	0.135	13	5.5	71.6	3.3	130	0.36
4.7	54	2400	13	0.1	0.135	13	5.5	71.6	6.6	52	0.14
47	54	2400	13	0.2	0.135	13	5.5	71.6	3.3	130	0.36
4.7	27	1200	13	0.2	0.135	13	5.5	71.6	1.7	1.6	0.004

With the modified Bichara method we calculate for ASR well 20 during cycle 3A, resetting the clock after each backflushing event, a flow reduction of about 9% for each day of continuous injection (Table 6). Maintaining the injection rate means that more pressure is needed, which leads to an increase of the impressed head in the well. Intuitively, this increase is proportional to the predicted flow reduction while ignoring temperature effects on viscosity, so that:

$$0.01 \, Q_t/Q_0 = f \, \varphi_0/\varphi_t \quad \text{so that} \quad \varphi_t = 100 \, f \, \varphi_0/(Q_t/Q_0) \tag{14}$$

where: φ_0, φ_t = the impressed head in the injection well without (significant) clogging, and with clogging at time t since start, respectively (m); f = empirical fit factor, being zero at $t = 0$, and otherwise >0.

Application of Equation (16) to the 3-day-long injection period from 27 October up to and including 30 October 2017 (Figure 3), yields an excellent overlap of calculated with measured φ_t (Table 6), with parameter settings of $f = 2$ (if $t > 0$) and $\varphi_0 = 10$ m. Table 6 also shows how much effect can be expected from reducing the injection rate, TSS input or augmenting the backflush frequency or borehole radius.

For ASR well 20 during cycle 3A, with an MFI value of 47 s/L², we calculate with the method of Buik and Willemsen [23] a clogging rate of 520 m/a or 1.42 m/day, assuming 100 days of injection a year (Table 7). This high clogging rate looks much smaller than the one obtained with Bichara's method (4.3 m versus 27 m in 3 days), but v_{CLOG} does not include the water level rise due to unsteady flow as does φ_t. Table 7 also shows how much effect can be expected from reducing the injection rate, MFI (by enhanced pretreatment) and equivalent full loading hours, or increasing the bore hole radius or implementing all measures to reduce the clogging rate.

3.4.3. Risk of Bioclogging

The results of calculating the number of bacteria (N) and the thickness of the biofouled layer (D_{BAC}) are presented for several scenarios in Table 8, and the bacterial growth curve is presented for selected cases in Figure 7. An unrealistic high D_{BACT} is predicted for scenarios A and E–G of the second Werribee trial, when an unlimited supply of assimilable carbon (and nutrients and O_2) is assumed, during a long uninterrupted injection run. The observed consumption level of BDOC in the aquifer (Δ_{BDOC}) is believed to represent the biodegradable fraction of DOC, which is then used for both respiration and biomass production. The concentration of Δ_{BDOC} (~0.2 mg C/L) is much lower than the NO_3 and PO_4 concentrations in the injectant (Table 5) and is, therefore, considered the growth-limiting factor.

Table 8. Predicted unlimited growth of bacteria (N_t–N_0) according to Equation (8) and carbon limited growth of bacteria (N_{tMAX}–N_0) according to Equation (11), and their clogging of open pore space on the borehole wall in terms of thickness of biofouled layer ($D_{BAC,t}$ and $D_{BAC,max}$; Equation (9)). Also shown is t_{MAX} (Equation (14); time since start of exponential bacterial growth to the maximum level of growth) assuming that carbon supply (with Δ_{BDOC} as indicator) for bacterial growth is the limiting factor. Explanations of parameters in Section 2.7. Red numbers = significant scenario changes.

Scenario	Data Input									
	Q_{IN} m³/day	L m	r_B m	ε Porosity	N_0 n/m³	T_D day	t day	V_{CELL} m³	t_{LAG} day	Δ_{BDOC} mg/L
Werribee well 20, A	1296	15.5	0.10	0.30	1.E+08	0.3	7.5	2.0E-18	0.5	0.2
Werribee well 20, B	1296	15.5	0.10	0.30	1.E+08	0.3	3	1.0E-18	0.5	0.2
Werribee well 20, C	1296	15.5	0.10	0.30	1E+09	0.3	3	1.0E-18	0.5	0.2
Werribee well 20, D	1296	15.5	0.10	0.30	1E+07	0.3	7.5	1.0E-18	0.5	0.2
Werribee well 20, E	1296	15.5	0.10	0.30	1.E+07	1	30	1.0E-18	0.5	0.2
Werribee well 20, F	2592	15.5	0.10	0.30	1E+07	1	30	1.0E-18	0.5	0.2
Werribee well 20, G	1296	15.5	0.20	0.30	1E+07	1	30	1.0E-18	0.5	0.2
Additional data input for all scenario's: f_{BOP} = 0.5; C_{CELL} (kg/m³) = 220										
Scenario	Calculated		No limitation		+ BDOC limitation					
	A_{OPEN} m²	V_{DIR} m/day	N_t n/m²	$D_{BAC,t}$ mm	t_{MAX} day	N_{tMAX} n/m²	$D_{BAC,max}$ mm			
Werribee well 20, A	2.9	133	6E+16	173	4.7	1E+14	0.29			
Werribee well 20, B	2.9	133	2E+12	0.003	5.0	2E+14	0.29			
Werribee well 20, C	2.9	133	2E+13	0.03	4.0	2E+14	0.29			
Werribee well 20, D	2.9	133	6E+15	8.67	6.0	2E+14	0.29			
Werribee well 20, E	2.9	133	1E+18	2074	17.2	2E+14	0.29			
Werribee well 20, F	2.9	266	3E+18	4148	17.2	4E+14	0.58			
Werribee well 20, G	5.8	67	7E+17	1037	17.2	1E+14	0.14			

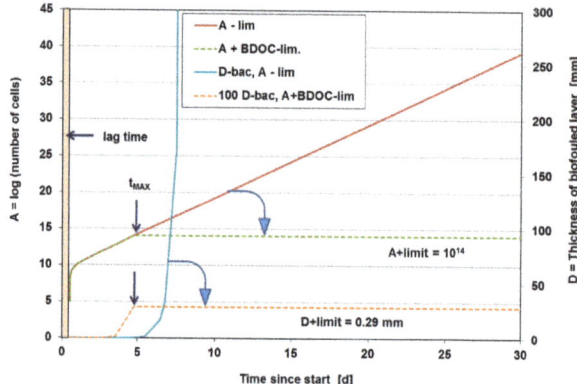

Figure 7. Predicted unlimited (-lim) and carbon limited (+BDOC-lim) growth of bacteria (plotted as log (number of cells)), and their clogging of open pore space on the borehole wall and behind, in terms of thickness of biofouled layer (D_{BAC}), for scenario A defined in Table 8. Note that right-hand scale for 100 D-bac has to be divided by 100. Carbon limitation calculated with observed biodegradable dissolved organic carbon (BDOC) decline during 1–2 days detention in aquifer. t_{MAX} = time since start of exponential bacterial growth to the maximum level of growth due to carbon limitation.

When we include growth limitation, then much less bacteria are accumulating in a much thinner biofouled layer: e.g., in case of Werribee scenario A, the calculated total number of bacteria decreases from 6.1×10^{16} to 1.0×10^{14} m^{-2} and the thickness of the biofouled layer declines from 173 to 0.29 mm, in case of BDOC limitation (Table 8).

These predictions are hampered by many imperfections due to among others: (i) inaccuracy of parameters, especially Δ_{BDOC}; (ii) accumulation of dead bacteria by lack of e.g., oxygen, (iii) accumulation of suspended particles other than bacteria, (iv) incomplete removal of biofouling and accumulated fines during backflushing, and (v) erosion by flowing water and predation by higher organisms.

The presented model calculations thus only serve the purpose of doing a sensitivity or risk analysis. What can we do to lower the risk of biological well clogging? The various scenarios in Table 8 reveal that this risk can be lowered by (i) further pretreatment to reduce N_0 (requires chlorination, advanced oxidation or reverse osmosis) and BDOC (requires granular activated carbon filtration or slow sand filtration), (ii) chlorination to reduce N_0 and extend t_{LAG} (this may raise BDOC and the capacity to oxidize aquifer minerals however), (iii) augmenting the frequency of backflushing (which reduces t), (iv) making wells with a larger A_{OPEN}, especially by drilling larger diameter holes, and (v) reducing the injection rate by augmenting the number of wells.

4. Discussion

Clogging is the enemy of managed aquifer recharge (MAR) operations [41,42], and this holds 'in extremis' for injection wells, especially in case of siliclastic aquifers. In Werribee, the results obtained during the first and the second ASR trial (up to and including cycles 1–3A), the first with and the second without additional filtration, confirm that well clogging is indeed cumbersome, requiring frequent backflushing (once per 1–3 days), incidental mechanical well regeneration and intensive monitoring of hydraulic heads. The ongoing research with the additional filtration steps in place (Figure 8) shows optimistic preliminary results, to be confirmed when injection phase 3B continues and more data become available.

Figure 8. (**Left**): Central manifold ASR System with temporary prefiltration unit that includes a 20 μm spin Klin disc (the 12 black cylinders) followed by a 1 μm bag filter (the 3 upright cylinders). (**Right**): clogging of the 1 μm bag filter after ~2 days of operation.

In this study, we focused on the 2 main causes of well clogging, i.e., physical clogging by suspended solids in the injectant, and biofouling, after evaluating the risks of 4 other causes of well plugging and the results of column studies by [7] with core material from well 20.

Hydraulic head trends indicate a more or less linear increase of clogging, which pleads for physical clogging as the main process [10]. However, injection runs were relatively short, which prevents biofouling to clearly manifest itself with the typical exponential progress and a typical H_2S smell during backflushing after a long storage period during which biomass will putrefy. Backflushing could be optimized by raising the pumping rate not 2 but 3 times the normal recovery pumping rate, by occasional backflushing several times, and by limiting the backflushing time to the period with a significant turbidity rise so as to reduce the waste water volume (Pyne, written communication). Chlorination has been applied during both pilots, but could also be optimized (see [42]) to reduce biofouling while keeping the formation of disinfection byproducts low.

We developed a simple method to deduce the composition and origin of the suspended solids removed during backflushing, based on readily available methods. A more direct method is desirable, however, and could consist of (i) TSS collection in the injectant from both filtration units (right panel in Figure 8), (ii) on-site filtration of a high volume sample during backflushing, and (iii) analysis of the material retained on the filters. For the injectant a lower minimum detection limit of TSS is needed (to be lowered from <1 to <0.1 mg/L), for the backflush water the additional analysis of loss on ignition of TSS is desired to estimate its organic matter content.

Microscopic analyses proved very useful to demonstrate the ubiquitous presence of diatoms, other algae, fine amorphous debris and bacteria in TSS of both the infiltration water and the backflushed water. Seasonal algae blooms should, therefore, be translated into an injection stop, possibly aided by on line turbidity measurements coupled to intelligent decision support and specific action protocols [43]. Another option is to utilize tangential columns [44] as an early warning system to prevent injection of e.g., an incidental TSS peak.

With the focus on well clogging by suspended particles and biofouling, we calculated these risks by three well clogging predictors: (i) the method of Buik and Willemsen [23] based on the MFI, well and hydrogeological parameters, (ii) a modification of Bichara's method [26] based on the ratio of the TSS input after a given infiltration time, to the open area of the external borehole wall, and (iii) a modification of the method of Huisman and Olsthoorn [27] which relates biofouling to the exponential growth of bacteria. Application of the first and second method is straightforward. The third is, however, handicapped by scarcity of relevant data on changes in biodegradable organic carbon (e.g.,

Δ_{BDOC}) during aquifer detention, and difficult to assess processes such as the lag time (period for bacteria to acclimatize) and growth limitation by either biodegradable carbon, nutrients or oxygen.

5. Conclusions

From the 6 main causes of injection well clogging, suspended particles and biofouling are the most common of ASR bore clogging [10,42]. This also holds for the ASR system in Werribee, as demonstrated by: (i) hydraulic data, geochemical and microscopic analysis of total suspended solids in the injectant and backflushed water, and (ii) 3 well clogging predictors, 2 of which were optimized for this study and are fit for general application. Clogging predictors yield an order of magnitude estimate of the clogging rate, but above all they serve the purpose of doing a sensitivity or risk analysis.

Based on these predictors the following recommendations can be given to reduce the risk of well clogging by suspended particles and biofouling: (i) reduce the *TSS* input by prefiltration (preferably by applying rapid followed by slow sand filtration), (ii) reduce the injection rate (may require additional wells), (iii) augment the backflush frequency and its performance, (iv) chlorinate or optimize the chlorination to reduce N_0 and extend t_{LAG} (this may raise BDOC, however; CWW already applies chlorination and occasionally a shock chlorination), (v) reduce BDOC (requires granular activated carbon filtration or slow sand filtration), and (vi) drill larger diameter holes. The strength of the predictors is that they yield insight into the relative effect of measures taken.

The order of recommendations presented above (i–vi) reflects the order of decreasing priority for the Werribee pilot. In case of a new pilot anywhere, we recommend the following general order of descending priority, with numbering as above: vi > ii > i > iii > iv > v. This order is subject to modifications depending on among others the outcome of the three presented clogging predictors.

Author Contributions: J.O. designed and supervised the pilot trials; P.S. assisted with the monitoring set-up, analyzed the data and wrote the paper with assistance by J.O.

Funding: This research was funded by the Australian Government Department of Environment and Energy in relation to The National Urban Water and Desalination Plan: West Werribee Dual Supply (Aquifer Storage and Recovery) Project and City West Water.

Acknowledgments: City West Water (CWW) gratefully acknowledges the support of the Australian Government in funding the West Werribee ASR Project. KWR thanks CWW for the assignment to assist in their ASR pilot studies, and for publication of results. Matthew Hudson (formerly CWW, now Southern Rural Water, Werribee) contributed to the project at an earlier stage. Two anonymous reviewers, David Pyne and Peter Dillon gave excellent suggestions to improve the manuscript.

Conflicts of Interest: The authors declare no conflict of interest.

Abbreviations

The following abbreviations are used in this manuscript:

AOC	assimilable organic carbon
ASR	aquifer storage and recovery
BDOC	biodegradable dissolved organic carbon
BGL	below ground level
BOD5	biological oxygen demand in 5 days at 20 °C
CWW	City West Water
D	aquifer thickness
D_{50}	median grain size diameter of aquifer
DOC	dissolved organic carbon
EC	electrical conductivity
L	length of well screen
LWF	Lower Werribee Formation
MAR	managed aquifer recharge
MFI	0.45 µm Membrane Filter Index
Mm^3	mega cubic meter

N_t, N_0	number of bacteria at time $t = t$ and $t = 0$
ORP	oxidation reduction potential
Q_{IN}	mean infiltration rate when well is recharging
r_B	radius of borehole
Q_0, Q_t	injection rate at time t since start and at start ($t = 0$), respectively
RE	recovery efficiency
S	aquifer storativity
SAR	sodium adsorption ratio
t	time since start (day)
t_{10}	injection period needed to reduce flow with 10% (day)
T	aquifer transmissivity
T_D	doubling time for bacterial population (day)
TDS	total dissolved solids
T_{EQ}	total amount of equivalent full load hours per year
t_{LAG}	lag time during which bacteria do not reproduce, e.g., due to chlorination (day)
t_{MAX}	time since start of exponential bacterial growth to maximum growth level (day)
TOC	total organic carbon
TSS	total suspended solids
v_{CLOG}	clogging rate
v_{DIR}	entrance velocity on borehole wall
α_L	longitudinal dispersivity
ε	porosity
$\Delta BDOC$	consumption of biodegradable DOC
φ_0, φ_t	impressed head in injection well without clogging, and with clogging at time $t = t$

References

1. Bixio, D.; Thoeye, C.; De Koning, J.; Joksimovic, D.; Savic, D.; Wintgens, T.; Melin, T. Water reuse in Europe. *Desalination* **2006**, *187*, 89–101. [CrossRef]
2. Kazner, C.H.; Wintgens, T.H.; Dillon, P. (Eds.) *Water Reclamation Technologies for Safe Managed Aquifer Recharge*; IWA Publishing: London, UK, 2012; p. 429.
3. Hudson, M.; Muthukaruppan, M. Meeting Melbourne's future demand of water using aquifer storage and recovery. In Proceedings of the 9th International Symposium on Managed Aquifer Recharge (ISMAR9), Mexico City, Mexico, 20–24 June 2016.
4. SKM. *West Werribee Dual Supply Project–Phase 2 ASR*; Unpublished Investigation Report for City West Water; City West Water: Footscray, Australia, 2013.
5. Stuyfzand, P.J.; Hudson, M. *Hydrogeochemical Evaluation of the Werribee ASR Pilot and Operational Trial*; Technical Report No. KWR-2016.107; KWR: Nieuwegein, The Netherlands, 2016; p. 91.
6. Crisalis International. *West Werribee Dual Supply Project: Assessment of Geochemical Data. From a Trial of Aquifer Storage and Recovery Using Class A reclaimed Water Injected and Recovered from the Werribee Formation*, Technical Report for SKM Pty Ltd. 2012; 56.
7. Barry, K.E.; Vanderzalm, J.L.; Page, D.W.; Gonzalez, D.; Dillon, P.J. *Evaluating Treatments for Management of ASR Well Clogging: Laboratory Column Study*; CSIRO Land and Water, Client Report to City West Water; City West Water: Footscray, Australia, 2015; p. 71.
8. Stuyfzand, P.J. *Water Quality Aspects and Clogging of the Werribee ASR Operational Trial*; Technical Report No. KWR 2018 071; KWR: Nieuwegein, The Netherlands, 2018; p. 90.
9. Dillon, P.; Vanderzalm, J.; Page, D.; Barry, K.; Gonzalez, D.; Muthukaruppan, M.; Hudson, M. Analysis of ASR clogging investigations at three Australian ASR sites in a bayesian context. *Water* **2016**, *8*, 442. [CrossRef]
10. Olsthoorn, T.N. *The Clogging of Recharge Wells Main Subjects*; Kiwa: Rijswijk, The Netherlands, 1982; Volume 72, p. 136.
11. Martin, R. (Ed.) Clogging Issues Associated with Managed Aquifer Recharge Methods. IAH Commission on Managing Aquifer Recharge, Australia. 2013. Available online: http://recharge.iah.org/recharge/documents/clogging-MAR-all.pdf. (accessed on 15 May 2019).

12. Houben, G.; Treskatis, C. *Water Well Rehabilitation and Reconstruction*; McGraw Hill: New York, NY, USA, 2007; p. 391.
13. Stuyfzand, P.J. Pyrite oxidation and side-reactions upon deep well injection. In Proceedings of the WRI-1010th International Symposiym on Water Rock Interaction, Villasimius, Italy, 10–15 June 2001; Volume 2, pp. 1151–1154.
14. Pavelic, P.; Nicholson, B.C.; Dillon, P.J.; Barry, K.E. Fate of disinfection by-products in groundwater during aquifer storage and recovery with reclaimed water. *J. Contam Hydrol.* **2005**, *77*, 119–141. [CrossRef] [PubMed]
15. Dillon, P.; Toze, S. (Eds.) *Water Quality Improvements during Aquifer Storage and Recovery*; Technical Report No. 91056F; American Water Works Assoc: Denver, CO, USA, 2005; p. 286.
16. Di Lorenzo, T.; Cifoni, M.; Fiasca, B.; Di Cioccio, A. Predictive ecological risk assessment of pesticide mixtures in the alluvial aquifers of central Italy: Towards more realistic scenarios for risk mitigation. *Sci. Total Environ.* **2018**, *644*, 161–172. [CrossRef] [PubMed]
17. Korbel, K.; Stephenson, S.; Hose, G.C. Sediment size influences habitat selection and use by groundwater macrofauna and meiofauna. *Aquatic Sci.* **2019**, *81*, 39. [CrossRef]
18. Hose, C.; Stumpp, C. Architects of the underworld: Bioturbation by groundwater invertebrates influences aquifer hydraulic properties. *Aquatic Sci.* **2019**, *81*, 20. [CrossRef]
19. AGT. *Aquifer Storage and Recovery Investigations for the West Werribee Area*; Technical Report No. 2009/934A; Australian Groundwater Technologies: Adelaide, Australia, 2010; p. 149.
20. Dudding, M.; Evans, R.; Dillon, P.; Molloy, R. *Report on Broad Scale Map of ASR Potential for Melbourne*; Technical Report No. WC02973; Sinclair Knight Merz (SKM) for Victorian Smart Water Fund: Canberra City, Australia, 2009; p. 46.
21. GHD. *Aquifer Storage and Recovery (ASR) Schemes; Groundwater Modelling of Existing and Potential ASR Schemes*, Unpublished Report by GHD and Groundwater Logic for CWW. 2017; 297.
22. ALS Environmental. Available online: https://www.alsglobal.com/au/ (accessed on 12 December 2018).
23. Buik, N.A.; Willemsen, A. Clogging rate of recharge wells in porous media. In *Management of Aquifer Recharge for Sustainability, Proceedings of the 4th Internat. Symp. on Artificial Recharge, Adelaide, Australia, 22–26 September 2002*; Dillon, P.J., Ed.; Balkema: Rotterdam, The Netherlands, 2002; pp. 195–198.
24. Schippers, J.C.; Verdouw, J. The modified fouling index, a method of determining the fouling characteristics of water. *Desalinisation* **1980**, *32*, 137–148. [CrossRef]
25. Dillon, P.; Pavelic, P.; Massmann, G.; Barry, K.; Correll, R. Enhancement of the membrane filtration index (MFI) method for determining the clogging potential of turbid urban stormwater and reclaimed water used for aquifer storage and recovery. *Desalination* **2001**, *140*, 153–165. [CrossRef]
26. Bichara, A.F. Clogging of recharge wells by suspended solids. *J. Irrigation Drainage Eng.* **1986**, *112*, 210–224. [CrossRef]
27. Huisman, L.; Olsthoorn, T.N. *Artificial Groundwater Recharge. Monographs and Surveys in Water Resources Engineering 7*; Pitman Advanced Publishing Program: Boston, MA, USA, 1983; p. 320.
28. Servais, P.; Anzil, A.; Ventresque, C. Simple method for determination of biodegradable dissolved organic carbon in water. *Appl. Environ. Microbiol.* **1989**, *55*, 2732–2734. [PubMed]
29. Del Giorgio, P.A.; Cole, J.J. Bacterial growth efficiency in natural aquatic systems. *Annu. Rev. Ecol. Syst.* **1998**, *29*, 503–541. [CrossRef]
30. Bratbak, G.; Dundas, I. Bacterial Dry Matter Content and Biomass Estimations. *Appl. Environ. Microbiol.* **1984**, *48*, 755–757. [PubMed]
31. Van Beek, C.G.E.M. Experiences with underground water treatment in the Netherlands. *Water Supply* **1985**, *3*, 1–11.
32. Stuyfzand, P.J.; Wakker, J.C.; Putters, B. Water quality changes during aquifer storage and recovery (ASR): Results from pilot Herten (Netherlands), and their implications for modeling. In Proceedings of the 5th International Symposiym on Management of Aquifer Recharge, ISMAR-5, Berlin, Germany, 11–16 June 2005; UNESCO IHP-VI, Series on Groundwater No. 13. 2006; pp. 164–173.
33. Emerson, W.W. A classification of soil aggregates based on their coherence in water. *Aust. J. Soil Res.* **1967**, *2*, 211–217. [CrossRef]
34. McKenzie, N.; Coughlan, K.; Cresswell, H. *Soil Physical Measurement and Interpretation for Land Evaluation*; CSIRO: Canberra, Australia, 2002; p. 390.

35. Maharaj, A. The Use of the Crumb Test as a Preliminary Indicator of Dispersive Soils. In Proceedings of the 15th African Regional Conference on Soil Mechanics and Geotechnical Engineering, Maputo, Mozambique, 18–21 July 2011; Quadros, C., Jacobsz, S.W., Eds.; IOS Press: Amsterdam, The Netherlands, 2011; pp. 299–306. [CrossRef]
36. Mohan, K.K.; Vaidya, N.; Reed, M.G.; Fogler, H.S. Water sensitivity of sandstones containing swelling and non-swelling clays, colloids and surfaces A. *Physiochem. Eng. Aspects* **1993**, *73*, 237–254. [CrossRef]
37. Muecke, T.W. Formation fines and factors controlling their movement in porous media. *J. Petrol. Techn.* **1979**, *31*, 144–150. [CrossRef]
38. Pavelic, P.; Dillon, P.J.; Barry, K.E.; Vanderzalm, J.L.; Correll, R.L.; Rinck-Pfeiffer, S.M. Water quality effects on clogging rates during reclaimed water ASR in a carbonate aquifer. *J. Hydrol.* **2007**, *334*, 1–16. [CrossRef]
39. Hijnen, W.A.M.; Van Der Kooij, D. The effect of low concentrations of assimilable organic material (AOC) in water on biological clogging of sand beds. *Water Res.* **1992**, *26*, 963–972. [CrossRef]
40. Pérez-Paricio, A.; Carrera, J. Operational guidelines regarding clogging. In *Artificial Recharge of Groundwater*; Peters, J.H., Ed.; Balkema: Rotterdam, The Netherlands, 1998; pp. 441–445.
41. Bouwer, H. Artificial recharge of groundwater: Hydrogeology and engineering. *Hydrogeol. J.* **2002**, *10*, 121–142. [CrossRef]
42. Pyne, R.D.G. *Aquifer Storage Recovery: A Guide to Groundwater Recharge Through Wells*, 2nd ed.; ASR Press: Gainesville, FL, USA, 2005; p. 608.
43. Lynggaard-Jensen, A.; Eisum, N.; Rasmussen, I. Real time monitoring and management of artificial recharge plants. In *ArtDemo: Reduction of Contamination Risks at an Artificial Recharge Demonstration Site in Denmark and Sweden*; Lynggaard-Jensen, A., Stuyfzand, P.J., Eds.; Technical Report No. EVK1-CT-2002-00114; Project Publication 5th Framework Programme of the EU Environment and Sustainable Development: Brussel, Belgium, 2018; p. 224.
44. Hijnen, W.A.M.; Bunnik, J.; Schippers, J.C.; Straatman, R.; Folmer, H.C. Determining the clogging potential of water used for artificial recharge in deep sandy aquifers. In *Artificial Recharge of Groundwater, Proceedings of the 3rd International Symp. on Artificial Recharge, Amsterdam the Netherlands 21–25 September 1998*; Peters, J.H., Ed.; Balkema: Rottherdam, The Netherlands, 1998; pp. 437–440.

© 2019 by the authors. Licensee MDPI, Basel, Switzerland. This article is an open access article distributed under the terms and conditions of the Creative Commons Attribution (CC BY) license (http://creativecommons.org/licenses/by/4.0/).

Article

Specific Types and Adaptability Evaluation of Managed Aquifer Recharge for Irrigation in the North China Plain

Shuai Liu [1], Weiping Wang [1,*], Shisong Qu [1], Yan Zheng [2] and Wenliang Li [1]

1. School of Water Conservancy and Environment, University of Jinan, Jinan 250022, China; 15053130651@sina.cn (S.L.); stu_quss@ujn.edu.cn (S.Q.); lwl17862903319@sina.cn (W.L.)
2. School of Environmental Science and Engineering, Southern University of Science and Technology, Shenzhen 518055, China; yan.zheng@sustech.edu.cn
* Correspondence: stu_wangwp@ujn.edu.cn; Tel.: +86-139-5316-2318

Received: 30 September 2019; Accepted: 14 February 2020; Published: 18 February 2020

Abstract: The North China Plain is the main grain production district in China, with a large area of well irrigation resulting in a large groundwater depression cone. In the 1970s and 1980s, small-scale managed aquifer recharge (MAR) projects were developed to recharge shallow groundwater, which played an important role in ensuring stable and high crop yields. MAR projects are divided into 10 types based on local water conservancy characteristics. The combined use of well–canal irrigation has been widespread in the Yellow River Irrigation District of Shandong Province for nearly 40 years, where canals play multiple roles of transporting and storing Yellow River water or local surface water, recharging groundwater and providing canal irrigation. Moreover, the newly developed open channel–underground perforated pipe–shaft–water saving irrigation system can further expand the scope and amount of groundwater recharge and prevent system clogging through three measures. Finally, an adaptability zoning evaluation system of water spreading has been established in Liaocheng City of Shandong Province based on the following five factors: groundwater depth, thickness of fine sand, specific yield, irrigation return flow, and groundwater extraction intensity. The results show that MAR is more adaptable to the western region than to the eastern and central regions.

Keywords: types of MAR for irrigation; Yellow River Irrigation District; adaptability zoning evaluation

1. Introduction

Managed aquifer recharge (MAR) is the intentional recharge of water to aquifers for subsequent recovery or environmental benefits, and MAR projects must achieve effective aquifer recharge under different terrains, hydrogeological conditions, water sources, and water demand characteristics [1]. As an effective water resources management measure, MAR has been widely used in many countries, especially in semi-arid and arid areas [2–4]. The North China Plain is the main grain-producing area in China. The average annual precipitation ranges from 500 to 900 mm, and the majority of the annual precipitation occurs from July to August. Agricultural irrigation requires a large amount of water and mainly relies on groundwater extraction, resulting in a large area of groundwater overexploitation. In the 1970s and 1980s, a variety of small-scale MAR projects were developed to recharge shallow groundwater, and they played an important role in ensuring stable and high yields of crops [5]. However, these small-scale MAR projects lack scientific review and evaluation.

In this study, 10 types of specific agricultural MAR in the North China Plain are summarized. Two methods have been developed in the Yellow River Irrigation District to apply and develop water spreading methods: a well–canal combination mode and an open channel-underground perforated pipe-shaft-water saving irrigation system. These two modes are both effective for sustaining a high grain yield and restoring groundwater overexploitation. Based on the special hydrogeological features of phreatic water (0–60 m) in the Yellow River Irrigation District of Shandong Province, a main aquifer of fine sand is determined to be suitable for water spreading instead of recharge through a well. However, the application of water spreading is influenced by many factors. Which area is suitable for it? Liaocheng City was selected to establish an adaptability zoning evaluation system for the MAR of water spreading, and five factors were evaluated: groundwater depth, thickness of fine sand, specific yield, irrigation return flow, and groundwater extraction intensity. Through the combination of GIS and the DRASTIC model, this adaptability zoning evaluation provides a scientific basis for the sustainable development of the MAR projects in the Yellow River Irrigation District [6–8].

2. Progress of MAR

2.1. MAR Application of Agriculture in the North China Plain

China has made many achievements in aquifer recharging, especially in industrial and agricultural production, urban water supply, and ecological protection [9–11]. The characteristics of MAR as a component of irrigation and drainage systems are as follows: small scale, short service life, low investment, fixed irrigation water sources, and crop planting structures. Considering these characteristics and the relationship between surface water and shallow aquifers, MAR is divided into three types: water spreading, well recharging, and a combination of both. MAR can be further divided into 10 types according to the specific farmland water conservancy project (Figure 1).

2.1.1. Water Spreading

The term "water spreading" refers to the release of water over the ground surface to increase the quantity of water infiltrating into the ground and percolating to the water table. Its characteristics are large influence areas, low investment, and high efficiency. It is divided into three types: field infiltration, infiltration pond, and infiltration ditch.

(a) Field infiltration: Field infiltration is achieved by check irrigation, flood irrigation, and large-scale winter irrigation, and it has the characteristics of steady water distribution and a wide range of influence. The amount of recharge is related to the irrigation quota and groundwater depth. ① Check irrigation: Water is diverted into fields with borders, and the thickness of the water layer is approximately 0.33 m. This method is carried out in winter wheat fields and white stubble land. Especially in coastal saline–alkali areas this method can be used to wash out soil salinity, dilute groundwater quality, and replenish groundwater. ② Flood irrigation: After water is injected into the recharge area through ditches, it is controlled by earthen embankments to make the water overflow. This method is suitable for flood plains with low slopes, old channels, sandy wastelands, woodlands, and orchards. ③ Winter irrigation with a large irrigation quota: In areas without check irrigation or flood irrigation conditions, large-quota winter irrigation can be used to recharge groundwater.

(b) Infiltration pond: Surface water is allowed to go through the unsaturated zone into the aquifer through natural ponds that are renovated and connected to conveyance canals. The characteristics of these ponds are low land occupation, a large amount of recharging, and a limited range of influence.

(c) Infiltration ditch: Infiltration ditches are the major method of recharging aquifers in North China. This method requires the selection of a reasonable ditch spacing. The infiltration water first forms a water peak along the ditch, then spreads from the ditch to both sides before reaching the center of the ditch.

(d) Ditch–underground permeable cement pipe-pond system: This system can be used in places with low permeability and a shortage of land. The diameter of the underground permeable cement pipe is more than 30 cm. A pond for desilting and lifting irrigation is built at a location 100 m away from the pipe head. When the pipe is buried as shallow as possible, it can also infiltrate the tilth topsoil and play an important role in irrigation.

(e) Tunnel–well: This system consists of four parts: river, tunnels, wells, and artificial water lifting facilities. A tunnel with a width of 0.8 m and a height of 1 m is connected to the river and located in the clay layer 7–8 m below the ground. Each tunnel is several kilometers long, and there is a well for water lifting every 30 m. During the flood season this system can increase the amount of recharging and raise the groundwater table rapidly, with a better infiltration effect than those of other facilities.

2.1.2. Well Recharging

(f) Seepage well: Water can flow directly into the aquifer through shallow wells. This system, which is suitable for areas with deeply buried gravel aquifers, is faster than water spreading and has low land occupation, although clogging frequently occurs.

(g) Shaft well: The vertical shaft is used to expose the soil layer so that the water is admitted directly into the sandy gravel aquifer with high permeability. However, the bottom silt needs to be cleared regularly. When combined with (b) infiltration pond, better efficiency is observed.

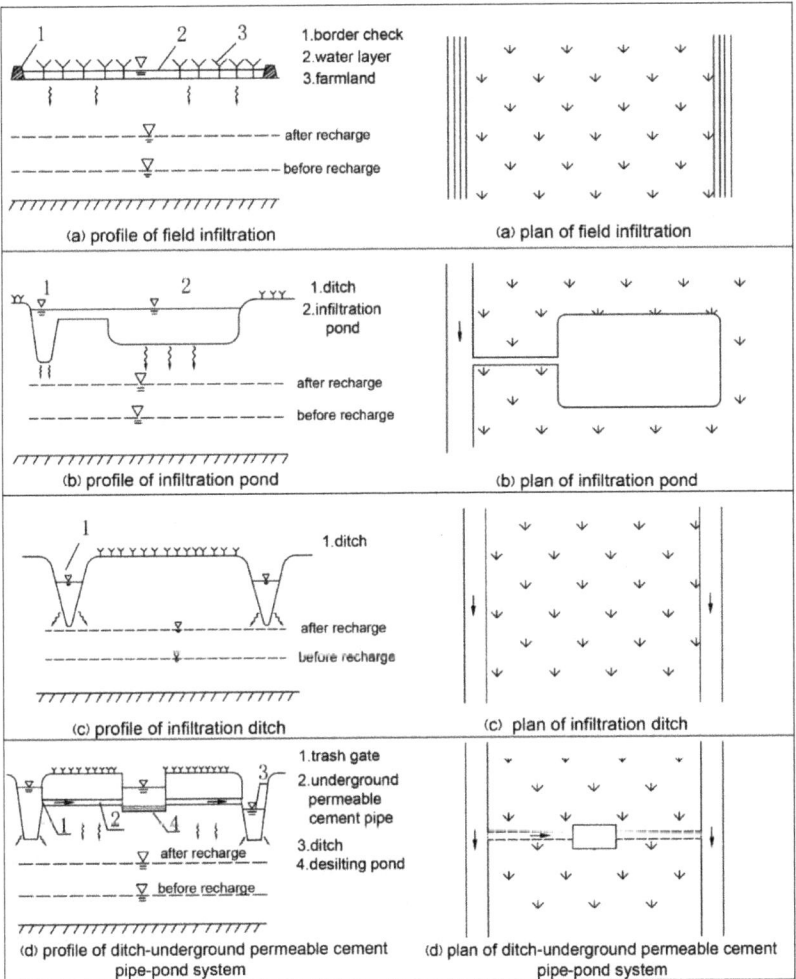

Figure 1. Cont.

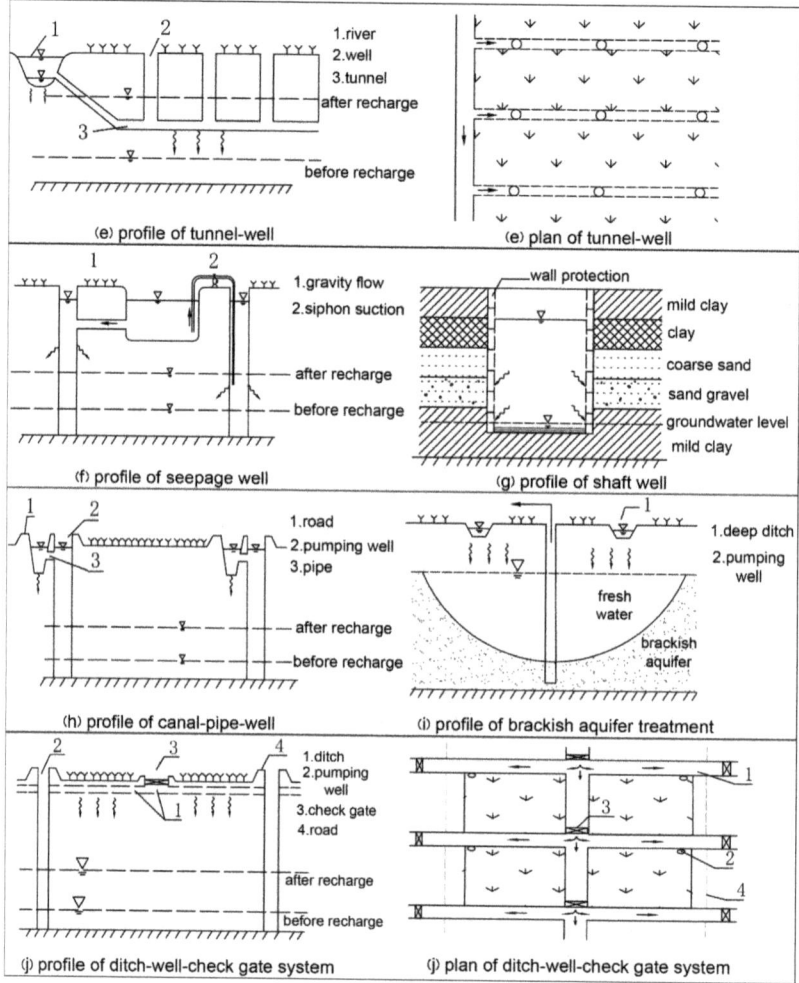

Figure 1. Specific types of MAR in agricultural irrigation.

2.1.3. Combination Methods

(h) Canal–pipe–well: The pumping well, which is connected to the canal by a pipe, lifts groundwater when the canal is waterless. The slope of the canal is gentle in the design, and the intake is more than 0.5 m above the bottom of the canal to reduce sedimentation.

(i) Brackish aquifer treatment: Brackish water is pumped through wells, and fresh water is diverted, stored, and infiltrated by deep ditches. The ditches can also be used for irrigation. Soil salinization can be controlled, and brackish aquifer can be desalinated.

(j) Ditch–well–check gate system: In plain areas, a number of check gates are added to ditches based on border checking, reticulated canal systems, and driven well groups. This system represents a farmland water conservancy system of diversion, storage, infiltration, water-saving, irrigation, and drainage.

2.2. Cases

2.2.1. Well-Canal Combination in the Yellow River Irrigation District

Overview of the Yellow River Irrigation District in Shandong Province

The Yellow River has a drainage area of 795,000 km^2 and a length of 5464 km, making it the second-longest river in China. The Lower Yellow River flows through six cities in Shandong Province and ends at the Bohai Sea in Lijin County, covering about one-third of Shandong Province [12]. The area has a warm temperate monsoon climate. The average annual rainfall is 606 mm, and the average annual evaporation is 1300 mm. Rainfall is mainly concentrated in the summer; thus, the area presents characteristics of spring and autumn droughts and summer floods. The Yellow River flood plain is the main part of the North China Plain. Due to the long-term geological movements and the sedimentary rhythm in the downstream area of the Yellow River, the formation is deep and the aquifer media are dominated by fine sand with a hydraulic conductivity of 0.3–2.5 m/day [13]. The area can be divided into three aquifers from top to bottom. The most important water supply for agriculture is phreatic water, with a buried depth of 0–60 m, a roof depth of 10–25 m, a thickness of 10–25 m, and a well yield of 6–10 m^3/h·m. The lithology of the unsaturated zone and the characteristics of the shallow aquifer provide favorable conditions for recharging aquifers and exploiting shallow groundwater. The Yellow River Irrigation District of Shandong province is shown in Figure 2. The designed irrigation area of the Yellow River of Shandong province is 16,667 km^2. The average annual water diversion for agriculture in the Yellow River Irrigation District is 4.5 billion m^3 (1995–2015). The main crops are winter wheat, corn, and cotton, with a multiple cropping index of 1.7. With the influence of inadequate local water resources, unreliable Yellow River water available in the low Yellow River flow years, and uneven distribution of water between upstream and downstream of the Yellow River Irrigation District, the shallow groundwater overexploitation area has reached 4500 km^2. In short, the Yellow River Irrigation District is facing drought, flooding, and sediment problems [14].

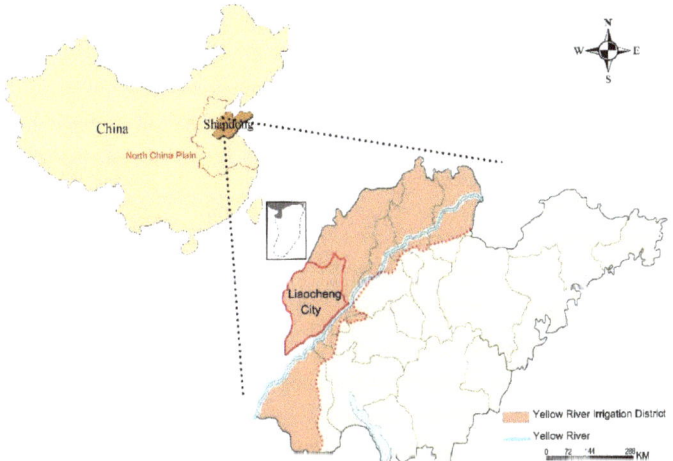

Figure 2. Yellow River Irrigation District of Shandong Province.

The Yellow River Irrigation District of Shandong Province is 16,667 km^2, and only 12% of this area uses gravity irrigation. Therefore, three irrigation-drainage modes have been established. (1) Irrigation and drainage systems are separated and use gravity flow. The canals of the irrigation system (including the main, branch, lateral, and sublateral canals) are above ground, while the corresponding ditches of the drainage system are underground. The drainage water enters the Bohai Sea through the river.

(2) Irrigation and drainage systems are separated, and water is pumped from the canal for irrigation and gravity drainage. The canals of the irrigation system and the ditches of the drainage system are underground. The ditches are deeper than the canals to ensure that the drainage water can flow into the river under the influence of gravity. (3) The well–canal combination, which is the most commonly used mode of irrigation and drainage, is a combination of irrigation and drainage systems.

Well-Canal Combination Mode

With the urbanization and industrialization of the Yellow River Irrigation District, the water distribution has shifted from agricultural irrigation to urban water supplies. Yellow River diversion occurs over a limited time; thus, although the Yellow River water is available, the flow is unreliable. Therefore, the well–canal combination mode is inevitably adopted to achieve a large-scale steady production increase in agriculture [15–17]. The well-canal combination mode is a double irrigation system, which has the advantages of recharging aquifers with river water through canal infiltration and irrigation return flow, and guaranteeing bumper harvests with wells (Figure 3). The canal has multiple functions of water delivery, storage, infiltration, irrigation by pumping, and drainage [18].

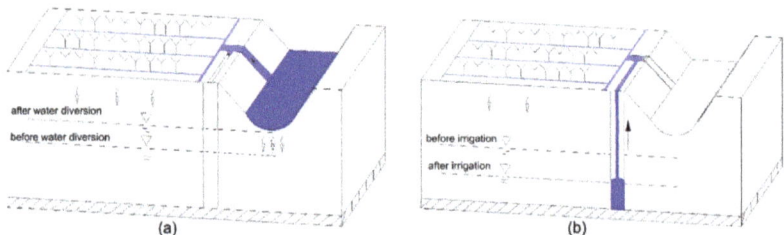

Figure 3. Well–canal combination mode (**a**) with Yellow River water diversion and (**b**) without Yellow River water diversion.

Historically, flood irrigation was widely used in the fields of the Yellow River Irrigation District; therefore, irrigation return flows became the main source of groundwater recharging. The irrigation efficiency and overall benefit of the Yellow River Irrigation District are low. At present, low-pressure pipeline irrigation has been widely used, resulting in a lower irrigation quota and irrigation return flow, a higher irrigation application efficiency, and lower groundwater pollution by soil leaching. The amount of groundwater recharged in this way is much less than that from flood irrigation. The method mainly depends on the leakage from canals after implementing the water saving measures, and does not pass through the plow layer and thus does not pollute the shallow groundwater. The wide use of the well–canal combination maintains the balance of diversion water and groundwater and has a good effect in regulating the groundwater table.

Figure 4 shows the water balance of the Yellow River Irrigation District with a 50% probability of precipitation. When the probability of precipitation is 50%, the total water demand of the main crops is 900 mm and the precipitation is 606 mm, of which 182 mm is infiltrated into the ground. Therefore, the water deficiency is 294 mm. The 269 mm of Yellow River water diversion can fill the water deficiency. The irrigation efficiency of Yellow River Irrigation District is 0.6. The average annual available local water can only guarantee 67% of the irrigation water requirement, and the rest needs to be provided by the Yellow River, of which the field irrigation volume is 161 mm and accounts for 60% of the total amount of Yellow River Diversion, and the remaining 108 mm accounts for 40% of the total, which can be pumped by well to irrigate crops for harvesting from the groundwater recharged through canal system infiltration and irrigation return flow. The Yellow River water diversion has played a key role in stable agricultural production and high yields. When the irrigation water ratio of well to canal is 1:0.93, the irrigation water in this area basically maintains a balance of supply and demand. In addition, other research has shown that when the irrigation water ratio of well to canal is at 1:0.78 in the People's Victory Canal Irrigation District of neighboring Henan Province, it can basically maintain the balance of supply and demand [19]. If water balance is reached in the water deficient area by Yellow River water diversion, this mode is influenced by irrigation efficiency. The irrigation water ratio of well to canal would decrease with increases in irrigation efficiency, resulting in a decreased capacity for groundwater regulation. So, the question is how to enlarge the groundwater recharge amount through Yellow River water diversion over a limited time [20–22].

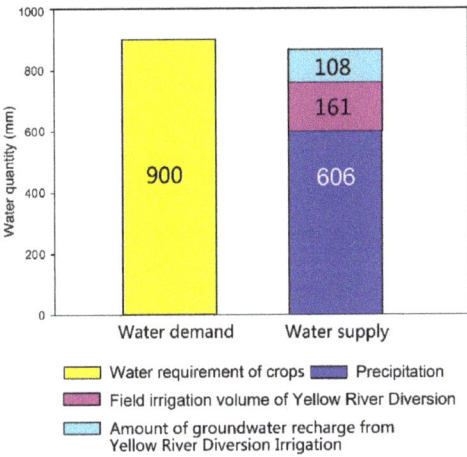

Figure 4. Water balance of the Yellow River Irrigation District at 50% probability of precipitation.

2.2.2. Open Channel–Underground Perforated Pipe–Shaft–Water Saving Irrigation System

With the large-scale promotion of water saving practices in the Yellow River Irrigation District, the flood irrigation method is no longer suitable. However, the canal system has already been formed. If new canals are dug, then they will occupy a large amount of cultivated land. Therefore, considering the existing irrigation-drainage system, a method of increasing the recharge rate of shallow groundwater overexploitation areas must be identified.

An open channel–underground perforated pipe–shaft–water saving irrigation system with high efficiency and ecology has been developed based on infiltration ditches, the combined tunnel–well mode, modern materials, and filtration technology (Figure 5). The underground perforated pipe is used to further increase the recharge rate and influence area, and to restore the groundwater overexploitation zone. The system consists of the Yellow River water source, open channels, prefilters, perforated pipes, dredging shafts, crops, irrigation systems, ambient groundwater systems, operation and monitoring facilities, equipment (e.g., electromagnetic flowmeters), and so on.

Three measures to prevent sediment blockage are set up in the recharging system. (1) A prefilter tank is provided at the canal head. (2) A 20-cm sand layer is placed around the pipe in the soil of the designed section, where the geotextile can prevent the outer sediments from entering into the pipe. The pipe has a certain slope and a dual function of water washing and seepage. (3) Shafts with well filters in the borehole are at the end of the recharge system, and they can not only receive the sediments but also contribute to the water seepage. The specific parameters of the recharge system are as follows: the bottom area of the prefilter tank is 4 m^2, and the underground perforated pipe is composed of plastic blind ditches with a diameter of 30 cm, a slope of 1/500, a pipe length of 200 m, and a shaft depth of 10 m. The parameters can be adjusted according to different regions and local conditions.

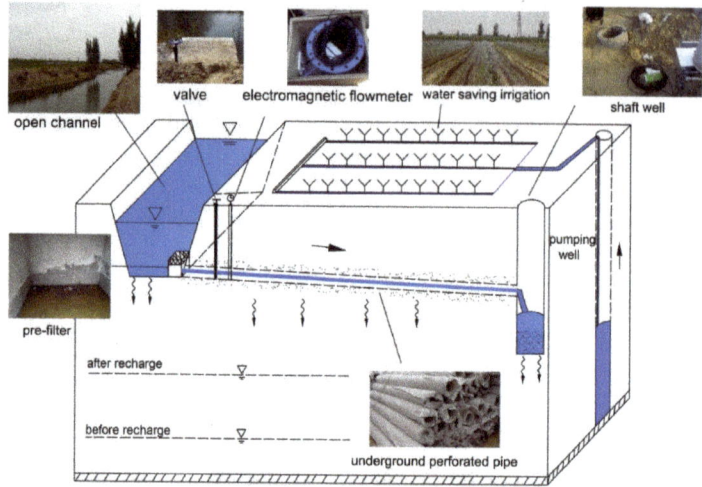

Figure 5. Open channel–underground perforated pipe–shaft–water saving irrigation system.

In the practical project, the system was conducted at a recharge area of 54,000 m² including three underground perforated pipes as subsystem and buried pipes spaced 90 m apart with 200 m lengths. According to the calculation by an analytical method, the project contributed a groundwater recharge amount of 52,310 m³, covering 60.7% of the total replenished water. The numerical simulation results showed that the groundwater table under the pipe rose 8.5 m throughout the simulation period. In the seven-day simulation period, the total infiltration amount of the open channel was 33,861 m³, and the total infiltration amount of the pipes was 56,224 m³, which accounted for 62.4% of the total infiltration amount [23]. In summary, this project has the advantages of no energy consumption, no land occupation, managed groundwater recharging, irrigation drainage, and water logging control, and it plays an important role in the recovery of groundwater funneling.

3. Adaptability Zoning Evaluation

3.1. Study Area

Liaocheng City is part of the North China Plain and is located in the northwestern part of Shandong Province and the western part of the Yellow River Basin, with good water diversion conditions (Figure 6). The terrain is flat, and the total area is 8715 km². Liaocheng City is a large agricultural city with an average annual precipitation of 560 mm, a cultivated land area of 6353 km², and an effective irrigation area of 4947 km². The groundwater depth of shallow aquifers as irrigation water sources is 0–60 m. Most rivers are approximately parallel and flow from the southwest to the northeast. The rivers have a crossing and repeated sedimentary structure in the vertical direction. In the horizontal direction, the land can be divided into an accumulation area, a flood plain alternating area, and an interstream area, wherein the aquifer media become finer, the thickness decreases, and the water abundance weakens. The aquifer is composed of fine sand with a thickness of 10–25 m, fine sand with a thickness of 5–10 m, and silt with a thickness of less than 5 m. The main irrigation water sources are groundwater, water diverted from the Yellow River, and a small amount of surface water. Over the past 60 years, the number of agricultural irrigation wells has increased rapidly to guarantee bumper harvests. Groundwater continues to be overexploited, and the groundwater table is declining in part of this area. Therefore, to ensure that the combined use of surface water and groundwater sustain crop yields and also groundwater storage levels, the types and densities of recharge mechanisms need to be adapted to the local intensity of water use and hydrogeological conditions. For these, zones were identified where water-spreading mechanisms were considered appropriate.

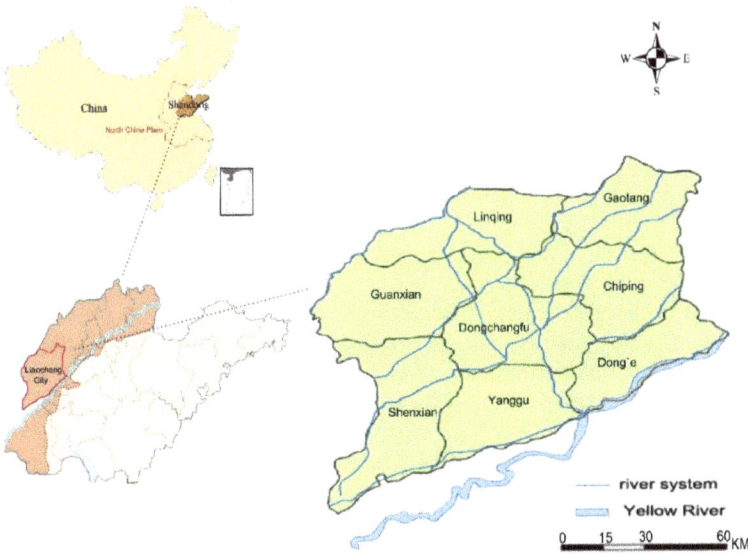

Figure 6. Location of Liaocheng City.

The constraint conditions for an adaptability zoning evaluation of MAR are as follows: (1) phreatic aquifers; (2) agricultural irrigation; (3) a Yellow River water recharge water source and a small amount of surface runoff during the flood season; and (4) the MAR project aim of a water spreading method, including indirect recharging methods such as field infiltration, infiltration ponds, ditches, and so on.

3.2. Evaluation Factors

An adaptability zoning evaluation means selecting a suitable recharge area to water spreading for irrigation, a process that is influenced by many factors. Based on the hydrogeological conditions of groundwater extraction, the evaluation focuses on the water-bearing characteristics of the vertical vadose zone during recharge. There are five factors for consideration: groundwater depth, thickness of the fine sand, specific yield, irrigation return flow, and groundwater extraction intensity.

Evaluation of groundwater vulnerability and recharge adaptability have similarities and differences in surface water infiltration. The process of the DRASTIC model can be used as a reference. The rating range for each evaluation factor is 1–5 points, for which the higher points correspond to a better adaptability of MAR for water spreading (Table 1). The overall score is calculated by equal weight.

Table 1. Grading of evaluating factors.

Factor	Level				
	1	2	3	4	5
groundwater depth (m)	<2	2~6	6~12	12~18	>18
thickness of fine sand (m)	<5	5~10	10~15	15~20	>20
specific yield	<0.05	0.05~0.07	0.07~0.09	0.09~0.11	>0.11
irrigation return flow	<0.01	0.01~0.1	0.1~0.2	0.2~0.35	>0.35
groundwater extraction intensity (10^4 m^3/a·km^2)	<9.1	9~11	11~14.6	14.6~18.3	>18.3

Because the five factors have different effects on groundwater recharge, they should be classified. The factors affecting the water storage capacity are groundwater depth, thickness of the fine sand, specific yield, and irrigation return flow (after decades of field experiments, abundant county-level data are available). The groundwater extraction intensity is equivalent to the user demand for recharge water. The lithology of the aquifer (0–60 m) is mainly fine sand, and the permeability coefficient is similar to that for a typical sand material. Therefore, the thickness of the fine sand can effectively represent the basic characteristics of Liaocheng City. Most data can be obtained from geologic reports and relevant departments, which are frequently used in groundwater resource assessments in the study area.

3.2.1. Groundwater Depth

Groundwater depth refers to the status of groundwater overexploitation. Statistical data for 2016 indicated that there was an area of 898 km^2 with a groundwater depth greater than 18 m, and this accounted for 10.3% of the total area. This area has a large demand for recharge, and the highest evaluation score of 5 points was observed when the aquifer had a sufficient recharge time and amount of water. The area with a groundwater depth between 12 and 18 m was 1043 km^2, which accounted for 12% of the total area, and the evaluation score was 4 points. The area with a groundwater depth between 6 and 12 m was 1240 km^2, which accounted for 14.2% of the total area, and the evaluation score was 3 points. The area with a groundwater depth between 2 and 6 m was 5517 km^2, which accounted for 63.3% of the total area, and the evaluation score was 2 points. The area with a groundwater depth less than 2 m was 16 km^2, which accounted for 0.2% of the total area. Such areas have a small demand for recharge, and the score of this area was 1 point (Figure 7).

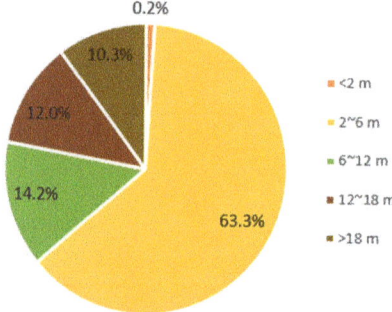

Figure 7. Corresponding areas of different groundwater depths.

3.2.2. Thickness of Fine Sand

Thickness of fine sand refers to the water abundance characteristics of shallow aquifers. Values are the cumulative thickness of fine sand (0–60 m). The existing data were obtained by a large number of geological surveys conducted by the relevant departments. The area with a thickness of fine sand less than 5 m accounted for 15.1% of the total area. The water storage space was not large, and the score is the smallest at one point. The areas with thicknesses of 5–10 m and 10–15 m accounted for 11.4% and 17.3% of the total area and presented scores of 2 and 3 points, respectively. Areas with thicknesses of 15–20 m and greater than 20 m accounted for 38.4% and 17.7% of the total area and presented scores of 4 and 5 points, respectively.

3.2.3. Specific Yield

Specific yield (μ) refers to the storage property of the formation (0–60m). A larger value of μ corresponds to a better water storage capacity. The corresponding values from 0.05 to 0.11 are based on the research results of the relevant departments. The specific yield can be obtained by the equation

$$\mu = \frac{\alpha \times P}{\Delta h}$$

where α is the recharge coefficient of precipitation, P is the precipitation (mm), and Δh is the change in water level (mm).

3.2.4. Irrigation Return Flow

Because the irrigation water quota is larger than individual rainfall under the same planting structure, groundwater depth, and uniform water distribution on farmland, irrigation return flow can better represent the hydraulic conductivity of unsaturated zones as the water spreading of MAR than with the recharge coefficient of precipitation.

$$\beta = \frac{\Delta h \times \mu}{Q}$$

where β is the irrigation return flow, Q is the amount of irrigation water (mm), Δh is the change in water level (mm), and μ is the specific yield of the area.

According to the above method, the relevant departments obtained the irrigation return flow through multiple irrigation experiments with an empirical value of 0.01-0.35. The larger the β is, the more permeable the unsaturated zone is.

3.2.5. Groundwater Extraction Intensity

Groundwater extraction intensity not only refers to water demand but also to impact on a groundwater system. The empirical values of groundwater extraction intensity for irrigation in the study area were 9.1~18.3 × $10^4 m^3/a \cdot km^2$. Groundwater extraction intensity was equal to the ratio of extraction volume to irrigation area.

3.3. Results

The maps of the five factors are as follows: (a) groundwater depth, (b) thickness of fine sand, (c) specific yield, (d) irrigation return flow, and (e) groundwater extraction intensity (Figure 8). The map is divided into 1~5 points based on the hydrogeological data.

Based on the scores of the five evaluation factors, ArcGIS (10.2, Esri, Beijing, China) was used to evaluate the applicability in the study area. Through spatial analyst of ArcGIS, the evaluation results were divided into five levels by their natural breaks (jenks): (1) unsuitable area, (2) small potential area, (3) general potential area, (4) medium potential area, and (5) high potential area. The zoning map for the adaptability zoning evaluation is obtained by ArcGIS in accordance with the spatial distribution of the recharge potential (Figure 9).

The results show that the western part of Liaocheng City is a suitable area for MAR. The overall score of the western region was higher because of the greater groundwater depth, thicker sand layer, and higher groundwater exploitation. Especially in Guanxian County, the groundwater depth was approximately 20 m, the thickness of the fine sand layer was the thickest, and the specific yield was the highest. This area is the high-potential area, while the medium-potential areas were near Linqing and Shenxian counties. The eastern part of Liaocheng City had a groundwater depth of 2 to 6 m. The thickness of the fine sand layer was relatively large, but the urgency of recharge was small. Therefore, the eastern part mainly had areas with less potential or that were unsuitable.

The main existing MAR project is the well–canal combination mode in Liaocheng City. The existing MAR project needs to be strengthened in the western region via water spreading methods, in which areas (4) and (5) are located at the end of the irrigation district of Liaocheng City. A new mode of the open channel–underground perforated pipe–shaft–water saving irrigation system should be extended in order to increase the groundwater recharge amount and expand the scope of groundwater recharge during the limited Yellow River water diversion period.

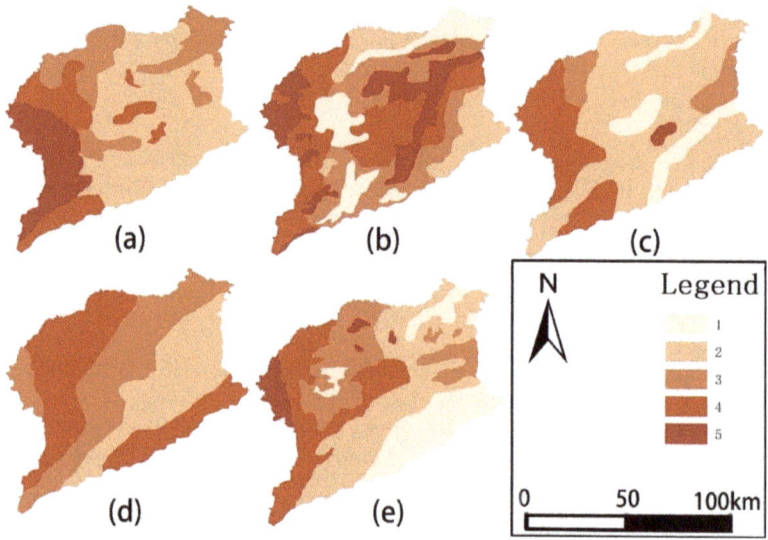

Figure 8. Maps of the five factors. (**a**) groundwater depth, (**b**) thickness of fine sand, (**c**) specific yield, (**d**) irrigation return flow, and (**e**) groundwater extraction intensity.

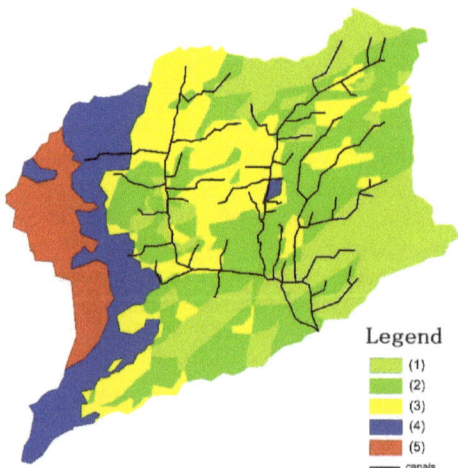

Figure 9. Adaptability zoning map of recharge and existing canal system in the Yellow River Irrigation District of Liaocheng City.

4. Discussion and Conclusions

In the 1970s and 1980s, various small-scale MAR projects were implemented to increase the amount of shallow groundwater in the North China Plain. Considering these characteristics and the relationship between surface water and shallow aquifers, MAR is divided into the three types: water spreading, well recharging, and a combination of both. MAR can be further divided into 10 forms according to the specific farmland water conservancy project (field infiltration, infiltration pond, infiltration ditch, ditch–underground permeable cement pipe–pond systems, tunnel–wells, seepage wells, shaft wells, canal–pipe–wells, brackish aquifer treatment, and ditch–well–check gate systems). These projects guarantee high grain yields and maintain the balance between recharging and extraction.

The Yellow River Irrigation District of Shandong Province is located in the lower reaches of the Yellow River. The effective irrigation area is mainly based on well irrigation. With the increase of crops, the water demand for agricultural irrigation has increased rapidly. Local groundwater can only provide half of the irrigation water, and Yellow River water diversion is performed to supplement irrigation. Large-scale overexploitation of shallow groundwater has occurred in some areas. The well–canal combination mode has been widely used in the Yellow River Irrigation District of Shandong Province for nearly 40 years. Because the well–canal mode has many advantages, it is applicable to the downstream, middle-stream, or upstream areas of the Yellow River Irrigation District. During water diversion, the water of the Yellow River is diverted by deep canals and relies on pumping irrigation and gravity drainage. As irrigation proceeds by pumping from canals, part of the water in the canals is recharged to the groundwater, which improves irrigation conditions. Without water diversion, the water demand can be satisfied by well irrigation. This mode has the advantage of recharging aquifers with river water and guaranteeing bumper harvests with wells. It also not only maintains high crop yields and basically guarantees the balance of exploitation and the recharging of shallow groundwater, but it solves the problem of aquifers being clogged by sediment from the Yellow River diversion because silt sediment at the canal head and sediment in the canal are dredged.

Based on field infiltration, infiltration ditches, and infiltration ponds, a new open channel–underground perforated pipe–shaft–water saving irrigation system was developed. The new system further expands the recharge scope and replenishment, and has three anti-blocking measures. The sustainable development of agriculture in the North China Plain is ensured by implementing the well–canal combination mode and adopting shallow groundwater recharge as the main line. The method of integrating MAR into agricultural facilities to form a farmland water conservancy system of water diversion, storage, infiltration, water savings, irrigation, and drainage is proposed to achieve the goal of comprehensively controlling droughts, floods, and salinization.

Liaocheng City was selected as the study area because of the distribution of aquifers, permeability of the unsaturated zone, and groundwater extraction intensity in the Yellow River Irrigation District of Shandong Province. An adaptability zoning evaluation system for water spreading was established based on the common modes of water spreading. Five factors were selected to reflect unsaturated zones and groundwater extraction: groundwater depth, thickness of fine sand, specific yield, irrigation return flow, and groundwater extraction intensity. The results show that MAR projects are adaptable to the western region and can resolve agricultural irrigation problems. The eastern and central regions have high groundwater tables, better diversion conditions for the Yellow River, and superior aquifer water storage capacities. However, these areas are not suitable for MAR projects due to their low groundwater extraction intensity. Thus, water diversion from the Yellow River and groundwater exploitation should be maintained in balance.

Author Contributions: This paper was composed by collaboration among all authors. Supervision, S.Q., Y.Z. and W.L.; Writing—original draft, S.L.; Writing—review & editing, W.W. All authors have read and agreed to the published version of the manuscript.

Funding: This study was supported by the Shandong Provincial Key Research and Development Project (2017GSF17121) and the Danish Development Agency (DANIDA) coordinated by the DANIDA Fellowship Center (DFC) through grant No. 17-M08-GEU.

Acknowledgments: The authors would like to acknowledge the editor and three anonymous reviewers for their valuable comments, which have greatly improved this paper.

Conflicts of Interest: The authors declare no conflict of interest.

References

1. Dillon, P.; Stuyfzand, P.; Grischek, T.; Lluria, M.; Pyne, R.D.G.; Jain, R.C. Sixty years of global progress in managed aquifer recharge. *Hydrogeol. J.* **2019**, *27*, 1–30. [CrossRef]
2. Changming, L.; Jingjie, Y.; Kendy, E. Groundwater Exploitation and Its Impact on the Environment in the North China Plain. *Water Int.* **2001**, *26*, 265–272. [CrossRef]
3. Ahmed, A.A. Using Generic and Pesticide DRASTIC GIS-based models for vulnerability assessment of the Quaternary aquifer at Sohag, Egypt. *Hydrogeol. J.* **2009**, *17*, 1203–1217. [CrossRef]
4. Saidi, S.; Bouri, S.; Dhia, H.B. Groundwater vulnerability and risk mapping of the hajeb-jelma aquifer (central tunisia) using a gis-based drastic model. *Environ. Earth Sci.* **2010**, *59*, 1579–1588. [CrossRef]
5. Hamza, S.M.; Ahsan, A.; Imteaz, M.A.; Rahman, A.; Mohammad, T.A.; Ghazali, A.H. Accomplishment and subjectivity of GIS-based DRASTIC groundwater vulnerability assessment method: A review. *Environ. Earth Sci.* **2015**, *73*, 3063–3076. [CrossRef]
6. Wang, W.P.; Dillon, P.; Vanderzalm, J. *New Progress in Aquifer Recharge Management between China and Australia*; The Yellow River Water Conservancy Press: Jinan, China, 2009; pp. 79–83.
7. Dillon, P. Future management of aquifer recharge. *Hydrogeol. J.* **2005**, *13*, 313–316. [CrossRef]
8. Asano, T.; Wassermann, K.L. Groundwater recharge operations in California. *J. Am. Water Works Assoc.* **1980**, *72*, 380–385. [CrossRef]
9. Cao, G.; Scanlon, B.R.; Han, D.; Zheng, C. Impacts of thickening unsaturated zone on groundwater recharge in the North China Plain. *J. Hydrol.* **2016**, *537*, 260–270. [CrossRef]
10. Min, L.; Shen, Y.; Pei, H. Estimating groundwater recharge using deep vadose zone data under typical irrigated cropland in the piedmont region of the North China Plain. *J. Hydrol.* **2015**, *527*, 305–315. [CrossRef]
11. Huang, T.; Pang, Z. Groundwater Recharge and Dynamics in Northern China: Implications for Sustainable Utilization of Groundwater. *Procedia Earth Planet. Sci.* **2013**, *7*, 369–372. [CrossRef]
12. Wang, Z.; Liu, C. Two-thousand years of debates and practices of Yellow River training strategies. *INT J. Sediment Res.* **2019**, *34*, 73–83. [CrossRef]
13. Li, X.; Zhong, D.; Zhang, Y.J.; Wang, Y.; Wang, Y.; Zhang, H. Wide river or narrow river: Future river training strategy for Lower Yellow River under global change. *INT J. Sediment Res.* **2018**, *33*, 271–284. [CrossRef]
14. Zhang, K.; Xie, X.; Zhu, B.; Meng, S.; Yao, Y. Unexpected groundwater recovery with decreasing agricultural irrigation in the Yellow River Basin. *Agric. Water Manag.* **2019**, *213*, 858–867. [CrossRef]
15. Xu, X.; Huang, G.H.; Qu, Z.Y. Integrating MODFLOW and GIS technologies for assessing impacts of irrigation management and groundwater use in the Hetao Irrigation District, Yellow River basin. *Sci. China* **2009**, *52*, 3257–3263. [CrossRef]
16. Sun, H.Y.; Wang, S.F.; Hao, X.M. An Improved Analytic Hierarchy Process Method for the evaluation of agricultural water management in irrigation districts of north China. *Agric. Water Manag.* **2017**, *179*, 324–337. [CrossRef]
17. Mao, W.; Yang, J.; Zhu, Y.; Ye, M.; Wu, J. Loosely coupled SaltMod for simulating groundwater and salt dynamics under well–canal conjunctive irrigation in semi-arid areas. *Agric. Water Manag.* **2017**, *192*, 209–220. [CrossRef]
18. Zhou, Z.M.; Zhou, K.; Wang, X.C. Influence of Irrigation Water-Saving on Groundwater Table in the Downstream Irrigation Districts of Yellow River. *Adv. Intell. Soft Comput.* **2012**, *138*, 79–83. [CrossRef]
19. Zhang, J.X.; Qi, X.B.; Magzum, M. Appropriate Well–canal Irrigation Proportion in Irrigation District Based on MODFLOW and GIS. *J. Irrig. Drain.* **2017**, *36*, 58–63. (In Chinese) [CrossRef]
20. Chen, L.L.; Sun, Y.; Yang, Y.G. Benefit analysis of well canal combined irrigation in Yellow River Diversion Irrigation Area of Henan Province. *Yellow River* **2011**, *33*, 92–93. (In Chinese) [CrossRef]
21. Qi, X.B.; Fan, X.Y.; Wang, J.L. Optimal allocation of water resources in well–canal combined irrigated area for high efficient utilization of water. *J. Hydraul. Eng.* **2004**, *10*, 119–124. (In Chinese) [CrossRef]

22. Liu, L.; Luo, Y.; He, C.; Lai, J.; Li, X. Roles of the combined irrigation, drainage, and storage of the canal network in improving water reuse in the irrigation districts along the lower Yellow River, China. *J. Hydorl.* **2010**, *391*, 157–174. [CrossRef]
23. Rong, Q.; Wang, W.; Qu, S.; Li, J.; Li, F.; Xu, Q. A MAR to address the water with high content of suspended solid with a case study in the Yellow River flood plain, China. *Agric. Water Manag.* **2017**, *182*, 165–175. [CrossRef]

© 2020 by the authors. Licensee MDPI, Basel, Switzerland. This article is an open access article distributed under the terms and conditions of the Creative Commons Attribution (CC BY) license (http://creativecommons.org/licenses/by/4.0/).

Article
Evaluation of MAR Methods for Semi-Arid, Cold Regions

Nasanbayar Narantsogt * and Ulf Mohrlok

Department of Civil Engineering, Geo and Environmental Sciences, KIT—Karlsruhe Institute of Technology, 76131 Karlsruhe, Germany; ulf.mohrlok@kit.edu
* Correspondence: ugeph@student.kit.edu or nnasan.4@gmail.com

Received: 22 August 2019; Accepted: 22 November 2019; Published: 2 December 2019

Abstract: Mongolia is a semi-arid, highly continental region with highly variable precipitation and river discharge. The groundwater aquifer located near Ulaanbaatar, the capital city of Mongolia, is the only one source for city water supply consumption, and it is important to ensure that groundwater is available now and in the future. The main watercourse near the capital city is the Tuul River, fed by precipitation in the Khentii Mountains. The semi-arid and cold environment shows high variability in precipitation and river discharge. However, due to absence of precipitation in winter and spring, the riverbed usually runs dry during these times of the year, and weather observations show that the dry period has been extending in recent years. However, in parallel with urban development, the extended groundwater aquifer has shown a clear decline, and the groundwater levels have dropped significantly. Therefore, a groundwater management system based on managed aquifer recharge is proposed, and a strategy to implement these measures in the Tuul River valley is presented in this paper. This strategy consists of the enhancement of natural recharge rates during the wet summer from the northern drainage canal, an additional increase in groundwater recharge through melting the ice storage in the dry period, as well as the construction of underground dams to accumulate groundwater and a surface water reservoir that releases a constant discharge in the outlet. To increase natural recharge rates of groundwater during the early dry period through the melting ice storage period, the MATLAB icing code, which was written for ice storage for limited and unlimited areas, was considered through finite element subsurface FLOW (FEFLOW) simulation scenarios as a water source in ice form on the surface. A study of the artificial permafrost of underground as an ice dam was processed in FEFLOW simulation scenarios for accumulating groundwater resources. The results of these artificial recharging methods were individually calculated, combined, and compared with the surface reservoir, which releases a constant discharge through the dam. In this paper, new ideas are presented involving managed aquifer recharge—MAR methods, and include application to aufeis, a mass of layered ice for groundwater recharge by melting. Additionally, the accumulation of groundwater using artificial permafrost is used as an underground dam. In addition, was considered recharging scenario only with constant release water amount from water reservoir also with all MAR methods together with reservoir combination.

Keywords: Ulaanbaatar; MAR; MATLAB; FEFLOW; artificial recharging scenarios

1. Introduction

The Tuul River, which flows through Mongolia's capital city Ulaanbaatar (UB), originates from the Khentii Mountains on the front side of the capital city. The water level of the Tuul River fluctuates according to an annual high-flow to low-flow cycle, with its average water flow of 26.6 m^3/s [1].

This river also recharges the Ulaanbaatar and upper aquifers that provide the city's water supply by 218 wells divided into nine groups of intake wells and 21 booster pump stations operated by a water supply agency. Daily domestic consumption fluctuates around 150–200 m^3/day and depending on the four seasons; see Figure 1. The water supply agency estimates that another 130 × 10^3 m^3/day of water is pumped from the aquifer in private wells by industries and individuals [2].

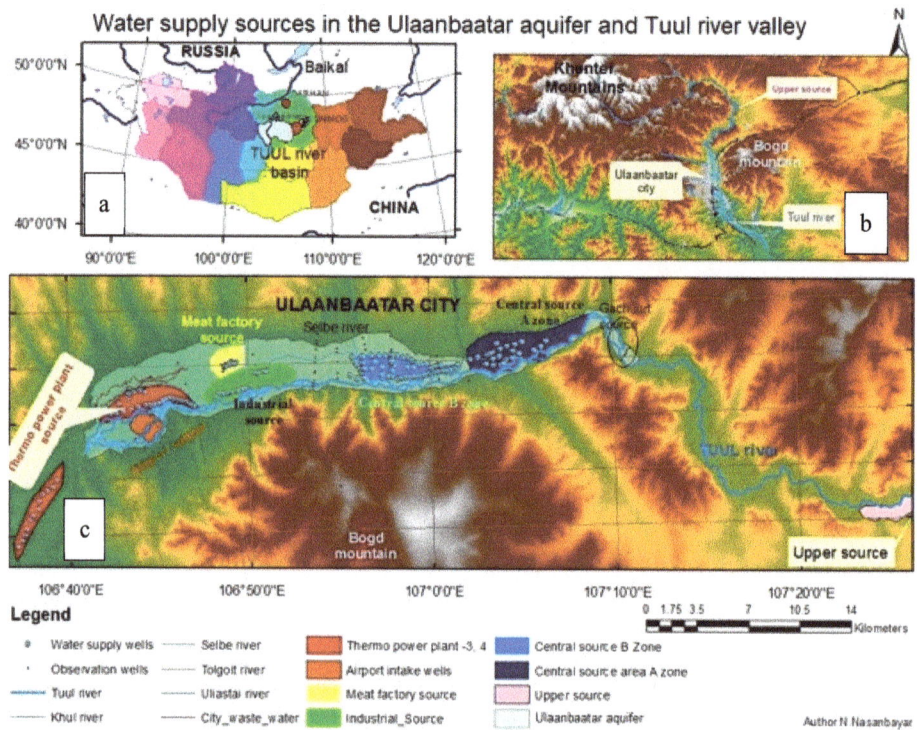

Figure 1. Groundwater aquifer south side of Ulaanbaatar city. (**a**). Hydrological drainage basins of Mongolia, (**b**). origin of Tuul River, (**c**). Ulaanbaatar aquifer and wells.

The joint meeting of the Mongolian and Russian mineral resources commissions estimated the useful capacity of groundwater resources of Ulaanbaatar aquifer at 264 × 10^3 m^3/day [3], without an upper source in the eastern side of the Ulaanbaatar aquifer (Figure 1), which pump groundwater around from 24 × 10^3 m^3/day to 48 × 10^3 m^3/day from aquifer [4].

As with natural recharge formation of groundwater resources, the main components are water entering in the aquifer as result of surface runoff loss and the capacitive reserves of the upper water-bearing layer. During the period from May to December, the aquifer takes recharge due to surface runoff loss. In this period, replenishment infiltrates from the surface water to capacitive reserves of the stored groundwater in the Ulaanbaatar aquifer [5].

In March and April, the groundwater table of water source aquifers decreases to minimum of 8 to 10 m from the surface, but it reaches maximum level of 2 m under the earth surface in the central area of the source in June, July, and August [6], see Figure 2. This table begins from a hydrological year where the wet season begins with melt water from snow and ice after a long, cold winter.

Water demand increases day by day with the development of industries and increasing population growth, but groundwater resources have decreased due to the climate change and excess water usage. In the last decade, during spring (April and May), the dry period the Tuul River flow continues longer than one month after ice breakup [7].

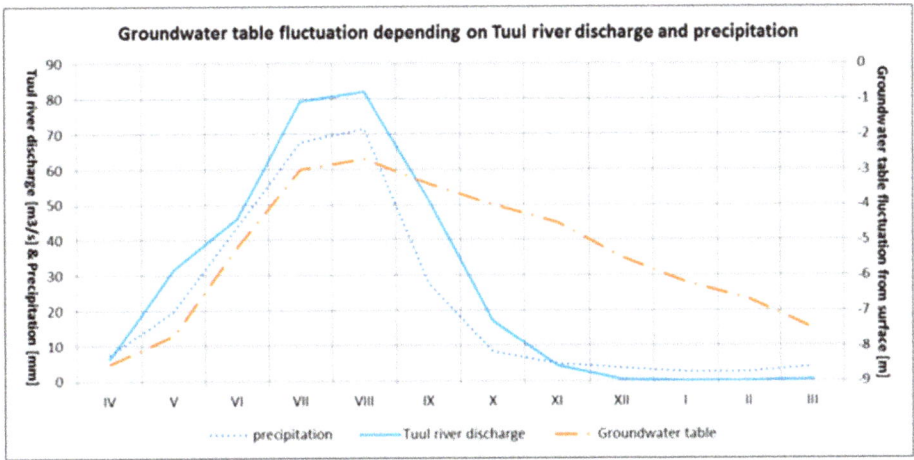

Figure 2. Table of groundwater fluctuation depending on Tuul River flow.

The following Table 1 shows a calculated balance of the outflow and inflow at the Tuul River valley and Ulaanbaatar aquifer, which is used for the water supply of the city. The Tuul River and tributaries discharge was taken as the inflow rate, precipitation, and evaporation volume averaged over many years. The average discharge amounts over the last ten years of Tuul River was taken as inflow including other inflow sources as surface water from the Selbe, Uliastai, and Khul Rivers they flow to the Tuul River through aquifer for water supply of Ulaanbaatar city.

Table 1. Ulaanbaatar city water exploration volume balance for water supply.

Months	IV	V	VI	VII	VIII	IX	X	XI	XII	I	II	III	Total
Balance (M·m³)	10.95	79.72	125.38	216.15	231.68	135.96	42.82	5.81	−4.23	−5.74	−5.56	−4.76	828.18
sum (M·m³)	848.47								−20.29				

The balance rate between surface water as inflow including precipitation and city consumption was estimated as following: from total inflow subtracted total pumped groundwater and added reused water from thermopower plants. see Table 1.

$$Q_{balance} = Q_{inflow} - Q_{pumped} + Q_{reused}$$

The above estimated calculation is based on the total water supply of the upper and central sources of Ulaanbaatar aquifers with the annual average flow rate of the Tuul River.

The water consumption of the whole aquifer of Ulaanbaatar, including domestic and thermal power plants, is shown above and is sorted by hydrologic year from April, when balance starts with a positive value [8].

The total amount of water consumption in Ulaanbaatar used for water supply in the dry, non-recharging season from surface water of Tuul River reaches about −20 × 10⁶ m³/year depending on the weather conditions and general precipitation in a given year. This balance between water consumption and recharging inflow from Tuul River was estimated with all pumped groundwater from all nine water sources, including an upper source.

The balance between incoming flow coming from the Tuul River to the aquifer and outcoming flow as an exploration rate by intake wells [9]. This means city consumes −20 × 10⁶ m³/year groundwater from aquifer without recharging from river. The balance estimation between groundwater income and pumped water for domestic water supply from the Ulaanbaatar aquifer source area without thermopower plants water supply consumption is -7.1×10^6 m^3/year water.

Therefore, there is an urgent need to solve this problem through the design of hydraulic structures for an underground dam, managed aquifer recharge, and the promotion of ice storage or the building of a surface reservoir dam that releases constant controlled outlet water.

In the next chapter, this problem is considered only for domestic consumption in the central source area of the Ulaanbaatar aquifer. A finite element subsurface FLOW (FEFLOW) simulation was taken in the upper and central source areas of the A zone, where there are only 23 wells for water supply; see Figure 3.

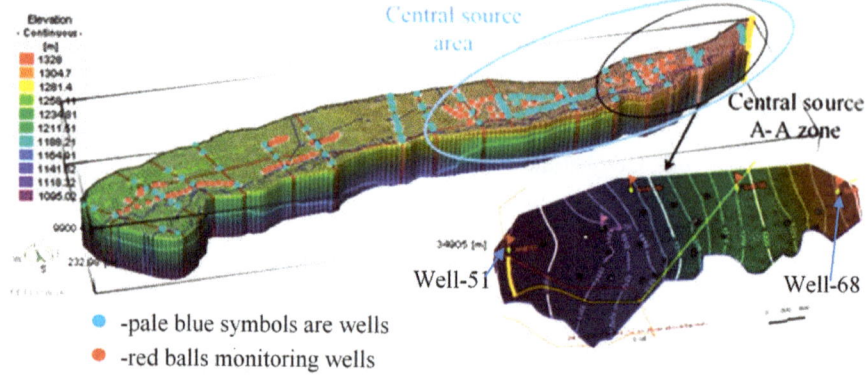

Figure 3. Ulaanbaatar aquifer, central source area, and the simulated A zone area.

The average date of the first ice formation on rivers is the third week of October. The freezing of the rivers starts from the end of October and lasts until the end of December. The ice cover lasts, on average, for 145 days. During the last 60 years, the annual mean of air temperature in Mongolia has increased by 1.66 °C, the winter temperature has increased by 3.61 °C, the spring–autumn temperature has increased by 1.4–1.5 °C, and the summer temperature has had no clear trend [10]. Temperature has rapidly increased in March, May, September, and November, and the ice regimes of the Mongolian rivers have therefore changed [2–4].

Ice phenology has shifted about 3–30 days in terms of freeze-up and break-up dates, and ice cover duration has shortened. Maximum ice thickness also decreased from the 1960s to 2000 [11].

A small river like Tuul is frozen to the bed for 2.5 months [3], and in mid-April, it has no ice cover with some places being dry bed without flow.

2. Materials and Methods

2.1. FEFLOW (Finite Element Subsurface FLOW and Transport System) Transit Model with Ice Scenarios

Nowadays, groundwater simulation and modeling are some of the main tools for groundwater aquifers [12], which visualize the situation and conditions of water in underground porous media for the protection of groundwater, as well as the restoration and development of aquifers. Groundwater level loggers collect more data, but because they cannot be manually controlled, only software modeling and processing provide an option that is both fast and accurate [13].

Every spring over the last decade, during the months of March and April, the Tuul River has dried out or has not flowed [14]. Therefore, we need to address on this problem by building complexes of hydraulic structures, establishing measures for flow control, and building artificial groundwater recharge systems like drainage or flooding areas near drinking water extraction wells. This study attempts to find suitable artificial groundwater recharging methods for the upper part of the central source groundwater aquifer, encompassing the water supply source area of the Tuul River valley inside Ulaanbaatar. The central source of drinking water supply system was established and put into operation in 1959. Drinking water is extracted from 93 deep well pumps with seven booster pumping stations; see Figure 4.

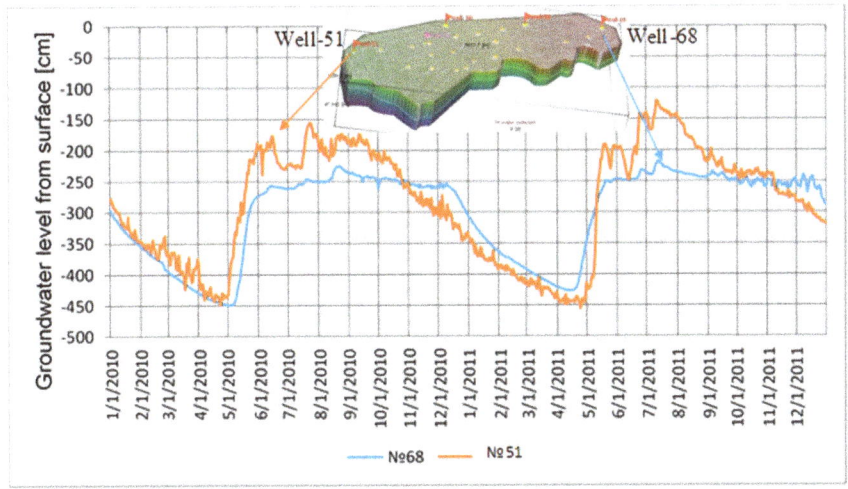

Figure 4. Table of groundwater fluctuation in Wells 51 and 68 in the simulation area.

The capacity of the source reserve is 114×10^3 m^3/day. Nowadays, water extraction is from 70–80 wells, and its volume reaches 87–90×10^3 m^3/day for supply to the capital city [4]; see Figure 3.

The central source of the A–A zone was simulated in FEFLOW simulation. The upper part of the central source intake area near Well 68 takes recharge from river surface water until January, when the Tuul River flows under ice cover. After that period, the Tuul River freezes while the riverbed bottom and groundwater continuously decrease until May. However, in the western side of Zone A, the intake wells area near Well 51 recharge comes not only from Tuul River but also from the Uliastai River; therefore, recharge takes place early in May. As shown in graph, the Uliastai River flow freezes early in November as a small river due to a groundwater decrease that begins in October, and the following recharge in May depends on melting water from both rivers; see Figure 5.

Figure 5. Icing dynamic process in the Uliastai River valley.

The groundwater recharging process runs as follows:

In the initial phase in which surface runoff is non-existent, maximum groundwater declines are observed from the end of April to early May. If the value of the groundwater table is same and above the elevation of the riverbed, it becomes possible for the river to flow downstream without recharge.

The icing phenomena is one of the most important parameters for semi-arid, highly continental climate conditions and plays significant role in the hydrological cycle and regime of Mongolia. The icing or ice cover of rivers and lakes takes place in the cold season, which occurs over five to six months, when the ice cover thickness reaches to 0.8–3.2 m. However, big mountain rivers with a greater slope and bigger perturbation boulders in riffle sections stay open for the whole year and do not completely freeze along the length.

Generally, for surface ice cover to become firmly established, the mean (depth-averaged) temperature of water must be less than 2 °C, the daily average temperature must be less than −5 °C [15], and the wind speed must be less than 5 m/s [11].

The Uliastai River, one of the tributaries of the Tuul River, originates from the Khentii Mountains, flows from the north east side to the south through Ulaanbaatar, and contributes to the Tuul River through the water supply wells of the central source for Ulaanbaatar.

Icing in the aufeis accumulates during winter along the streams and the river valley in northern Mongolia, which is dominated by semi-arid, highly continental regions. In the Uliastai River, building ice lasts from the middle of October until the end of December, and the melting process starts from the end of March and ends in April.

The icing dynamics depend on the groundwater fluxes and discharges alongside of the main channel. The spring leakes through drainage channel, builds ice sheets over the frozen riverbed, where the main stream flows under ice cover. The ice generating process and icing dynamics are studied from the middle of October to the end of December.

The main stream flows by the main channel under ice cover. The side spring streams over the top of the ice sheets. This phenomenon is called icing or aufeis. Icing or aufeis consist of sheets of stratified ice formed by freezing consecutive water leaks [16].

The water flows over existing ice layers. It forms through the upwelling of groundwater discharge or manmade drainage channels, where groundwater discharge is blocked by ice, perturbing the steady-state condition and causing a small incremental rise in the local water table until discharge occurs along the bank and over the top of the previously formed ice [16].

In the beginning of November, river flow freezes from side benches, with the spring discharge no longer extending energy because it had frozen first. After the riverbed completely freezes and takes ice cover, the spring discharges leak from under the ice or ice hummock while the groundwater head and pressure flows increase over frozen ice sheets to create the next ice sheet.

A groundwater flux from drainage canal flows on surface as spring and creates the next ice sheet, which fills the lower ravines and smoothest horizontal by ice sheet.

The groundwater flux alongside the river drains through drainage canal-built icing phenomena, while springs' flow beds are blocked by ice and the main stream in the main riverbed flows under the ice cover. In this section, the Uliastai River gains streams of almost one-third of the flow of the discharge abstracts to groundwater, and groundwater comes from springs.

Thus, phenomena create side spring leakage from under frozen soil, leakage which flows over frozen soil and ice-covered streams, fills ravines and lower lands, creates ice sheets on the ice until the river valley gains the same level of ice. After that, the average daily temperature decreases to under −20 °C, and then the groundwater leakage discharge decreases the head and pressure of groundwater flux in underground flows. Additionally, drained groundwater, spring discharge from the surface, and groundwater decreases influence the quantity of groundwater volume. The ice thickness measurements from the 15th to the 30th of December showed that the thickness of the aufeis sheets have not increased; see Figure 4.

In this way, groundwater leakage, like drained water or springs over frozen soil and ice cover, creates stratified ice sheets over other ice sheets. In the Uliastai River, there are spring discharges of 32

L/s—a small amount of water that nonetheless build ice thicknesses in some places up to 1 m thick. Some of the rivers make ice sheets of thicknesses of several meters. The decline of the ice storage depends on the quantity of spring discharge. The end of building aufeis is usually in the last days of December or the first days of January—after the longest December night ends and the colder days of the year start. In a larger river, such as the Tuul, is possible to create an aufeis until February. An aufeis typically starts to melt during summer and finishes by the end of April (or sometimes, the beginning of May), and it will often form in the same place each year [7].

From these icing ideas, a MATLAB code was written and used for the management of aquifer recharge for the central source of the A zone, which accumulates ice storage due to the use of melt water for recharge in the dry season [17].

The aufeis code was written as one mole of water exchanging energy with cold air and ice, extending velocity of in both the x and y directions could widen in an ellipsoidal way due to the Darcy–Weisbach law. The aufeis spreading dynamics of one mole thick water sheet over ice sheets decrease the extension size while decreasing air temperature day by day from −5 to −30 °C [17].

The result shows that length of ice spread is 2329 m, with a width of 458 m and thickness of 1.54 m. The code for ice storage was simplified with the same slope along flat area. The MATLAB code presents the ice extending dynamics of spreading water with pipeline levees, and from leakage point to pipeline levee of 1500 m. The length of ice storage is only 1500 m; see Figure 6. The results show that the length of ice spread is 1500 m, with a width of 400 m and thickness of 2.72 m, thicker than the unlimited area. In the following figure, you can see two ice storages: unlimited and unlimited area with underground pipeline levee, see Figure 7. These ice storages will change recharging boundary condition in FEFLOW model as a one-month early recharge in April by melt water from north side, where the Tuul River flows in May [8].

The relation between the estimated quantity of the MATLAB ice code and the FEFLOW simulation characterizes the boundary condition change in the northwest side of simulated area. The groundwater recharge from the river starts in May, when the Tuul River flows again. However, in the simulation, the aufeis-stored ice was shown to melt earlier in April and to recharge groundwater. The aufeis changes the northwest boundary condition so that the recharge starts from the beginning of April. The drainage canal begins to flow with water from May, when the Tuul River flows again; see Figure 6. The eastern boundary groundwater fluctuation taken from Well 68 was not found to change. However, the downwards flow direction of the northwest boundary condition was found to change with recharge from the ice melt water; see Figure 6, blue line.

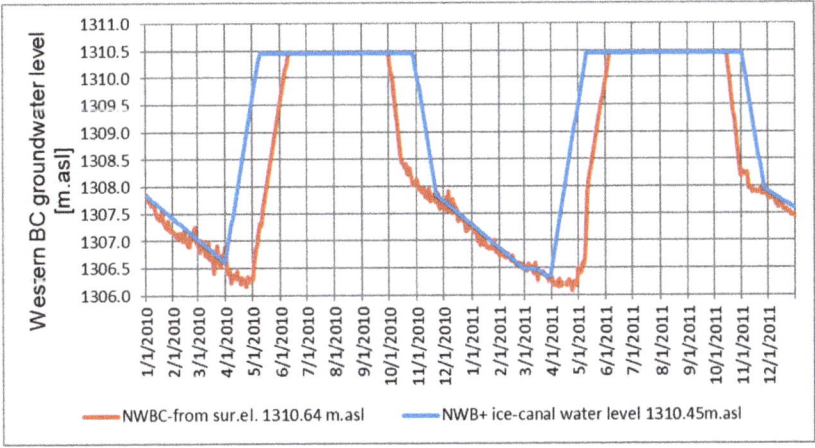

Figure 6. Recharging time difference on northwest boundary (NWBC) with ice storage (blue), only from drainage canal (red).

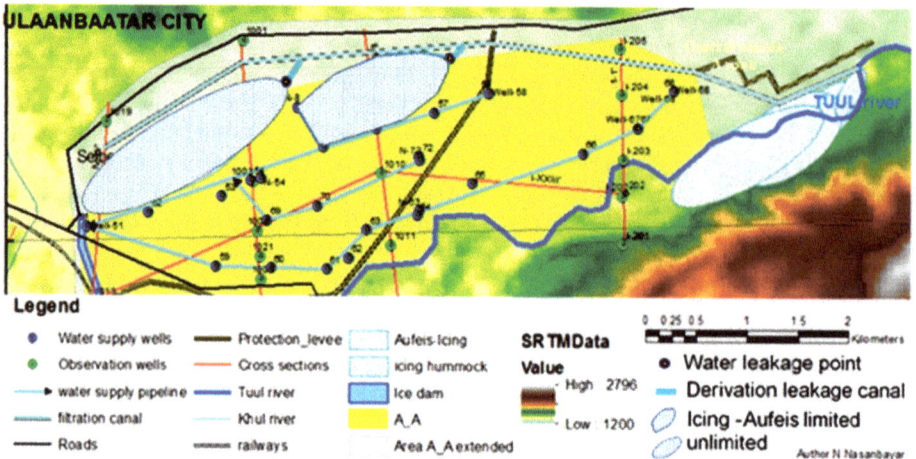

Figure 7. The water supply pipelines connecting wells.

One of the artificial recharging groundwater resources is the temporary and spatial redistribution of surface runoff, which in the beginning of winter results in ice formation. The surface runoff flows over the northern side of the central section A zone through filtration channel will release water at the end of canal and, thus, it again creates more ice sheets over frozen ones.

There are three ways of promoting ice creation in cold regions in winter: on the ground surface, in the underground open pit or channel, and on the river bed. Of these, surface and underground ice creation were considered in the FEFLOW simulation scenarios. See Figure 7.

The rate of water release is 1 m^3/s at the beginning of November until the middle of December, and creates ice storage when the average temperature decreased under −5 °C.

From the above figures, it is shown that the northern drainage canal allows us to accumulate 1 m^3/s flow water for a month, from November to the middle of December, and 3.9×10^6 m^3 water can be stored on the surface. However, with losses from evaporation and winter fog over frozen ice sheets during the melting season, evaporation loss allowed only half of this quantity—about 2×10^6 m^3—to accumulate for recharging groundwater. Here melted water from ice storage recharges groundwater in the central source A–A zone from the northern side, from the beginning of April until May.

Ice accumulation from November to the end of December also recharges groundwater while water flows through the drainage canal. It is then transferred to the artificial regime with the subsequent supply of water to canals, functioning together until the period of ice formation, provided there is ice accumulation from surface water on the end of the canal.

The underground dam will be built in the first layer until reaching the natural permafrost layer 5–10 m from the surface [18,19], considered as an ice wall on the western boundary line.

2.2. FEFLOW Steady Model with Water Reservoir Scenario

A preliminary analysis, as presented in this study, was to identify low-cost MAR implementation measures adapted to the specific natural conditions of the Northern Mongolia. Thus, the coldness of cold weather can be used to keep water in ice form and as a water resource in the winter season in addition to being used during the dry season, characterized by low flow, by melting ice when the rivers have dried out.

The accumulated ice would recharge the groundwater in the dry season from March to May by melting, and the riverbed would be dry without water or cover ice.

To find the total additional water resource recharged by the northern drainage canal and ice storage, the abstraction rate should be increased until the available maximum rate, when groundwater drawdown decreases under the filter screen at the bottom of wells and soaks up air. For calculation of the potential maximum abstraction quantity in the central source A zone without MAR methods, it should be estimated using FEFLOW simulation until some well groundwater drawdown reaches to the bottom of the well screen, pumping air instead of water. In this way, groundwater fluctuation ranges can be established in FEFLOW.

The groundwater fluctuation graph of monitoring well N8 show that by increasing daily extraction to 60,986 m^3/day, it was not possible to pump water while the groundwater drawdown was under the bottom screen. However, in the middle of group of wells, groundwater level is in extreme drawdown, and only 45,121 m^3/day is possible. Therefore, the maximum abstraction rate of this area is 45,121 m^3/day, which means 16,469,165 m^3/year = 16.5 × 10^6 m^3 of water per year.

Nowadays, the exploitation rate is 20,329 m^3/day (average of 2009–2011), 7.5 × 10^6 m^3/year corresponding to half of the possible maximum abstraction.

The recharge quantity with MAR method and then the maximum abstraction rate increase until 70,133 m^3/day, which corresponds to 25.6 × 10^6 m^3/year.

The difference between the above maximum abstraction simulations shows that it is possible to increase the groundwater resource extraction rate to around 25 × 10^3 m^3/day using MAR methods. That means approximately 912,500 m^3/year = 9.125 × 10^6 m^3 additional groundwater during wet season could be kept as reserve in the upper part of this central source area A–A zone.

The percentage for recharge groundwater sources in this area was estimated by the sum of all methods of each MAR method, and is taken as 100% for all together.

The percentage of the increased amount of water reserve by each method, after combination of the MAR methods, from increased groundwater sources is as follows.

- Drainage canal brings 73.5% of artificial recharge in 6.7 × 10^6 m^3 water (1 m^3/s discharge water through the drainage canal from May to November 86,400 m^3/day × 184 day = 15,897,600 m^3 = 15.9 × 10^6 m^3).
- Ice storage keeps water in ice form in 21% or 1.9 × 10^6 m^3 (1 m^3/s discharge water for ice storage would be 5.2 × 10^6 m^3, but fog and evaporation during icing and melting bring about huge loss).
- Ice walls delay groundwater flux and maintain stability in the groundwater table, leading to retention of an additional 5.65% of groundwater, which is about 516 × 10^3 m^3 in water amount [17].

All of these FEFLOW simulations were simulated in a transient model, where the river surface water level increases in the wet season over the riverbed, and decreases in the dry season under the dry riverbed. The table of the eastern and western boundary groundwater fluctuates under the surface, depending on the recharge from the river for whole year.

Another alternative simulation scenario is for the water release from the surface water reservoir with a dam, with a constant q rate of 26.6 m^3/s and creation of table for constant groundwater water. In this case, the model is taken with the eastern and western boundary using a constant groundwater level and hydraulic head and, also, in the southern boundary as the Tuul River with a constant level value for groundwater recharge. In the FEFLOW simulation, the maximum abstraction rate reached 90,288 m^3/day. Following this, the simulation produced an error report indicating that the aquifer has no groundwater; see Figure 8.

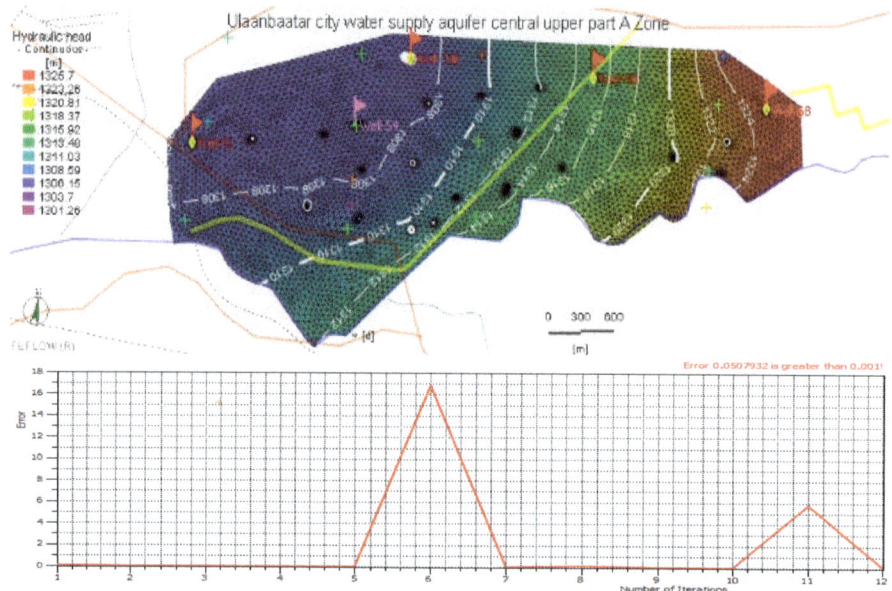

Figure 8. The groundwater contours and simulation error by pumped 166,700 m³/day water.

This means the flow control by surface reservoir accumulates flood water and, additionally, the recharge groundwater quantity for the central source of A zone is 16.5×10^6 m³/year.

The flow control upper reservoir releases a constant outflow throughout the year, and it means it also recharges the central source of B zone, occupying a larger area twice in size, with 50 intake wells. The constant release rate for the whole central source area that is recharged from the reservoir would be 49.5×10^6 m³/year. An additional 350×10^6 m³ of accumulated water in the reservoir would be reserved for use as a freshwater resource [20].

All these simulated calculations involved recharging water from only reservoir. When we are simulating a combination of artificial recharging groundwater sources and reservoir release for surface water recharge, then the additional recharged groundwater would be 167,700 m³/day; see Figure 9.

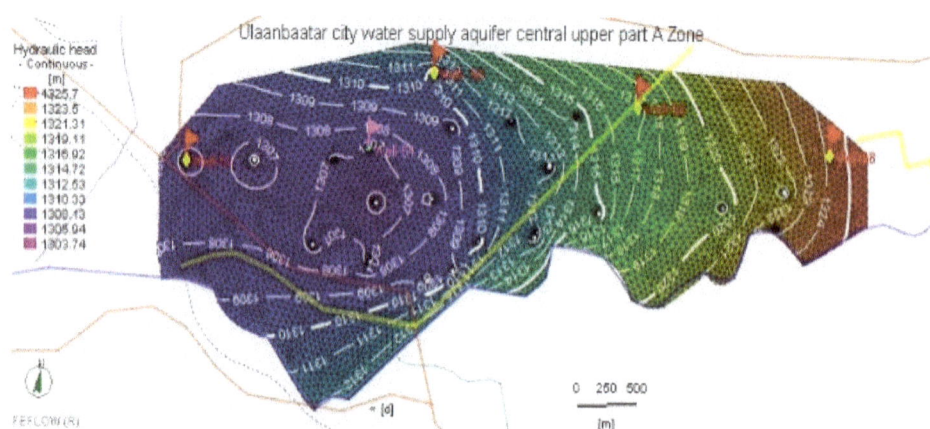

Figure 9. The groundwater simulated contours by a combination of managed aquifer methods and recharge from Tuul River outflowing from reservoir.

These MAR methods and the reservoir recharge system include the following simulation scenarios.

- The northern drainage canal recharge groundwater resource from the northern side in wet season;
- Ice storage for groundwater recharge by melting ice in the dry season in early April with Tuul River flow starting in May;
- The underground ice dam which accumulates groundwater backside up to the permafrost layer 5–10 m [18,19];
- The surface reservoir with constant release discharge into Tuul River, which recharge this area from the southern side over the whole year.

In this combination of scenarios, the maximum possible abstraction rate is increased up to 167,700 m^3/day after simulation. The maximum possible abstraction quantity per year reaches up to 44.7 × 10^6 m^3/year.

The FEFLOW simulation was only run for the central source of A zone. When we take the second B zone with 50 wells, this maximum abstraction rate would be increased, and at least doubled or 89.5 × 10^6 m^3/year due to water recharging because the B zone is twice as large as the A zone. If the Tuul River flows yearly with constant discharge, then river surface water will be recharged for industrial sources, and that leads to increases in the additional groundwater resource. This will increase the flow from groundwater sources to 149 × 10^6 m^3/year and create natural conditions for discontinuous Tuul River flow to assist in avoiding a situation of shortages in the fresh water supply for Ulaanbaatar city.

3. Analyses of FEFLOW Model

Aufeis is an icing method that brings more recharge water in the middle of the study area for its other constituent subsections as well as an ice wall, increasing the backside of the western boundary. The drainage canal filtrates through the northern side and the recharge occupies the whole area from the east to the west boundary.

The combination of all methods includes an ice wall, icing, and drainage canal. The drainage canal filtrates water from east to west along the north side of the study area, and the rest are released from it as water, since early winter creates icing, see Figure 10. Thus, this icing method is performed after drainage canal filtration has ceased altogether.

As seen in the following figure, recharge from the drainage canal begins in May and ends in November, with icing recharge beginning in the middle of March and lasting until May, and the ice wall holding groundwater undergoing the reverse, from March until November.

In nature, we can reserve more water in ice form as aufeis to help in building some hydraulic structures, such as drainage canals or underground dams. Both of these structures can help us to reserve water stores underground, as well as keep them on the surface in ice form.

From October to December, surface water naturally accumulates and is kept in ice form from river flow in the semi-arid, highly continental region of the study area. During the subsequent dry season, these sources increase and recharge potential water availability by melting water sources.

One possibility is to reserve water resources to regulate groundwater flux control in highly continental cold regions to eliminate dry riverbeds and keep primary source rivers, such as the Tuul River, continuously flowing during low flow periods is by melting ice blocks.

From the following graph, it can be seen that the northern drainage canal will allow accumulation 1 m^3/s flow water for a month from the beginning of November to mid-December, creating 3.9 × 10^6 m^3 water for surface storage.

However, due to losses from evaporation and winter fog over frozen ice sheets and, also, evaporation loss during the melting season, only half of this quantity, or about 2 × 10^6 m^3, will actually accumulate and be available for groundwater recharge.

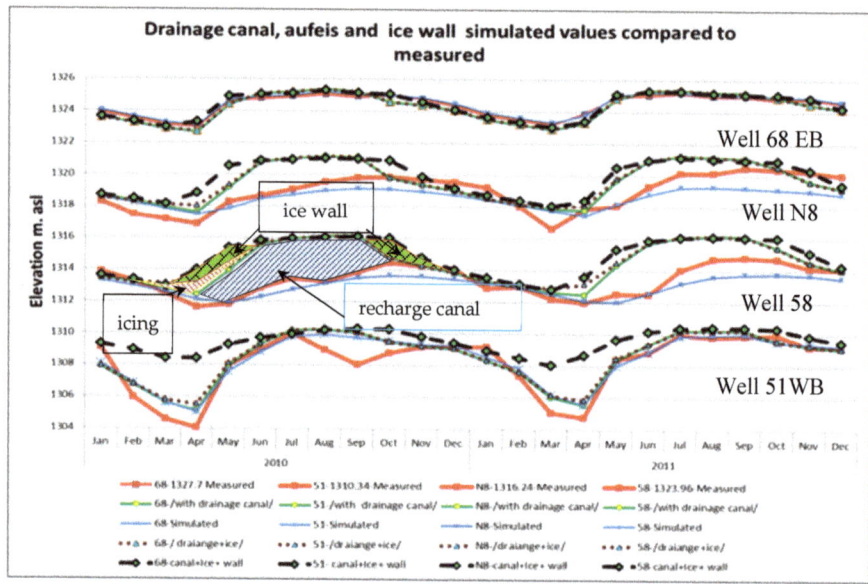

Figure 10. The combination of all variations in simulated area compared to the measured groundwater level.

For this FEFLOW simulation, the maximum difference between the simulation in natural conditions and the simulation in each managed aquifer recharge scenario is taken into account. The result of the FEFLOW simulation in each MAR method scenario demonstrates the following results:

- Single ice wall on the western boundary in natural condition without drainage canal increases groundwater resources in 516×10^3 m^3/year;
- The drainage canal through the northern side filtrates groundwater at around 1×10^6 m^3/year;
- The icing on the end of the drainage canal recharges groundwater resources in the aquifer corresponding to 1.6×10^6 m^3 water;
- The aquifer additional recharge groundwater quantity of all combined MAR methods is 2.55×10^6 m^3/year [8], see Figure 10.

4. Results and Conclusions

In recent decades, as the city develops and expands, the consumption of the domestic and industrial water supply of the city is increasing intensively, but the availability of the water supply, both now and in the future, has become a pressing issue. Therefore, in this paper, possible variants for recharging the groundwater resource are considered using FEFLOW simulation, and compared with the results for modeling in the upper part of Ulaanbaatar aquifer.

The simulation variants of the recharging methods of managed aquifer are northern drainage canal, which recharges groundwater from the opposite side such as for the Tuul River, icing or aufeis, which keeps water in ice form, brings it from the end of the wet season through winter to the dry season, and increases groundwater sources, by melting. In the end of the upper part of the aquifer in the western side, building an underground dam which accumulates groundwater on the backside. All these variants were used in FEFLOW simulation, and the recharged quantity was calculated separately, and combined ones demonstrated the following results.

- A single ice wall on the western boundary in natural conditions without drainage of canal increases groundwater resources by 516×10^3 m^3/year.

- The drainage canal through the northern side filtrates groundwater around 1×10^3 m^3/year;
- The icing on the end of the drainage canal recharges groundwater resources in the aquifer with 1.6 million m^3 water;
- The aquifer additional recharge groundwater quantity of all combined MAR methods is 2.55×10^6 m^3/year [8].

The additional recharged groundwater difference by increased daily abstraction rate from this aquifer both with and without managed aquifer recharge would be 9.16×10^6 m^3/year.

When the FEFLOW simulation scenario has taken into account the surface water reservoir, which releases constant discharge flow, then the calculations involving recharge from only reservoir separately and also the combined simulation with MAR methods demonstrate the following results.

- The recharge quantity from the Tuul River with constant discharge during an entire year is 16.5×10^6 m^3/year;
- The combination MAR and surface water from reservoir outflow corresponds to a recharge of 44.7×10^6 m^3/year.

These simulation results are from the upper part with only 23 wells of central source area, which has a total of 72 wells. When the simulation is based on all the Ulaanbaatar aquifers and additional recharged groundwater resource from Tuul River flow with constant runoff, the results is at least 3 times more than this small area or 150×10^6 m^3/year.

The simulation of both transient (MAR methods) and steady reservoir outflow recharge methods show that artificial recharge methods of groundwater are not enough for additionally recharged quantity compared with surface reservoir outflow recharging. However, a combination of both methods demonstrates that use of artificial recharging groundwater sources in the aquifer is more effective.

Using coldness and icing in a semi-arid, highly continental region such as Mongolia can increase groundwater resources. The nonconventional artificial recharging methods specified in this paper, including ice storage, can assist in groundwater recharge in dry seasons, when there is no flow and dry beds in river systems.

The Tuul River reservoir will be located upstream from the aquifer and can accumulate 350×10^6 m^3 in water [20], creating a more humid environment in the region of Ulaanbaatar. A positive side effect of increased evaporation from the open water is that air humidity will be increased, which will lead to artificial precipitation and deposition of atmospheric particles from air pollution and, thus, helps to improve air quality in the city and its surroundings.

Groundwater flux disturbed by the presence of an ice wall comes out on the surface during spring, spreading, evaporating, and losing some of its mass. This creates fog in winter and turbidity in air when a cold high-pressure cyclone dominates Ulaanbaatar city. The turbidity of air corresponds to windy surroundings and clears air pollution.

Following the MAR method, water storage in the form of ice in semi-arid, cold, highly continental regions would also produce coldness during the dry season and increase the inner continental hydrologic cycle through melting and evaporation. In the dry season, melting ice evaporates and increases precipitation. Many forest fires occur every year in Mongolia during the dry season due, at least partially, to the lack of precipitation and the lower air humidity. Therefore, ice-keeping methods would help in keeping the environment green, environmentally close to the natural process and improve human and natural habitats.

The aufeis blocks, involving icing in the river valley, are increasing continental cycles by evaporating melted water and producing coldness [21]. Thus, it helps in growing of vegetation and increasing precipitation in the dry period. During the fall months, from September to October, the soil moisture increases with more rain, and during spring (March and April), there is more humidity in the air and more rain. Meltwater is a common source for rivers and enables their continuous flowing [22] as well as the recharging of groundwater aquifers.

Author Contributions: U.M., First editor and reviewer of my (N.N.) dissertation structure. This article structured by his (U.M.) recommendation which is part of my (N.N.) dissertation in groundwater modelling and management.

Funding: This research received no external funding.

Conflicts of Interest: The authors declare no conflict of interest.

References

1. Davaa, G. *Surface Water Resources in Mongolia*; Interpress LLC: Ulaanbaatar, Mongolia, 2008.
2. GIM—Geo-Ecology Institute Mongolia. *The Research Works Report of Tuul River Water Reserves Decreases Reason; Protection Provision UB*; GIM: Ulaanbaatar, Mongolia, 1997.
3. RIBES-Research Institute on Building Engineering Studies. *Technical Report of Engineering Survey TOM-2*; RIBES-Research Institute on Building Engineering Studies: Moskow, Russia, 1979.
4. GIM—Geoecology Institute of Mongolia. *The Current Situation, Location, Reserves and Future Concept of Domestic and Industrial Water Supply for Ulaanbaatar City*; GIM: Ulaanbaatar, Mongolia, 2000.
5. Gombo, D.; Magsar, E. *Hydrological Changes in the upper Tuul River Basin*; Juuljinii St. 5, Ulaanbaatar-46; Institute of Meteorology and Hydrology: Ulaanbaatar, Mongolia, 2004.
6. GIM—Geoecology Institute of Mongolia. *The Ecological Assessment of Tuul River*; GIM: Ulaanbaatar, Mongolia, 1999.
7. IMHM—Institute of Meteorology and Hydrology Mongolia. *Weather Parameters Data of Ulaanbaatar, 1986–2006*; IMHM: Ulaanbaatar, Mongolia, 2006.
8. Nasanbayar, N. Icing Mar Method for Semi-Arid, Highly Continental Region (an Example at Ulaanbaatar city, Mongolia). *Int. J. Recent Sci. Res.* **2019**, *10*. [CrossRef]
9. SIWRMM. *Strengthening Integrated Water Resources Management in Mongolia*; Tuul River Basin Integrated Water Resources Management Assessment Report; SIWRMM: Ulaanbaatar, Mongolia, 2012.
10. Punsalmaa, B.; Nyamsuren, B.; Buyndalai, B. Trends in River and Lake Ice in Mongolia. AIACC Working Paper No. 4. 2004. Available online: http://www.start.org/Projects/AIACC_Project/working_papers/Working20Papers/AIACC_WP_No004.pdf (accessed on 30 November 2019).
11. Ashton, G.D. Deterioration of Floating Ice Covers. *J. Energy Resour. Technol.* **1985**, *107*, 177–182. [CrossRef]
12. Kresic, N. *Hydrogeology and Groundwater Modeling*; CRC Press: Boca Raton, FL, USA, 2006; p. 807.
13. MIKE Powered by DHI. Available online: http://www.feflow.com (accessed on 1 January 2012).
14. Tsujimura, M.; Ikeda, K.; Tanaka, T.; Janchivdorj, L.; Erdenchimeg, B.; Unurjargal, D.; Jayakumar, R. Groundwater and Surface Water Interactions in an Alluvial Plain, Tuul River Basin, Ulaanbaatar, Mongolia. Sciences in Cold and Arid Regions. 2013, pp. 126–132. Available online: http://www.scar.ac.cn (accessed on 30 November 2019).
15. Ashton, G.D. Thin ice Growth. *Water Resour. Res.* **1989**, *25*, 564–566. [CrossRef]
16. Froehlich, W.; January, S. River Icings and Fluvial Activity in Extreme Continental Climate: Khangai Mountains, Mongolia. 1982. Available online: http://pubs.aina.ucalgary.ca/cpc/CPC4-203.pdf (accessed on 30 November 2019).
17. Nasanbayar, N. Icing phenomena for Managed aquifer recharge (MAR) and It's FEFLOW simulation result. *Ulaanbaatar PMAS* **2019**, *59*, 229.
18. Dashjamts, D. Geotechnical problems of construction on permafrost in Mongolia. *Sci. Cold Arid Reg.* **2013**, *5*, 0667–0676. [CrossRef]
19. Dashjamts, D. Permafrost and geotechnical investigations in Nalaikh Depression of Mongolia. *Sci. Cold Arid Reg.* **2015**, *7*, 438–455.
20. Monhydroconstruction LLC. *The Results Report of Feasibility Study Multipurpose Dam Built on the Tuul River*; Monhydroconstruction LLC: Ulaanbaatar, Mongolia, 2007.
21. Grayson, R. Asian Ice Shields and Climate Change. *World Placer J.* **2010**, *10*, 21–45.
22. Emerton, L.; Erdenesaikhan, N.; de Veen, B.; Tsogoo, D.; Janchivdorj, L.; Suvd, P.; Enkhtsetseg, B.; Gandolgor, G.; Dorjsuren, C.; Sainbayar, D.; et al. *The Economic Value of the Upper Tuul Ecosystem, Mongolia*; World Bank: Washington, DC, USA, 2010.

© 2019 by the authors. Licensee MDPI, Basel, Switzerland. This article is an open access article distributed under the terms and conditions of the Creative Commons Attribution (CC BY) license (http://creativecommons.org/licenses/by/4.0/).

Article

New Methods for Microbiological Monitoring at Riverbank Filtration Sites

Yasmin Adomat [1],*, Gerit-Hartmut Orzechowski [1], Marc Pelger [1], Robert Haas [2], Rico Bartak [2], Zsuzsanna Ágnes Nagy-Kovács [3], Joep Appels [4] and Thomas Grischek [1]

1. Division of Water Sciences, University of Applied Sciences Dresden, 01069 Dresden, Germany; gerit.orzechowski@htw-dresden.de (G.-H.O.); marc.pelger@mailbox.tu-dresden.de (M.P.); thomas.grischek@htw-dresden.de (T.G.)
2. DREWAG NETZ GmbH, 01067 Dresden, Germany; Robert_Haas@drewag-netz.de (R.H.); Rico_Bartak@drewag-netz.de (R.B.)
3. Budapest Waterworks Ltd, 1134 Budapest, Hungary; zsuzsanna.nagy-kovacs@vizmuvek.hu
4. microLAN B. V., 5145 Waalwijk, The Netherlands; joep.appels@microlan.nl
* Correspondence: yasmin.adomat@htw-dresden.de; Tel.: +49-351-462-3944

Received: 19 November 2019; Accepted: 18 February 2020; Published: 20 February 2020

Abstract: Water suppliers aim to achieve microbiological stability throughout their supply system by regular monitoring of water quality. Monitoring temporal biomass dynamics at high frequency is time consuming due to the labor-intensive nature and limitations of conventional, cultivation-based detection methods. The goal of this study was to assess the value of new rapid monitoring methods for quantifying and characterizing dynamic fluctuations in bacterial biomass. Using flow cytometry and two precise enzymatic detection methods, bacterial biomass-related parameters were monitored at three riverbank filtration sites. Additionally, the treatment capacity of an ultrafiltration pilot plant was researched using online flow-cytometry. The results provide insights into microbiological quality of treated water and emphasize the value of rapid, easy and sensitive alternatives to traditional bacterial monitoring techniques.

Keywords: online flow-cytometry; enzymatic activity; riverbank filtration; ultrafiltration; ATP

---

1. Introduction

Riverbank filtration (RBF) systems are operated in many countries for the public and industrial water supply due to their efficient removal of pollutants such as microorganisms [1,2]. During RBF, the removal of bacteria, viruses, and protozoa in surface water is attained trough filtration, sorption, and grazing processes besides die-off. Such processes can be influenced by the aquifer material composition, hydraulic gradient, temperature, redox conditions, organic/inorganic nutrients, and travel time in the aquifer [3–5].

By measuring the microbiological characteristics such as bacterial biomass concentration or enzymatic activities, these interactions between surface water and groundwater can be determined. The access of limited nutrient supply (e.g., organic carbon, nitrogen or phosphorus) or environmental conditions (temperature, inhibitory substances) can lead to complex interactions between various microbes [6]. As a result, unwanted changes in microbiological water quality such as an excessive growth of bacteria can lead to a degradation of drinking-water quality and operational problems [7,8].

The World Health Organization (WHO) specified that water which enters the distribution system must be microbiologically safe and ideally should also be biologically stable, meaning microbiological water quality must be maintained from the point of drinking water production up to the point of consumption [9,10]. To ensure safe and effective water treatment, distribution, and consumption, reliable procedures for characterizing and monitoring waterborne microbes need to be carried out by water suppliers on a regular basis.

Water produced at RBF sites is commonly monitored for the absence of pathogen indicator organisms like *Escherichia coli (E.coli)*, total coliforms (TC), enterococci and *Clostridium* using cultivation-based methods with targeted growth media [4]. Additionally, heterotrophic plate counts (HPC) with non-specific media are frequently assessed. Although there is no evidence of a link between HPC results and health risk, it is of major importance to assess data of microbiological growth during drinking-water treatment, and to detect changes in bacterial concentration and composition of monitored water [5,6,11]. Since the HPC method was first introduced in the 1800s as a public health indicator, science has advanced. Hence, HPC monitoring became more useful as an operational rather than a health-based indicator [9]. At present, within water-treatment facilities in Germany, The Netherlands, and Hungary the HPC method is used for validation and verification of drinking-water treatment processes. Abnormal changes in HPC indicate problems in the treatment process and appropriate actions are essential to ensure that the problem is identified and eliminated [9].

Although the HPC method was introduced more than 100 years ago and enhancements regarding general performance and interpretation of data were developed, additional work concerning sample incubation times, temperatures, and acceptable critical thresholds is required [12]. Despite time-intensive laboratory procedures and incubation, the HPC method only detects a fraction of bacterial cells in water samples. This is due to the fact that only 0.1%–1.0% of bacteria species present in aquatic samples is culturable under laboratory conditions which was confirmed in various studies [5,13,14]. Furthermore, an estimation of the percentage of subpopulation of heterotrophic bacteria as well as a differentiation of which of these subpopulations include potential pathogens is not possible using HPC techniques [5].

In the past decade, significant advances in rapid cultivation-independent techniques, mostly fluorescence-based methods, have been developed. These methods focus on direct measurements of indigenous bacterial growth or enzymatic activity. Examples include optical methods (e.g., flow-cytometry, FCM) which count suspended particles in water samples and are able to differentiate between bacteria and abiotic particles based on e.g., 3D scanning or chemical staining techniques (e.g., SYBR®Green or propidiumiodide) [14,15]. Also, on-site sensors measuring indirect indicators of microbiological fluctuations such as adenosine triphosphate (ATP) concentration or specific enzymatic activities have been developed in the past decade [16–18].

In this study, FCM and two enzymatic detection methods (ALP, *alkaline phosphatase*) were used to analyze the total microbiological water quality after different treatment processes in samples of three RBF treatment sites. The two main objectives were to assess the applicability of each method in routine monitoring programs and, to compare the methods with each other as well as with HPC data. Additionally, an ultrafiltration pilot plant was monitored using online FCM with the goal to assess the pilot plant's performance and to test if a continuous measurement of bacterial removal is possible.

2. Materials and Methods

2.1. Site Description

Samples were collected from continuously and discontinuously operated sample taps from two RBF sites around Dresden: D1 and D2, Csepel Island, Hungary (CI), and various RBF wells at Szentendre Island (SI), Hungary. D1 and D2 are situated on the floodplain of the Elbe River around Dresden, the state capital of Saxony, Germany. D1 has a total capacity of 72,000 m^3/day, 111 vertical siphon wells as well as 36 single-operated wells [19]. The waterworks operates two separate treatment trains: a RBF treatment train and a managed aquifer recharge (MAR) treatment train. After well extraction, the RBF and MAR water is aerated and filtered using granular activated carbon (GAC) and disinfected with chlorine before it is distributed as drinking water. Water in D2 is abstracted from three siphon well galleries with 72 vertical wells with a total capacity of 36,000 m^3/day (approx. 65%–80%

bank filtrate) and post-treated by cascade aeration, GAC filtration and disinfection with chlorine dioxide. In Hungary, 756 horizontal as well as vertical wells on CI and SI are operated by Budapest Waterworks Ltd with a maximum capacity of 1.0 million m^3/day and an average supply of about 456,000 m^3/day [20]. Post-treatment is performed in Cl using ventilation, coagulation (aluminium sulphate), ozonation, sand filtration, ultraviolet (UV) radiation (temporary), and disinfection with chlorine.

2.2. Sample Collection and Microbiological Characterization

Prior to sampling at discontinuously operated taps, a disinfection-step with ethanol (96% Merck KGaA, Darmstadt, Germany) and flame sterilization (propane/butane gas, 1350 °C) were applied followed by a 3 min flushing interval before samples were collected. Samples were collected into 5 mL (polypropylene rack tube, Corning, New-York, USA), 15 mL (polypropylene rack tube, VWR, Radnor, USA), and 1 L (Borosilicate glass with polypropylene screw cap, VWR; Radnor USA) sterile sample bottles. Afterwards, each sample was transported to the laboratory for analysis within 12 h.

Intracellular ATP (ATPi) was determined using a luminesce-based Clean-Trace ATP water test kit (3M, St. Paul USA). Based on an enzymatic reaction (firefly luciferase), total (ATPt) and extracellular ATP (ATPe) were measured in relative light units (RLU) as per the manufacturer's instructions. ATPi was calculated from ATPt and ATPe values. Using a fully automatic operating system called BACTcontrol (microLAN, Waalwijk, Netherlands), total enzymatic activity (TEA) was analyzed by measuring the specific activity of ALP as an indicator for the presence of bacteria. Prior to each measurement, the water sample was pumped at 0.2–1 mL/s through a 0.45 µm ceramic filter into a reactor chamber. While constantly stirring, the concentrated water sample was incubated for 20 min at 45 ± 0.1 °C. During incubation, the enzymatic activity of ALP was detected in methylumbelliferone (MUF, in pmol MUF/(min·100 mL) by a fluorimeter which was pre-calibrated using a standard concentration of 1000 nM MUF.

Flow-cytometry analysis was carried out with a BactoSense (Sigrist, Switzerland) flow-cytometer equipped with a 488 nm solid-state laser and an optional continuous/discontinuous sampling port. Sample volumes of 260 µL were drawn at a flow rate of 200–400 mL/min and mixed with fluorescent stain (SYBR®Green, propidiumiodide). After incubation (10 min, 37 °C), samples were analyzed (FL1 channel at 525 nm, FL3 channel at 721 nm) using fixed gates to separate cells and background signals and additionally to distinguish between so-called high (HNAP) and low (LNAP) nucleic acid content cells.

2.3. Ultrafiltration Pilot Plant and Online Flow-Cytometry (FCM) Measurements

The ultrafiltration pilot plant was operated at D1 with a treatment capacity of 20 m^3/h. Using either Elbe River water or flocculated Elbe River water as feed supply, water was pumped directly into a storage tank (1.9 m^3) through a supply pipe. Ultrafiltration was processed by using two membrane modules consisting of polyvinylidene difluoride (PVDF) with pore sizes of 20 nm (UF 1, Pall Corporation, Port Washington, USA) and 18 nm (UF 2, inge GmbH, Greifenberg, Germany) and operated at a flux of 40–80 L/(m^2 h) at 1.5 bar.

Online FCM sampling was realized using an automatic programmed magnetic-valve system (Figure 1) consisting of three sampling ports (Feed, Permeate 1, and Permeate 2). Water samples were drawn in a bypass which was controlled using a programmed Delphin-EMS-system (Delphin Technology AG, Gladbach, Germany) and the online sampling setting of the BactoSense flow-cytometer. Sample analysis was carried out applying the same method which is outlined in Section 2.2.

The cleaning process of the membrane modules was adjusted every 30 min by backwashing. UF 1 was cleaned using a combination of air and water at a flux of 5 m^3/h for 60 s. While backwashing air was added in the membrane direction, water was added in the reverse direction. Afterwards, a 45 s forward flush processing step in the direction of filtration at 7 m^3/h was performed. UF 2 was pre-cleaned by air flushing for 10 s followed by air–water backwashing in the flow direction for 50 s. The cleaning was completed with a 15 s forward flush step at 7 m^3/h. Taking into account backwash cycles every 30 min to avoid a sampling of backwash water, samples were drawn by flushing the FCM system for 2 min and analyzed within a cycle of 105 min according to Table 1.

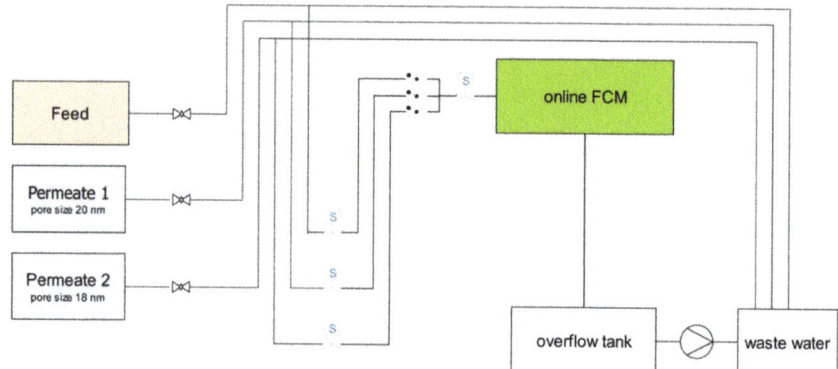

Figure 1. Scheme of online flow-cytometry (FCM) sampling system.

Table 1. Of water sample analysis within a measurement cycle of 105 min.

Time in min	Process	Sample
0–5	backwash	
5-35	sampling and analysis	Feed
35–40	backwash	
40–70	sampling and analysis	Permeate 1
70–75	backwash	
75–105	sampling and analysis	Permeate 2

2.4. Data Analysis

Statistical data processing was carried out using MS Excel and OriginLab. All microbiological data analysis were carried out using the provided device-specific software.

3. Results

3.1. Correlation of New Methods and Conventional Cultivation-Based Methods during Riverbank Filtration (RBF) and Drinking-Water Treatment

To elucidate the correlation of FCM, ALP-TEA and ATP values a serial dilution was applied with ultrapure water (autoclaved at 121 °C and 15 min, V75, SysTech, Pegnitz, Germany) and bottled mineral water (Evian, France). FCM and ALP-TEA provided positive bacterial counts for all water samples with an average maximum Pearson correlation coefficient of $R = 0.80$ for the intact cell count (ICC). There was a high correlation between ICC and ALP-TEA (Table 2). This was also observed for the total cell count (TCC)/ATPt ratio ($R = 0.78$) and the ALP-TEA/HNAP ($R = 0.76$).

Table 2. Correlation coefficients of FCM, ALP-TEA and ATP methods.

	ATPi	ATPt	HNAP	ICC	TCC	ALP-TEA
ATPi	1					
ATPt		1				
HNAP	0.87	0.62	1			
ICC	0.76			1		
TCC		0.78			1	
ALP-TEA	0.32	0.57	0.76	0.80	0.78	1

The sample that contained 5% bottled drinking water showed a slight increase by 16.3% ICC with FCM and ATPt which is not in accordance with the ALP-TEA decrease by 18.3 pmol/min (Figure 2). Also, ALP-TEA/ATP standard deviation in 100%-ultrapure water are fluctuating around ± 10 pmol/min which may be caused either by background noise or still intact cells with a high amount of intercellular ALP. Suprisingly, a higher correlation coefficient with ALP-TEA/ATPt (R = 0.57) than ALP-TEA/ATPi (R = 0.32) was determined given the fact that the ALP-TEA amount refers only to living organisms and, therefore, a better ALP-TEA/ATPi would have been expected. Further to note is the higher correlation of ATPi/HNAP (R = 0.87) than ATPi/ICC (R = 0.76) ratio. These values are in accordance to [21] since HNAP refers to the ratio of large cells to small cells and could therefore contain higher levels of intracellular ATP.

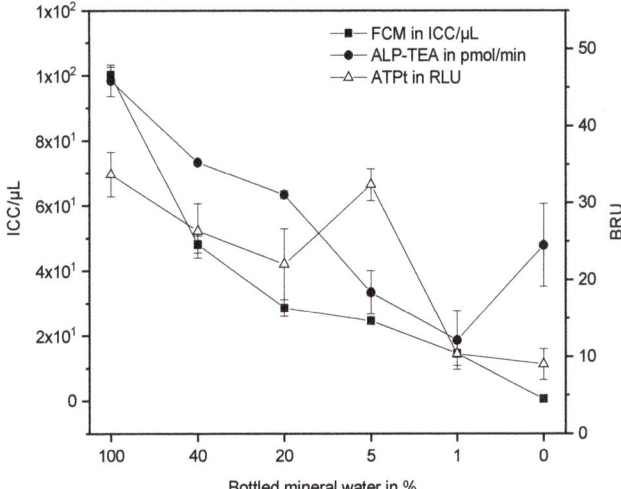

Figure 2. Correlation of FCM (n = 5), total adenosine triphosphate (ATPt, n = 3) and alkaline phosphatase-total enzymatic activity (ALP-TEA, n = 3) for measuring total biomass related units (BRU), samples from bottled mineral water (Evian) were diluted with ultrapure water from the same bottle, based on a method described in [13]. FCM samples were stained with SYBER®Green and propidiumiodide, error bars indicate standard deviation on samples.

To further demonstrate potential links between FCM and ALP-TEA related data, results of the CI, D2 and D1 sampling campaigns are given in Figure 3. Unfortunately, ALP-TEA sampling was only possible in CI due to logistical problems in D1 and D2. 1.5×10^2 ICC/µL and 20.2% HNAP were detected in RBF samples with corresponding ALP-TEA values of 70 pmol/min in CI. In D2 and D1 similar values of 3.7×10^2 ICC/µL (D1) and 1.4×10^2 ICC/µL (D2) were measured in RBF samples with an average HNAP portion of 20.9% and 20.4%. Regarding Elbe River water and RBF, BRU in D1 decreased by 97.0% (1.53 log units).

Biomass concentration increased further in open aeration towers by 4%, to 1.5×10^2 ICC/µL, 86 pmol/min and 23.4% HNAP in CI, and to 4.1×10^2 ICC/µL and 21.6% HNAP in D2 due to process-related oxygen and external biomass entries and a rainfall event (CI).

Ozonation eliminated most intact organisms to 2.0×10^1 ICC/µL and 25 pmol/min in CI based on its high redox potential, but at the same time this process provides organic substrate: dead cell material to still intact organisms. This conjuncture is also confirmed when comparing the lysed cell gates in Section IV and Section V in Figure 4 with 2.3×10^2 desolate cell count (DCC)/µL (aeration) and 2.0×10^2 DCC/µL (ozonation). Also, the wide spread of standard deviation results, regarding HNAP, indicates that the amount of present cells is below detection limit since standard deviation values surpass the mean value by 40.6%. As a result, biomass increased tenfold in the sandfilter effluent in CI.

This relation however, could not be observed with the BACTcontrol method where ALP-TEA slightly decreased to 23 pmol/min. BRU in D1 and D2 decreased by 46% (D1) and 51.6% (D2) indicating that sandfilters act as a microbiological barrier.

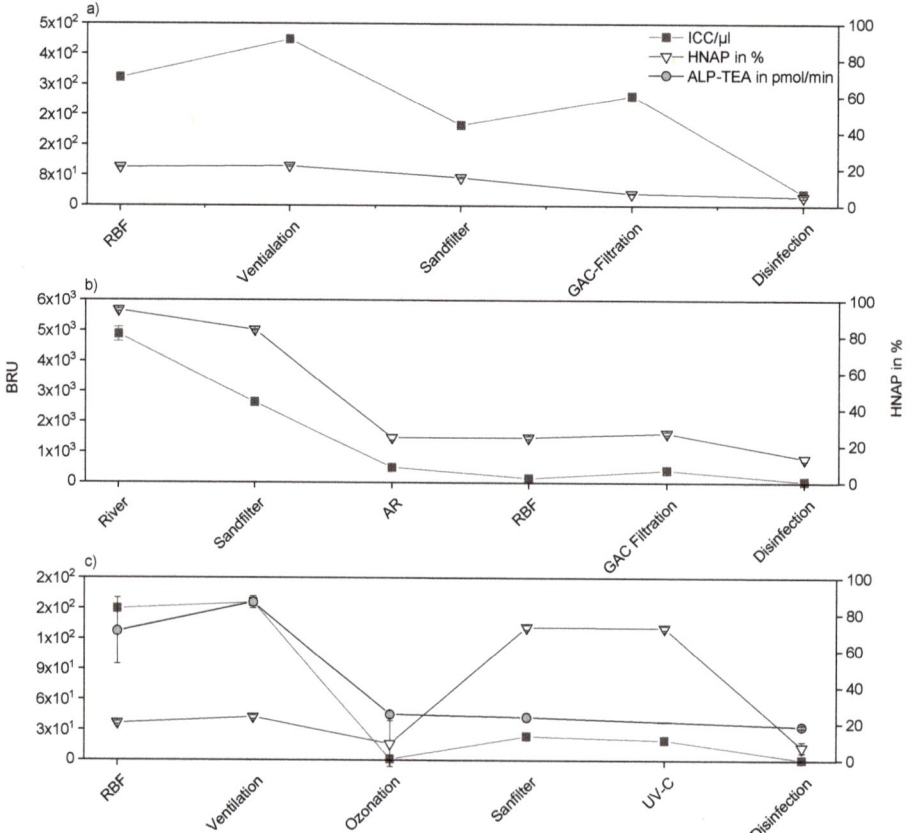

Figure 3. Comparison of FCM intact cell count in (**a**) D2, (**b**) D1 and (**c**) CI, (ICC, n = 5) with total enzymatic activity (ALP-TEA, n = 3 only CI), Sampling Dates: 9 May 2019 (CI), 15 May 2019 (D2) and 5 June 2019 (D1), FCM samples were stained with SYBER® Green and propidiumiodide, error bars representing standard deviation for those points.

Also, no significant impact of UV disinfection in CI on ICC as well as HNAP and LNAP amount was observed. ICC decreased by 16% to 2.0×10^2 ICC/µL and HNAP by 0.33%. UV-C treatment only causes damages of the bacterial genome but has no impact on bacterial cell membranes which is also confirmed in Section II and Section III in Figure 4 [22]. GAC filter effluents show an increase of BRU by 63.5% (D1) and 56.4% (D2). For GAC filter effluents in D1, BRU increase may be explained by an additional aeration process after water extraction and/or biological active GAC filters. The cell amount of 2.7×10^2 ICC/µL in D2 with a decrease of HNAP by 56.4% at the same time in the GAC filter effluent may indicate that certain strains, especially HNAC, are more adept at colonizing the filter surface and may be able to out-compete LNA cells [23].

Final disinfection using chlorine leads to an ICC decrease by 95.0% to 1.0×10^1 ICC/µL (CI), by 92.1% to 4.0×10^1 ICC/µL (D1) and by 88.6% to 3.0×10^1 ICC/µL (D2). HNAP decreased by 90.1% to 7.2% (CI), by 50.8% to 13.3% (D1) and by 25.4% to 5.0% (D2). ALP-TEA was detected in CI with 18 pmol/min which is in accordance with other studies [16,22]. Interestingly, the HNAP portion in

D2 decreased by 85.2% and the LNAP amount increased correspondingly. This may indicate that ClO_2 damages HNAC membranes faster and more effectively than LNAC membranes which was also observed in several water samples from a previous study [22]. Alternatively, there may be a correlation between perception-events and rising biomass in GAC effluents.

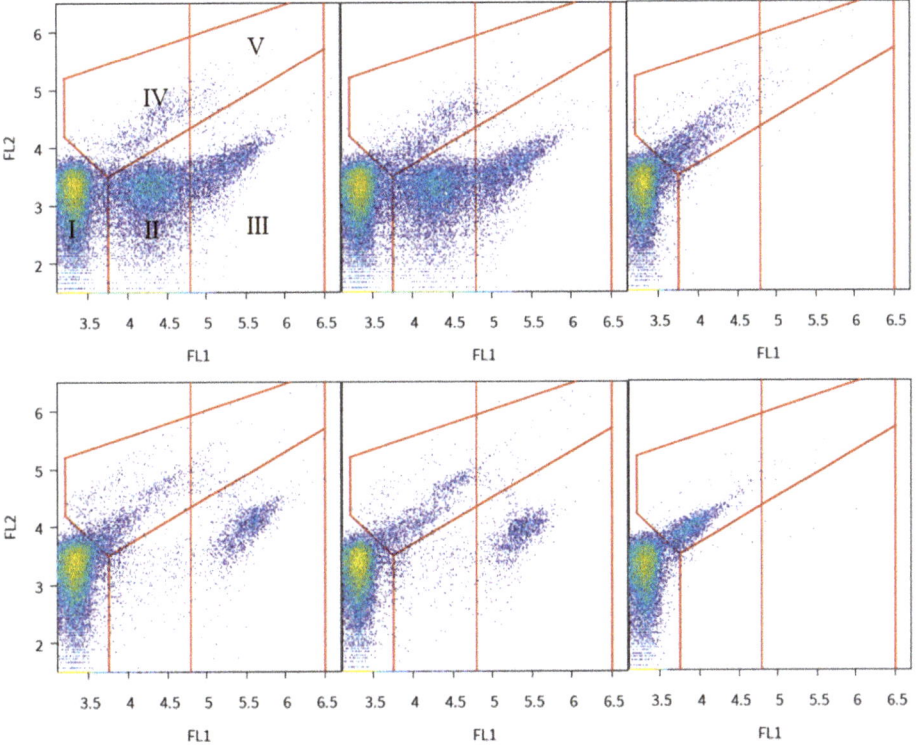

Figure 4. Dot-plots of water treatment trains in CI, Line 1: RBF, aeration, ozonation; Line 2: sandfilter, ultraviolet (UV-C) disinfection, assignments of Sections I to III: Background Signal, Intact LNAC, Intact HNAC, Sections IV-V: Lysed Cells, x-axis: Flourescence Signal 1 (FL1), y-axis: Flourescence Signal 2 (FL2).

The results shown in Figure 5 confirm the correlation between the attenuation of microorganisms and travel time in the aquifer. ICC decreased by 1.10 log units to 1.5×10^7 ICC/µL in well BF1 while ICC in well BF4 with a travel time of 100 ... 220 days decreased by 1.61 log units to 4.7×10^2 ICC/µL which is in accordance with the D2 results of average retention of 1.22 log units (not all sampling days are shown in Figure 5). The limit of detection using the HPC method is also demonstrated in Figure 5. While 220 MPN/mL and 48 MPN/mL were determined in Danube River water, no colonies could be detected in water from the RBF wells. These results suggest that new rapid microbiological methods (e.g., FCM) could be powerful tools for monitoring general microbiological water quality during treatment and distribution, as well as for the design and optimization of RBF site operations.

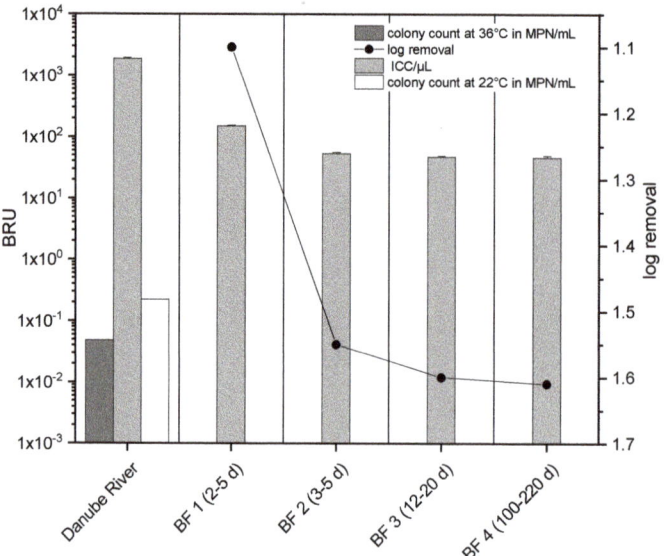

Figure 5. ICC monitored in various RBF wells in Budapest and their travel times in days, BF 1 = Csepel water plant, well No. 1, BF 2 = Tahi 1, well No. 5, BF 3 = Tahi 2 well No. 5, BF 4 = Szigetujfalu, well No. 7 with error bars rep-resenting standard deviation for those points, FCM samples were stained with SYBER®Green and propidiumiodide, HPC data were provided by Budapest Water Works Ltd. Modelled travel times may overlap depending on Danube River water levels.

3.2. Ultrafiltration Pilot Plant and Online FCM Measurements

During operation of the ultrafiltration pilot plant, bacteria and other particulate matter were efficiently retained independent of feed-water quality. Figure 6 shows online FCM measurement results of flocculated river water and river water as feed water. TCC in flocculated river water was fluctuating between 3.8×10^3 and 1.2×10^3 TCC/µL. The 30 cycles from 9 November to 12 November showed overall stable TCC of 2.1×10^3 TCC/µL but also daily fluctuations, and two biomass peaks (cycle 9 and cycle 29).

TCC was considerably higher in Elbe River water with an average amount of 9.4×10^3 TCC/µL with daily fluctuations from 4.4×10^4 to 1.9×10^3 TCC/µL (Figure 7) and biomass peaks at 27 April (cycle 12) and 3 May (cycle 11) caused by a rainfall event on April 26th, and due to several rainfall events around 1 May and 4 May. LNAP increases during rainfall and decreases during dry periods by 74.5% (not shown in Figure 6) as well as the TCC amount which was also documented in two studies of Besmer et al. [24,25]. TCC in Permeate 1 range from 2.9×10^3 to 4.0×10^1 TCC/µL, and in Permeate 2 from 2.6×10^3 to 3.7×10^0 TCC/µL dependent on the feed water quality in the corresponding cycle but are in a normal range due to inevitable bacterial regrowth after treatment. Cell numbers in the range from 10^1 to 10^2 cells/µL of diverse bacterial dynamics in similar water habitats such as riverbank filtrate or spring water were reported to be normal in previous studies [13,24,26]. However, the authors are not aware of any online FCM long-term studies that provide data of a similar ultrafiltration pilot setup.

With respect to the influence of feed water quality no significant difference in permeate quality using river water or flocculated river water as Feed (*t*-test, n = 44 (Elbe River), n = 59 (flocculated Elbe River water) $p > 0.05$) was reported, demonstrating the high performance of ultrafiltration in terms of microbiological removal of bacteria. Moreover, all online FCM results revealed higher amounts of TCC (3.6 fold) in Permeate 1 in comparison to Permeate 2 due to the fact that the membrane units in this pilot study differ in membrane area per module and pore size [19]. Hence, unit 1 with a pore size of 20 nm and 60 m^2 is more permeable to microorganisms than unit 2 with 18 nm and 55.7 m^2.

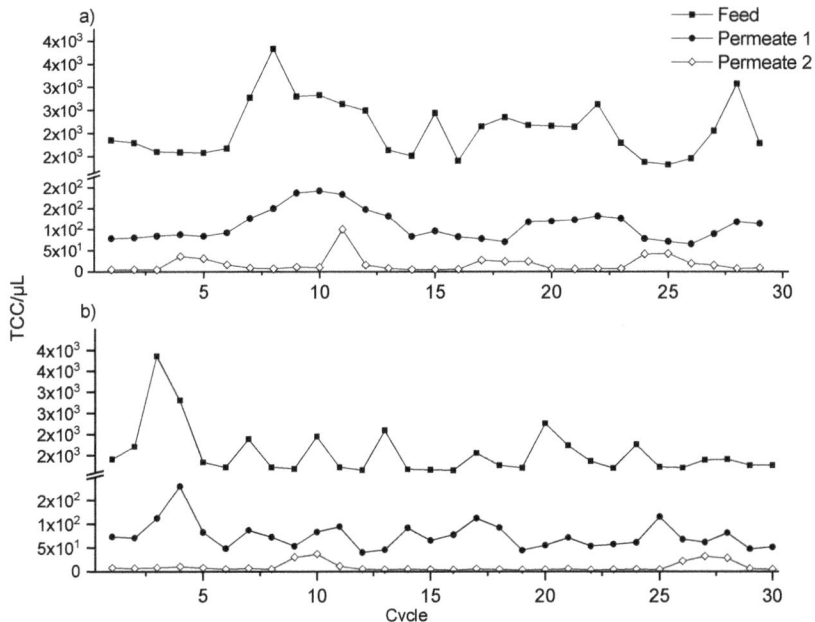

Figure 6. Continuous determination of total cell count (TCC) in flocculated Elbe River water (feed), Permeate 1 and Permeate 2, (**a**) 9–12 November 2018, (**b**) 19–22 November 2018), samples were stained with SYBER® Green.

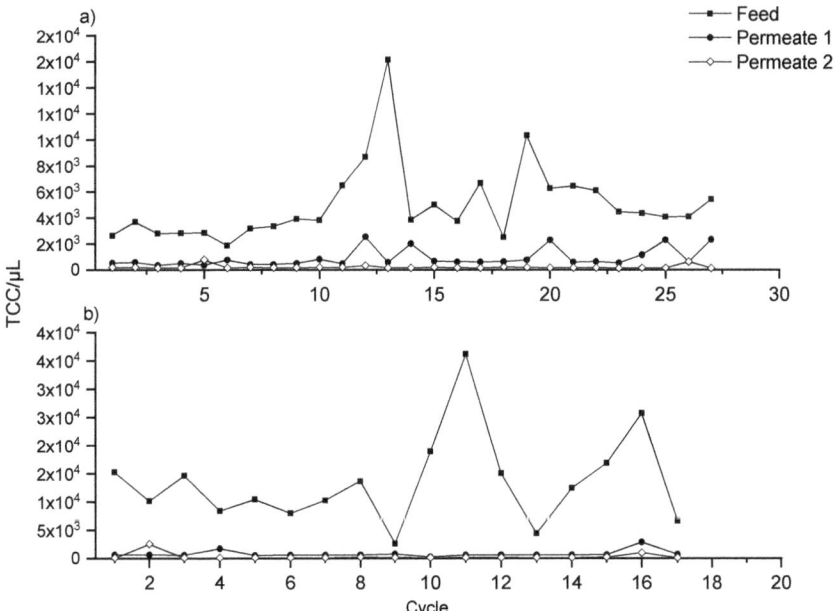

Figure 7. Continuous determination of total cell count (TCC) in Elbe River water (feed), Permeate 1 and Permeate 2, (**a**) 26 April–29 May 2019, (**b**) 2 May–4 May 2019, samples were stained with SYBER®Green and propidiumiodide.

When investigating the cut-off values of Permeate 1 and Permeate 2 (Figure 8) median microbiological removal rates of 2.41 and 1.19 log units were observed in flocculated river water feed and Elbe River water feed in Permeate 2, whereas log removal efficiencies of only 1.28 and 0.92 log units were achieved in Permeate 1. Such differences were also reported by Haas et al., [19] regarding ATPt results at the same pilot setup. Log units in Elbe River water were in addition 28.1% (Permeate 1) and 50.6% (Permeate 2) lower than in flocculated river water due to the membrane flux being influenced stronger by particular matter (e.g., dissolved iron or manganese). This leads to severe fouling problems, particularly biofouling and organic fouling on the membrane surface [27].

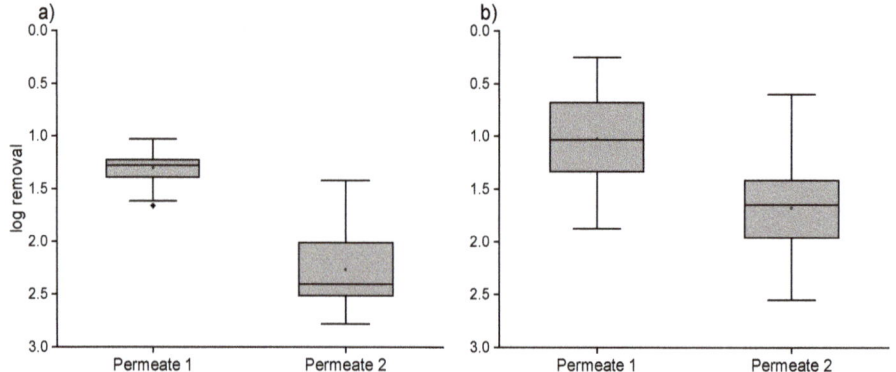

Figure 8. Differences in log removal rates of Permeate 1 and Permeate 2 in (**a**) flocculated Elbe river water t-test, n = 59, $p < 0.05$, n = 59 and (**b**) Elbe River water, t-test, n = 44, $p < 0.05$.

4. Discussion

In this study, the applicability of FCM and two enzymatic detection methods (ATP, ALP-TEA) for monitoring water-quality parameters at RBF sites was investigated. Despite a good correlation between FCM and ALP-TEA values, several differences were observed. In general ALP-TEA values correspond better to FCM results. This may be due to the fact that ATPi amount varies across living organisms and species, and is inter alia dependent of physiological states, especially HNA cells contain a higher ATPi amount than LNA cells [28]. Furthermore, ATPi is calculated from ATPt and ATPe results which were measured with a hand-held device where no collection of a fixed sample volume was possible. Therefore, values were fluctuating which is proven by the standard deviation on the triplicate samples.

The results of the dilution series (Figure 2) and also ALP-TEA values in CI (Figure 3) differentiate from the ICC trend as well as from the ATPt results. This could be caused by the nature of intracellular enzymatic methods. Intracellular enzymatic activity is generally bound to enzymatic concentration which is mostly dependent on bacterial state of growth while no information about species, size or number of a general population pattern is obtained.

ALP-TEA activity is especially high in the exponential stage of growth [13]. The probability of exponential growth in smaller bacterial communities is lower due to its limited number of microbiological species which is confirmed in Figure 2. Besides concentration, enzymatic processes are also dependent on specific reaction conditions which can be disturbed by interfering substances such as iron and manganese compounds. This was observed during ATP measurements in RBF well water samples (not shown in Figure 5) which usually contain higher concentrations of dissolved iron and manganese.

Additionally, FCM results may provide false positive signals due to staining limitations. SYBER®Green binds to any source of DNA including higher animals and plants. Although dividing cells and background signals through gates will remove most errors in quantitation, an overestimation of particle numbers must be always taken into account when assessing FCM results. Moreover, a recent study revealed that propidiumiodide-based viability staining can significantly overestimate DCC due to the presence of extracellular nucleoid acid (eNA) biofilms [29]. False positive results with propidiumiodide dye have been also associated with high membrane potential or might be influenced on physiological processes other than membrane damage in earlier studies [30,31].

Furthermore, the removal of microorganisms was correlating with travel time, proven by FCM and HPC measurements for various RBF wells in Budapest. The observed log removal rates for bacteria during RBF were in agreement with average values of 1.5 ... 3.5 [32]. Despite the short travel time of 2 ... 5 d in BF1, the removal rate of 1.10 log units is only slightly lower than values found at other RBF sites [14,15]. No colony counts were detected by HPC or ALP-TEA (not shown in Figure 3), indicating the limitations of these methods, whereas by FCM and ATP methods a low limit of detection was proven and a high efficiency to assess microbiological dynamics during RBF.

Changes in BRU concentration after aeration processes were observed in D1 and CI, both showing a slight increase in cell concentration by 4% and 13%. Here the aeration is operated as open cascade towers and microbiological growth is stimulated by external biomass entries, especially seasonal pollination. Further research is needed since analyzed data of the aeration processes suggest a connection between bacterial growth and pollination. This may be based upon the assumption that, especially on days without rainfall events, the growth was more intense. Alternatively, cell growth in aeration towers D1 and CI could be caused by FCM dyes on pollen entries. This may lead as previously mentioned to an over estimation of fluorescence signals.

Furthermore, differences in microbiological removal efficiency during sand filtration were observed at all sites. Previous ozonation provides dead cell material, a carbon source, to still intact microorganisms which are able to pass the sandfilter, and are therefore responsible to BRU increase in the filter effluent. In D1 and D2 however, sand filtration operated as a microbiological barrier dependent on cell size, cell morphology, motility, and membrane surface chemistry [33]. Additionally, microbiological removal efficiency is dependent on the sandfilters general conditions such as the thickness of the uppermost biofilm (*Schmutzdecke*), sand composition, or maturation of microbiological community in the *Schmutzdecke* [34].

In terms of BRU increase in the D2 GAC filter effluent and the corresponding HNAC decrease, it is presumed that certain strains, especially HNAC adapt more at colonizing the filter surface. In a pilot study by Shirey et al. [23], the community structure of heterotrophic bacteria associated with three GAC and two anthracite filters was examined over 12 months. Besides the diversity of bacterial community structures in prefiltered water, media composition, and depth, time was also a significant factor influencing bacterial communities within the filters. After the initial acclimation and colonization period had passed, overall community structure became less variable [23]. To clarify whether those cells are composed of a high HNAP amount, a similar pilot study using online FCM is necessary to prove this assumption.

Final disinfection results in a strong BRU decrease which was observed at all sites. Interestingly, in comparison to CI and D1 a decrease by 25.5% HNAP was observed in D2. This was likely due to the fact that ClO_2 was used as a disinfectant instead of Cl_2. Previous studies revealed cell membranes of HNAC bacteria were damaged much faster than those of LNAC bacteria during treatment with ClO_2 while only small differences were observed during treatment with chlorine and chloramine, and no difference was observed for ferrate treatment [22]. Thus, the considerably high HNAP decrease of 85.2% in D2 may be due to ClO_2 treatment.

Results from the ultrafiltration pilot plant study confirmed that online sensors are of advantage in terms of microbiological monitoring. Online measurements allow real-time detection of stable phases which are missed by grab sampling or incorrectly characterized. Optionally, online FCM can be coupled with online enzymatic activity probes (e.g., BACTcontrol). Besides ALP-TEA enzymatic activities of EC (*β-galactosidase*) and *E. coli* (*β-glucoronidase*) can be determined that may speed-up the detectability of incidents which impair microbiological water quality and safety [16]. More detailed evaluation regarding advantages of microbiological online sensors is given in previous studies [16,24–26].

Ultrafiltration proved to operate as an efficient barrier against microorganisms. During online FCM measurements, the average TCC values measured in river water as well as in flocculated river water are in normal range of surface water [35]. All feeds indicate an apparent bacterial fluctuation as well as one event linked to rainfall (cycle 13 on 27 April). According to a local meteorological station in D1, a rainfall event of 16.3 mm was recorded on 26 April in the evening [36]. This event is in accordance with the observed TCC peaks. The second event indicates a strong TCC increase of 65.9% (cycle 11, 3 May) which is likely caused by several entries of different origin (e.g., shipping or leisure activities) into the Elbe River due to a national holiday event on 1 May, and minor rainfall events including 3 May (0.2 mm) [36].

To underline those assumptions, abiotic parameters such as EC, pH, or dissolved oxygen should have been measured directly in feed waters which was beyond being integrated into the experimental setup. Diurnal fluctuations of abiotic parameters in river water have been investigated in previous studies [24,37,38]. Microbiological changes in surface waters may therefore be linked to photosynthetic and respiratory metabolic processes as well as to changes in precipitation and dissolution of ions which was inter alia assumed by Besmer et al. 2014 [24].

While the authors are not aware of previous studies regarding online FCM during ultrafiltration, the TCC amount detected in permeates are in accordance to typical values of riverbank filtrate or spring water. The daily patterns in Figures 6 and 7 were in accordance with microbiological dynamics in feed water. The occurrence of TCC in the permeates was probably due to bacterial regrowth after treatment [26,27].

Despite general membrane characteristics such as pore size and membrane surface and thus differences in permeate quality were observed (Figure 8), median removal rates in river water (0.92 log units Permeate 1 and 1.19 log units Permeate 2) was less than in flocculated river water (1.28 log units Permeate 1 and 2.41 log units Permeate 2). Jadoun et al. 2018 suggested that biofouling promotes the establishment of populations of water-borne pathogens on membrane surfaces. This was proven in monitoring studies using both culture-based and quantitative polymerase chain reaction (qPCR) methods in which the ability of microorganisms to establish rapid biofilm formation and persist on the membrane surfaces was demonstrated [27].

Considering the average amount of TCC in Elbe River water which is five times higher than in flocculated river water, the impact of biofouling on the membrane surface would most likely affect microbiological removal efficiency. The survival of biofilm-forming bacteria on the membrane surface despite in situ backwash and chemical treatment highlights the importance of optimization of the ultrafiltration process e.g. by applying RBF as a pretreatment step to ultrafiltration. In addition, if coupling RBF and ultrafiltration, fewer chemicals would be needed with regards to disinfection and membrane backwashing.

5. Conclusions

The results in this study provide insights into microbiological quality of treated water and emphasize the value of rapid, easy, and sensitive alternatives to HPC-based monitoring techniques. Despite a good parameter correlation, in particular in between ICC and ALP-TEA results, several divergences were observed. These were probably caused by method-based limitations such as staining techniques (FCM) or interfering ions (ATP, ALP-TEA). Also, due to different parameters defining bacterial concentration being measured, a direct comparison between these variables was rather difficult. The sets of microbiological data of RBF and other water treatment-related samples, nevertheless, enhance the understanding and improve the assessment of microbiological dynamics during drinking water treatment. However, further research, especially long-term studies to cover different seasons at RBF sites, are of vital importance to underline the results gathered in this study.

The online FCM data during ultrafiltration revealed diurnal BRU fluctuations, likely in response to nutrient concentration and abiotic parameters in feed water. These rapid changes should be considered when water is monitored by grab sampling. At present, little is known about online FCM during UF. Further long-term data combined with abiotic parameter monitoring and optional enzymatic activity are, therefore, required.

The presented methods may serve as possible alternatives in the future for the assessment of the quality of RBF water. However these methods are not yet accredited, hence measurement results are informative. Nonetheless, the article is of interest to all drinking-water suppliers that operate RBF systems to consider improving general microbiological monitoring.

Author Contributions: Y.A. reviewed previous literature and prepared the article draft, T.G. initiated the research and acquired the funding, Y.A., G.-H.O. and M.P. carried out measurements, R.H., R.B. and Z.Á.N.-K. organized sampling campaigns and operation of treatment facilities, J.A. introduced ALP-TEA measurements and provided the device, and all authors reviewed the final manuscript. All authors have read and agreed to the published version of the manuscript.

Funding: All primary data was collected within the AquaNES project. This project has received funding from the European Union's Horizon 2020 Research and Innovation Program under grant no. 689450.

Acknowledgments: The authors gratefully acknowledge support from the DREWAG NETZ GmbH, Budapest Waterworks Ltd, Hungary, microLAN B. V.

Conflicts of Interest: The authors declare no conflict of interest. The founding sponsors had no role in the design of the study; in the collection, analysis or interpretation of data; in the writing of the manuscript; or in the decision to publish results.

Abbreviations

The following abbreviations are used in this manuscript:

ALP	Alkaline phosphatase
ATP	Adenosine triphosphate
MAR	Managed Artificial Recharge
BRU	Biomass Related Unit
DCC	Desolate Cell Count
FCM	Flow cytometry
HNA	High Nucleic Acid
HNAP	High Nucleic Acid Percentage
ICC	Intact Cell Count
LNA	Low Nucleic Acid
LNAP	Low Nucleoid Acid Percentage
RBF	Riverbank filtration
TC	Total coliforms
TCC	Total Cell Count
TEA	Total Enzymatic Activity

References

1. Ray, C.; Jain, R. *Drinking Water Treatment: Focusing on Appropriate Technology and Sustainability*, 2011th ed.; Springer: Dordrecht, The Netherland; New York, NY, USA, 2011; ISBN 978-94-007-1103-7.
2. Grischek, T.; Schoenheinz, D.; Worch, E.; Hiscock, K.M. Bank filtration in Europe—An overview of aquifer conditions and hydraulic controls. In *Management of Aquifer Recharge for Sustainability*; Balkema Publishers, Swets & Zeitlinger: Lisse, The Netherlands, 2002; pp. 485–488.
3. Ray, C.; Soong, T.W.; Lian, Y.Q.; Roadcap, G.S. Effect of flood-induced chemical load on filtrate quality at bank filtration sites. *J. Hydrol.* **2002**, *266*, 235–258. [CrossRef]
4. De Vera, G.A.; Wert, E.C. Using discrete and online ATP measurements to evaluate regrowth potential following ozonation and (non)biological drinking water treatment. *Water Res.* **2019**, *154*, 377–386. [CrossRef] [PubMed]

5. Allen, M.J.; Edberg, S.C.; Reasoner, D.J. Heterotrophic plate count bacteria—What is their significance in drinking water? *Int. J. Food Microbiol.* **2004**, *92*, 265–274. [CrossRef] [PubMed]
6. Prest, E.I.; Hammes, F.; van Loosdrecht, M.C.M.; Vrouwenvelder, J.S. Biological Stability of Drinking Water: Controlling Factors, Methods, and Challenges. *Front. Microbiol.* **2016**, *7*, 45. [CrossRef]
7. Sun, H.; Shi, B.; Lytle, D.A.; Bai, Y.; Wang, D. Formation and release behavior of iron corrosion products under the influence of bacterial communities in a simulated water distribution system. *Environ. Sci. Process Impacts* **2014**, *16*, 576–585. [CrossRef]
8. Szewzyk, U.; Szewzyk, R.; Manz, W.; Schleifer, K.H. Microbiological safety of drinking water. *Annu. Rev. Microbiol.* **2000**, *54*, 81–127. [CrossRef]
9. World Health Organization. *Guidelines for Drinking Water Quality: Incorporating First Addendum*; World Health Organization: Geneva, Switzerland, 2006.
10. Rittmann, B.; Snoeyink, V.L. Achieving biologically stable drinking water. *Am. Water Works Assoc. J.* **1984**, *76*, 106–114. [CrossRef]
11. Bartam, J.; Cotuvo, J.; Exner, M.; Fricker, C.; Glasmacher, A. *Heterotrophic Plate Counts and Drinking-Water Safety*; IWA Publishing on behalf of the World Health Organization: London, UK, 2003.
12. Van Nevel, S.; Koetzsch, S.; Proctor, C.R.; Besmer, M.D.; Prest, E.I.; Vrouwenvelder, J.S.; Knezev, A.; Boon, N.; Hammes, F. Flow cytometric bacterial cell counts challenge conventional heterotrophic plate counts for routine microbiological drinking water monitoring. *Water Res.* **2017**, *113*, 191–206. [CrossRef]
13. Hammes, F.; Berney, M.; Wang, Y.; Vital, M.; Köster, O.; Egli, T. Flow-cytometric total bacterial cell counts as a descriptive microbiological parameter for drinking water treatment processes. *Water Res.* **2008**, *42*, 269–277. [CrossRef]
14. Stadler, P.; Loken, L.C.; Crawford, J.T.; Schramm, P.J.; Sorsa, K.; Kuhn, C.; Savio, D.; Striegl, R.G.; Butman, D.; Stanley, E.H.; et al. Spatial patterns of enzymatic activity in large water bodies: Ship-borne measurements of beta-D-glucuronidase activity as a rapid indicator of microbial water quality. *Sci. Total Environ.* **2019**, *651*, 1742–1752. [CrossRef]
15. Højris, B.; Christensen, S.C.B.; Albrechtsen, H.-J.; Smith, C.; Dahlqvist, M. A novel, optical, on-line bacteria sensor for monitoring drinking water quality. *Sci. Rep.* **2016**, *6*, 23935. [CrossRef] [PubMed]
16. Appels, J.; Baquero, D.; Galofré, B.; Ganzer, M.; van den Dries, J.; Juárez, R.; Puigdomènech, C.; van Lieverloo, J.H. *Safety and Quality Control in Drinking Water Systems by Online Monitoring of Enzymatic Activity of Faecal Indicators and Total Bacteria*; IWA Publishing: London, UK, 2018; pp. 171–195. ISBN 978-1-78040-869-9.
17. Vang, Ó.K.; Corfitzen, C.B.; Smith, C.; Albrechtsen, H.-J. Evaluation of ATP measurements to detect microbial ingress by wastewater and surface water in drinking water. *Water Res.* **2014**, *64*, 309–320. [CrossRef] [PubMed]
18. Park, S.J.; Hong, J.T.; Choi, S.J.; Kim, H.S.; Park, W.K.; Han, S.T.; Park, J.Y.; Lee, S.; Kim, D.S.; Ahn, Y.H. Detection of microorganisms using terahertz metamaterials. *Sci. Rep.* **2014**, *4*, 4988. [CrossRef] [PubMed]
19. Haas, R.; Opitz, R.; Grischek, T.; Otter, P. The AquaNES Project: Coupling Riverbank Filtration and Ultrafiltration in Drinking Water Treatment. *Water* **2019**, *11*, 18. [CrossRef]
20. Nagy-Kovács, Z.; László, B.; Simon, E.; Fleit, E. Operational Strategies and Adaptation of RBF Well Construction to Cope with Climate Change Effects at Budapest, Hungary. *Water* **2018**, *10*, 1751. [CrossRef]
21. Vital, M.; Dignum, M.; Magic-Knezev, A.; Ross, P.; Rietveld, L.; Hammes, F. Flow cytometry and adenosine tri-phosphate analysis: Alternative possibilities to evaluate major bacteriological changes in drinking water treatment and distribution systems. *Water Res.* **2012**, *46*, 4665–4676. [CrossRef] [PubMed]
22. Ramseier, M.K.; von Gunten, U.; Freihofer, P.; Hammes, F. Kinetics of membrane damage to high (HNA) and low (LNA) nucleic acid bacterial clusters in drinking water by ozone, chlorine, chlorine dioxide, monochloramine, ferrate(VI), and permanganate. *Water Res.* **2011**, *45*, 1490–1500. [CrossRef]
23. Shirey, T.B.; Thacker, R.W.; Olson, J.B. Composition and stability of bacterial communities associated with granular activated carbon and anthracite filters in a pilot scale municipal drinking water treatment facility. *J. Water Health* **2012**, *10*, 244–255. [CrossRef]
24. Besmer, M.D.; Weissbrodt, D.G.; Kratochvil, B.E.; Sigrist, J.A.; Weyland, M.S.; Hammes, F. The feasibility of automated online flow cytometry for in-situ monitoring of microbial dynamics in aquatic ecosystems. *Front. Microbiol.* **2014**, *5*, 265. [CrossRef]

25. Besmer, M.D.; Epting, J.; Page, R.M.; Sigrist, J.A.; Huggenberger, P.; Hammes, F. Online flow cytometry reveals microbial dynamics influenced by concurrent natural and operational events in groundwater used for drinking water treatment. *Sci. Rep.* **2016**, *6*, 38462. [CrossRef]
26. Egli, T.; Zimmermann, S.; Schärer, P.; Senouillet, J.; Künzi, S.; Köster, O.; Helbing, J.; Montandon, P.-E.; Marguet, J.-F.; Khajehnouri, F. Automatische Online-Überwachung. Bestimmung der Bakterienzahl im Roh- und Trinkwasser: Resultate aus der Praxis. *Aqua Gas* **2017**, *97*, 52–59.
27. Jadoun, J.; Mreny, R.; Saad, O.; Azaizeh, H. Fate of bacterial indicators and Salmonella in biofilm developed on ultrafiltration membranes treating secondary effluents of domestic wastewater. *Sci. Rep.* **2018**, *8*, 18066. [CrossRef] [PubMed]
28. Greenstein, K.E.; Wert, E.C. Using rapid quantification of adenosine triphosphate (ATP) as an indicator for early detection and treatment of cyanobacterial blooms. *Water Res.* **2019**, *154*, 171–179. [CrossRef] [PubMed]
29. Rosenberg, M.; Azevedo, N.F.; Ivask, A. Propidium iodide staining underestimates viability of adherent bacterial cells. *Sci. Rep.* **2019**, *9*, 6483. [CrossRef]
30. Kirchhoff, C.; Cypionka, H. Propidium ion enters viable cells with high membrane potential during live-dead staining. *J. Microbiol. Methods* **2017**, *142*, 79–82. [CrossRef]
31. Yang, Y.; Xiang, Y.; Xu, M. From red to green: The propidium iodide-permeable membrane of Shewanella decolorationis S12 is repairable. *Sci. Rep.* **2015**, *5*, 18583. [CrossRef]
32. Nagy-Kovács, Z.; Davidesz, J.; Czihat-Mártonné, K.; Till, G.; Fleit, E.; Grischek, T. Water Quality Changes during Riverbank Filtration in Budapest, Hungary. *Water* **2019**, *11*, 302. [CrossRef]
33. Becker, M.W.; Metge, D.W.; Collins, S.A.; Shapiro, A.M.; Harvey, R.W. Bacterial Transport Experiments in Fractured Crystalline Bedrock. *Groundwater* **2003**, *41*, 682–689. [CrossRef]
34. Ranjan, P.; Prem, M. Schmutzdecke—A Filtration Layer of Slow Sand Filter. *Int. J. Curr. Microbiol. Appl. Sci.* **2018**, *7*, 637–645. [CrossRef]
35. Wang, Y.; Hammes, F.; Boon, N.; Egli, T. Quantification of the Filterability of Freshwater Bacteria through 0.45, 0.22, and 0.1 µm Pore Size Filters and Shape-Dependent Enrichment of Filterable Bacterial Communities. *Environ. Sci. Technol.* **2007**, *41*, 7080–7086. [CrossRef]
36. ReKIS-Viewer. Available online: http://141.30.160.224/fdm/rekisViewer.jsp#menu-5 (accessed on 15 November 2019).
37. Vogt, T.; Hoehn, E.; Schneider, P.; Freund, A.; Schirmer, M.; Cirpka, O.A. Fluctuations of electrical conductivity as a natural tracer for bank filtration in a losing stream. *Adv. Water Resour.* **2010**, *33*, 1296–1308. [CrossRef]
38. Hayashi, M.; Vogt, T.; Mächler, L.; Schirmer, M. Diurnal fluctuations of electrical conductivity in a pre-alpine river: Effects of photosynthesis and groundwater exchange. *J. Hydrol.* **2012**, *450–451*, 93–104. [CrossRef]

© 2020 by the authors. Licensee MDPI, Basel, Switzerland. This article is an open access article distributed under the terms and conditions of the Creative Commons Attribution (CC BY) license (http://creativecommons.org/licenses/by/4.0/).

MDPI
St. Alban-Anlage 66
4052 Basel
Switzerland
Tel. +41 61 683 77 34
Fax +41 61 302 89 18
www.mdpi.com

Water Editorial Office
E-mail: water@mdpi.com
www.mdpi.com/journal/water